Casting of Near Net Shape Products

Casting of Near Net Shape Products

Proceedings of an International Symposium on Casting of Near Net Shape Products.

Co-sponsored by:

The Metallurgical Society
Iron and Steel Society
The Mining and Metallurgical Institute of Japan
The Iron and Steel Institute of Japan
Benelux Metallurgy
The Metallurgical Society of CIM
The Australasian Institute of Mining and Metallurgy
Deutsche Gesellschaft fur Metallkunde
Societe Francaise de Metallurgie
The Institute of Metals and Materials Australasia

Held at the Fall Extractive and Process Metallurgy Meeting, November 13-17, 1988, in Honolulu, Hawaii.

Edited by

Y. Sahai
J.E. Battles
R.S. Carbonara
C.E. Mobley

A Publication of The Metallurgical Society, Inc.

A Publication of The Metallurgical Society, Inc.
420 Commonwealth Drive
Warrendale, Pennsylvania 15086
(412) 776-9024

Printed in the United States of America
Library of Congress Catalogue Number 88-62072
ISBN Number 0-87339-079-2

PREFACE

The papers in this bound volume were presented at the Fall Extractive and Process Meeting, International Symposium on Casting of Near Net Shape Products held in Honolulu, Hawaii, November 13-17, 1988. The casting of near net shapes is an age-old technology, and a variety of processes exist for casting such products. However, this symposium was particularly focused on the continuous casting of thin section (sheet-like) products. In the recent past, a significant development and research effort has been directed toward a better understanding of these processes. Thus, most steel and non-ferrous metal producers and research laboratories around the world are conducting work directed to the continuous casting of thin section products. This symposium was intended to bring the researchers and engineers working in this area together to discuss both the theoretical and practical aspects of the major casting technologies for the production of ferrous and non-ferrous sheet-like products.

With few exceptions, the manuscripts appear in this volume essentially in the form originally prepared by the authors. This decision to publish several of the manuscripts without extensive revision was taken because of time shortage and for rapid dissemination of information.

On behalf of the organizing committee, I wish to express my sincere appreciation and gratitude for the time and effort contributed by many authors to make this symposium and this volume a success. We are also grateful to the plenary lecturers, specially invited lecturers, and to the session chairmen for accepting our invitations. Finally, our sincere thanks to The Metallurgical Society staff for their assistance and to Ms. E. Davidson of the Ohio State University for her extensive secretarial assistance.

Y. Sahai
The Ohio State University
Columbus, Ohio

July 1988

TABLE OF CONTENTS

AFTER DINNER TALK

PLENARY LECTURE

FUNDAMENTAL ASPECTS III: MODELING (continued)

CASTING TECHNIQUES I: PROCESS CONTROL

LUNCHEON LECTURE

CASTING TECHNIQUES II: SPRAY CASTING

CASTING TECHNIQUES III: GENERAL

PLENARY LECTURE

FUNDAMENTAL ASPECTS IV: ELECTROMAGNETIC CONFINEMENT

CASTING TECHNIQUES IV: THIN SLAB CASTING

CASTING TECHNIQUES V: ELECTROMAGNETIC CONFINEMENT

CASTING TECHNIQUES VI: STRIP CASTING

STRUCTURE CONTROL

MELT TREATMENT

PLENARY LECTURE

Session Chairman: **Y. Sahai**
Ohio State University
Columbus, Ohio

FUNDAMENTAL ASPECTS OF THE CONTINUOUS

CASTING OF NEAR-NET-SHAPE STEEL PRODUCTS

J.K. Brimacombe and I.V. Samarasekera

The Centre for Metallurgical Process Engineering
The University of British Columbia
Vancouver, B.C. Canada V6T 1W5

Abstract

In the continuous casting of thin steel slabs and strip, the introduction of molten metal into the mould cavity and the extraction of heat present formidable challenges. This paper focuses on heat extraction from a fundamental standpoint and attempts to provide a framework for evaluation of the different near-net-shape casting processes. The heat-extraction rates necessary to meet the production requirements of an integrated steel plant or minimill are shown to be about 6 to 22 MW and casting speeds must be in the range of 10 to 40 m/min. Measured heat-transfer coefficients of 1.3 to 2.6 kW/m^2°C for twin-belt casters are similar to values close to the meniscus of a conventional slab caster. Consequently the cooling rate (20 to 100°C/s) and structure are like those obtained within 25 mm from the surface of a continuously cast slab. Less information is available on heat-transfer coefficients for a twin-roll caster; comparative measurements of secondary dendrite-arm spacing suggests that the heat fluxes in the twin roll caster are probably somewhat higher than heat fluxes measured at the meniscus of a billet mould. The heat-extraction capabilities of twin-belt and twin-roll casters have been assessed, in the light of production requirements, with the aid of a computer model. Surface quality and shape defects associated with near-net-shape casting processes are reviewed; and, finally, thermal distortion of heat-extraction devices is discussed.

Casting of Near Net Shape Products
Edited by Y. Sahai, J.E. Battles,
R.S. Carbonara and C.E. Mobley
The Metallurgical Society, 1988

Introduction

Since its implementation on a global scale, conventional continuous casting of semi-finished shapes (slabs, blooms and billets) has been driven by the twin goals of quality and productivity. Both have been achieved to a large extent since links between quality (freedom from cracks and macrosegregation, correct shape and cleanness) and machine design/operating practice have been established. The same can be said for productivity although operating problems like break-outs remain an abiding difficulty, particularly as casting speeds are increased. Thus in many respects, the challenge to conventional continuous casting is the ongoing implementation of existing knowledge more than the creation of new knowledge.

The maturity of conventional continuous casting and the large pool of knowledge supporting it provide an excellent base for the development of the next generation of continuous-casting processes to produce thin slabs ($\sim$ 20-70 mm thick), strip ($\sim$ 5-20 mm thick) and thin strip (< 5 mm thick). The need for such processes has never been greater than at present in the steel industry of the industrialized nations which must reduce production costs to meet competition from alternative materials and from low-cost producers in developing nations. The key is to cast a near-net shape which requires a minimum of hot or cold work to achieve the desired product.

The successful development of near-net-shape casting depends critically on an understanding, and on the application, of fundamental knowledge particularly of heat and fluid flow. The feeding of liquid uniformly into a near-net-shape channel and the extraction of heat at a rate which meets productivity requirements, but does not result in crack formation, are at the heart of the new processes under development. Unfortunately very little information exists in the literature on fluid flow in near-net-shape casting relative to heat transfer.

Thus in this paper, emphasis is placed on the heat flow aspects which are examined in detail from a fundamental viewpoint. The analysis begins with a consideration of the heat-extraction rates to be expected in near-net-shape casting based on scanty literature reports and from our own measurements of mould heat transfer close to the meniscus in conventional billet and slab moulds. Having established the level of heat extraction, at least approximately, its effect on production rate, structure and surface crack formation is examined. It then becomes possible to establish fundamental requirements of different heat-extraction devices and to assess current designs.

Types of Heat-Extraction Device

Currently there are six basic types of heat-extraction devices in near-net-shape casting processes under development world-wide:

[1] Single roll (melt drag and spray deposition)
[2] Twin roll
[3] Twin belt
[4] Wheel-belt (billets)
[5] Single belt (mould-trough train and spray deposition)
[6] Convex mould/conventional equipment

These are shown schematically in Fig. 1. The application of the different heat-extraction devices has been reviewed by Dancy (1) and others (2-4). Table I shows the section size, casting speed and feeding technique employed in the various installations under development. Thus it is seen that the single- and twin-belt devices are being utilized for the casting of thin slabs having a thickness ranging from 10 to 70 mm, while thinner sections are being cast with single- and double-roll devices. The thinnest sections are cast using a single roll.

With one exception, all the heat-extraction devices shown in Table I move continuously in a rolling or linear action and are unlubricated in the conventional sense with oil or mould flux. The exception is the Schloemann-Siemag mould which is reciprocated and lubricated with mould flux to prevent sticking between the solidifying steel and the heat-extraction surface. Thus sticking, excessive friction between the steel and heat-extraction device, and consequent stress generation, which may lead to surface quality problems (cracks, depressions), are serious potential difficulties in near-net-shape casting. These questions will be taken up later in the paper.

Heat-Extraction Requirements of Near-Net-Shape Casting Machines

Although the advantages of continuously casting thin steel slabs or strip are obvious, the process has another potential problem of having to operate at higher speeds to maintain the production rate of a given steelmaking operation. The relationship between casting speed and thickness of cast section for an assumed section width of 1 m, a casting time of 80 min and different heat sizes is shown in Fig. 2, taken from Cygler and Wolf (2). It is seen that for a heat size of, say, 240 tons, typical of an integrated steel operation, and an assumed section thickness of 10 mm, a single casting machine would have to operate at about 40 m/min. In the case of a minimill, having a heat size more typically of 60 tons, the same section would need to be cast at 10 m/min. These casting speeds may be compared to 0.5 - 1.5 m/min for a conventional bloom/slab caster and 1.5 - 4.0 m/min for a billet caster. The high speeds required for near-net-shape casting machines can be reduced by the installation of multiple units but only to a certain extent owing to the concomitant penalties of cost and complexity of operation.

To achieve these production rates, the casting machine must be capable of extracting the superheat ($\sim$ 25°C) and latent heat of solidification of the steel (271 kJ/kg) together with a fraction of its sensible heat (cooled to, say, 1200°C). This capability amounts to about 490 kJ/kg of cast steel, or at a production rate of 3 tons/min (2.73 tonnes/min), must be about 22 MW, as can be seen in Fig. 3. For a heat size of 60 tons, cast in 80 min, the heat-extraction requirements would be one-fourth as large, or roughly, 5.5 MW. A serious concern, then, is whether or not in the relatively short dwell times of the steel in the heat-extraction device, at the high casting speeds, sufficient heat can be extracted to solidify and cool the metal.

Heat Extraction Rates in Continuous Casting of Thin Slab/Strip and Slabs/Billets

The rates of heat extraction, measured or estimated, in near-net-shape casting studies are summarized in Table II. Thus it is seen that the heat-transfer coefficient reported for twin-belt casters varies from 1.29 to 2.58 kW/m^2°C which would give rise to a heat flux of about 1.5 to 3.1 MW/m^2 (temperature driving force of 1200°C assumed). This range is very close to that observed near the meniscus of a conventional slab casting mould lubricated with mould flux. Takeuchi and Brimacombe (20), for example, report a heat flux of 2.6 MW/m^2 at the meniscus of a slab mould; the value decreases to about 1.8 MW/m^2 at 80 mm below the meniscus. Consequently the cooling and solidification rates and structure (secondary dendrite-arm spacing) to be expected in twin-belt casting processes will be similar to the values currently obtained in the first 25 mm (from the surface) of a continuously cast slab. From Table II, cooling rates in twin-belt casters are 20 to 100°C/s, estimated from the secondary dendrite-arm spacing which ranges from 20 to 150 μm. The solidification constant reported (for $s = k\sqrt{t}$) is about 2.58 $mm/s^{1/2}$ similar to that for a conventional slab mould. The low heat-extraction rates are likely related to the small ferrostatic head as well as to belt distortion.

For twin- and single-roll casters, the cooling rates are higher, about 10^2-10^4°C/s, again estimated from secondary dendrite-arm spacing measurements typically of 1-10 μm close to the surface of the strip. Cox[11] has reported heat-flux data, determined from measurements made by Nishioka et al. (9) with the aid of a one-dimensional, heat-flow model of a twin-roll caster having an assumed roll temperature of 480°C. The results are shown in Fig. 4. The heat flux is seen to vary from 1.63 to 7.3 MW/m^2 and is inversely dependent on dwell time between the rolls. The upper figure is similar to values measured close to the meniscus in billet moulds (21). Thus the secondary dendrite arm spacing in the subsurface region of continuously cast billets should be similar to that in thin strip, cast on twin- or single-roll casters, if Nishioka's data is representative of heat-flux typically obtained on such machines. Secondary dendrite-arm spacing measurements have been made on continuously cast billets with carbon contents ranging from 0.29% to 0.57% to a depth of 2 mm from the surface and the magnitude of the former ranges from 13 ∿ 35 μm. From measurements of mould temperatures obtained during the casting of these particular billet samples heat fluxes at the meniscus have been computed (21). This data has been employed in a heat-flow model for the solidifying billet from which cooling rates during solidification have been determined. For a region 1.0 mm from the surface, the cooling rates range in magnitude from 50°C/s to 130°C/s. These preliminary calculations suggest that since the cooling rates reported for twin- and single-roll casters are higher, based on some-what smaller secondary dendrite-arm spacing it is likely that heat-fluxes in these casting machines are higher than the upper figure of 7 MW/m^2 reported for billet casters.

The heat-extraction conditions in static chill-block experiments are closer to those of the twin-roll than of the twin-belt caster, as can be seen in Table II.

To check on the consistency of the data in Table II, a mathematical model based on one-dimensional heat flow and solidification, was run to predict local cooling and solidification rates as a function of time with values of heat flux ranging from 2.5 MW/m^2 (twin-belt caster) to 10.0 MW/m^2 (twin-roll, short dwell times). The cooling rate at the surface of the steel is shown as a function of time in Fig. 5. The predicted rates are 10^2 to 10^4°C/s for the first second of cooling. Over this period the shell will grow to a thickness of 0.5 to about 2 mm depending on the heat flux, as shown in Fig. 6. The cooling rate is relatively constant within the first 1 mm of the steel as can be seen in Fig. 7. It is important to note that the very high cooling rates of 10^3 ∿ 10^4°C/s prevail for extremely short times (< 0.2 s) and the average values over the first second are of the order of 10^2 ∿ 10^3°C/s. It appears from the data in Table II that the cooling rates are slightly higher than the average values calculated. Toyama et al. (6), for example, estimated a cooling rate of 10^3-10^4°C/s at 0.5 mm from the surface of the steel. This suggests that heat fluxes in twin-roll casters may be higher than 10.0 MW/m^2. Although calculations were not done for the longer dwell times of the twin-belt casters, predicted values certainly would be in the range of 10 to 100°C/s as reported.

The solidification rate predicted by the model does not conform to that given by the square-root law with the constants in Table II but no reliable values have been reported for short dwell times. Most of the solidification constants in Table II have been determined for twin-belt casters with dwell times of 45 to 60 s. Indeed for short dwell times, of say less than 6 s, our measurements and calculations in billet moulds reveal that the shell thickness, s, increases linearly with time, t, as follows (21)

$$s = 0.64\ t \qquad (1)$$

This is an expected result since during this early solidification period, the extraction of heat is dominated by the resistance of the mould/steel gap. The resistance offered by heat conduction through the shell does not become significant until the shell is about 5-6 mm thick. Consequently a nearly constant solidification rate may be expected for the casting of strip less than 10 mm thick depending on the heat-extraction rate.

Assessment of Current Heat Extraction Devices

A critical component of the assessment of heat-extraction devices for near-net-shape casting is determining whether the heat-extraction capabilities of a given device meet casting speed requirements set by current production standards. Thus it is important to evaluate the various processes from this stand-point, and to examine other factors such as dimensional stability of the device, uniformity of cooling, and associated quality problems.

Heat Extraction Capability

The processes for the direct casting of thin sections falls into two categories based on heat extraction capability. From Table II, it appears that the lowest heat-transfer and cooling rates are associated with twin/single belt processes and processes based on moving moulds such as the Alusuisse caterpillar type or the stationary oscillating mould in the Schloemann-Siemag process. The casting speeds are in the 3 $\sim$ 13 m/s range whilst the section sizes cast vary from 12 $\sim$ 75 mm thick. These casting speeds barely meet production standards for the thinnest sections in this range of thickness; consequently these processes appear more suitable for the casting of thin slab (20-75 mm) which subsequently will have to be finish hot rolled. At this point in time, it appears that the Schloemann-Siemag process (22) has a distinctive edge over the twin-belt processes because the former utilizes existing technology and is an incremental step forward relative to the latter which is based on more radical technology. However the SMS process may be limited to a casting speed of about 6 m/s owing to friction between the solidifying shell and mould. The moving belts present less of a problem in this regard and are capable of higher speeds.

The second category of processes for the casting of thin sections have high heat-extraction rates, in excess of that measured in the meniscus region of billet casters, and are based on single or double rolls. From the data of Cygler and Wolf (2) given in Fig. 2, the casting speed requirements for two section sizes, 5 and 10 mm, and two heat sizes 60 ton and 240 ton, have been calculated and are presented in Table III. It appears that a four-strand caster would be required for the integrated mill in order to implement a design that would be suitable for the casting of both section sizes unless very high casting speeds are attainable.

To determine whether these casting speed requirements could be met by a double-roll caster, a one-dimensional, finite-difference heat-flow model has been employed to predict the shell thickness as a function of dwell time.

Owing to the lack of directly applicable heat transfer data, the heat-transfer measurements of Takemoto (23) and co-workers for steel solidifying on a stationary copper chill have been employed. During solidification of steel on a moving roll, relative motion between the two is unlikely; hence the heat-transfer rates measured by Takemoto et al. (23), which compare well with those of Nishioka et al. (9), should provide a good estimate of the heat extraction between the steel and rolls at short dwell times. This data and the results of the calculation are given in Fig. 8. The dwell times and the solid shell thickness on exit from the roll bite have been calculated for twin rolls of diameter 400, 600 and 800 mm with a freeboard of 50 mm and are presented in Table IV. For the 5-mm section, solidification would essentially be complete within the roll bite when casting with 800-mm diameter rolls whilst the 10-mm section would be very nearly fully solid. Complete solidification within the roll bite may be undesirable owing to the large increase in roll forces (10) which could give rise to cracking and shape problems but bulging below the rolls is also a problem in the case of incomplete solidification. The calculations also show that for large heat sizes such as 240 tons, an effective way of decreasing the number of strands and the concomitant capital cost would be to increase the roll diameter since dwell times increase significantly. These preliminary calculations have demonstrated that it is possible to achieve satisfactory solidification rates in a twin-roll heat-extraction device which, together with containment and cooling below the rolls, if solidification is incomplete, would make such an operation viable from the standpoint of productivity.

For very thin strip (thickness less than 3 mm) the melt drag process appears to be most suitable, and Allegheny-Ludlum (5) are now commercially producing alloy strip with this technique. The use of the single-roll casters for thicker sections appears less viable owing to the need to successfully extract heat from two surfaces to complete solidification in a reasonable period of time. If the surface not in contact with the roll were to be shielded from the atmosphere to prevent oxidation and cooled by a containment device such as a belt or a shoe then the heat-transfer rates at that cooling surface would be considerably lower than at the roll and greater problems could be encountered with unequal cooling, strip shape and quality than if double rolls were employed.

Surface Quality and Shape Defects

The major impetus for the direct casting of strip is to produce material that can be cold rolled with a minimum of hot reduction. The shape and surface quality of the cast product must meet stringent requirements set by current standards for flat-rolled products. If excessive surface conditioning were required, the process would lose its economic viability owing to the large surface area that would have to be inspected and conditioned.

A comprehensive evaluation of the surface and internal quality of near-net-shape cast material cannot be conducted since information on this topic in the literature is sparse. However, an attempt has been made to summarize the available information and it is presented in Table V. For single-roll processes casting thin sections, (< 1 mm), thickness variations across the width could become significant and constitute a shape problem. Mismatch in the cooling rates between the upper and lower surfaces, if excessive, could cause surface cracks to occur (24). In twin-roll processes, sticking in the rolls has been identified as a problem in casting aluminum (26-28), and defects such as edge cracks and heat lines have been found to be prevalent during casting (26-28). For steel, based on studies on an experimental caster, it is apparent that solidification in the roll bite could lead to high roll separating forces and surface cracks (10). It also seems evident that crack formation will increase with the casting of 0.10% carbon steel owing to the δ-γ transformation and concomitant shrinkage.

During casting of thicker sections employing twin-belt casters, transverse cracking has been reported and attributed to frictional forces which are generated between the belt and the steel due to differential cooling and shrinkage (17). Belt distortion, if excessive, can also contribute to poor flatness of the as-cast slab (30). During development of the Schloemann-Siemag process employing a semi-conventional mould, transverse cracks, corner cracks and longitudinal surface depressions were prevalent. These problems have apparently been minimised through improvements to mould design, oscillation characteristics and mould operation (22).

Dimensional Stability of Heat-Extraction Device

In the direct casting of steel strip, dimensional stability of the heat-extraction device assumes paramount importance owing to its possible effect on the air gap between the steel and the device, which dominates heat transfer, and also to its impact on the shape of the cast product. Since the major impetus for the development of a near-net-shape casting process lies in the potential for eliminating all or many of the hot rolling steps, it is important to briefly examine the requirements currently imposed on the shape of hot strip. Figure 9 shows the widthwise variation in the thickness of the strip and the limitations on strip crown denoted by Cr_{25} (difference in sheet thickness at the center and at 25 mm from the strip edge) for a variety of applications (32). Ideally strip crown should not exceed 60 μm if it is to successfully replace conventionally rolled material. With the advent of hydraulic gauge control systems, the variation of strip thickness along its length has been reduced to ± 3 μm. Thus it is evident that the requirements on the shape of the cast strip and on the associated heat-extraction device are extremely stringent if the strip is to be directly cold rolled. If one or two passes of hot rolling are interposed between the caster and the cold mill then the dimensional control becomes less stringent but does not diminish in importance.

In the hot rolling of steel strip, the variation in thickness across the width is produced as a result of bending of the work rolls in response to the large roll separating forces coupled with the thermal crown resulting from the temperature gradients in the steel work rolls. The cyclic fluctuations in temperature, due to intermittent contact with the strip, are restricted to a surface layer approximately 1 mm deep (33) and do not contribute measurably to the thermal crown (34). The latter is affected by the steady-state temperature distribution in the bulk of the roll an example of which is shown in Fig. 10 (35). The sharp reduction in crown close to the edge, from a maximum value of 80 μm at the center, is due to the strip half width (B/2) being considerably less than the half width of the work roll, ($Z_o/2$).

When rolls are employed for the direct casting of strip, and solidification is avoided in the roll bite, the rolls are subject only to the development of a thermal crown. As an example, for a 400-mm diameter copper roll with a cooling channel located 50 mm below the surface, the steady-state average temperature during a revolution at the hot face and at the cooling channel is 110.95 and 46.67°C, for an average heat flux of 500 kW/m^2. The thermal crown at the center of the rolls for this case is 1082 μm decreasing to zero at the ends if the width of the sheet being cast is somewhat less than the roll width. Increasing the roll diameter to 800 mm increases the thermal crown at the midface to 2209 μm owing to the increase in the circumferential length. Thus it is clear that considerable attention has to be paid to the design, geometry and alignment of the rolls, and the location of the cooling channel if variations in shape and heat transfer across the width are to be minimised.

For the direct casting of thin slabs utilizing a Hazelett type machine, if the cast section is greater than 20 mm thick, the required tolerances on shape may be attained by finish rolling. However belt distortion can cause serious variations in local heat transfer. Itoyama et al. (16) claim that in the thin slab caster developed at Kawasaki Steel, belt distortion is minimised by force cooling the belts with water flowing at high velocity. They observed that austenitic stainless steel belts (20) gave rise to fluctuations in surface flatness due to the higher thermal expansion coefficient of stainless steel as compared with plain carbon steel. In the direct casting of aluminum employing a horizontal Hazelett-type machine, the rapid change in temperature of the belt as it enters the region of the molten metal pool causes considerable belt distortion, non-uniform contact and varying heat transfer rates. This problem has been significantly reduced by belt preheating techniques, which involve applying steam to the cooling side of the belt before it contacts molten metal (31). It also has been reported that for the casting of steel with a Hazelett-type caster, the mean belt temperature can be maintained below 190°C and the belt tension required to keep the belts taut does not exceed the yield strength of the material (36). The life of belts varies from eight hours to two weeks which is low relative to mould life in a continuous billet or slab casting machine.

Summary

This paper explores the fundamental aspects of near-net-shape casting, particularly heat-extraction, because it profoundly influences the production rate, cast structure and surface quality of the cast section. Six basic types of heat-extraction devices are currently under development for casting near-net-shapes: single roll, twin roll, twin belt, wheel belt, single belt and convex mould. To achieve the production rates of an integrated steel company or minimill, casting speeds will have to be considerably higher (10-40 m/min) than currently achieved with conventional continuous casting machines (0.5 - 4.0 m/min).

To achieve the production rates stated the heat-extraction device must be capable of removing between 6 and 22 MW if the thin section is to be fully solidified at these speeds. Measured heat-transfer coefficients reported for twin-belt casters vary from 1.3 to 2.6 $kW/m^2{}^\circ C$, which are very similar to those close to the meniscus of a conventional slab caster. Consequently the cooling rates (20 to 100°C/s) and secondary dendrite-arm spacing are like those obtained in the first 25 mm of a continuously cast slab. The secondary dendrite-arm spacing of thin sections cast on single or twin-roll machines have been reported to be in the range 1 $\sim$ 10 μm which have been associated with cooling rates of $10^2 \sim 10^4$°C/s; heat-transfer coefficient data for these machines is sparse. Secondary dendrite-arm spacing near the surface (0 $\sim$ 3 mm) of conventionally cast billets has been found to be in the range 15 $\sim$ 30 μm and the calculated cooling rates are of the order of $10^2 \sim 10^3$°C/s. This suggests that heat-extraction rates in single- and twin-roll devices are probably higher than the upper limit of 7 MW/m^2 measured at the meniscus of billet casters.

Surface quality problems resulting from excessive friction are a concern because with the exception of the convex mould (SMS Concast), all the heat-extraction devices move continuously in a rotary or linear manner and are unlubricated in a conventional sense. For single roll processes thickness variations across the width and surface cracking have been reported to occur, whilst for twin-roll processes complete solidification within the roll-bite has been associated with surface cracking. During twin-belt casting transverse cracking and poor flatness may result due to friction between the steel and the belt, and belt distortion.

The dimensional stability of the heat-extraction device is of paramount important because of stringent requirements on the shape of the cast section if it is to replace conventionally rolled material. Thus the development of thermal crown in roll-based processes and belt-distortion in Hazelett type machines are a concern. Thus considerable attention would have to be paid to the design, geometry and cooling of the heat-extraction device in order to ensure dimensional stability and uniform heat-extraction in the transverse direction.

Acknowledgements

The authors wish to express their appreciation to Dr. R. Babu for his efforts. The financial support of NSERC is gratefully acknowledged.

References

1. T.E. Dancy: "The Direct Casting of Steel Strip", Report prepared for Industrial Materials Research Institute, 1986.
2. M. Cygler and M. Wolf: "Continuous Strip and Thin Slab Casting of Steel - An Overview", Iron and Steelmaker, 1986, Vol. 13, No. 8, pp. 27-33.
3. M.C. Flemings and N.J. Grant: "Technology Development of Thin Strip Metal Casting - Feasibility Study", Report prepared for the U.S. Department of Energy, Office of Industrial Programs (Contract No. DE-AS07-83ID12444), January 1984.
4. W. Reichelt, W. Kapellner and R. Steffen: Near-Net-Shape Casting of Flat Products", Metallurgical Plant and Technology, 1988, Vol. 11(2), pp. 18-35.
5. R.K. Pitler: "Direct Casting of Coilable Ferrous Alloy Strip", N.J. Grant Symposium on Processing and Properties of Advanced High Temperature Alloys, M.I.T., June 1985, ASM Metals/Materials Technology Series.
6. N. Toyama, H. Abo, H. Yoshimura and H. Arai: "Direct Strip Casting of 18-8 Stainless Steel by Unequal Diameter Twin-Roll Method", Trans. ISIJ, 1986, Vol. 26, pp. B-149 - B-150.
7. A. Kasama, S. Mizogouchi, K. Miyazawa, M. Ito and T. Sugai: "A Consideration on the Rapid Solidification of Steels - Solidification of Steel by Twin-Roll Caster", Fifth Intl. Iron and Steel Congress, Washington, D.C., Steelmaking Proceedings, 1986, Vol. 69, pp. 849-853.
8. F. Kogiku, M. Ozawa, K. Shibuya, H. Shishido and T. Kan: "Rapid Solidification Process of High-Silicon Steel Strips and Their Characteristics", Trans. ISIJ, 1986, Vol. 26, pp. B-141 - B-142.
9. S. Nishioka, H. Tsuchida, Y. Kitagawa, Y. Fujita, H. Sato and M. Ohtsuki: "Solidification Behavior in the Twin-Roll Casting Process", Tetsu-to-Hagane, 1985, Vol. 71, No. 10, pp. A241-244.
10. K. Kawakami, J. Inagaki, S. Nishioka, H. Sato and M. Ohtsuki: "Research on Twin Roll Casting Process", Fifth Intl. Iron and Steel Congress, Washington, D.C., Steelmaking Proceedings, 1986, Vol. 69, pp. 861-870.
11. J.H. Cox: "Near Net Shape Casting of Flat Products - State of the Art", Private Communication.
12. C. Yoshida, K. Taniguchi, T. Nakagawa, M. Sudo and T. Nozaki: "Characterization of Rapidly Solidified Cast Iron and High-Carbon Steel", Fifth Intl. Iron and Steel Congress, Washington, D.C., Steelmaking Proceedings, 1986, Vol. 69, pp. 855-859.
13. T. Kusakawa, H. Yamamoto, Y. Yanagi, K. Endo and K. Oka: "Characteristics of Iron and Steel Thin Sheet Made by Strip Casting", Fifth Intl. Iron and Steel Congress, Washington, D.C., Steelmaking Proceedings, 1986, Vol. 69, pp. 871-875.
14. B.C. Whitmore and J.W. Hlinka: "Continuous Casting of Low-Carbon Steel Slabs by the Hazelett Strip-Casting Process", TMS-AIME Open Hearth Proceedings, 1969, Vol. 52, pp. 40-45.
15. Y. Sugitani, M. Nakamura, Y. Shirai, T. Okazaki and M. Yoshihara: "Solidification and Heat Transfer Phenomena in the Twin Belt Caster", Trans. ISIJ, 1986, Vol. 26, pp. B-153 - B-154.
16. S. Itoyama, H. Nakato, T. Nozaki, Y. Habu and T. Emi: "Development and Solidification Characteristic of Horizontal Twin-Belt Type Thin Slab Caster", Fifth Intl. Iron and Steel Congress, Washington, D.C., Steelmaking Proceedings, 1986, Vol. 69, pp. 833-838.

17. K. Ushijima, T. Okazaki, M. Yoshihara, S. Shiode and M. Koide: "Thin Steel Slab Continuous Casting Process", Fifth Intl. Iron and Steel Congress, Washington, D.C., Steelmaking Proceedings, 1986, Vol. 69, pp. 827-833.
18. J. Inagaki, Y. Takada, K. Nakaoka and K. Yamamoto: "Microstructural Changes of Steels Induced by Rapid Solidification and Their Applications", Trans. ISIJ, 1986, Vol. 26, pp. B-143 - B-144.
19. M.M. Wolf: "A Review of Published Work on the Solidification Control of Steel in Continuous Casting Moulds by Heat Flux Measurement", Trans. ISIJ., 1980, Vol. 20, pp. 718-724.
20. E. Takeuchi and J.K. Brimacombe: "The Formation of Oscillation Marks in the Continuous Casting of Steel Slabs", Metall. Trans. B, 1984, Vol. 15B, pp. 493-509.
21. I.V. Samarasekera and J.K. Brimacombe: "Heat Extraction Capability of Continuous-Casting Billet Moulds", Proc. Philbrook Memorial Symposium on the Fundamentals of Iron and Steelmaking, Toronto, ISS, 1988.
22. H.F. Marten and H. Fastert: "Latest Developments and Operating Results of the High-Speed Strip-Casting Machine at SMS Schloemann-Siemag AG", 44th Electric Furnace Conference, Dallas, Texas, 1986.
23. H. Takemoto, K. Ohnishi, K. Nagai, H. Nishiyama and K. Kunugiza: "Cooling and Solidification Characteristics of Thin Slab Casting by Twin Roll Caster", Transactions of ISIJ, 1986, Vol. 26.
24. J.F. Grubb, D.B. Love, A. Murthy and J.D. Nauman: "Casting Specialty Steel Strip", Near Net Shape Casting, Published by the Iron and Steel Society, 1987, pp. 51-57.
25. A.R.E. Singer and W.N. Jenkins: "An Assessment of Spray Forming as a Potential Continuous Route to Steel Strip", ibid., pp. 87-91.
26. H. Westengen and K. Nes: "Twin Roll Casting of Aluminum: The Occurrence of Structure Inhomogeneities and Defects in As-Cast Strip", Reference details not available.
27. R. Iricibar and I. Jin: "Roll Aspects of Twin Roll Casting", Reference details not available.
28. M.J. Bagshaw, J.D. Hunt and R.M. Jordan: "A Steady State Model for Roll Casting", Proceedings of the Third Conference on Modeling of Casting and Welding Processes, Engineering Foundation of AIME, January, 1986, pp. 135-147.
29. H.S. Marr, G.D. Spenceley, V. Ludlow and J. Peace: "Horizontal Continuous Casting of Thin Slabs at the British Steel Corporation - Teesside Laboratories", Near Net Shape Casting, published by the Iron and Steel Society, 1987, pp. 21-30.
30. T. Nozaki, S. Itoyama, N. Bessho, T. Fujii and K. Nakanishi: "Heat Extraction from Thin Slab and Concept of its Caster Developed by Kawasaki Steel Corporation", Journee's Siderurgiques ATS 1986.
31. D. Jaffrey, I. Dover and L. Hamilton: "Recent Developments in Continuous Casting Technology: An Overview", Metals Forum (7), 1985.
32. Y. Miyake, I. Yarita, K. Hamada, Y. Hiroshe, Y. Ogawa and K. Toyoshima: "Development of Hot Rolling Technology for Improving Strip Profile and Flatness", Kawasaki Steel Technical Report, No. 12, 1985, pp.1-13.
33. C. Devadas and I.V. Samarasekera: "Heat Transfer During Hot Rolling of Steel Strip", Ironmaking and Steelmaking, 1986, Vol. 13, No. 6, pp. 311-321.
34. W.L. Roberts: "Frictional Effects in Hot Rolling", Hot Rolling of Steel, Marcel Dekker Inc., New York, 1983, pp. 845-853.

35. C. Ueda: "Calculation of Thermal Crown of a Hot-Rolling Roll at Kakogawa Works", Private Communications.
36. C.J. Petry and J.F.B. Wood: "Hazelett Twin-Belt Thin Slab Casting and Downstream Processing Alternatives", Reference details not available.

Table I. Partial Listing of Heat-Extraction Devices Under Development for the Continuous Casting of Near Net Steel Shapes (adapted from R. Littlewood, Stelco)

Heat-Extraction Device	Metal Feeding	Company/ Research Institution	Section Dimensions Width x thickness (mm)	Casting Speed (m/min)
Single roll	Melt drag	Allegheny-Ludlum	300 x (1-2)	
		Armco/Westinghouse	75 x (0.7-3)	(460)
	Melt drag	Battelle/National	1800 x (0.25-4.5)	
		Thyssen	< 1 mm	
	Spray deposition	Swansea		
	Melt drag	Nippon Metal	315 x (1-3)	
Twin roll		Nippon Steel Co.	260 x 1	
	Top feed	NKK	100 x (2-5)	2-25
	Top feed	Kawasaki	200 x (0.1-0.4)	180-720
	Inclined feed	Kobe	270 x (1-2)	5-10
		Hitachi Zosen	350 x (6-60)	7 max
	Top feed	BSC	400 x (1-25)	
		CSM	300 x (1-25)	
		IRSID	200 x (1-25)	
	Inclined feed	Waseda University	100 x (1-1.5)	4-14
		LTV (roll within roll)	250 x (2-5)	7.5
		Bethlehem/Armco/ Weirton/Inland	30 x 2	
Single belt	Mould-trough train	BSC	500 x 50	10-13
	Spray deposition	Mannesmann Demag	600 x 20	
Twin belt	Inclined	Sumitomo (Hazelett)	600 x (20-80)	15 max
	Horizontal	Kawasaki	(100-150)x(10-30)	0.7-12.5
		Kawasaki	600 x (40-50)	> 15 (planned)
	Inclined	Bethlehem/USS	(1220) x (12-25)	(stopped)
		(Hazelett)	1800x (12-15)	9-12
	Inclined	Nucor (Hazelett)	1300 x 40	
	Inclined	Krupp (Hazelett)	200 x (20-70)	10
	Vertical	Nippon Steel	600 x (50-80)	4
Wheel-belt	Inclined	Southwire	3110 sq.mm	14
(billets)	Top feed	Hitachi-Korf	30600 sq.mm	3-6
Convex mould/ conventional equipment		Schloemann-Siemag (SMS Concast)	300 x 50	6 max

Table II. Heat-Extraction, Cooling Rates and Structure Reported Under Thin Slab/Strip Casting Conditions

Heat Extraction Device/Company/ Comments	Section Dimensions (mm)	Casting Speed (m/min)	Dwell Time (s)	Superheat (Cast Temp.) (°C)	Steel Grade	Heat-Transfer Coefficient (kW/m² °C)	Cooling Rate, R (°C/s)	Solidification Constant mm/s$^{1/2}$	Dendrite Arm Spacing, S (μm) 1° or 2°	Reference
Single roll										
Allegheny Ludlum (227 kg)	330 x 1	-	-	-	304	-	5000 (0.2 mm) 800 (0.63 mm)	-	5-10(2°)	Pitler[5]
Twin roll										
Nippon Metal (unequal diam. rolls)	-	-	-	50-70	304	-	~10^3-10^4 (est.)	-	9(2°) 0.5 mm from lower surface	Toyama et al.[6]
Nippon Steel (Lab., 100 kg)	200 x 1	30			304	1.25	10^3-10^4 (est.) 10^2-10^3 (est.)	-	2-5(2°) < 0.2 mm from surface 4-8(2°) 1.7 mm from surface	Kasama et al.[7]
Kawasaki	(0.12-0.5) thick	3-12	0.01 (est.)	-	Si steel	(7) (est.)	10^3-10^4	-	1-7(1°)	Kogiku et al.[8]
Nippon Kokkan K.K.	(2-17.0) thick	25.0-0.6	0.4-16.3	-	-	7.3-1.4		1.55-2.07	40 1.3 mm from surface	Nishioka et al.[9,10] (Cox[11])
Kobe (Lab., 15 kg)	-	5-10	-	-	Cast iron	-	10^3 (surface)	-	7(2°) at surface	Yoshida et al.[12]
Waseda Univ.	100 x (1-1.5)	4-14	-	(1250)	Cast iron 18-8 Stainless	-	10^2-10^3 (est.)	-	-	Kusakawa et al.[13]
Twin belt										
Bethlehem	(75) thick	1.8-2.5	71	-	Low C(AK)	0.6	-	-	-	Whitmore & Hlinka[14]
Sumitomo(simulated with single belt)	45 x 45	Static	up to 60 s	-	Low C(AK)	2.73 (0-2 s) 1.13 (> 2 s)	~ 10^2 (surface)	2.58 (same as skin of conv.slab)	50-150 1-20 mm from surface	Sugitani et al.[15] (Cox[11])
Kawasaki	100 x 30	1.9-6.0	30	24-70	0.002-0.16%C Stainless Si steel	1.4(avg) 0.8-2.1 (upper side)	~ 20	2.32 1.29-2.58 (upper side)	20-60(2°) S_{II} = 146 $R^{-0.386}$	Itoyama et al.[16]
Sumitomo (50 tonne ladle)	600 x (20-80)	4-6	45	(1550-1580)	Low C(AK) SiK High Mn(AK) Stainless	0.93	~ 25 (surface)	2.58	S_{II} = 158 $R^{-0.36}$	Ushijima et al.[17]
Laboratory Studies										
Split Plate Mould (Static) NKK	(1-20) thick	Static	-	50	Carbon Stainless Low C(AK)	-	10-10^3	-	S_{II} = 181 $R^{-0.36}$ 3.8 (2°), 1mm 9.0 (2°), 5mm 17.2 (2°), 17mm	Inagaki et al.[18,10]
Chill Block, Kobe (Caterpillar block simulation)	75 x (3-40)	Static	-	(1590)	0.8% C	2.51	10^3 (surface)		5-6 (2°) at surface	

Table III. Casting Speed Requirements to Meet Production Rates for an Integrated Operation and Minimill

Heat Size	240 t			60t
No. Strands	1	2	4	1
Gauge (mm)	Casting Speed m/min			
5	80	40	20	20
10	40	20	10	10

Table IV. Shell Thickness on Exit from Rolls in a Twin-Roll Device

Casting Speed m/min	10			20		
Strip Thickness (mm)	10			5		
Roll Diameter (mm)	Dwell Time (s)	Roll Speed (r.p.m.)	Shell Thickness on Exit (mm)	Dwell Time (s)	Roll Speed (r.p.m.)	Shell Thickness on exit (mm)
400	0.99	8.0	2.25	0.50	16.0	1.25
600	1.76	5.3	3.40	0.88	10.6	2.10
800	2.54	4.0	4.30	1.27	8.0	2.70

*Calculations have been performed for a freeboard 50 mm below the top of the rolls.

Table V. Surface and Shape Defects in Near Net Shape Cast Product

Heat Extraction Device	Metal Feeding	Product	Quality Problems	Causes	Remedy
Single Roll	Melt drag	300 x (1-2) 304 stainless	Upper surface cracking[24] Upper surface porosity transverse thickness variation Edge tear[5,24]. Strip buckle.	Mismatch in speed of coiling and casting[5].	Control of casting parameters, casting wheel design changes, shrouding.
	Spray deposition		Transverse thickness variation[25]. High porosity[25].	Non-uniform spray deposition[25]. Low spray density[25].	Development of scanning atomisers.
Twin Roll	Top feed	100 x (2 ~ 25 mm) Experimental 0.2% Carbon steel	Surface cracks[10].	High roll separating[10] forces, non-uniform solidification, large unbending strains at bonding point.	Prevent solidification in roll bite.
Twin Roll	Horizontal feeding	Aluminum	Edge cracking[26]. Intrusions on strip[26].	Heavy deformation in roll bite. Fire cracking of work rolls.	
	Horizontal feeding	Aluminum	Edge cracking[27]. Sticking[27].		
	Horizontal feeding	Aluminum	Heat line Formation[28].	Cast material leaves[28] roll bite partially molten.	Improve rate of heat[28] removal.
Twin Roll	Unequal diameter	18-8[6] stainless steel	Edge crack and flash.	Molten metal penetration in the gaps between upper roll and nozzle.	Control casting conditions
	Inclined feeding		Longitudinal crack in lower side.	Delayed solidification locally.	
Single belt	Mould trough train	0.3% carbon steel 300 x 50	Hot tearing[29].	Mismatch in contraction of steel and mould and extensive friction between the two[29].	·
Twin belt	Inclined	Steel 600 x (20-80)	Transverse cracks[17] 1 mm deep ~ 30 mm long. Hair cracks[17]. 'Dam marks' on narrow face[17].	Friction between belt and slab surface during cooling of shell. Belt coating thickness irregularities. Discrepancy of dam blocks.	Set belt speed higher than pinch roll speed. Uniform coating of lubricant.
Twin belt	Horizontal	Steel (.16% Carbon) 100 x (25 ~ 40)	Ripple marks on narrow face similar to cold shuts in horizontal casting[30]. Fluctuations in surface flatness[30].	Stationary narrow sides. Thermal distortion of the belt.	Employ plain carbon steel belts instead of stainless steel.
	Horizontal	Aluminum	Surface, liquation porosity and cracking[31].	Uneven distortion of belts.	Belt preheating.
Convex mould/ conventional equipment		Steel	Transverse cracks, surface laps, corner cracks and breakouts, longitudinal depressions and cracks[22].	Problems associated with mould geometry in area of mould funnel[22].	Improved mould geometry, increase in speed with decreased cooling, optimization of mould oscillation[22].

[1] Single roll

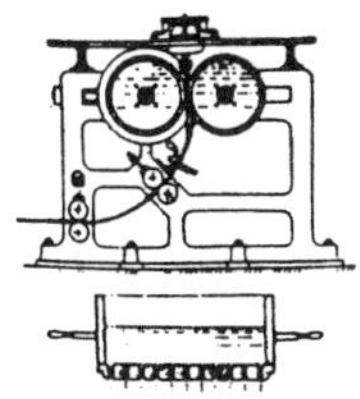

[2] Twin roll

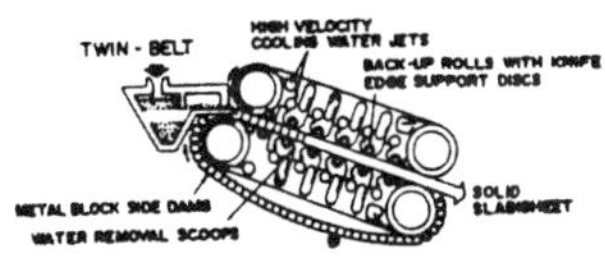

[3] Twin belt

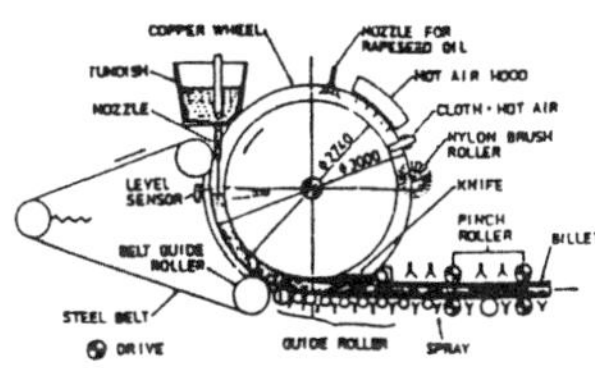

[4] Wheel-belt

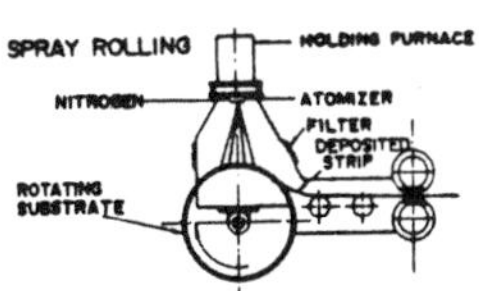

[5] Single belt/spray

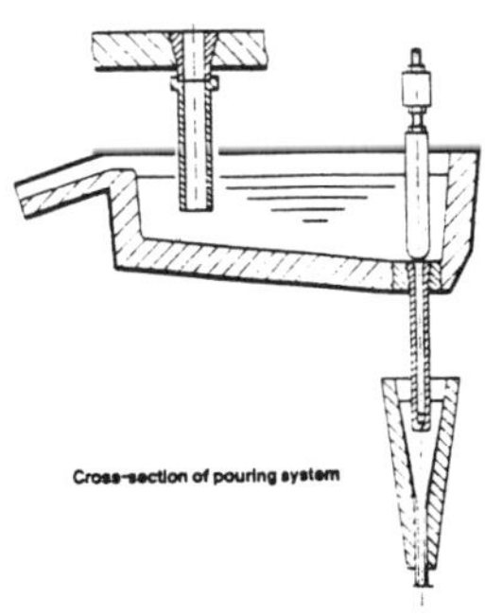

[6] Convex mould (SMS)

Figure 1 - Schematic diagrams of near-net-shape casting processes.

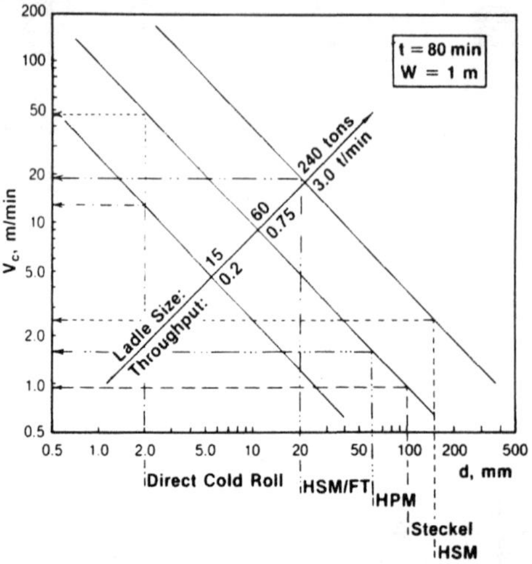

Figure 2 - Casting speed versus product thickness (width of 1 m) as a function of the required throughput rate for a casting time of 80 minutes, and the range of applicable rolling processes[2].

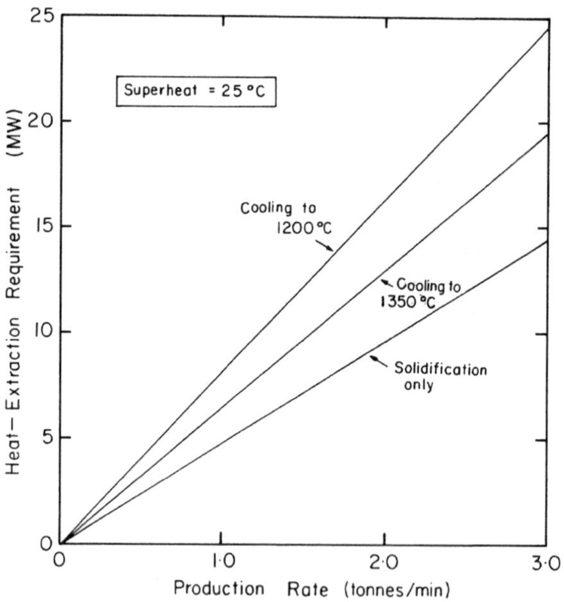

Figure 3 - Heat extraction requirements for different production rates.

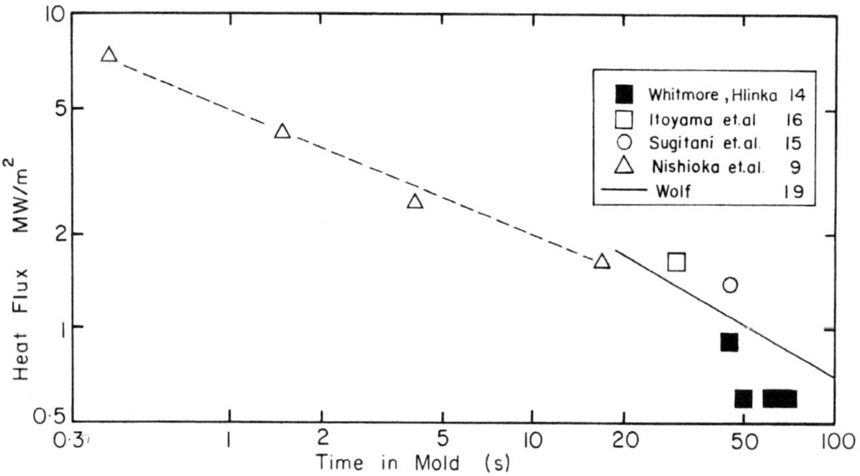

Figure 4 - Heat flux in heat extraction device as a function of dwell time[11].

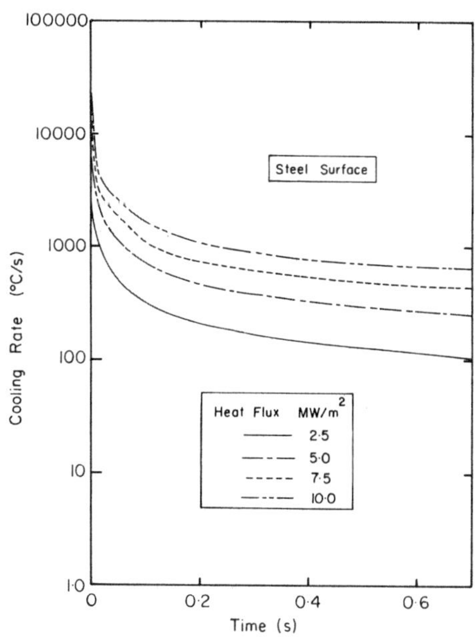

Figure 5 - Cooling rate at the surface of the strand as a function of time.

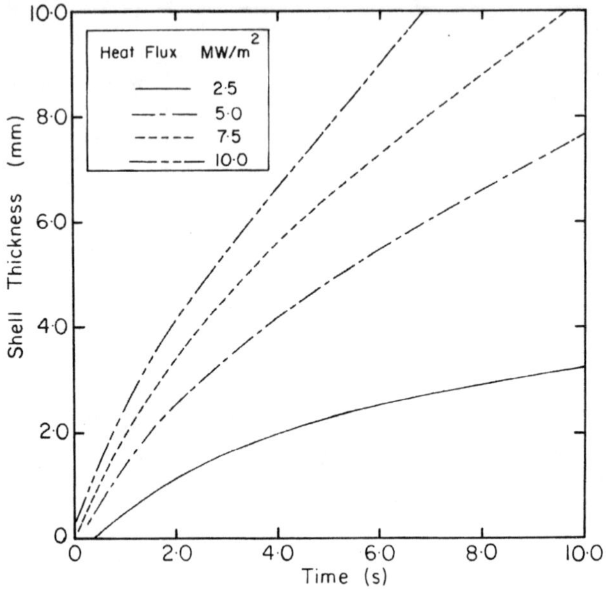

Figure 6 - Shell thickness profiles for different heat extraction rates in the mould.

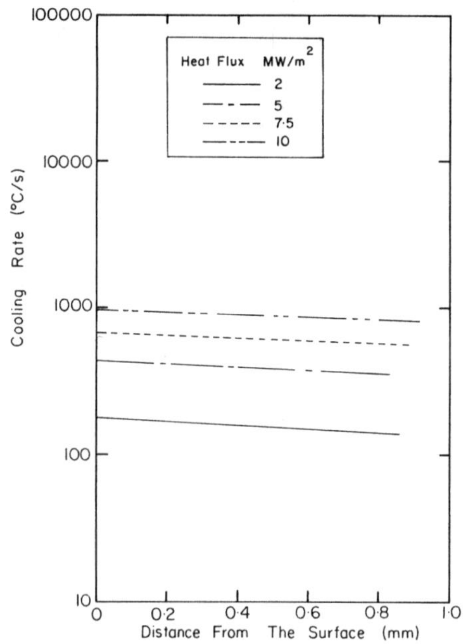

Figure 7. Calculated cooling rate as a function of distance from the surface.

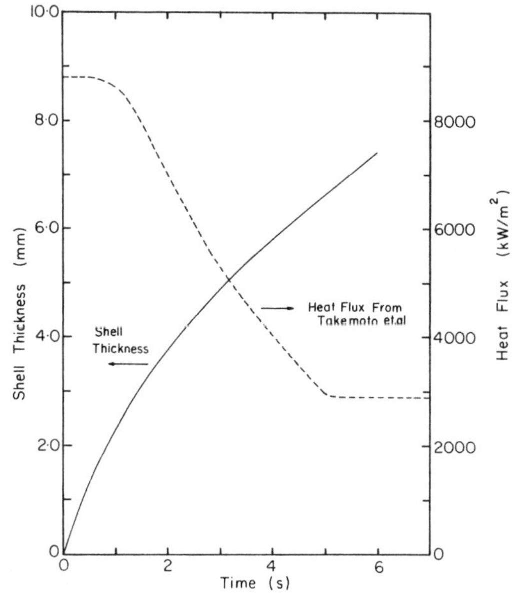

Figure 8 - Predicted shell thickness profile for heat-extraction rates obtained from Takemoto's data.

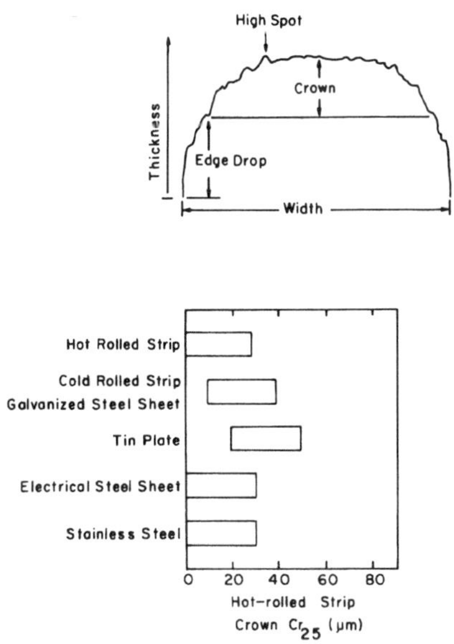

Figure 9. Restrictions on strip crown for different applications.

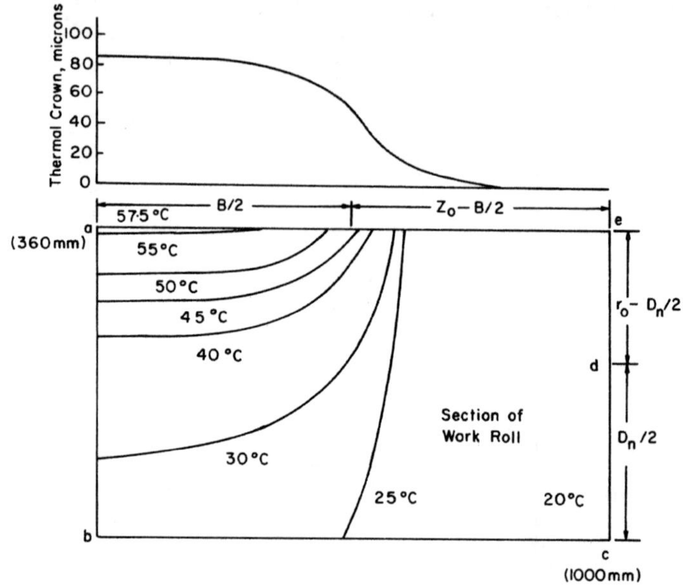

Figure 10 - Thermal crown and roll temperature distribution in a steel work roll in a hot strip mill.

FUNDAMENTAL ASPECTS I:

STRIP CASTING

Session Chairman: **G.R. St. Pierre**
Ohio State University
Columbus, Ohio

PLANAR FLOW CASTING OF LOW CARBON STEEL STRIP

R. Sussman*, R. Williams** and D. Follstaedt*

*Armco Inc., 703 Curtis Street, Middletown, OH 45043.
**Westinghouse Electric Corp., 1310 Beulah Road
Pittsburgh, PA 15235.

Abstract

In the planar flow casting process, metal is teemed out of a nozzle designed to be delivered to the substrate at approximately the rate that can be solidified on the substrate. This paper will describe experimental research into the casting of low carbon steel strip by the planar flow casting method. Particular features of the casting process will be discussed especially those influencing fluid flow and solidification. The experiments are being performed on a large water-cooled, single-wheel caster. Mathematical modeling of fluid flow and solidification is used to supplement the experimental effort. Water modeling of flow from the pouring box and nozzle is also essential to determine the critical flow parameters.

Casting of Near Net Shape Products
Edited by Y. Sahai, J.E. Battles,
R.S. Carbonara and C.E. Mobley
The Metallurgical Society, 1988

THREE-DIMENSIONAL MATHEMATICAL MODELING OF FLUID FLOW AND HEAT TRANSFER ASPECTS IN TUNDISHES FOR STRIP-CASTING

Y. Sahai* and R. C. Sussman**

*Dept. of Materials Science and Engineering, The Ohio State University, 116 West 19th Avenue, Columbus, OH 43210.
**Armco Inc., 703 Curtis Street, Middletown, OH 45043.

Abstract

A three-dimensional mathematical model has been developed to predict the melt flow in strip casting tundishes. These tundishes are smaller than the typical tundishes employed in continuous casting of thicker cross-section products. However, the optimum flow characteristics of molten metal from these tundishes to the substrate are extremely important for the production of good quality of strip. In this paper, three-dimensional model and the computed results will be presented. The effect of the tundish design on flow control will be discussed.

Casting of Near Net Shape Products
Edited by Y. Sahai, J.E. Battles,
R.S. Carbonara and C.E. Mobley
The Metallurgical Society, 1988

A MATHEMATICAL MODEL OF THE STRIP CASTING PROCESS

Ernesto Gutierrez-Miravete

School of Engineering and Science
The Hartford Graduate Center
Hartford, CT 06120

Abstract

The formulation of a mathematical model of the strip casting process is described and selected results are discussed. Emphasis is made on the single chill process but comments are also included regarding the twin chill operation. The model is derived by combining the theories of lubrication and capillarity with standard methods for the solution of phase change problems. The examples discussed include the casting of low carbon steel strip by the melt overflow process, the production of superalloy ribbons by the planar flow melt spinning process, and the production of aluminum sheet by a single chill process. The model uses as input data the values of the operating variables and the physical properties. The microstructure of the resulting strip is calculated and the predictions are in reasonable agreement with experimental measurements.

Casting of Near Net Shape Products
Edited by Y. Sahai, J.E. Battles,
R.S. Carbonara and C.E. Mobley
The Metallurgical Society, 1988

Introduction

The possibility of producing metal sheets directly from the melt has long attracted the attention of metallurgists (1), (2). However, the earliest trials apparently were not very successful and the idea was temporally abandoned. The possibility transformed into a reality during the last few decades as wide sheets of low melting point alloys were successfully cast. During the last ten years or so the steel industry has invested a great deal of effort on research towards the production of thin steel sheets (3). The main objective of strip casting operations is the continuous production of controlled gage metal sheets directly from the melt in a single operation. Considerable savings in metal processing costs are expected to result from industrial scale strip casting when compared with the more conventional sheet producing techniques based on the mechanical working of thick sections.

There are at least three essential elements in any strip casting installation. First, a furnace or molten metal holder is required to supply molten metal with the desired temperature and chemistry in a continuous or semicontinuous fashion. Next one needs a tundish in order to be able to control the cleanliness of the melt and also to streamline the flow on its way towards the casting slot. Finally there is the casting mold where the solidification takes place. The mold usually moves and it consists of one or two quenching surfaces with lateral containment to avoid spilling. A means of capturing the solid strip as it emerges from the caster may also be required. Despite the apparent simplicity successful production of good quality strip is not easy and the situation is particularly complicated in the case of the high temperature melts such as steel. Steel strip a few thousands of micron thick is yet to be produced on an industrial scale in spite of intense efforts. As mentioned before, strip casting of low melting point alloys has been more successful. For example, aluminum alloy sheets almost 2 m wide and 7 mm thick have been routinely cast at rates of about 500 kg/s on an industrial scale since the late 1960's (4). It should be noted that this success became possible, at least in part, thanks to the availability of modern instrumentation and computers. The much higher temperatures involved in the case of steel melts still pose a formidable challenge. However some progress has been reported in the case of specialty steels (5). The evidence indicates that the simplicity of the process is more apparent than real and that very close control of the process is required in order to succeed.

The thermal and chemical conditions in the holding furnace and particularly in the tundish must be carefully controlled. One must have a good quantitative understanding of the fluid flow phenomena in the tundish in order to be able to control the supply of molten metal to the mold. Finally, the cooling conditions in the strip casting mold must also be carefully controlled, in order to be able to obtain the desired solidification microstructure in the resulting strip or sheet.

The objective of this paper is to describe a simple mathematical model of the solidification process in the strip casting mold. The formulation is followed by a discussion of results obtained for three different strip casting systems.

Formulation

A common feature of strip casting processes is the thinness of the melt in the casting mold. The small section is required in order to produce the desired solidification rates. Additionally, the melt layer is usually much elongated in the casting direction with the result that the fluid flow phenomena in the casting mold approach a condition known as thin film flow. It is well known that the flow in thin liquid films can often be represented by the theory of lubrication. The theory of lubrication is based on the following assumptions: (i) since the liquid film is thin, the pressure can be assumed constant across the thickness, (ii) since the film is long, the component of velocity in the thorough thickness direction can be neglected in comparison with the component along the length of the film, and (iii) the inertia forces are negligible in comparison with the viscous forces. The introduction of these assumptions leads to considerable simplification in the analysis of fluid flow in thin liquid films. Analytical solutions of the Navier-Stokes equations become possible and these have been available for over one hundred years (6).

The theory of lubrication was combined with a solidification model to produce a mathematical representation of the twin roll rapid solidification process (7). Velocities, temperatures and cooling rates were calculated and good agreement with experiment has been found (8). The model presented in the present paper was inspired by that in (7) and (8) and represents a generalization of the early model. As mentioned above, strip casting molds consist of either one or two quenching surfaces. Melt spinning and melt overflow are examples of the first case while twin roll casting exemplifies the second. In the case of strip casting molds with a single quenching surface, the film surface opposite to the melt-chill interface is actually a free surface. Such free surface is relatively unconstrained and its motion must be controlled in order to obtain steady operating conditions. The theory of capillarity (9) was developed to analyze the shapes of free liquid surfaces. The main result of the almost two hundred years old theory is an expression relating the pressure difference across a free liquid surface to the local curvature of the interface. More than a century after its invention, the theory of capillarity was combined with the theory of lubrication to develop a mathematical model of the problem of dragging a liquid film by a moving plate. The problem is a common occurrence not only in the movie industry where a uniform layer of photosensitive emulsion is deposited upon a photographic film base (10), but also in various other contexts (11).

The theory of capillarity, the theory of lubrication and a solidification model have been combined to produce a mathematical representation of the casting of thin foils (about 50 micron) by the planar flow melt spinning process (12),(13). The particular solidification model employed is the so-called enthalpy method (14). Since the model is quite simple and also because the details of the formulation are readily available, only the final resulting equations will be shown here. Then, for the problem of solidification of thin strip on a single chill and with a free liquid surface one obtains:

$$\Psi = \begin{cases} V_x\, y/(Q_v/w) , & \text{for } y < y_s \\ [V_x\, y_s + A_2(y^3-y_s^3)/3 + A_1(y^2-y_s^2) + A_o(y-y_s)]/(Q_v/w) , & \text{for } y > y_s \end{cases} \quad (1)$$

$$H_1''' = [(Q_v/w) - V_x(H_1+y_s) - \sigma_T\, T_x(H_1^2/2\mu)]* [3\ \mu/H_1^3\sigma] \quad (2)$$

$$A_2 = [V_x(H_1+y_s) - (Q_v/w) + \sigma_T\, T_x(H_1^2/2\mu)]* [3/2H_1^3] \quad (3)$$

$$A_1 = \sigma_T\, T_x/\mu - 2(H_1+y_s)\, A_2 \quad (4)$$

$$A_o = V_x - \sigma_T\, T_x\, y_s/\mu - (y_s^2 - 2(H_1+y_s)y_s)\, A_2 \quad (5)$$

$$H_1''' = d^3H_1/dx^3 \quad (6)$$

$$\sigma_T = d\sigma/dT \quad (7)$$

$$T_x = dT/dx \quad (8)$$

where

- x is the distance along the casting direction,
- y is the distance along the thickness of the film,
- Ψ is the stream function,
- Q_v/w is the volumetric flow rate per unit width
- V_x is the velocity of the moving chill
- H_1 is the thickness of the liquid layer
- y_s is the thickness of the solidified layer
- μ,σ are, respectively, the viscosity and surface tension of the melt, and
- T is the temperature

Note that the solidification heat transfer model is needed in order to evaluate the solidified thickness and also the temperature at the free surface of the melt. The solution of the solidification equations must be carried out using numerical techniques such as the enthalpy method (14). Equations (1)-(8) have been used to model the production of Ni-based superalloy foils by the melt spinning process (12),(13) and the microstructures predicted were in good agreement with metallographic observations. The same equations have been used in combination with an empirical solidification model to simulate planar flow casting of 200 micron thick aluminum foils (15). Also good agreement was found between predictions and experiment. The calculations in (13) and (15) neglected the

Marangoni effect, however, it has been shown that this assumption is appropriate in many cases (12). A set of equations analogous to Eqns. (1)-(8) can be readily derived for strip casting in molds with cooling from both sides. The resulting set of equations for twin roll quenching can be found in (7). In the next section a selection of computed results obtained from the model described will be presented.

Discussion

The model has been used to analyze the solidification process during strip casting for three distinctly different systems. The process parameter values employed for the calculations are representative of those used in existing strip casting installations. The examples have been selected because they represent a wide range of strip thicknesses, material properties and operating conditions. It is shown that the predictions of the model are in reasonable agreement with experiment. The examples selected for study are: (i) casting of 400 micron thick low carbon steel strip by the melt overflow process (16), (ii) casting of 40 micron thick superalloy ribbons by the planar flow melt spinning process (12), and (iii) casting of 7000 micron thick aluminum sheets by the single roll process.

Figure (1) shows the computed fluid flow pattern inside the meniscus formed in the melt overflow process. The figure shows the melt overflowing from the pouring lip and impinging on the chill which is moving from left to right. Values of the operating parameters and physical properties have been derived from the information in (16). The streamline pattern shows two distinct flow regions inside the puddle. Next to the moving chill the streamlines are compressed and more or less uniformly spaced. This is a result of the solidification and the viscous drag exerted on the melt by the partially solidified ribbon. The flow in this region is along the casting direction. Above this region of through flow there is a larger, slower moving region where the melt recirculates inside the molten puddle. Fluid flow patterns like the one shown in figure (1) are a common occurrence in lubrication and thin film flow systems (17). The computed temperature field associated with the flow map shown in the previous figure is shown in figure (2). As expected, the temperature gradients are concentrated in the vicinity of the chill and decrease gradually from the point of impingement. Furthermore the temperature in the bulk of the puddle is quite uniform and very close to the pouring value. The calculations show that the actual location of the free surface is closely related to the solidification rate (see equation(2)). Therefore, it should be possible to monitor the progress of the solidification process by careful observation and control of the meniscus. This is yet to be attempted. Once the temperature field has been computed other useful quantities are easily derived. For example, the local cooling rate in this case is calculated to vary from about 700,000 degrees C/s for solidification at the chill-strip interface down to about 1000 degrees C/s for solidification at the top surface of the strip. This suggests that the size scale of the microstructure (i.e. dendrite arm spacing or inclusion diameter) increases markedly across the strip thickness. This microstructure gradient is frequently observed in practice.

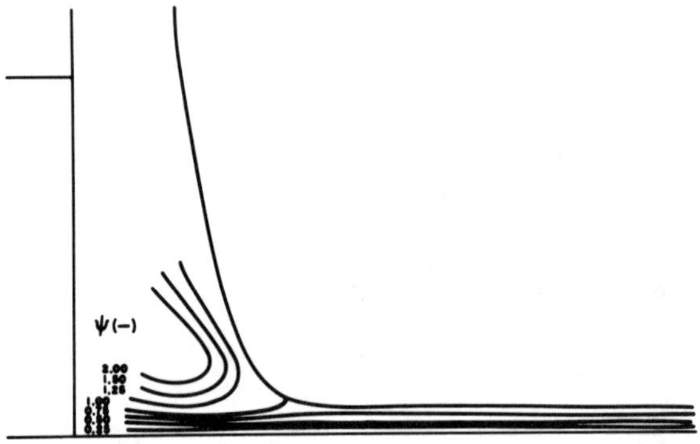

Figure (1). Fluid flow map in the meniscus

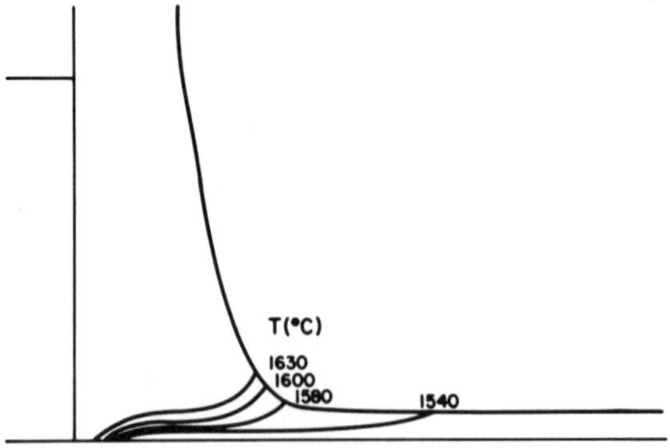

Figure (2). Temperature field in the meniscus

The general behavior shown in figures(1) and (2) has also been obtained for the planar flow melt spinning of a niquel-based superalloy ribbon (12), (13). Measured secondary dendrite arm spacings across the thickness of actual ribbons agree reasonably well with the predictions of the model. Agreement was obtained not only regarding the average size of the arms (about 0.5 micron) but also concerning the microstructure gradient across the ribbon thickness. Furthermore, there is experimental evidence that the fluid flow inside the melt spinning puddle is similar to the one predicted by the model. Additional discussion and specific references about all this can be found in (12) and (13).

It was decided to investigate the feasibility of producing 7000 micron thick aluminum sheets by the single chill process using the mathematical model. Although these sheets are routinely cast using the twin roll system (4) there does not seem to be a comparable single chill installation. For the calculations, the values given in (4) were used. The results indicate that the resulting strip would cool during solidification at a rate of the order of 200 degrees C/s . This value is in very good agreement with the observations reported in (4), which is to be expected since both processes work with the same melt thickness. Moreover the results also indicate that in order for the process with a single chill to be successful, the thickness of the liquid film entering the casting mold should not be more than a few mm thicker than the desired strip. However, it should be noted that thick liquid films are more prone to instabilities and this has not yet been accounted for in the calculations.

In summary, the proposed model of the strip casting process appears capable of predicting the behavior of the melt during processing. Predictions are possible for a wide range of operating conditions and physical properties with a very modest effort. Furthermore, the predictions are in reasonable good agreement with the limited experimental evidence available. Now that large efforts are in progress for the development of strip casting processes for steel it would be prudent to consider the possibilities of the mathematical modeling approach.

References

1. Bessemer, H., U.S. Patent No. 49053, July 25, 1865.

2. Lang, E.M., U.S. Patent No. 112054, February 21, 1871.

3. Cramb, A.W., "New Developments in the Continuous Casting of Steel. Part V: Strip Casting", Ironmaker and Steelmaker, **15** (5)(1988), 65-68.

4. Nes, E. and Slevolden, S., "Casting and Annealing Structures of Strip Cast Aluminium Alloys", Aluminium, 55(5)(1979), 319 -324.

5. Allegheny Ludlum Corporation, U.S. Patent No. 4475583, October 9, 1984.

6. Reynolds, O., "On the theory of Lubrication and its applications to Mr. Beauchamp Tower's experiments, including an

experimental determination of the viscosity of olive oil", Philosophical Transactions of the Royal Society of London, 177 (1886), 157-234.

7. Miyazawa, K. and Szekely, J., "A Mathematical Model of the Splat Cooling Process using the Twin-Roll Technique", Metallurgical Transactions A, 12A (1981), 349-358.

8. Miyazawa, K., Choh, T. and Inouye, M., "Cooling and Rolling Characteristics of the Twin-Roll Rapid Solidification Process", Transactions of the Japan Institute of Metals, 24(10) (1983), 696-703 and 781-788.

9. Finn, R., Equilibrium Capillary Surfaces (New York, Springer-Verlag, 1986), 1-16.

10. Landau, L. and Levich, B., "Dragging of a Liquid by a Moving Plate", Acta Physicochimica of the U.R.S.S., XVII(1-2)(1942) 42-54.

11. Ruschak, K.J., "Coating Flows", Annual Reviews of Fluid Mechanics, 17 (1985), 65-89.

12. Gutierrez, E.M. and Szekely, J., "A Mathematical Model of the Planar Flow Melt Spinning Process", Metallurgical Transactions B, 17B (1986), 695-703.

13. Gutierrez-Miravete, E., The Mathematical Modeling of Rapid Solidification Processing, NASA Contractor Report 179551, (National Aeronautics Space Administration, Lewis Research Center, Cleveland, OH 44135). Copies available on request from the author or from the National Technical Information Service, Springfield, Virginia 22161. pp. 59-103.

14. Crank, J., Free and Moving Boundary Problems (Oxford, Clarendon Press, 1984), 217-282.

15. Yu, H., "A Fluid Mechanics Model of the Planar Flow Melt Spinning Process under Low Reynolds Number Conditions", Metallurgical Transactions A, 18A (1987), 557-563.

16. Hackman, L.D. and Gaspar, T., "Producing Strip by Melt Overflow Process", Industrial Heating, LIII (1)(1986), 36-38. Additional information obtained during telephone conversations is gratefully acknowledged.

17. Schlichting, H., Boundary Layer Theory, 7th edition, (New York, McGraw-Hill Book Company, 1979), 116-122.

FUNDAMENTAL ASPECTS II: MODELING

Session Chairman: **K. Blazek**
Inland Steel Company
East Chicago, Indiana

THE ROLE OF THE MOLD-METAL INTERFACE

IN NET-SHAPE ALUMINUM CASTING

John T. Berry

ACIPCO Professor and Departmental Head
Department of Metallurgical and Materials Engineering
THE UNIVERSITY OF ALABAMA

Abstract

The development of our understanding of the role of the mold-metal interface in solidification heat transfer is critical to both modeling and controlling solidification processing operations, especially in net shape casting. Associated with this interface is the formation of an air-gap, which has been both discussed and measured by a variety of researchers. The present paper reviews and compares some of the important findings in this area and reports on some recent work in the author's laboratory which refers to net shapes produced by investment casting. The author concludes that the current absence of a rigorous thermo-mechanical mold of the zones surrounding the mold-metal interface precludes progress in modeling and control applications.

Casting of Near Net Shape Products
Edited by Y. Sahai, J.E. Battles,
R.S. Carbonara and C.E. Mobley
The Metallurgical Society, 1988

Introduction

Unlike many heat transfer dependant industrial processes, for example that of the heating and air conditioning of work spaces, the process of metallic alloy solidification, whether applied to net-shaped castings or simple ingot or continuously-cast configurations, is largely an open-loop type process. In other words, metallurgists have yet to successfully close the feed-back loop in processes of this nature.

As our knowledge of microstructural evolution during solidification is improved and as suitable criteria functions emerge which will facilitate the evolution of specific microstructures, effort must be turned to the task of bringing the solidification process truly under control

In order to facilitate this and especially in view of the time constraints of many aspects of the process of solidification, a better understanding has to be developed of the role played by the mold-metal interface in heat transfer processes involved. This paper discusses some of the generic aspects of heat transfer in this region.

Heat-Transfer at the Mold-Metal Interface

Veynik(1) and subsequently Hillier(2) have attempted to classify the types of thermal contact that exist in solidification processing by using a modified Biot Number

$$Bi = \frac{\text{conduction thermal resistance}}{\text{contact thermal resistance}} = \frac{h.X}{K}$$

For the modified number, the expression may be applied to each side of the interface, to the casting (Bi_1) and to the mold (Bi_2). Under steady state conditions, a state which does not necessarily apply in solidification processing, the ratio of the temperature differences which exist on either side of the interface δT_1 and δT_2 (pertaining to the casting and the mold temperature fields respectively) may be related to ratio of the Biot numbers (The Veynik Number? Ve)

$$Ve = \frac{Bi_1}{Bi_2} = \frac{\delta T_1}{\delta T_2}$$

Hillier(2) discusses four cases which appear below:

Case	Bi_1	Bi_2	Significant Temp. Difference
1	<<1	>>1	δT_2
2	<<1	<<1	ΔT
3	>>1	>>1	δT_1, δT_2
4	>>1	<<1	δT_1

ΔT in the above is actually the temperature difference across the mold-metal interface itself.

Case 1 adequately describes metallic solidification in a sand, or particulate refractory mold, while Case 2 describes metallic solidification in a metallic mold, where considerable thermal resistance is exhibited (air-gap or thickly coated chill). Case 3 might describe metallic solidification, where an uncoated mold, affording only moderate thermal resistance, is in contact with a metallic solidification system. (See Fig. 1) Case 4 is, of course, more appropriate to the freezing of injection molded polymeric melts. The first three cases are appropriate to net-shape casting.

The thermal contact conduction problem is generally described in terms of a heat transfer coefficient (h) or its inverse, the thermal contact resistance. In many approximate numerical analyses, the value of h is held constant during solidification. However, the time or more properly, the surface temperature dependance of h is sometimes incorporated into the computational scheme.

Clearly h will depend upon the true nature of the interface concerned. This aspect has been comprehensively discussed by Ho and Pehlke(3) These workers mention three factors as being important in the formation of the all important air-gap which often forms at this interface:

(i) surface interactions
(ii) transformations in the mold or the metal
(iii) geometric effects

They also cite work upon the effects of applied pressure(4) on the solidification heat transfer of aluminum in a thick-walled cylindrical steel mold enclosure (essentially squeeze casting). Figure 2, after Nishida and Matsubara(4), indicates how under the action of an applied pressure, the h value will rise to a maximum after pressure application but will then fall off as soon as the solidified shell is sufficiently thick, and thus strong enough, to resist the applied pressure. After this point, h falls rapidly, as the conformity of contact is destroyed by the receding away of the solidified shell. Ho and Pehlke(H/P) attribute this to the eventual dominance of thermoelastic behavior over plastic behavior of the shell. Quite independently, Lodhia, Colwell and Berry(5) have reached a similar conclusion, a numerical study of the solidification of a vertical column of a 60/40 PbSn alloy against a 70/30 brass water-cooled chill. Comparing their computations with the experimental findings of Durham(6), they infer that this effect may be interpreted as the increase in the effective air-gap size as the skin thickens and recedes, resisting the metallostatic pressure as outlined above (Fig. 3).

That the heat-transference across the mold-metal interface depends upon the geometry concerned becomes especially clear on examining the work of Nishida, Droste and Engler(7) and further upon comparing this work with that of H/P(3). The geometry used by the latter group represents an almost one-dimensional system (insulated column-top chill) whereas that of Nishida et alia (NDE) is considerably more complex (flat plate and cylindrical shapes formed by metallic mold walls). Although only the h(t) curves of NDE show an immediate increase, generally within 5-10 seconds after pouring, both sets of curves show distinct "knees" in the subsequent downward swing of h with increasing time, see Figs. 4, 5, 6. However, the knees occur at times of the order of 15-10 seconds for NDE's flat plates (pure aluminum and Al-13.2% Si) and just beyond 40 seconds for their cylindrical casting surfaces in these materials, whereas the results of H/P (pure Al) indicate that the knee should occur at considerably greater times, of the order of five minutes. The distinct differences in the experiments are principally those of geometry, as mentioned, although there appears to be differences in the materials used as the mold or chill surface. NDE also discuss important effects of mold constraint.

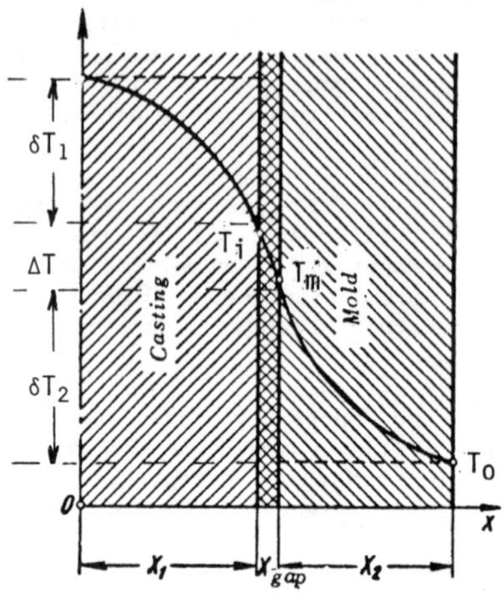

Figure 1. Temperature distribution for a metallic material cast in a metal mold. (Case 3) (After Veynik)

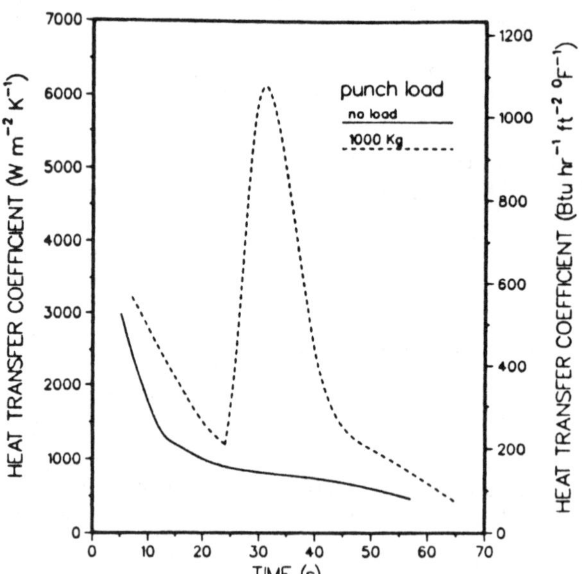

Figure 2. Variation of interfacial heat transfer coefficient with time, pure aluminum in metal mold, with and without applied load. (After Nishida and Matsubara)

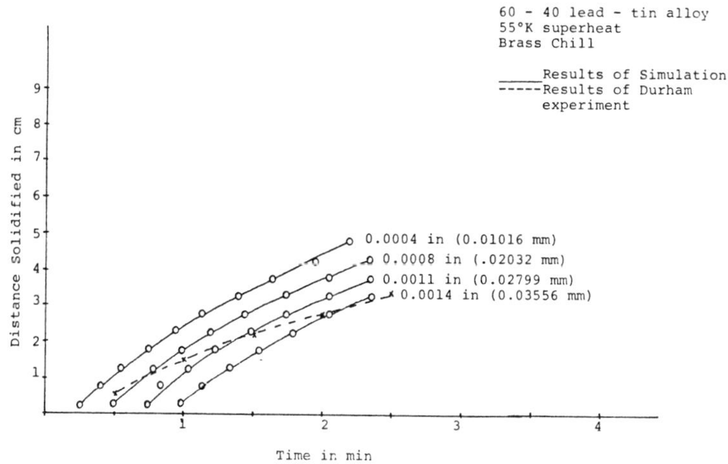

Figure 3. Variation of interfacial heat transfer coefficient with time for various air-gap thickness values. (After Lodhia, Colwell, and Berry) Superimposed is experimentally derived data of Durham. (60-40 Pb-Sn on water-cooled 70-30 brass)

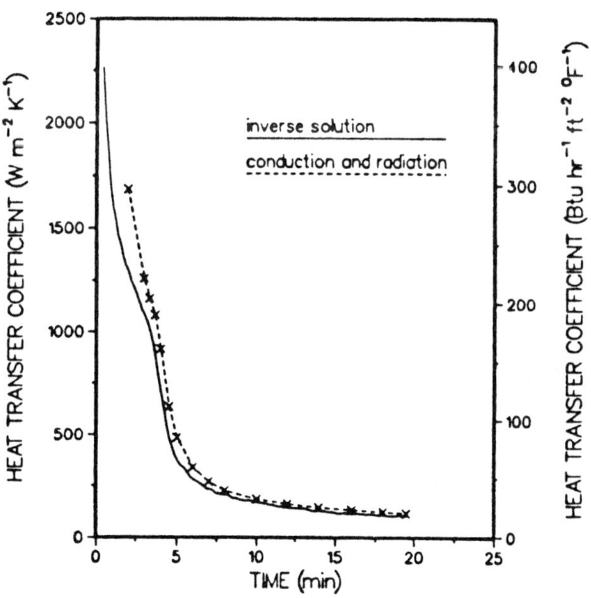

Figure 4. Variation of interfacial heat transfer coefficient with time. (Aluminum on water-cooled copper chill, after Ho and Pehlke)

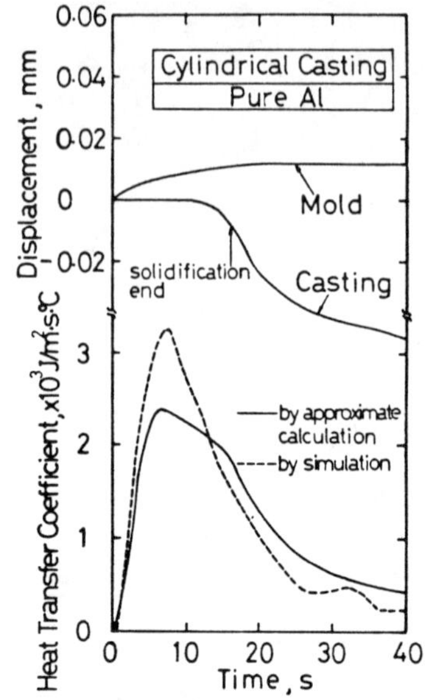

Figure 5. Variation of interfacial heat transfer coefficient with time, pure aluminum in metal mold (Cylindrical casting, after Nishida, Droste, and Engler)

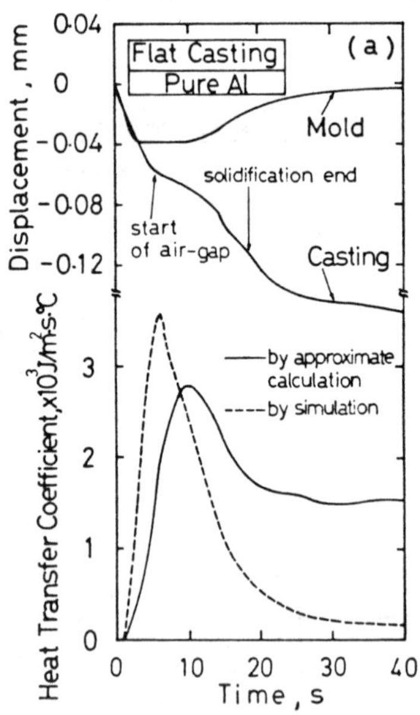

Figure 6. As Figure 5 but for plate casting. (After Nishida, Droste, and Engler)

Turning to net shape cast alloys which solidify in seemingly rigid but more refractory mold enclosures (sand and investment castings) some particularly complex situations are seen to occur. The early work of Gittus(8), which seems to have been overlooked by many workers, points out that special differences exist between the more easily distorted green sand molds and the more rigid dry sand variety. Working with gray irons of near eutectic carbon equivalent, this work indicated that although green sand mold cavities may initially enlarge under the effect of metallostatic pressure, the eventual drying out of the mold plus the expansion associated with the graphite precipitation during the eutectic arrest, promotes good thermal contact. Consequently, casting "swelling" - a considerable departure from net-shape - can occur with incorrectly prepared green-sand molds.

In dry sand in contact with such casting media, Gittus implies that good thermal contact will occur at an earlier stage in the solidification. Pehlke and coworkers(9) have documented a variety of metal-sand mold combinations of simple geometry and have indicated clearly for which combinations air-gaps are likely to be of significant size.

Recently, the present writer and his colleagues had the occasion to examine the question of heat transfer into investment shell molds containing a molten aluminum casting medium. Solidification times were measured in a number of disc-shaped castings of varying thickness.(10) Mold (shell) wall thickness, mold preheat and the superheat of the metal were also variables in the experiments. A one-dimensional finite-difference model of the heat transfer was set up in order to predict the effects of the above variables. Separately determined values of the thermal conductivity of the shell materials(11) were used in the numerical experiments. Initial experiments assumed good thermal contact of the mold-metal interface - something frequently held (although not exclusively) for castings in non-metallic molds of refractory oxides (the shells were of a siliceous nature, dried and fired along normal lines.)

The initial results of the simulation were extremely disappointing and in spite of adjustment to cover all possible circumstances for convective heat-transfer at the outer (mold-air) interface the models were "solidifying" much too early! Subsequent discussion convinced the team that the rigid monolithic mold of well-bonded partially sintered material was acting in a way more similar to a plate cast in a metallic mold or die, rather than as an aluminum-sand casting. Computational experiments were undertaken which examined two alternative ways of describing the heat transfer away from the solidifying metal. The first, a simplistic approximation treats the entire environment of the metal (air gap, shell and surrounding air) as an equivalent convective boundary(12). The second, perhaps more realistic, follows the model for the heat-transfer coefficient at an air gap described by Ho and Pehlke(3). The description of the interfacial contact provided by these workers suggests that the "gap" forming might be divided into two parts a "macro-gap" (the air-gap itself) and a "micro-gap" (characterized by the roughness of the mold material.) Accordingly simulations were running using an air-gap thickness made up of a micro-gap of the order of the average particle size of the investment and a macro-gap of a variety of magnitudes. Figure 7 shows some typical results(14). Separate computational experiments were run to study the extent of the effect of the variation of the heat transfer coefficient at the outer (mold-air) interface(14).(Figure 8) From these experiments it was inferred that air-gap formation is not only present but is probably of some importance in the investment casting of aluminum.

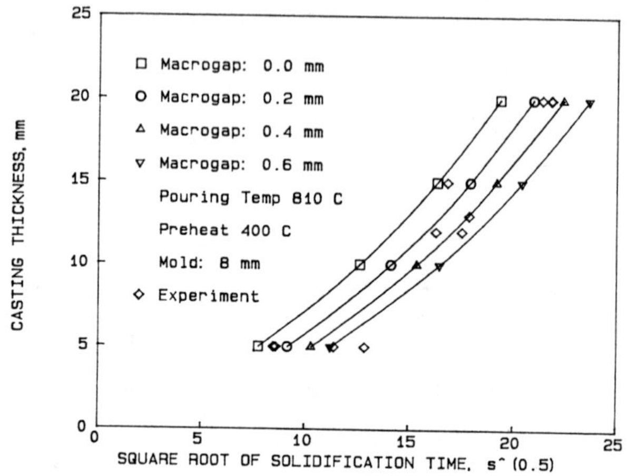

Figure 7. Effect of air-gap size on solidification times of aluminum investment castings of various thicknesses. (After Huang)

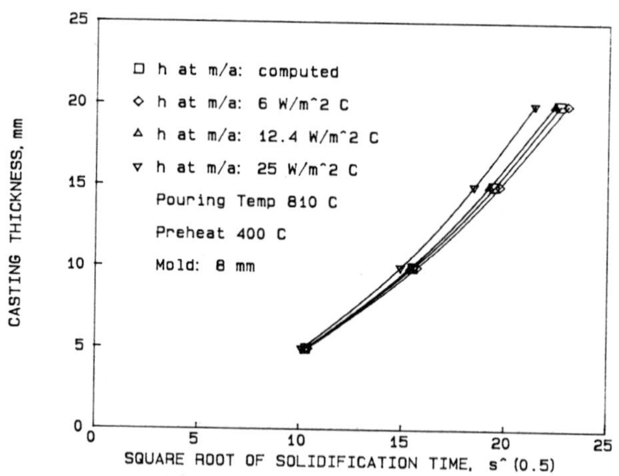

Figure 8. Effect of mold-air interfacial heat transfer coefficient of aluminum investment casting of various thicknesses. (After Huang)

Concluding Discussion

From the preceding discussion of experiments undertaken by various teams over a number of years, it is abundantly clear that interfacial heat transfer between the casting and the mold is of an immensely complicated nature. Nonetheless, certain definitive works have attempted to develop consistent rationales. Although only two publications of those usually cited by workers in this area have contained anything approaching a quantified treatment of air-gap formation(7,13), no one has, as yet, assembled a more generalized numerical description. This will be a necessary condition for genuine progress in the control of both casting dimensions and microstructure and the eventual closing of the control loops currently left open in the system of casting solidification.

Acknowledgements

The work on the solidification of investment castings has been supported by the National Science Foundation Grant No. DMC-8612019, which forms part of the data-base exploration phase of the joint UA-UM CADCAST Project. This experimental and computational work was performed by Mr. Heng Huang, a graduate assistant in the author's department.

REFERENCES

1. A. I. Veynik, *Special Casting Methods*, ASME, New York, 1962.

2. M. J. Hillier, *Casting Science*, NASA Report N67-36119, 1967.

3. K. Ho and R. D. Pehlke, *Transactions American Foundrymen's Society*, 1984 (92), 587-598.

4. Y. Nishida and M. Matsubara, *British Foundryman*, 1976 (69), 274-278.

5. A. Lodhia, C. Colwell, and J. T. Berry, *Computer Aided Design for Casting and Solidification Technology Project Progress Report*, Georgia Institute of Technology, 1984.

6. D. Durham, *Welding Journal* (Welding Research Supplement), 1979 (58), 301s-306s.

7. Y. Nishida, W. Droste, and S. Engler, *Metallurgical Transactions*, 1986 (17B), 833-844.

8. J. H. Gittus, *BCIRA Journal of Research and Development*, 1954 (5), No. 5, 1.

9. R. D. Pehlke et alia, *Computer Aided Design for Casting and Solidification Technology Project Progress Report*, University of Michigan, 1980.

10. H. Huang and J. T. Berry, "Solidification Heat Transfer in Aluminum Investment Castings," (Paper to be presented at ASME Winter Annual Meeting, Chicago, Illinois, 28 November - 2 December, 1988).

11. X. Z. Xheng, J. Hartley, J. T. Berry, and H. Huang, "Thermal Conductivity of Investment Shells and Core Materials," (Paper in preparation, 1988).

12. C. S. Kanetkar, D. M. Stefanescu, N. El-Kaddah, and I. G. Chen, *Proceedings of Solidification Processing 1987*, Sheffield, U.K., 404-407.

13. A. I. Veynik, *Thermodynamics for the Foundryman*, Maclaren, London, 1968.

14. H. Huang, "Computer Simulation of Solidification in Investment Casting," (M.S. thesis, The University of Alabama, 1988).

FIGURES

Figure 1. Temperature distribution for a metallic material cast in a metal mold. (Case 3) (After Veynik)

Figure 2. Variation of interfacial heat transfer coefficient with time, pure aluminum in metal mold, with and without applied load. (After Nishida and Matsubara)

Figure 3. Variation of interfacial heat transfer coefficient with time for various air-gap thickness values. (After Lodhia, Colwell, and Berry) Superimposed is experimentally derived data of Durham. (60-40 Pb-Sn on water-cooled 70-30 brass)

Figure 4. Variation of interfacial heat transfer coefficient with time. (Aluminum on water-cooled copper chill, after Ho and Pehlke)

Figure 5. Variation of interfacial heat transfer coefficient with time, pure aluminum in metal mold (Cylindrical casting, after Nishida, Droste, and Engler)

Figure 6. As Figure 5 but for plate casting. (After Nishida, Droste, and Engler)

Figure 7. Effect of air-gap size on solidification times of aluminum investment castings of various thicknesses. (After Huang)

Figure 8. Effect of mold-air interfacial heat transfer coefficient of aluminum investment casting of various thicknesses. (After Huang)

A TWO DIMENSIONAL TIME-DEPENDENT THERMAL MODEL

FOR THE SOLIDIFICATION OF SHEET ON MOVING SUBSTRATES

J. P. Papai and C. E. Mobley
Department of Metallurgical Engineering
The Ohio State University

ABSTRACT

A two dimensional, time-dependent finite difference program has been developed to compute the thermal fields in the solidifying sheet and the moving substrate in sheet casting processes. This program has been used to model the solidification of copper on plain carbon steel substrates and to study the effects of sheet width-to-substrate width and the substrate cooling conditions on the copper sheet solidification history and temperature profiles. The use of the thermal model as a tool for designing sheet casting systems is discussed.

Casting of Near Net Shape Products
Edited by Y. Sahai, J.E. Battles,
R.S. Carbonara and C.E. Mobley
The Metallurgical Society, 1988

INTRODUCTION

A number of ribbon and sheet casting processes involve the controlled flow of a liquid onto a moving heat sink substrate, such as an internally cooled drum, where the liquid layer moves with the substrate and is solidified. Following partial or complete solidification, the solidified sheet product leaves the substrate. A simplified schematic drawing of such a sheet casting system is shown in Figure 1. The sheet is in contact with the substrate for a period known as the contact time, $t_{contact}$, and the substrate surface is exposed to the environment for the period known as the noncontact time, $t_{noncontact}$. The time for one complete revolution, t_{cycle}, is the sum of the contact time and the noncontact time, i.e.,

$$\pi D/V = t_{cycle} = t_{contact} + t_{noncontact} \qquad [1]$$

where D = the surface diameter of the drum,
and V = the surface velocity of the drum.

The thermal histories and profiles in the solidifying sheet and the drum are dependent on a number of factors, the principal ones being:

a. the thermophysical properties of the sheet and drum materials,
b. the dimensions of the drum and cast product,
c. the initial temperatures of the liquid flowing onto the drum, the drum itself, the drum internal coolant, and the environment surrounding the drum,
d. the contact and noncontact times,

and e. the heat transfer coefficients characterizing the heat fluxes at the cast sheet-drum interface, the drum-coolant interface, and the drum-environment interface.

A two dimensional, time dependent finite difference computational program, which incorporates all of the above factors, has been developed to calculate the thermal fields in the solidifying sheet-like products and the substrate in sheet casting processes.

MODEL DESCRIPTION

The computational program, referred to as the model, is a 2-D, time dependent, finite difference heat transfer code similar to the program known as CAST as described by Walther(1). The heat of fusion for the solidifying material is handled via the enthalpy formulation method(2). The model considers a radial slice through the width of the drum and the sheet. The radial slice is divided in rectangular elements, as shown in Figure 2. The centerline of the sheet and drum are assumed to coincide, i.e., the cast sheet is centrally located on the drum surface. Thus, only one-half of the cast sheet and the drum are included in the computational model. The elements within the drum next to the drum-sheet interface are smaller than those within the body of the drum to gain better temperature resolution in the temperature calculations. The elements within the drum near the location of the sheet free edge are also of smaller volume than elsewhere to better define any thermal edge effects. Each volume element in the sheet and the drum is assigned an initial temperature and thermal conductivity. The program algorithm calculates the heat flow into (or out of) each element via the Fourier heat flow relation:

$$J = - k(dT/dx) \quad [2]$$

where J = the heat flux,
k = the thermal conductivity,
and (dT/dx) = the thermal gradient.

The system is discretized, as shown in Figure 3, so that the heat flow between adjacent elements may be calculated through equations such as:

$$J_{2-1} = (T_2-T_1)(A_{2-1})\ \Delta t/\ (1/h + x_1/k_1 + x_2/k_2) \quad [3]$$

where J_{2-1} = the heat flux into element 2 from element 1 in time step Δt.
(T_2-T_1) = the temperature difference between the two elements,
A_{2-1} = the area of contact between elements 2 and 1,
Δt = the time step
h = the heat transfer coefficient between the two elements if an interface exists at that location,
x_1 = the element half dimension of element 1,
and k_1 = the thermal conductivity of element 1 at its known temperature T_1.

The heat flow into an element in a given time step is divided by the volume of that element to calculate the change in the heat content per unit volume. The heat content per unit volume is then divided by the element's specific heat to obtain the temperature change of the element for that particular time step. These calculations are done iteratively to determine the thermal history of each element. The computational method is stable so long as the time step is sufficiently small relative to the size of the elements selected. The model sets the time step to comply with the absolute stability requirement for the particular element dimensions selected. The model starts with the 2-D slice at the initial temperatures specified by the input to the model. The drum slice is then "exposed" to the liquid and is in contact with the solidifying sheet for the contact time. The drum is then in contact with the environment along itsouter surface and the internal coolant for the noncontact time. The temperatures of the environment and the internal coolant are assumed constants in this model.

The cycle, i.e., contact time + noncontact time, is then repeated for a user specified number of times. The model computes the temperature of each element in the sheet material and the drum as functions of time. The model also provides the data as to the time to complete solidification of each element in the cast sheet.

The model currently includes a data base of the relevant thermophysical properties of liquid and solid aluminum, magnesium, and copper as cast sheet materials and aluminum, copper and plain carbon steel as drum materials.

The typical computation time for a system containing about 600 elements in the sheet and drum and for 20 complete cycles is about 15 minutes CPU time using a VAX 8550 computer.

EXAMPLE OF MODEL COMPUTATIONS

The above described model was used to compute the thermal profiles associated with the casting of copper sheet on a plain carbon steel drum for the casting conditions given in Table 1.

TABLE 1. INPUT TO SHEET CASTING MODEL

Sheet Material:	Copper
Drum Material:	Plain Carbon Steel
Sheet Thickness:	0.1 cm
Drum Thickness:	0.5 and 1.0 cm
Sheet Width:	20 and 100 cm
Drum Width:	35 and 125 cm
Liquid Superheat:	10 C
Initial Drum Temperature:	30 C
Internal Coolant Temperature:	30 C
Environment Temperature:	30 C
Sheet-Drum Heat Transfer Coefficient:	1.0 to 3.0 cal/cm^{2}C sec
Drum-Coolant Heat Transfer Coefficient:	0.05 to 0.3 cal/cm^{2}C sec
Drum-Environment Heat Transfer Coefficient:	0.005 cal/cm^{2}C sec
Contact Time:	0.1 sec
Noncontact Time:	1.9 sec

The particular heat transfer coefficients used in the model calculations were selected from values reported in the literature for the processes of chill block melt spinning and melt extraction[3-10].

Influence of Selected Variables

Sheet and Drum Width

To identify the influence of the sheet and drum widths on the time to complete solidification of the cast copper sheet on the plain carbon steel drum, the input values of Table 1 were used and the heat transfer coefficients for the sheet-drum interface and the drum-coolant interface were selected to be 2.0 and 0.2 cal/cm^{2}C sec, respectively. The drum thickness was set at a value of 1.0 cm. The time to complete solidification of the 0.1 centimeter thick copper sheet along the outer 10 centimeters of the sheet width for various sheet and drum width combinations is shown in Figure 4. The three quantities defined in Figure 4 are the time to solidify the edge of the sheet, the time to solidify the center of the sheet , and the difference between those two limiting solidification times. These computations indicate that the edge of the sheet solidifies before the center of the sheet and that the difference between the two limiting solidification times is relatively independent of the sheet and drum widths over the range of values considered (i.e., for those cases where the sheet width is 20 cm or greater and the drum width exceeds the sheet width by 15 or more centimeters).

Sheet-Drum and Drum-Coolant Heat Transfer Coefficients

The influences of the two heat transfer coefficients, the sheet-drum heat transfer coefficient and the drum-coolant heat transfer coefficient, on the solidification of the 0.1 cm thick copper sheet were determined for two given thickness steel drums, a 1.0 cm thick drum, referred to as the thicker drum, and a 0.5 cm thick drum, referred to as the thinner drum. For these computations, the sheet and drum widths were set at 20 and 35 cm, respectively. The time to solidify at the middle of the sheet, at the edge of the sheet, and the solidification time difference based on these two positions are shown in Figures 5, 6, and 7, respectively as functions of the various heat transfer coefficient values for the sheet-drum and the drum-coolant heat transfer coefficients for the 1.0 centimeter thick drum case. As shown in Figures 5 through 7, while the times to complete solidification of both the center and the edge of the copper sheet decrease with increasing heat transfer coefficients at either interface, i.e., for both the sheet-drum and/or the drum-coolant interfaces, the difference in solidification time between the sheet center and edge is more strongly influenced by the heat transfer coefficient for the drum-coolant interface than the sheet-drum interface for this particular casting system.

The time to solidify at the sheet middle, the sheet edge, and the solidification time difference between these two positions are shown in Figures 8, 9, and 10, respectively as functions of the two heat transfer coefficients for the 0.5 cm thick steel drum. The data shown in Figures 8 through 10 are similar to that shown in Figures 5 through 7, however, it should be noted that the thinner (i.e., 0.5 cm thick) drum presents a smaller solidification time difference than the thicker (i.e., 1.0 cm thick) drum for the same combination of heat transfer coefficients at the sheet-drum and drum-coolant interfaces.

As anticipated, it was also noted that the heat transfer coefficients at both the sheet-drum interface and the drum-coolant interface must exceed given values to ensure that the 0.1 cm thick copper layer solidifies in a time less than or equal to the contact time of 0.1 second.

The time to solidify at any position within the copper sheet increased with accumulative time, i.e., as the drum warmed up from its initial temperature to its quasi-steady state condition. The time required to reach the quasi-steady state condition can be defined as the number of cycles when the time to solidify the center of the sheet does not vary within a specified amount (e.g., solidification time is constant to within 1 millisecond of the steady state value). The number of cycles required for the copper on steel system to reach quasi-steady state as influenced by the heat transfer coefficients and the drum thickness is given in Table 2.

As indicated in Table 2, the time for the system to reach quasi-steady state is primarily a function of the drum-coolant heat transfer coefficient and the drum thickness and is relatively independent of the sheet-drum heat transfer coefficient for this particular collection of parameters. For the same combination of heat transfer coefficients, the thinner drum reaches the quasi-steady state conditions faster than the thicker drum.

TABLE 2: Number of Cycles to Reach Steady State

$h_{Sheet-Drum}$ \ $h_{Drum-Coolant}$	0.05	0.1	0.2	0.3
	Thicker (1.0 cm) Drum			
1.0			14	11
1.2		22	14	11
1.6		22	14	11
2.0		22	14	11
2.5	35	22	13	11
3.0	35	22	13	11
	Thinner (0.5 cm) Drum			
1.0		10	6	5
1.5		10	6	4
2.0	17	10	6	4
2.5	18	10	6	4
3.0	18	10	6	4

The heat transfer coefficients listed in Table 2 are given in cgs units, i.e., cal/cm^2 C sec.

SUMMARY AND CONCLUSIONS

A two dimensional, time dependent, finite difference computational model has been developed for predicting the thermal fields in sheet casting systems. Application of this model to the case of 0.1 cm thick copper sheet cast on an internally cooled plain carbon steel drum indicated the following:

1. The time to complete solidification at any position in the copper sheet decreases with increasing values of the sheet-drum and drum-coolant heat transfer coefficients.

2. Thinner drums provide increased solidification times compared to thicker drums.

3. The time difference between completion of solidification at the center and edge of the copper sheet is decreased by increasing the heat transfer coefficients at both the sheet-drum interface and the drum-coolant interface.

4. For a rapid approach to the quasi-steady state condition, the drum should be as thin as feasible and the drum-coolant heat transfer coefficient should be large.

REFERENCES

1. M. K. Walther, "Solidification Analysis Using CAST," Modern Casting, 74, No. 3, (1984), 18-22.
2. J. Crank, "How to Deal with Moving Boundaries in Thermal Problems'" Numerical Methods in Heat Transfer, J. Wiley & Sons, (1981), 177-200.
3. S. C. Huang and H. C. Fiedler, "Effects of Wheel Surface Conditions on the Casting of Amorphous Metal Ribbons," Met Trans A, 12A, (June 1981), 1107-1112.
4. A.G. Gillen and B. Cantor, "Photocalorimetric Cooling Rate Measurements on a Ni-5 wt% Al Alloy Rapidly Solidified by Melt Spinning," Acta. Metall. 33, (October 1985), 1813-1825.
5. L. Katgerman and P.J. Van Den Brink, "A Mathematical Model for Melt Spinning of Crystalline Alloys," Proceedings 4th Int. Conf. on Rapid Quenched Metals, Sendai, (1981), 61-64.
6. L. Katgerman, "Computer Simulation of Cellular Growth during Rapid Solidification of Binary Aluminum Alloys," Modeling of Casting and Welding Processes II, Henniker, N.H., (1983), 135-143.
7. M. J. Tenwick and H. A. Davies, "The Mechanism of Ribbon Formation in Melt-Spun Copper and Copper-Zirconium," Rapid Quenched Metals, (1985), 75-78.
8. M. Yukumoto, K. Shibuya, T. Kan, amd Y. Ito, "Analysis of the Solidification Process in the Roller Quenching Method," Rapid Quenched Metals, (1985), 91-94.
9. H. Soderhjelm and L. Mandal, "On the Possibilities of Making Multiple layer Ribbons by Single Roller Melt Spinning," Rapid Quenched Metals, (1985), 107-110.
10. H. Jones, "Experimental Methods in Rapid Quenching from the Melt," Treatise on Materials Science and Technology, 20, Ultrarapid Quenching of Liquid Alloys, Academic Press, (1981), 1-72.

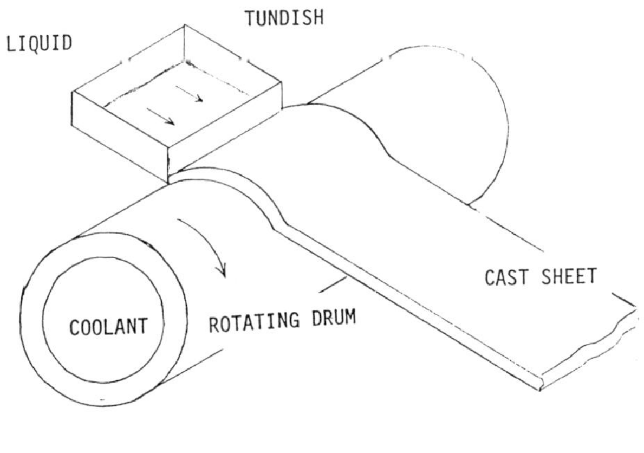

FIGURE 1. Schematic of sheet casting process.

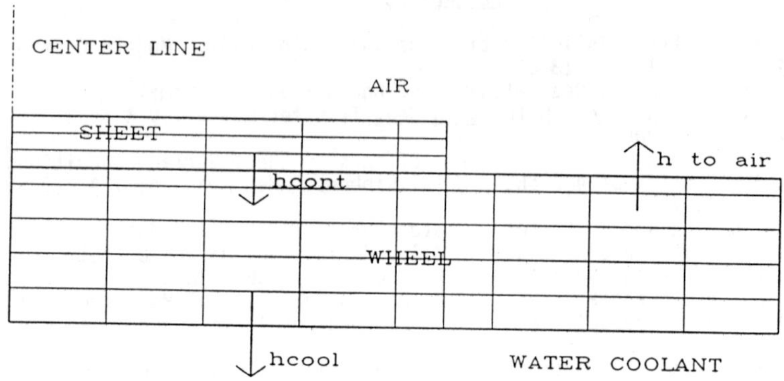

FIGURE 2. Radial slice of drum and casting are divided into rectangular elements. Heat transfer coefficients are illustrated. Only half of system need be simulated due to symmetry.

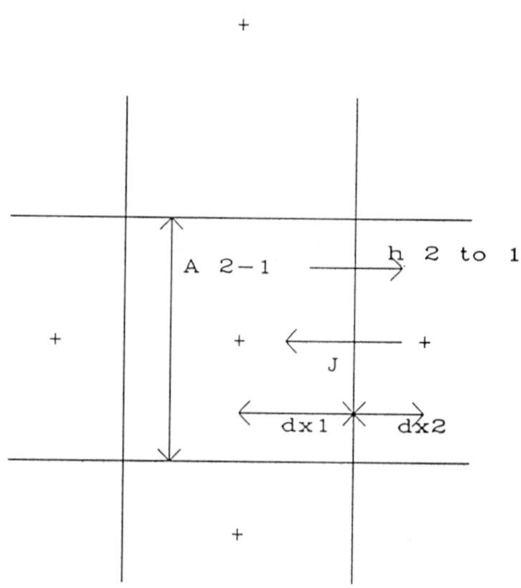

FIGURE 3. Schematic illustration of the quantities that determine the heat flow, J, between two elements including the distances, area of contact, and heat transfer coefficient.

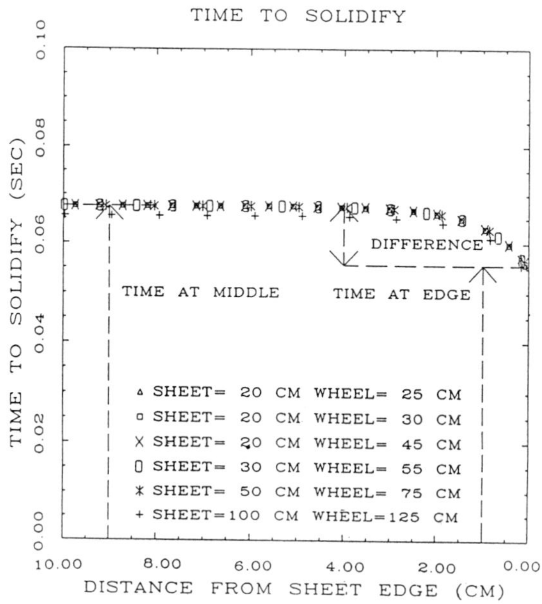

FIGURE 4. Time to solidify as a function of the distance from the sheet edge for the same heat transfer coefficients and different widths.

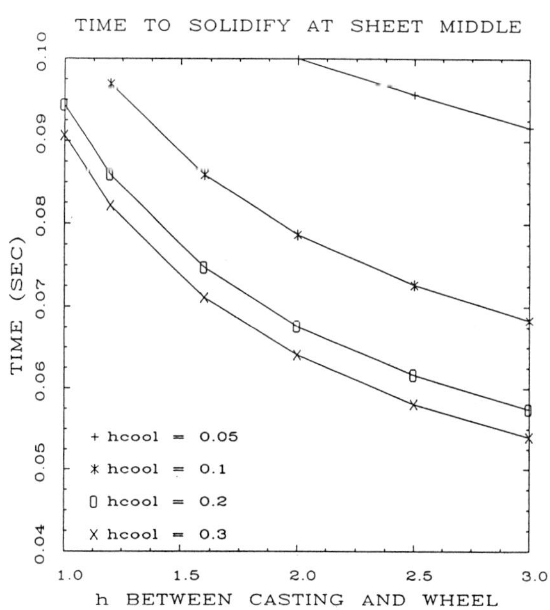

FIGURE 5. Time to solidify at sheet middle for thicker drum.

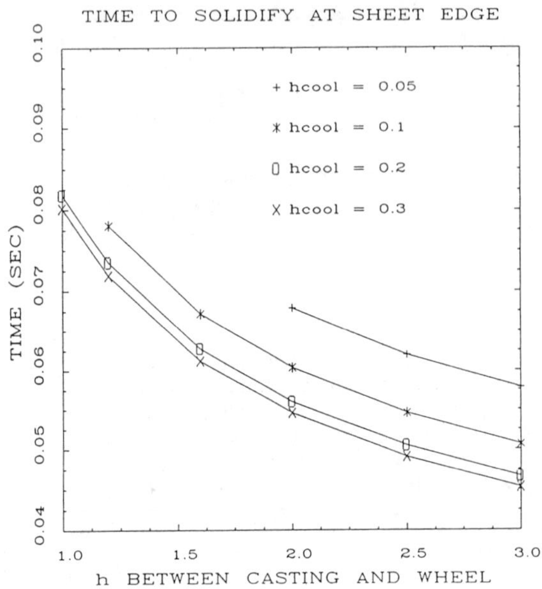

FIGURE 6. Time to solidify at sheet edge for thicker drum.

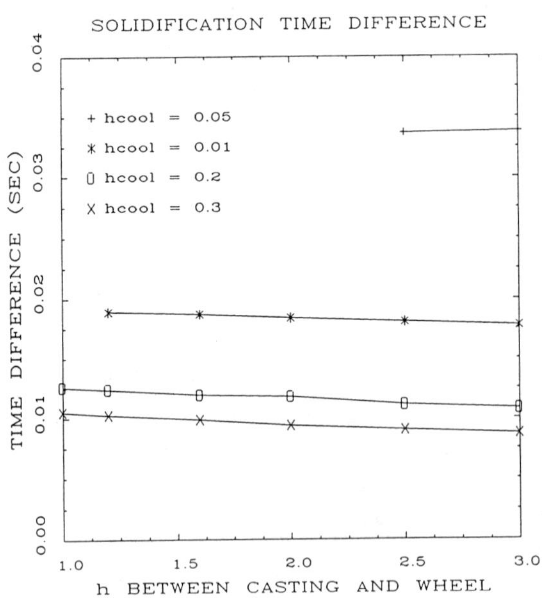

FIGURE 7. Time difference in solidification for thicker drum.

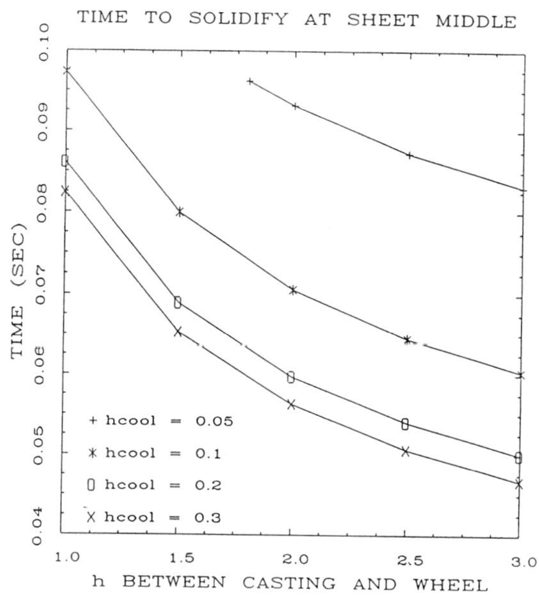

FIGURE 8. Time to solidify at sheet middle for thinner drum.

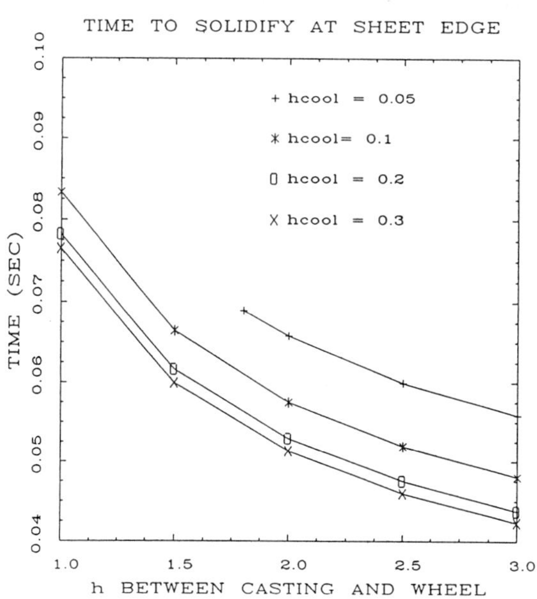

FIGURE 9. Time to solidify at sheet edge for thinner drum.

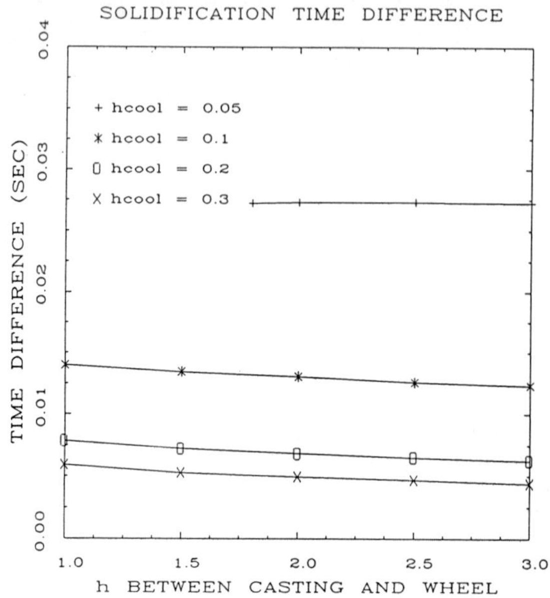

FIGURE 10. Difference in time to solidify between middle and edge of sheet for thinner drum.

ON HEAT AND FLUID FLOW PHENOMENA IN SHALLOW TUNDISHES

- THE EFFECT OF MAGNETIC FLOW CONTROL

O.J. Ilegbusi and J. Szekely

Department of Materials Science and Engineering
Massachusetts Institute of Technology
Cambridge, MA 02139 USA

Abstract

A model is developed to predict fluid flow and heat transfer in a generic shallow tundish system. Emphasis is placed on the influence of magnetic flow control on these phenomena. Allowance is made for a sloping bottom to approximate tundish systems of the type employed in belt and roll casters.

Casting of Near Net Shape Products
Edited by Y. Sahai, J.E. Battles,
R.S. Carbonara and C.E. Mobley
The Metallurgical Society, 1988

Introduction

A key feature of many near net shape casting operations employing belts, single or twin rolls, is the use of a relatively shallow tundish, into which molten metal is being fed from a ladle, and which in turn supplies the feed to the caster [1,2]. While these tundishes will vary in size with the actual application, a typical "generic" tundish may be considered to be about 1 m long, 0.5 m wide and a few cm deep.

It has been appreciated in a qualitative sense that the distribution of the molten metal may play an important role in affecting the performance of the casting system. However, up to the present time, very little mathematical modelling work has been done to describe the transport phenomena in these systems.

In this paper, we shall draw on previous tundish modelling experience [3-6] to examine the flow behavior of a typical shallow tundish system together with the means available for controlling the flow maldistribution.

While the actual calculations will be presented for one fixed geometry, it is thought that the results will provide a general indication regarding the behavior of this whole class of tundish systems.

Formulation

Let us consider a typical shallow tundish system, such as sketched in Fig. 1. It is seen that molten metal is being fed in at the upper right hand corner and exits through a slot, near the top left hand corner. Such an arrangement would correspond to many belt casting and single roll casting arrangements [1,2].

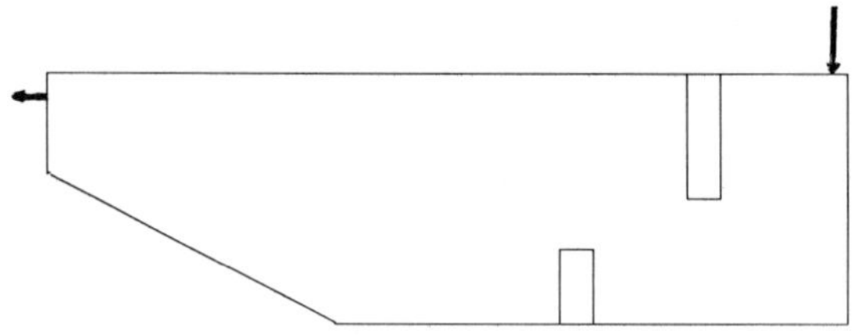

Figure 1 Schematic sketch of a typical shallow tundish system

In case of twin roll casting systems, the tundish geometry would be somewhat different, but a relatively small diameter inlet stream and an exit through a slot would be still typical features of the overall flow arrangement.

We should remark here that while a dam and a weir are being shown in Fig. 1, some calculations will be presented for systems in the absence of such devices. In addition, we shall explore the effect of imposing a magnetic field on the system.

It should be noted, further, that "fluid mechanics common sense" will suggest, even without considering any differential equation, that these shallow tundishes will be very liable to produce quite severe flow maldistribution, because of a relatively short distance between a "point inlet" and a "line sink". These features, coupled with the presence of very large surfaces retarding flow, are likely to produce strong recirculating loops. Expressed in other words, the original function of tundishes, envisioned for conventional continuous casting system, as acting as a buffer, is not likely to be fully fulfilled by these shallow tundish designs.

In the light of prior experience with tundish type systems, such as in [3-8], the mathematical statement of the problem is readily given as follows:

Equation of Continuity:

$$\nabla \cdot \rho \underline{u} = 0 \tag{1}$$

Equation of Motion:

$$\underline{u} \cdot \nabla \rho \underline{u} = -\nabla P + \nabla \cdot (\mu_{eff} \nabla \underline{u}) + F_B + F_E \tag{2}$$

in which μ_{eff} is the effective viscosity, F_B is the buoyancy force and F_E is the electromagnetic (Lorentz) force.

Thermal Energy Balance:

$$\underline{u} \cdot \nabla \rho T = \nabla \frac{\mu_{eff}}{\sigma_T} \nabla T \tag{3}$$

in which σ_T is the Prandtl number.

Equation of Tracer Transport:

$$\frac{\partial c}{\partial t} + \underline{u} \cdot \nabla \rho C = \nabla \frac{\mu_{eff}}{\sigma_c} \nabla c \tag{4}$$

where σ_c is the Schmidt number.

The auxilliary relations employed are as follows:

$$F_B = \rho_o \, g[1-\beta(T-T_o)] \tag{5}$$

where ρ_o,To are the reference density and temperature, respectively, and β is the coefficient of thermal expansion.

$$F_E = \underline{J} \times \underline{B} \tag{6}$$

where $\underline{J}$ is the induced current density and $\underline{B}$ is the magnetic field. From Maxwell's relations, the components of F_E for a fixed magnetic field B_o oriented in the y-direction can be expressed as:

$$F_x = -\kappa E_z B_o - \kappa u B_o^2 \tag{7}$$

$$F_y = 0 \tag{8}$$

$$F_z = \kappa E_x B_o - \kappa w B_o^2 \tag{9}$$

where κ is the electrical conductivity and E_x, E_z are the x- and z-components of the electric field $\underline{E}$, respectively, which are obtained from the relations:

$$\underline{E} = -\nabla\Psi \tag{10}$$

and

$$\nabla^2\Psi = B_o(\nabla \cdot u) \tag{11}$$

The effective viscosity is given as:

$$\mu_{eff} = \mu_l + \mu_t \tag{12}$$

where μ_l is the laminar viscosity and μ_t is the turbulent viscosity deduced from:

$$\mu_t = 0.09\rho K^2/\epsilon \tag{13}$$

where K is the turbulent energy and ϵ is the rate of dissipation of this energy. K and ϵ are in turn deduced from transport equations similar to (2) as detailed in reference [9].

The boundary conditions employed are as follows:

(a) Velocity is specified at inlet.

(b) A no-slip condition at all solid surfaces.

(c) Zero shear at the free surface and symmetry planes.

(d) Stefan-Boltzmann radiative heat loss at the free surface.

(e) Thermal insulation at all surfaces except the free surface.

(f) Electrical insulation at all surfaces.

Computed Results

The input parameters used in the computation are summarized in Table 1; these are thought to be within the typical range of shallow tundish operations. We should note that the flow is turbulent near the inlet stream, while the flow will be largely laminarized elsewhere in the system. Within this general framework, we shall consider three specific cases, namely:

(i) the absence of flow control

(ii) flow control through dams and weirs

(iii) magnetic flow control, through the imposition of 1 kilogauss field in a direction perpendicular to the main flow direction.

Let us consider these results in detail.

Table 1: Principal Input Parameters

Tundish length	1m
Tundish width	0.6m
Tundish height	0.2m
Inlet nozzle diameter	4cm
Outlet slot height	3cm
Inclination of bottom (near exit) to horizontal	25^0
Inlet flow velocity	0.25 m/s
Inlet flow temperature	1873K

(i) Behavior in the Absence of Flow Control

Figures 2-6 show the velocity fields in the absence of flow control. Figure 2 depicts a strong recirculating flow field near the plane of the incoming metal stream (y/b = 0.1). Figure 3 shows a plan view of the velocity field close to the bottom of the tundish at z/h = 0.2, while Figures 4, 5 and 6 show the corresponding plots just below the plane of the exit slot (z/h = 0.65), in the plane of the exit slot (z/h = 0.75) and above the exit slot (z/h = 0.85), respectively. These plots are quite remarkable, because they show very definite flow maldistribution. It is thought that this maldistribution would be a characteristic feature of these systems, quite independent of the geometry. One should also note that as seen in Figure 5, there appears to be a dead zone in the plane of the exit slot, which seems to be fed by a melt stream coming from the upper region of the tundish.

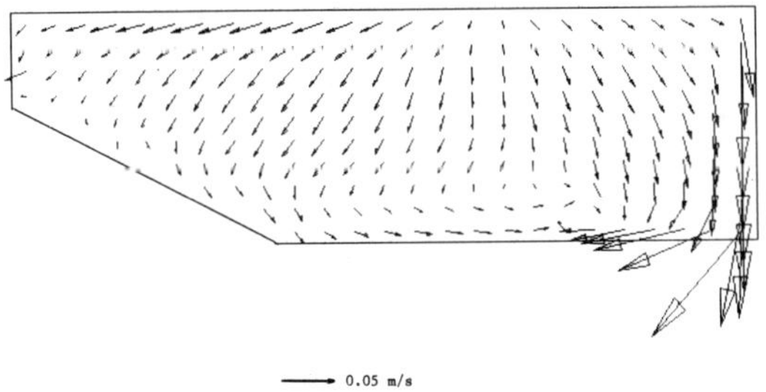

Figure 2 Velocity field at plane y/b = 0.1 for no flow control

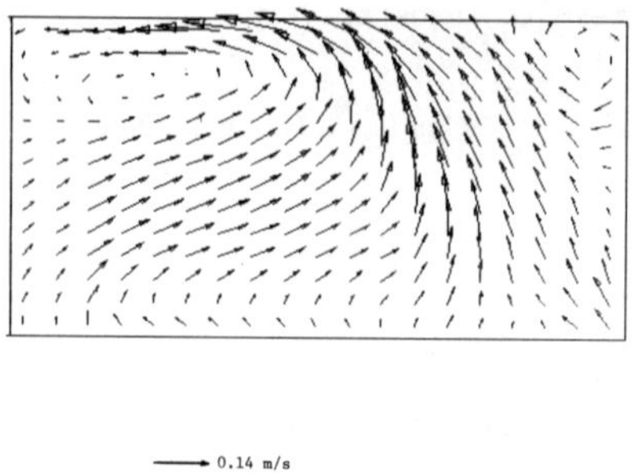

Figure 3 Velocity field at z/h = 0.2 for no flow control

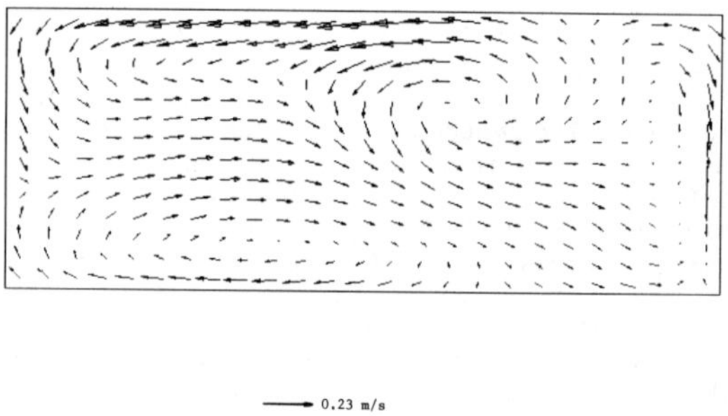

Figure 4 Velocity field at z/h = 0.65 for no flow control

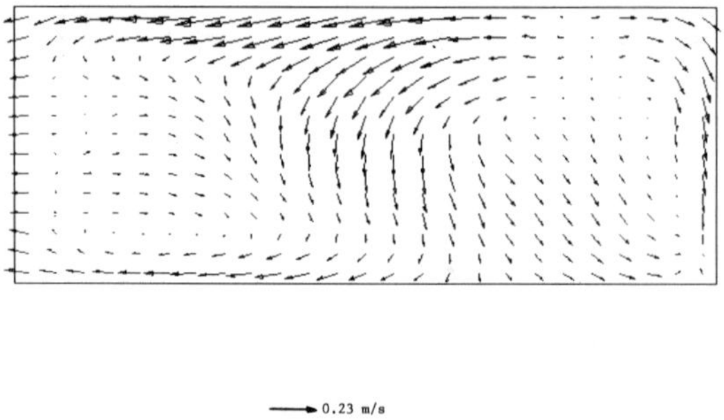

Figure 5 Velocity field at z/h = 0.75 for no flow control

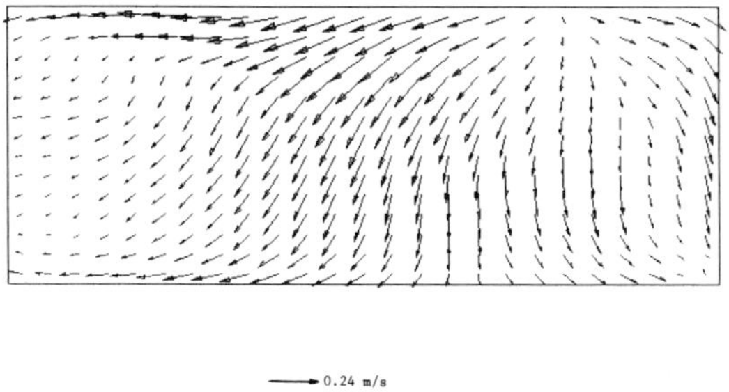

Figure 6 Velocity field at z/h = 0.85 for no flow control

The thermal energy balance equation stated in the formulation, enables us to calculate the temperature profiles within the system. Figure 7 shows the temperature profiles in the vertical plane of the inlet stream, while Figure 8 shows the map of the isotherms in the plane of the exit slot. It is seen that there is a significant vertical stratification, which is to be expected, because of the laminar nature of the flow. As seen in Figure 8, there are some lateral variations in temperature as well.

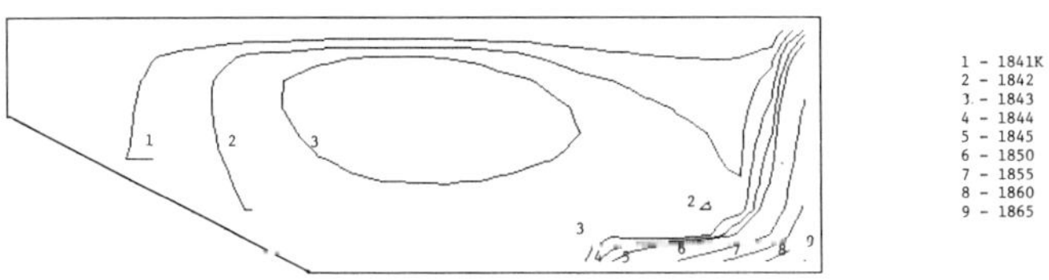

Figure 7 Temperature profiles at y/b = 0.1 for no flow control

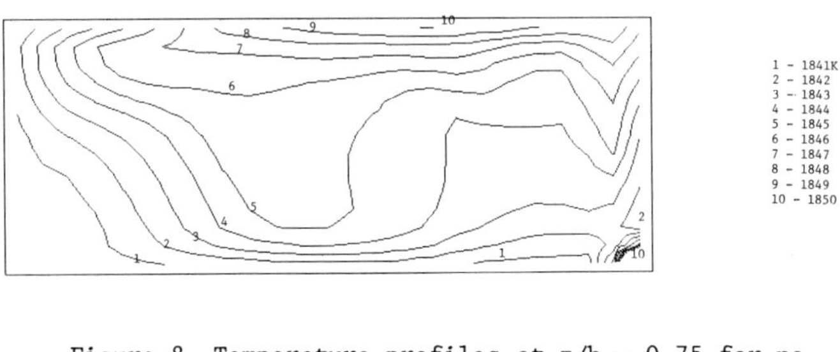

Figure 8 Temperature profiles at z/h = 0.75 for no flow control

(ii) The effect of flow control by a dam and a weir

Figures 9, 10 and 11 show some selected computed results, depicting the effect of a dam and a weir arrangement, as an attempt to control the flow. It is seen, consistent with earlier findings reported by the authors [6], that a dam and weir arrangement will not be very effective in controlling the flow. As discussed in a previous paper, this behavior is attributable to the fact that the flow field is dominated by the "point inlet" and the "line sink".

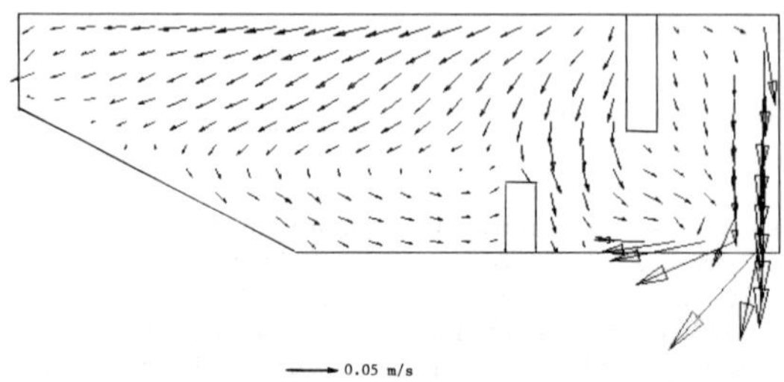

Figure 9 Velocity field at y/b = 0.1 for flow control with a dam and a weir

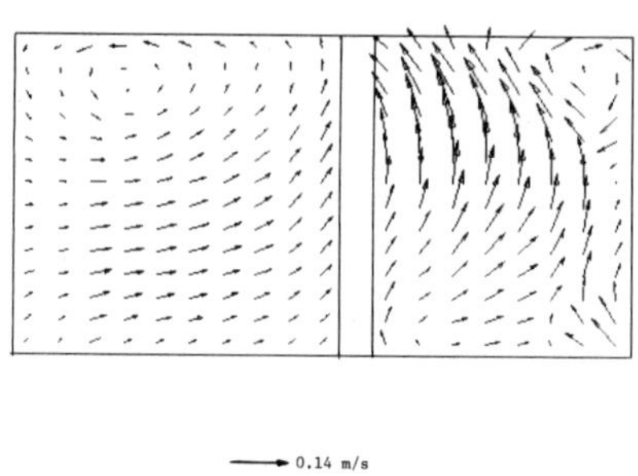

Figure 10 Velocity field at z/h = 0.2 for flow control with a dam and a weir

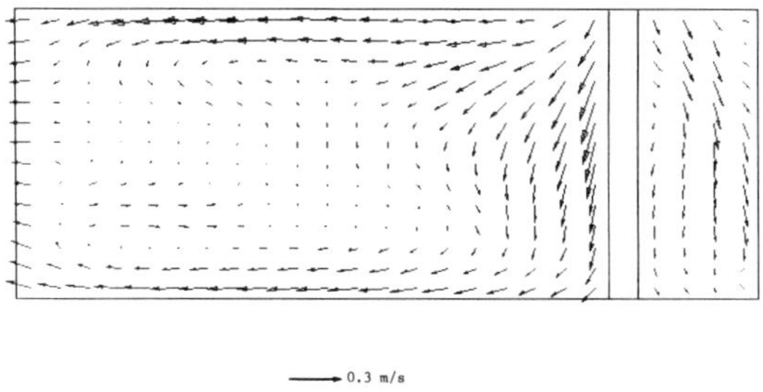

Figure 11 Velocity field at z/h = 0.75 for flow control with a dam and a weir

Figures 12 and 13 show the computed temperature profiles at y/b = 0.1 and z/h = 0.75, respectively; and these are seen not to be much affected by this flow control arrangement.

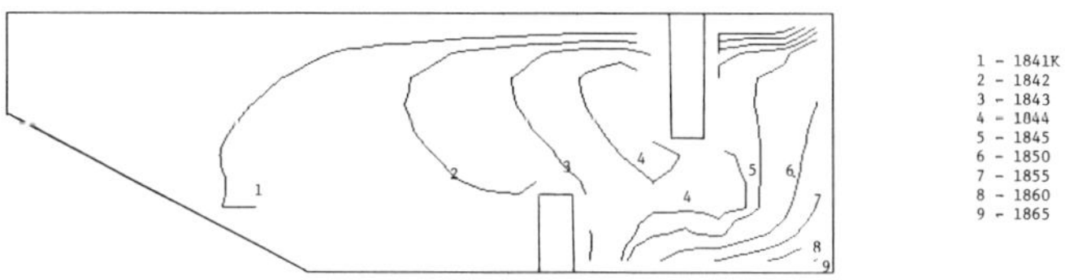

Figure 12 Temperature profiles at y/b = 0.1 for flow control with a dam and a weir

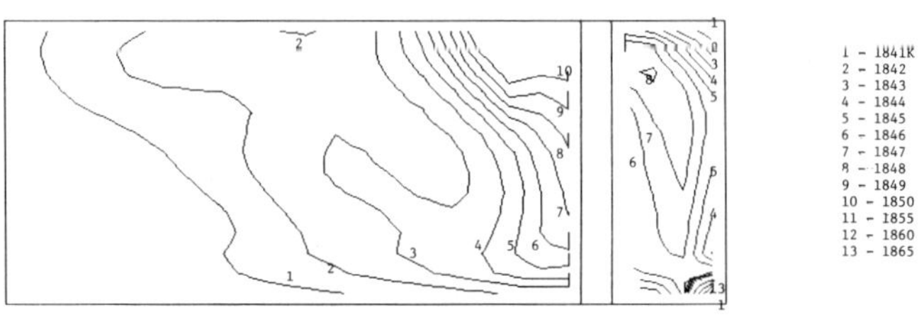

Figure 13 Temperature profiles at z/h = 0.75 for flow control with a dam and a weir

(iii) Magnetic Flow Control

The following Figures 14-16 show the effect on the velocity field of imposing a magnetic field, of 1 kilogauss strength, perpendicular to the principal direction of the flow. It is seen that this field will act to "straighten out" the flow and to eliminate the recirculating loops.

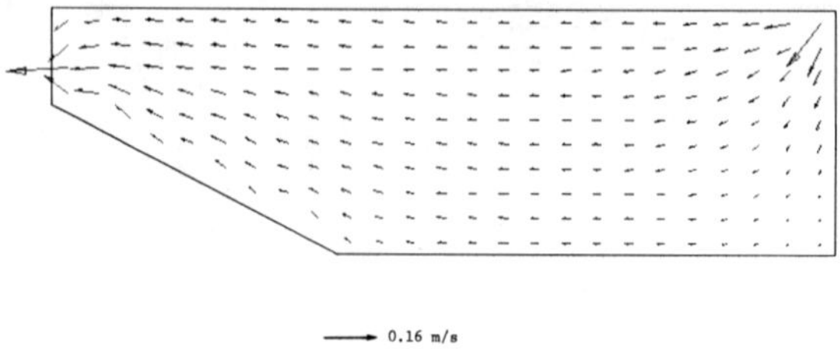

Figure 14 Velocity field at y/b = 0.1 for flow control with a magnetic field of 1 kilogauss

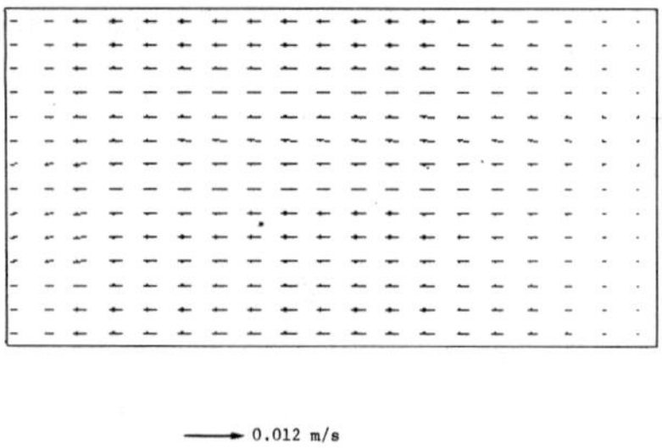

Figure 15 Velocity field at z/h = 0.2 for flow control with a magnetic field of 1 kilogauss

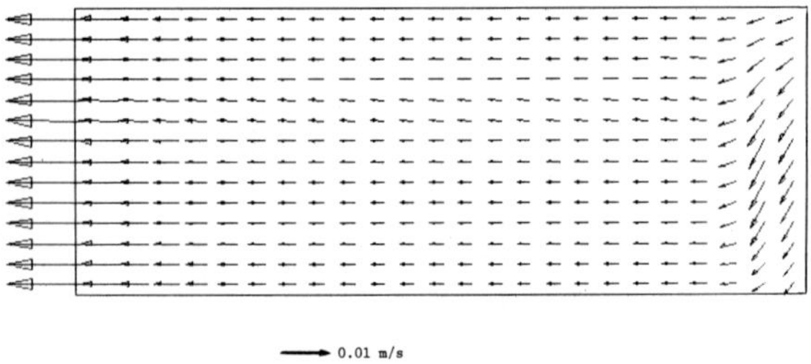

Figure 16 Velocity field at z/h = 0.75 for flow control with a magnetic field of 1 kilogauss

Figures 17 and 18 show the corresponding computed temperature profiles, which are seen to be much more regular, than those in the absence of magnetic flow control.

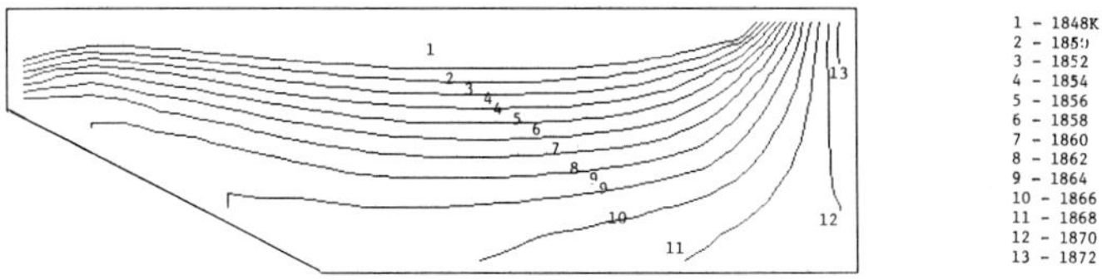

Figure 17 Temperature profiles at y/b = 0.1 for flow control with a magnetic field of 1 kilogauss

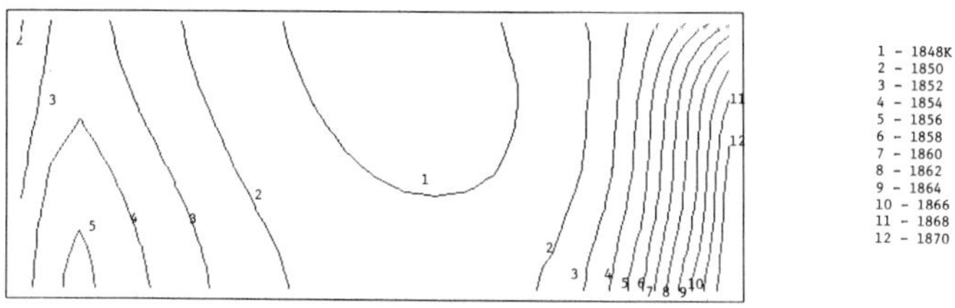

Figure 18 Temperature profiles at z/h = 0.75 for flow control with a magnetic field of 1 kilogauss

Tracer Dispersion

The use of tracer dispersion tests is a very convenient way of characterizing the behavior of continuous flow systems. Figure 19 provides a summary of the flow characteristics of the three cases considered in this paper, in terms of tracer dispersion. It is readily seen that very marked short-circuiting or by-passing would occur both in the absence of flow control, and when dams or weirs are used. In contrast, the use of a suitably oriented magnetic field could produce a close approximation to the ideal plug flow behavior.

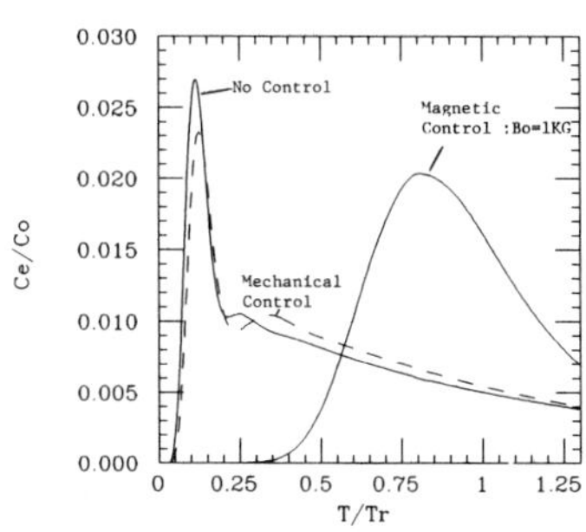

Figure 19 Tracer concentration at the exit against time for various tundish arrangements

Concluding Remarks

A mathematical representation has been developed for the velocity field and the temperature field in a "generic tundish system" of the type likely to be used in belt and roll casting arrangements. The principal findings of the calculations may be summarized as follows:

(1) Shallow tundishes are not very effective in evening out the flow and serious flow maldistribution is likely to be encountered in most applications. This flow maldistribution, with strong recirculating patterns will occur in both the horizontal and the vertical planes of the system.

(2) The use of dams or weirs is not likely to be very effective either, simply because the flow will be dominated by the combined effects of the inlet and the exit regions.

(3) The results presented in the paper for a shallow tundish with a sloping bottom and a relatively wide outlet slot (e.g. corresponding to a belt caster) appear to be quite similar in nature to some of our earlier calculations for shallow tundishes [6], with different geometries (e.g. point exit, non-sloping bottom, narrow slow exit). This suggests that the conclusions concerning flow maldistribution should be quite generally valid.

(4) The heat transfer calculations were somewhat idealized, in that no allowance has been made for either the transient conductive loss into the lining or for slag cover or radiation shields. Nonetheless, these calculations have shown the existence of significant temperature gradients within the system; the heat flow model presented here could be readily refined and could also incorporate the effect of auxiliary heating, either by induction or by a plasma torch. Such auxiliary heating may be very effective in controlling the superheat.

(5) Perhaps the most important finding reported in this paper is that the use of a relatively modest magnetic field (e.g. 1 kilogauss) would enable one to provide almost complete laminarization of the flow and a very close approach to plug flow. Such flow control may be very effective in providing for a smooth operation of these shallow tundish systems.

REFERENCES

1. J.P. Birat, "Manufacture of Flat Products for the 21st Century," Proc. Conf. on Restructuring Steel Plants for the Nineties, London: Institute of Metals (1986), 140.

2. J.D. Naumann and D.B. Love, "Controlling the Shape of Cast Stainless Steel Strip," presented at International Symposium on Near Net-Shape Casting of Strip, 89th Annual Meeting of CIM, Toronto (May 1987).

3. J. Szekely, O.J. Ilegbusi and N. El-Kaddah, Journal of Physico-Chemical Hydrodynamics 9 (3/4), 453 (1987).

4. J. Szekely and O.J. Ilegbusi, "Magnetic Streamlining and Flow Control in Tundishes," U.S. Patent File No. 87151 (1987).

5. O.J. Ilegbusi and J. Szekely, "The Effect of an Externally Imposed Magnetic Field on the Velocity, on Tracer Dispersion and on the Flotation of Inclusion Particles in Tundishes," Ironmaking and Steelmaking, in press (1988).

6. O.J. Ilegbusi and J. Szekely, "Fluid Flow Phenomena and Tracer Dispersion in Shallow Tundishes," Steel Research, in press (1988).

7. Y. He and Y. Sahai, "Fluid Dynamics of Continuous Casting Tundishes: Mathematical Modelling," Proc. 5th International Iron and Steel Congress, Washington, DC, Book 1. Warrendale, PA: Iron and Steel Society, AIME (1986).

8. Y. Sahai and R. Ahuja, Ironmaking and Steelmaking 13, 241-247 (1986).

9. B.E. Launder and D.B. Spalding, Mathematical Models of Turbulence, London: Academic Press (1972).

NOTATION

b	Half width of tundish
$\underline{B}$	Magnetic field
B_o	Constant magnetic field in y direction
c	Tracer concentration
c_e	Tracer concentration at exit
c_o	Tracer concentration at inlet
$\underline{E}$	Electric field
E_x, E_z	X and Z components of $\underline{E}$
F_B	Buoyancy force per unit volume
F_E	Electromagnetic force per unit volume
g	Gravitational acceleration
h	Height of tundish
K	Specific turbulent energy
t	Time
t_r	Nominal residence time
T	Temperature
$\underline{u}$	Velocity vector
x	Longitudinal coordinate (origin at exit plane)
y	Transverse coordinate (origin at central plane)
z	Vertical coordinate (origin at bottom plane)
ϵ	Rate of dissipation of turbulent energy per unit volume
κ	Electrical conductivity
ρ	Density
σ_c	Schmidt number
σ_t	Prandtl number
ψ	Electric potential
μ_{eff}	Effective viscosity
μ_i	Molecular viscosity
μ_t	Turbulent viscosity

ROLE OF PROCESS SIMULATION IN CONTROL

OF CASTING PROCESSES

Sulekh C. Jain

General Electric Aircraft Engines
One Neumann Way
Cincinnati, Ohio 45215

Abstract

Total quality control (TQC) is now the chief concern of many progressive U. S. manufacturers. TQC, rather than being concerned principally with inspection and rejection after the fact, is an approach that calls for prediction, detection, correction and prevention. Thus TQC simply does not mean just monitors and sensors and statistical process control (SPC) but also front-end simulation of the manufacturing process, sensitivity analysis of various process elements, equipment and variables, determination of processing windows, number, location and accuracy ranges of these sensors, quality function and quality policy deployment etc. TQC today is not only just designing and making a product but also intelligent processing of materials (IPM) to manufacture a product right the first time and consistently thereafter. Process simulation via computer aided engineering (CAE) is a roadmap to TQC and plays a major role. Many CAE tools have been developed or are being developed for several unit manufacturing processes. Time has come when these process simulation tools can be used not only for analyzing and designing the manufacturing process but also to design the control of the process and equipment and thus assist in inspection and quality assurance functions "Up-Front". This paper describes the status, scope and use of several of these tools in casting process and post-casting process such as heat treatment etc.

THE EFFECT OF NONUNIFORM CONTACT ON THE SOLIDIFICATION OF ALUMINUM SHEET ON COPPER SUBSTRATES

J. P. PAPAI and C. E. MOBLEY
The Department of Metallurgical Engineering
The Ohio State University
Columbus, Ohio 43210

ABSTRACT

The extent of physical contact between a liquid cast onto a solid substrate may vary spacially as a result of entrapped gases, substrate debris, and the surface profile of the substrate. In modeling the solidification of sheet casting processes, spacial variations of the physical contact may be treated as equivalent variations in the interface heat transfer coefficient. The effects of nonuniform interfacial heat transfer coefficients on the solidification of aluminum on copper substrates are presented. A two dimensional, time dependent, finite difference computation with spacially varying interface coefficients is used to model the sheet solidification problem. The shape of the liquid-solid interface and the occurrence of defects in the sheet are discussed as they relate to the spacial variation of the interface heat transfer conditions.

Casting of Near Net Shape Products
Edited by Y. Sahai, J.E. Battles,
R.S. Carbonara and C.E. Mobley
The Metallurgical Society, 1988

INTRODUCTION

A number of researchers[1-6] studying metallic ribbon and sheet casting processes have indicated that the heat transfer across the sheet-substrate interface may vary both spacially and temporally in a given process. The variation in the value of the heat transfer coefficient at the sheet-substrate interface has been attributed to entrapped gases between the sheet and substrate, lack of wetting or complete physical contact between the liquid and substrate, debris on the substrate, and dimensional changes during cooling of the cast sheet and substrate. Other workers[2,7-10] have noted the effects of the nonuniform contact and/or heat transfer on the microstructures of cast ribbon and sheet. Several patents have issued which recognize and/or address the significance of nonuniform contact between cast sheet and the substrate as it influences sheet quality[11, 12]. To ascertain the influence of a periodic spacially varing extent of physical contact or equivalently, a periodic spacially varying heat transfer coefficient at the solidifying sheet-substrate interface, a two dimensional, time-dependent computational program was developed to model the solidification of aluminum sheet on a copper substrate.

MODEL

The problem being simulated is the solidification of a fixed thickness stagnant layer of aluminum on a planar copper substrate with a periodically varying heat flux along the aluminum-copper interface. The model incorporates alternating regions of poor and good thermal contact along the aluminum-copper interface with each type region being of fixed linear dimension. The spacing or wavelength, B, of the repetitive poor-good contact regions is defined as the distance between the centers of neighboring regions of poor contact, i.e.,

$$B = W_{poor} + W_{good} \qquad [1]$$

where W_{poor} = the width of the region of poor thermal contact

and W_{good} = the width of the region of good thermal contact

The length (and aerial) fractions of poor and good thermal contact, f_{poor} and f_{good}, are defined as:

$$f_{poor} = W_{poor}/B \qquad \text{and} \qquad f_{good} = W_{good}/B \qquad [2]$$

with the resultant identity:

$$f_{poor} + f_{good} = 1 \qquad [3]$$

Thus the configuration of the aluminum-copper planar interface may be characterized through either of the two variable sets, W_{poor} and W_{good}, or B and f_{good}.

The regions of poor and good thermal contact are characterized by the prescribed heat transfer coefficient, h_{poor} and h_{good}, respectively. It is assumed that the heat transfer coefficient within a given region is a constant (i.e., h_{poor} does not depend on position within the poor thermal contact area or time). The ratio of the values of the heat transfer coefficient in the good region to that of the poor region is defined as r, i.e.,

$$r = h_{good}/h_{poor} \qquad [4]$$

The surface of the copper substrate away from the casting interface is taken to be in contact with a constant temperature (i.e., 50 C) coolant. A constant value heat transfer coefficient equal to 10 cal/cm^2 C sec is assigned to the copper-coolant interface. The copper substrate, of thickness s, is assigned an initial homogeneous temperature of 50 C.

The exposed surface of the solidifying aluminum sheet is assumed in contact with a constant temperature (i.e., 30 C) environment and the heat transfer coefficient at the aluminum-environment interface is set at 0.001 cal/cm^2 C sec. The aluminum layer, of thickness d, is initially in the liquid state with an homogeneous temperature of 710 C (i.e., the initial superheat is 50 C).

A schematic drawing of the aluminum-copper system and the various regions with their associated heat transfer coefficients is shown in Figure 1.

A two-dimensional, time dependent explicit finite difference method(13) was used to compute the thermal fields in the aluminum layer and the copper substrate as functions of time. The enthalpy method was used to include the heat of fusion in the solidifying aluminum layer(14). Heat balances were computed for a rectangular array of volume elements as shown schematically in Figure 2. Incremental time steps were chosen in the computations to meet the stability criteria to insure stable solutions. Prior to using the computational model to study the effect of inhomogeneous thermal contact on the solidification of the aluminum layer, the finite difference model was used to calculate temperature distributions for a two dimensional body with a prescribed perimeter temperature (i.e., a Dirichlet boundary condition) without a phase change and the results compared with available analytical solutions. In all the test cases, the results of the model were in agreement with the analytical solutions.

The finite difference model was used to study the effect of inhomogeneous thermal contact on the solidification of aluminum on a copper substrate. Input values to the model are given in Table 1.

Table 1. Input Values

Thickness of aluminum layer, d:	0.025, 0.050, or 0.10 cm
Thickness of copper substrate, s:	0.50 cm
Initial temperature of liquid aluminum:	710 C
Initial temperature of copper substrate:	50 C
Temperature of environment:	30 C
Temperature of substrate coolant:	50 C
Heat transfer coefficient at copper-coolant interface:	10 cal/cm^{2}C sec
Heat transfer coefficient at metal-environment interface:	0.001 cal/cm^{2}C sec
Poor-good contact spacing distance, B:	0.04 or 0.08 cm
Fraction of good thermal contact, f_{good}:	.1, .2, .33, .5, .67, or.8
Heat transfer coefficients at the aluminum-copper interface:	See Table 2

The sets of values used for the heat transfer coefficients at the regions of good and poor thermal contact along the aluminum-copper interface are listed in Table 2. The corresponding values of r, the ratio of the good to poor heat transfer coefficient, are also listed in Table 2.

Table 2: Heat Transfer Coefficients at the Al-Cu Interface Used in the Model Simulations

h_{good}	h_{poor}	$r = h_{good}/h_{poor}$
5.0	0.01	500
2.0	0.01	200
5.0	0.50	10
1.0	0.10	10
0.5	0.05	10
5.0	1.00	5
1.0	0.20	5
0.5	0.10	5
5.0	2.50	2
2.0	1.00	2

All the heat transfer coefficients listed in Table 2 are given in units of cal/cm^2 C sec.

The values of the heat transfer coefficients listed in Table 2 and used in the model silmulations are comparable to the heat transfer coefficient values reported in the literature for thin section aluminum castings on metallic substrates [15].

COMPUTATIONAL RESULTS

The results of a typical computation run are shown in Figure 3, in the form of the position of the liquid-solid interface in the aluminum layer for several selected times. The particular input variables associated with the Figure 3 results are:

$d = 0.025$ cm　　$s = 0.5$ cm

$h_{good} = 1.0\ cal/cm^2$ C sec　　$h_{poor} = 0.1\ cal/cm^2$ C sec

$W_{good} = 0.064$ cm　　$W_{poor} = 0.016$ cm

$B = 0.08$ cm　　$r = 10$　　$f_{good} = 0.8$

The x-axis represents the distance from the center of the poor contact region and the y-axis represents the thickness or depth of the aluminum layer measured from the planar copper-aluminum interface.

Two computed parameters are used to characterize the influence of the inhomogeneous thermal contact on the solidification of the aluminum layer. These two parameters are:

1. the thickness solidified over the center of the poor contact region at the time the region in the center of the good contact region completes solidification, identified with the symbol M with units of centimeters.

and

2. the ratio of the time to complete solidification at a position at the center of a poor contact region to the time to complete solidification at the center of a good contact region, i.e., t_{poor}/t_{good}.

M, the thickness solidified over the center of a poor contact region when the center of the good contact region just completes solidification, is shown as a function of the fraction good contact, f_{good}, for B = 0.04 cm and d = 0.025 cm and for r values from 2 to 500 in Figure 4. The ratio of t_{poor} to t_{good} is shown in Figure 5 as a function of f_{good} for the various r values for the case with B = 0.04 cm and d = 0.025 cm.

M is plotted as a function of f_{good} and the various r values for B = 0.08 cm and d = 0.025 cm in Figure 6. Figure 7 shows t_{poor}/t_{good} as a function of f_{good} and r for the same case as used in Figure 6.

Figures 8 and 9 show M and t_{poor}/t_{good} ,respectively as functions of f_{good} and r for the case of B = 0.08 cm and d = 0.05 cm.

DISCUSSION

As indicated in Figures 4 through 9, the dimensional and thermal characteristics of the casting-substrate interface influence the planarity of the liquid-solid interface. As the factor d/B (i.e., the ratio of the sheet thickness to the thermal inhomogeneity spacing) increases, the planarity of the liquid-solid interface improves. A comparison of Figure 6, with d/B = 0.31, to Figures 4 and 8, where d/B = 0.62 shows that M is greater for the smaller d/B case. Computations for cases with d/B equal to or greater than 1.0 indicate that M approaches the value of d to within about 1%.

The influence of B or the scaling parameter d/B on the planarity of the liquid-solid interface is also shown in the plots of t_{poor}/t_{good} versus f_{good} (i.e., Figures 5,7 and 9). The parameter t_{poor}/t_{good} is greater than 1.3 for the cases where d/B is 0.31, but was 1.0 to 1.01 when d/B = 1.25, independent of the particular r values considered. Thus, as the ratio d/B increases, the planarity of the liquid-solid interface improves.

The effects of the r value (i.e., $r = h_{good}/h_{poor}$) on the planarity of the liquid-solid interface are also evident in Figures 4 through 9. With all other variables held constant, as r increases M decreases and t_{poor}/t_{good} increases. Increasing values of r generally lead to a less planar liquid-solid interface, as inferred by smaller values of M and larger values of t_{poor}/t_{good}.

The influence of f_{good} on M and t_{poor}/t_{good} depends on the particular value of r considered, but generally appears as a skewed sine type function with M and the time ratio having values of 1.0 at the limiting conditions f_{good} equal to 0 and 1. The value of f_{good} at which M and the time ratio have their minimum or maximum values, respectively, depend on the r value and the specific B value.

It should be noted that the computed thermal fields are based solely on "macroscopic" heat flow considerations and the model described herein does not include any phase nucleation and microscopic growth considerations. Thus, one would not anticipate that this model would predict grain morphology relations within the cast aluminum sheet.

SUMMARY

An explicit two dimensional, time dependent finite difference model has been applied to the problem of solidification of a fixed thickness layer of aluminum on a copper substrate with periodically varying heat transfer along the planar aluminum-copper interface. The computational results of this model indicate that:

1. When d/B approaches unity the liquid-solid interface is nearly planar as it approaches the aluminum free surface for all combinations of variables studied.

2. When d/B is less than unity the liquid-solid interface is nonplanar with:
   a. M, the thickness solidified over the center of the poor contact region when the region over the center of the good contact region has just completed solidification decreasing as $r = h_{good}/h_{poor}$ increases

   and b. t_{poor}/t_{good}, the time to freeze at the center of the poor region divided by the time to freeze at the center of the good contact region, increasing as r increases.

REFERENCES

1. T. R. Anthony and H. E. Cline, "Dimensional Variations in Newtonian-quenched Metal Ribbons Formed by Melt Spinning and Melt Extraction," J. Appl. Phys., 50(1), (January, 1979), 245-254.
2. O. Chopra and P. Niessen, "Solidification of Lead and Lead Alloys in Continuous Drum Sheet Casting," Z. Metallkde., 66, (1975), 10-16.
3. H. Muhlbach, G. Stephani, R. Sellger, and H. Fiedler, "Cooling Rate and Heat Transfer Coefficient during Planar Flow Casting of Microcrystalline Steel Ribbons," Inter. Jour. of Rapid Solidification, 3, (1987), 83-94.
4. E. Vogt, "On the Heat Transfer Mechanism in the Melt Spinning Process,", Inter. Jour. of Rapid Solidification, 3, (1987), 131-146.
5. R. Sellger, W. Loser, and W. Neumann, "Heat Flow Model of Rapid Solidification with Non-homogeneous Thermal Contact," J. Materials Science, 19, (1984), 2145-2152.
6. P. Cremer and J. F. Wadier, "Analysis of Casting Conditions of Amorphous Ribbons," Rapidly Quenched Metals V, (1985), 83-87.
7. S. C. Huang and A. M. Ritter, "Structure Development in Melt-Spun Superalloy Ribbons," General Electric Report No. 83CRD015, (January, 1985).
8. H. A. Davies, "Metallographic Studies of Rapidly Quenched Superalloy Ribbons," General Electric Report No. 82CRD318, (December, 1982).
9. H.H. Liebermann, R.E. Maxwell, R. W. Samshey, and J. L. Walter, " Process and Structural Aspects of Melt-Spun Nickel-Base Superalloy Ribbons," General Electric Report No. 82CRD335, (January, 1983).
10. C. E. Mobley, R.E. Maringer, and L. Dillinger, "Surface Features of Rapidly Quenched Filaments," Rapid Solidification Processing, Principles and Technologies, Claitor's Publishing Division, (1978), 222-229.
11. E. Mizikar, W. M. Wojcik and K. Li, "Method of Producing Steel Strip of Uniform Thickness by Direct Casting," U.S. Patent 3,345,738, October, 1967.
12. W. M. Wojcik and E. A. Mizikar, "Method of Direct Strip Casting on a Coated Drum," U.S. Patent 3,540,517, November, 1970.
13. M.K. Walther, "Solidification Analysis Using CAST," Modern Casting, 74, No. 3, (1984), 18-22.
14. J. Crank, "How to Deal with Moving Boundaries in Thermal Problems," Numerical Methods in Heat Transfer, J. Wiley & Sons, (1981), 177-200.
15. J. Papai and C. Mobley, Heat Transfer Coefficients for Solidifying Systems, Report No. ERC/NSM-87-13, Engineering Research Center for Net Shape Manufacturing, (August, 1987), 1-50.

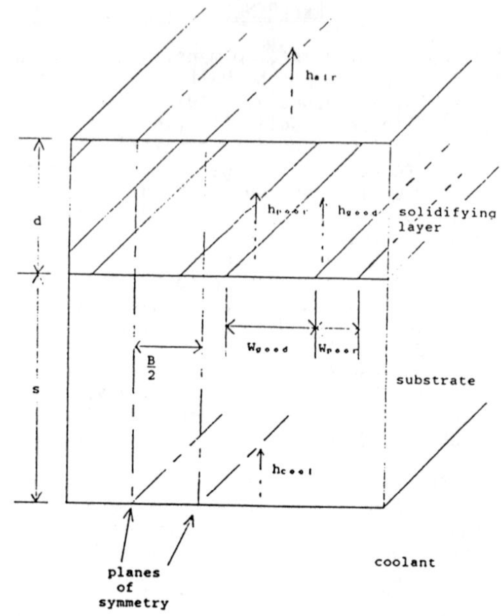

FIGURE 1. Situation being simulated, schematic.

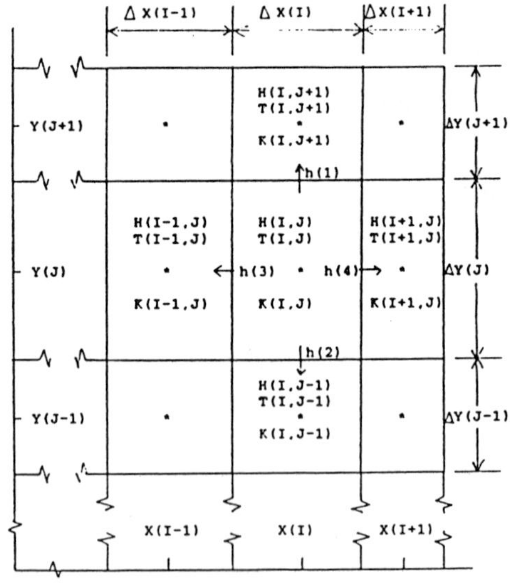

FIGURE 2. Schematic of finite difference elements.

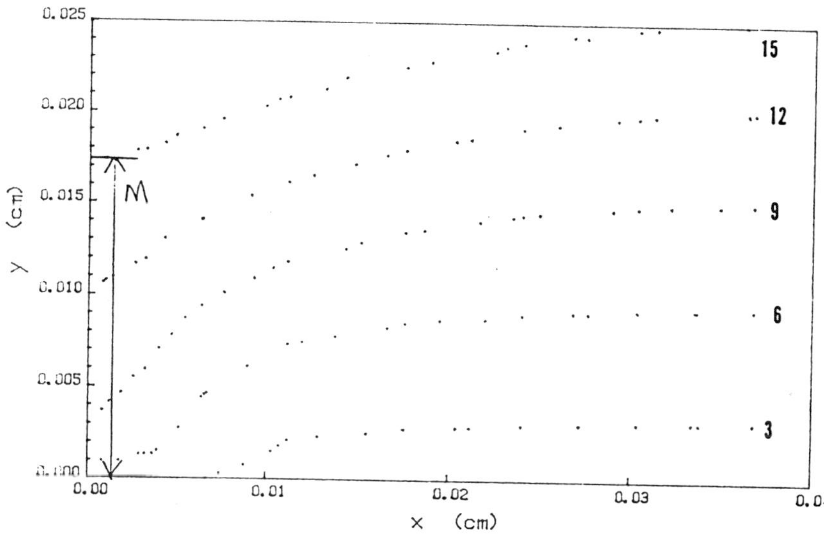

FIGURE 3. Results of the example simulation, where the vertical axis represents the distance from the substrate and the horizontal axis is the distance from the center of the poor contact region. The times beside the calculated solidification interface positions are in milliseconds. M is the depth of solidification when the position over the center of the good contact region has just solidified. For this simulation, M= .0174 cm.

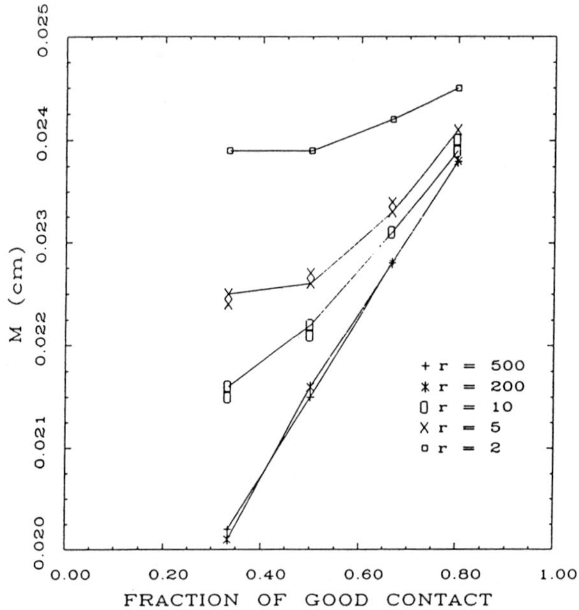

FIGURE 4. M as a function of fraction of good contact for several values of r. B_o=.04 cm d=.025 cm.

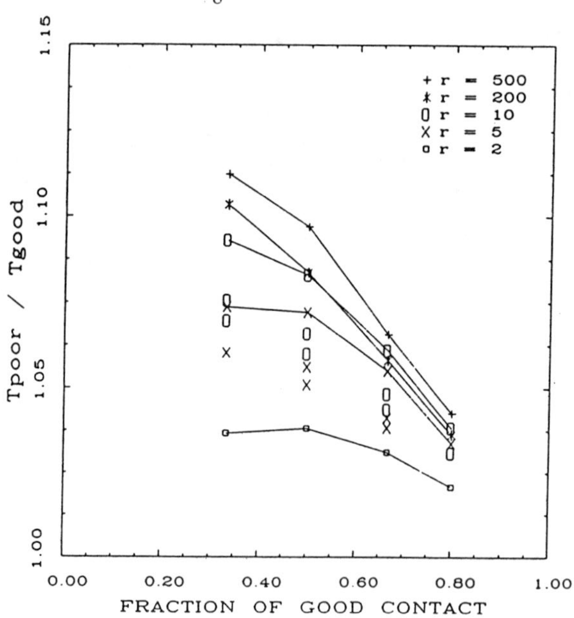

FIGURE 5. T_{poor} / T_{good} as a function of fraction of good contact for several values of r. B_o=.04 cm d=.025 cm

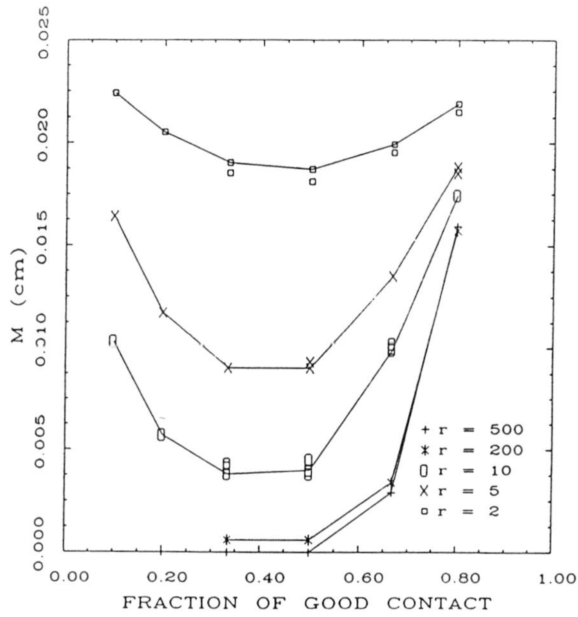

FIGURE 6. M as function of f_{good} for several values of r. B=.08 cm d=.025 cm.

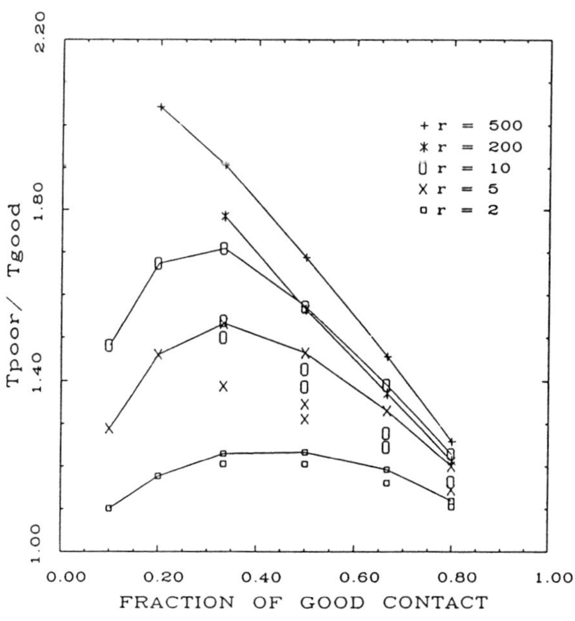

FIGURE 7. T_{poor}/T_{good} as function of f_{good} for several values of r. B=.08 cm d=.025 cm.

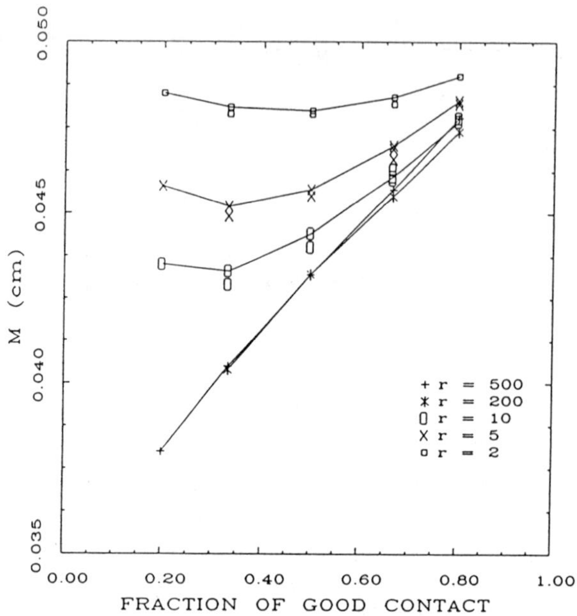

FIGURE 8. M as a function f_{good} for several values of r B=.08 cm d=.05 cm.

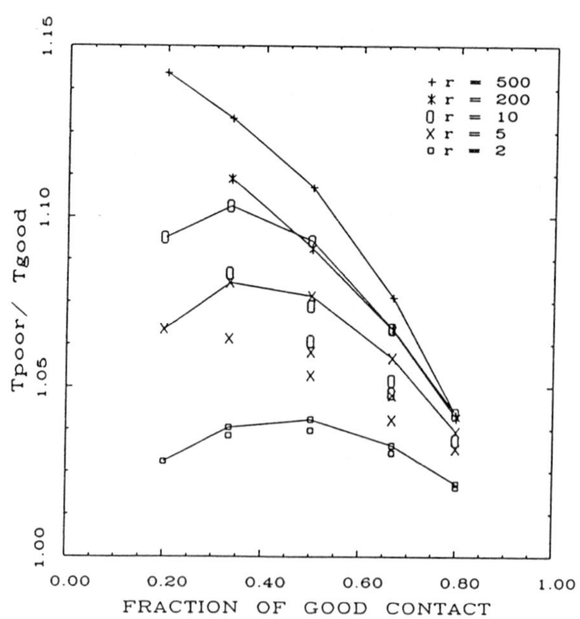

FIGURE 9. T_{poor} / T_{good} as function of f_{good} for several values of r. B=.08 cm d=.05 cm.

AFTER DINNER TALK

Introduction by: **C.E. Mobley**
Ohio State University
Columbus, Ohio

CONTINUOUS CASTING PROCESSES

(PAST, PRESENT AND HOPES)

Robert B.Pond. Sr.

Matetrials Science and Engineering Department
Johns Hopkins University
Baltimore, Md. 21218

Abstract

Before launchings into a journey through the various techniques of continuous casting it is advantageous to stop and consider a few definitions. If we are going to discuss CONTINUOUS CASTING we should have common definitions of these words.

Continuous casting processes do not necessarily result in continuous castings any more than a continuous production of automobiles results in one exceptionally long vehicle. We do have continuous casting of discontinuous products. If the time between the casting of the discontinuous products is measured in microseconds then we feel comfortable in saying that the process is continuous.. But what it the time period is in minutes or hours or days? When do we stop calling it a continuous process and refer to it as a static casting process?

We need also to look at the word CASTING. A casting is generally looked upon as an article resulting from throwing or casting material into a mold. But does the material being cast have to be consolidated or even condensed? If the material entering the mold is a vapor which converts to a liquid in the cavity or even condenses directly to a solid do we consider the product as a casting?

It is fun to consider what the largest continuous castings are if we accept the broadest definition of these terms. I contend that it would be snow. Certainly the generation of the snow flakes is continuous and although the products are discontinuous they form together to produce the net shape of a blanket covering the earth.

The word MOLD also needs to be examined. In his review paper on continuous casting Easton (1) recited the various molds described in U.S. patents which included sliding, oscillating, vibrating wheel or roll, chain, belt or strip, furnace, liquid and whirling molds made of materials such as copper, graphite, brass, iron, steel, mercury, salt and various refractories.

Casting of Near Net Shape Products
Edited by Y. Sahai, J.E. Battles,
R.S. Carbonara and C.E. Mobley
The Metallurgical Society, 1988

In this discussion I will use the broadest definition of continuous casting and as I review some of the advances that have been made in the field I will try to classify the techniques described as belonging to one of three classes. The first and perhaps most understood class includes those castings which are contained in a mold in three dimensions. Typical of such castings are sand castings and die castings. The second class includes those castings in which a mold contacts the cast material in only two dimensions or those castings which have one free surface. These castings are typified by chill block melt spun fibers. The third class includes those castings in which no mold is used but forces (such as viscosity and surface tension) within the metal control the terminal shape of the part.

Probably the earliest practice of continuous casting is found in the manufacture of shot in shot towers. Such a tower still stands as a museum in Baltimore. It was built in 1828, and at 246 feet, it is the tallest remaining shot tower in the world. In practice the molten lead flowed through orifices or ports in a "sieve" and of necessity had to solidify before falling into water catches at ground level. This continuous casting of discontinuous products falls into the third class of castings where surface tension forces shape the part.

Virtually all solidification involves motion of fluids with viscosities of only a few centipoise which are transformed to solids having viscosities of the order of 10 to the twentieth power in various time periods which generally depend on the section size. The teaching of Lord Raleigh (1) on jets and the excellent demonstrations of Boys (2) show that the instability o jets allows the surface tension forces to take over and disintegrate the stream as seen in a demonstration of the water telephone.

Although Boys (2) gives a more elaborate description of the water telephone it can be appreciated with the use of only a paper cup, a bit of aluminum foil, a wash bottle and a music box works. Water squirted from the wash bottle emanates as a smooth jet which breaks up due to the instability of the jet and surface tension forces at some distance from the orifice of the wash bottle. Raleigh insinuated that if the jet could be solidified before breakup a water wire would be generated. If the unbroken stream is impinged on a piece of aluminum foil stretched over the mouth of a paper cup a hiss like sound is produced. If the impingement is made further along the stream where it is broken into a continuous array of droplets then a high pitched note is generated. It the performing music box works is held against the orifice of the wash bottle the smooth stream is disintegrated at the orifice into a parade of jets of various length so that their impingement on the foil produces an amplified tune of the music box.

Without the advantage of the work of Raleigh and Boys, Edward Small (3)invented a "apparatus for making wire solder" in 1882. Small's device consisted of orifices through which molten metal was extruded and solidified into the wires later imagined by Raleigh. This moldless casting is easily demonstrated by pneumatically squirting a tiny jet of tin into the air to produce a round continuous wire.

A principal interest in continuous casting has been centered in the area of steel casting. Such work was pioneered by Bessemer (4) during the 1850's and 1860's. The many techniques and inventions following Bessemer's efforts have primarily been of the first class wherein the molten metal is contained on all sides. Many of the improvements have been concerned with the total containment.

While these endeavors were going on a number of other continuous casting processes were introduced. Most of these techniques are of the second class wherein there is always a free surface. The principle of all of these techniques is easily demonstrated in chill block melt spinning by using a rotating cake pan and a jet of molten tin. In the same class of castings there have been many inventions and improvements in the continuous casting of built up layers by atomization and vapor deposition and in the manufacture of flakes or "splats". Such techniques are easily demonstrated by use of the Cowing spinning cup(5).

There have also been many inventions and improvements of the third (or moldless) class. Some of the newest of these improvements involve levitated casting (6) and the continuous casting of hollow metal microspheres(7)

About fifty patents and discoveries are covered to illustrate the progress of continuous casting. The importance of the interrelation of fluid mechanics and solidification mechanics is stressed. A hope is offered that progress in continuous casting will continue if this interrelation is understood and taken advantage of.

References

1. Rufus Easton, "Continuous Casting: Review and Outlook", Journal of Metals, August 1957

2. J.W. Strutt (Lord Raleigh) The Theory of Sound, Dover Publications, New York, 1945

3. C.V. Boys, Soap Bubbles, Dover Publications, New York, 1959

4. H. Bessemer, inventor, U.S. Patent No. 409,053 "Steel Caster" July 1865

5. P. Cowing, inventor, U.S. Patent No. 809,671 "Shot Making Machine", 1906

6. P.R.Roberts, inventor, U.S.Patent No. 4,735,252 "System for Reforming Levitated Molten Metal into Metallic Forms"April 1988

7. T. Wang, M.C.Lee, C. Schilling, & G.O. Ladner, inventors, Patent Pending, "Producing Refractory Microballoons", NASA Tech Briefs, Nov. 1986

PLENARY LECTURE

Session Chairman: **J.E. Battles**
Argonne National Laboratories
Argonne, Illinois

THE DEVELOPMENT OF THIN STEEL SLAB CASTING PROCESS

M. Yoshihara

Steel Making Reseach Laboratory
Iron & Steel Making Reseach Department
Iron & Steel Research Laboratories
Research & Development Division
Sumitomo Metal Industries, Ltd.
Amagasaki, Japan

Abstract

Since 1983, as one of the continuous casting processes of 'Near Net Shape' slab, the thin slab casting process for steel has been developed in Sumitomo Metal Industries, Ltd. using a twin belt caster.

The casting tests of 600 mm width slab and 1320 mm width slab were carried out by a 50 tons pilot plant at Kashima Works.

The techniques to cast 50 mm thickness by 1320 mm width slab at casting speed 5 m/min have been developed. The cast slabs have been confirmed to be acceptable for commercial grade quality by hot rolling.

Casting of Near Net Shape Products
Edited by Y. Sahai, J.E. Battles,
R.S. Carbonara and C.E. Mobley
The Metallurgical Society, 1988

Introduction

In the recent years, the thin slab continuous casting process for steel has been developed in many places arround the world. This process has a number of advantages over the conventional continuous casting process.

(1) The number of rolling passes can be decreased.
Therefore, initial cost and energy cost at hot rolling process can be decreased.

(2) Machine construction cost of caster can be saved.

Omission of hot rolling process is most effective to decrease hot rolling machine construction cost and hot rolling energy. However, for this purpose, the slabs of the thickness below about 5 mm must be cast, and, in this case, there will be many problems to be solved in productivities and slab quality.

The other hand, in the case of omitting roughing mill, 30 to 50 mm thickness slabs must be cast, therefore, there will be few problems to be solved compared with 5 mm thickness slab casting process.

In Sumitomo Metals, since 1983, this slab casting process for steel has been developed using a twin belt type caster, and it has been possible to cast 50 mm thickness and 1320 mm width slabs.

This paper described the developed techniques for casting, the slab qualities and the properties of hot rolled products.

Facilities

The layout of the pilot plant is shown in Fig. 1 schematically, and main specifications are shown in Table 1.

(1) Ladle
The maximum capacity is 50 tons.

(2) Large tundish
The maximum capacity is 10 tons. The immersion nozzle with sliding gate used for controlling the pouring rate is equipped at the bottom of the large tundish.

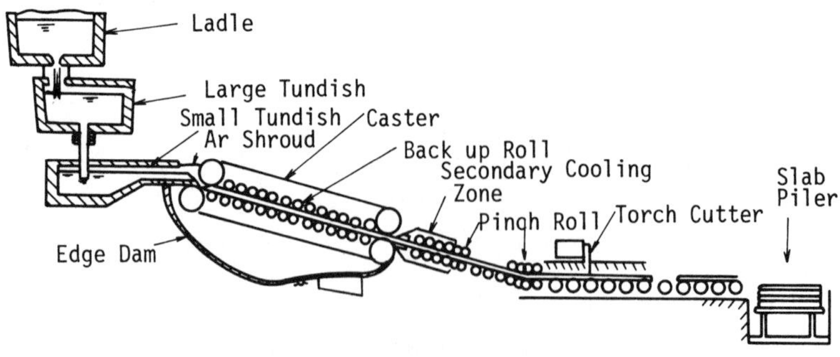

Figure 1 - Layout of the TSC pilot plant

Table I. Specifications of pilot plant

		Case a	Case b
Caster type		Hazellet twin belt caster	
Plant length		69 m	
Casting speed		max. 15 m/min.	
Slab size	Thickness	50 mm	
	Width	600 mm	1,320 mm
Caster angle		0 - 15°	
Caster length		2.9 m	4.2 m
Belt cooling		High speed water film cooling	
Secondary cooling		Water spray cooling	
Torch cutting speed		max. 2 m/min.	

(3) Small tundish

The maximum capacity is 1 ton. A refractory spout for pouring the molten steel into the caster is installed.

(4) Caster

The configuration of the caster is shown in Fig. 2 schematically.

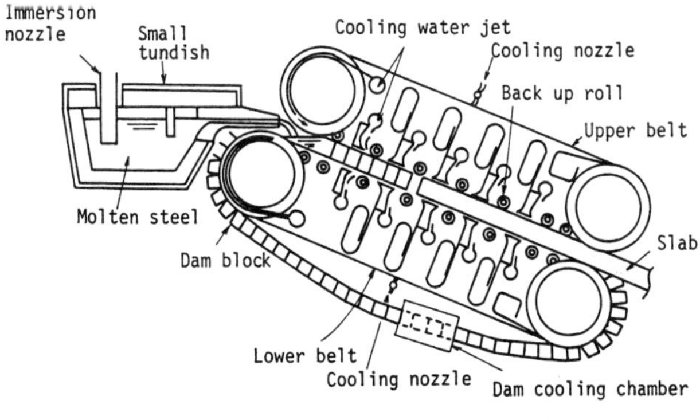

Figure 2 - Schematic diagram of twin belt caster

Mold cavity is formed by movable belts and side dam blocks. The belts are made from steel, the thickness of which are 1.1 mm to 1.4 mm. And the belts are cooled by high velocity water film which flow velocity is about 15 m/sec.

The side dam blocks which are moved with the belts are made from low alloy steel.

(5) Secondary cooling zone

Water sprays similar to conventional continuous caster are adopted to prevent the bulging of slabs.

The length of cooling zone is about 6 m.

(6) Pinch rolls

Pinch rolls are driven by an A.C. motor, which is synchronized with a caster motor.

(7) Torch cutter

Propane gas and high pressure oxygen gas are used for slab cutting. The cutting speed is maximum 2 m/min.

(8) Slab piler

Slab of the 10 m length are piled automatically and the maximum piling height is 800 mm.

Operation

Molten steel is tapped into the ladle from a 250 tons B.O.F.. The temperature and chemical compositions of molten steel in the ladle is controlled by Ar gas bubbling treatment, and it is poured into the caster through two tundishes.

The casting conditions is shown in Table II and Table III.

Table II. Casting conditions

Item	Case a	Case b	Note
1) Heat size	30 ton	50 ton	
2) Steel grade	Middle carbon silicon killed Low carbon Al-kilied		* Table-III
3) Casting size (thickness x width)	50 mm x 600 mm	50 mm x 1,320 mm	
4) Casting temp. Middle carbon Si-killed steel Low carbon Al-killed steel	 1,530°C - 1,555°C 1,540°C - 1,570°C		
5) Casting speed	max. 6 m/min.*		
6) Pouring method	Overflow		
7) Pool level detector	Optical method		
8) Casting start method	Automatic start		* Fig.-3
9) Pool level control	Automatic control		* Fig.-4
10) Belt	Steel belt		

* Stable casting is up to 5 m/min.

Table III. Chemical composition (Results)

Steel Grade	Composition (wt %)				
	C	Si	Mn	P	S
Low C Al killed	0.03 ~ 0.09	0.02 ~ 0.05	0.19 ~ 0.32	0.021 ~ 0.036	0.003 ~ 0.009
Middle C Si killed	0.16 ~ 0.19	0.22 ~ 0.36	0.46 ~ 0.62	0.023 ~ 0.034	0.006 ~ 0.009

(1) Automatic casting start up system

The diagram of the automatic casting start up system is shown in Fig. 3. In this system, pouring rate is adjusted to the target value automatically, and the belts and pinch rolls are moved at the equivalent speed to the pouring rate.

At first, the pouring rate from the large tundish to the small tundish is measured by road cells installed at the bottom of the small tundish. The pouring rate is adjusted to the target value by controlling the opening of the sliding gate of the large tundish.

Next, starting up casting speed is calculated and set to the equal value to the pouring rate. Fig. 3 shows that, for a short period of time, a target casting speed of minus 1 m/min is necessary to raise the pool level to the target position from the position of the top of the dummy bar set before casting.

This automatic start up system is necessary to prevent the problems of the overflow at the caster inlet or the breakout at the caster outlet.

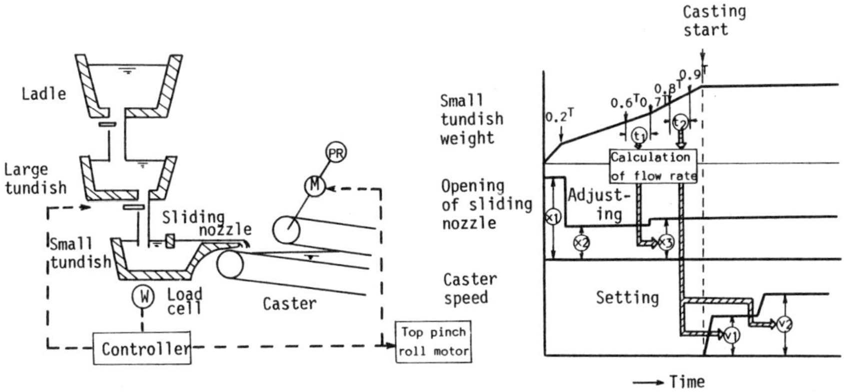

Figure 3 - Schematic diagram of automatic casting start system

(2) Automatic pool level control system

During the constant casting state, the pool level in the caster is controlled automatically. This system, as shown in Fig. 4, has been developed for stable casting and improvement of slab surface quality due to reduce fluctuation of the pool level.

The pool level in the caster is detected by an optical method using a CCD camera, and the pool level is adjusted to the target position by controlling the caster speed and pinch roll speed. At the same time, caster and pinch roll speed are automatically adjusted to the target value by controlling the pouring rate, which is also automatically by controlling the opening of the sliding gate of the large tundish.

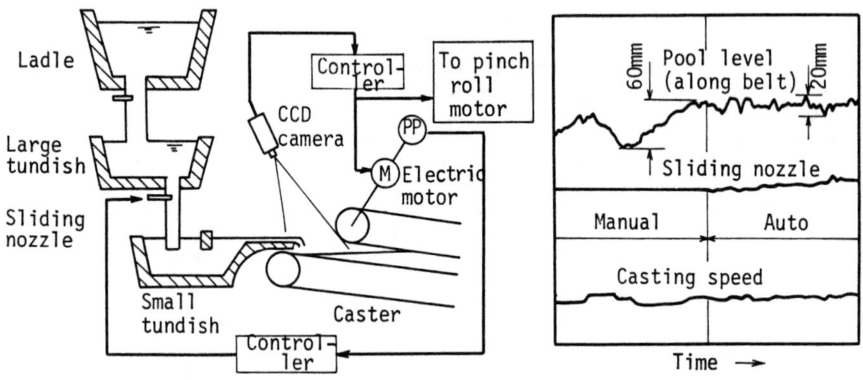

Figure 4 - Automatic pool level control system and result

(3) Countermeasures of pool level fluctuation

Periodic fluctuations of the pool level in the caster were observed at high speed casting more than 4 m/min. It was considered that this fluctuations caused by 'resonance phenomena' in the caster due to the fluctuations of pouring rate, belt deformation, and bulging of the slab in the secondary cooling zone.

This fluctuation of the pool level is improved by selecting the proper time constant of pool level control system, reducing the belt deformation and the bulging of the slab. The reduction of the bulging is achieved by decreasing the supporting length in the secondary cooling zone.

As shown in Fig. 5, after these countermeasures, the casting speed could be increased up to 5 m/min.

(4) Belt deformation

The belts are deformed by heat of the molten steel, and this deformation causes the defects of the slab surface qualities as well as instable operation of casting.

The belt deformation could be reduced to one third by improvement of the belt coating, as shown in Fig. 6.

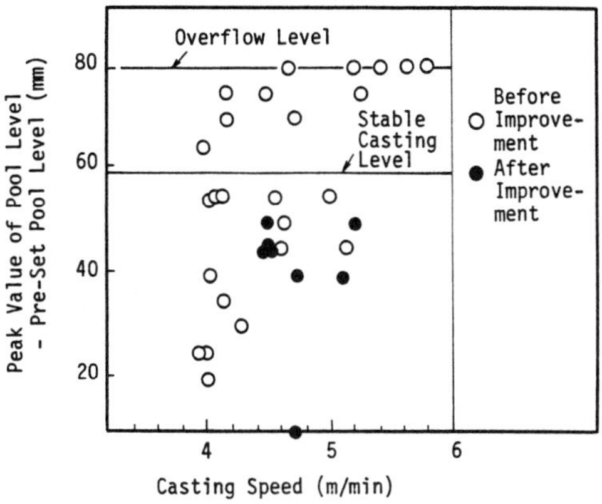

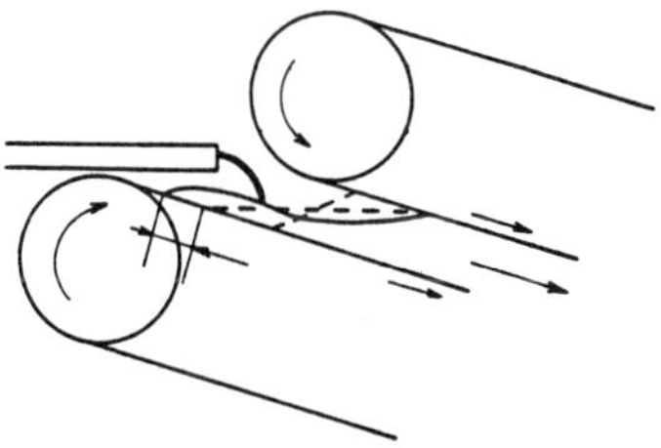

Figure 5 - Influence of casting speed on pool level fluctuation.

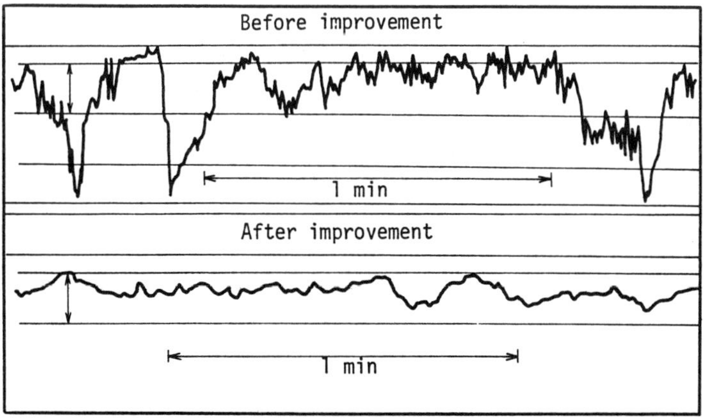

Figure 6 - Effect of belt coating on belt deformation

Slab Qualities

At the early stage of the casting test, many kinds of defects were experienced, such as transverse cracks, longitudinal cracks, hair cracks, and cold shut, and so forth.

After many investigations, the cause of these defects were made clear and these problems were solved.

(1) Transverse cracks

These cracks were about 1 mm depth and 30 mm length. The relation between transverse cracks and the ratio of caster and pinch roll speed is shown in Fig. 7.

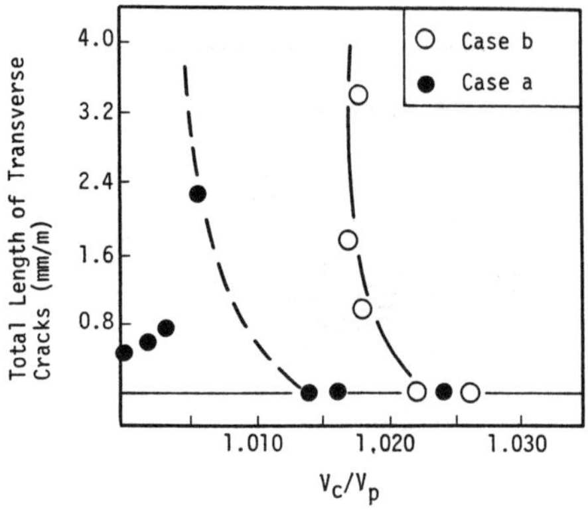

Figure 7 - Transverse cracks

From Fig. 7, it is clear that, to prevent these cracks, the caster speed must be slightly higher than the pinch roll speed.

It is considered that these cracks are caused by the friction between belt and slab surface due to the thermal shrinkage of the slab, and that the difference of the critical value between 600 mm width slab and 1320 mm width slab was caused by the difference of caster length, that is, 1320 mm width caster was longer than 600 mm width caster. Therefore, the friction force of 1320 mm width caster between belt and slab is larger than that of 600 mm width caster.

(2) Longitudinal cracks

The relation between longitudinal cracks and the belt coating is shown in Fig. 8. As shown in Fig. 8, the longitudinal cracks are remarkably reduced by the optimum coating on the belts, especially, it was able to prevent perfectly these cracks for middle carbon Si killed steel.

It is considered that these cracks was caused by un-uniform solidification and thermal shrinkage to transverse direction of the slab. Therefore, these cracks are frequently caused in 1320 mm width slabs compared with 600 mm width slab.

It is important to select the belt coating material in order to uniform cooling of slabs, lubrication between belt and slabs, and reducing the belt

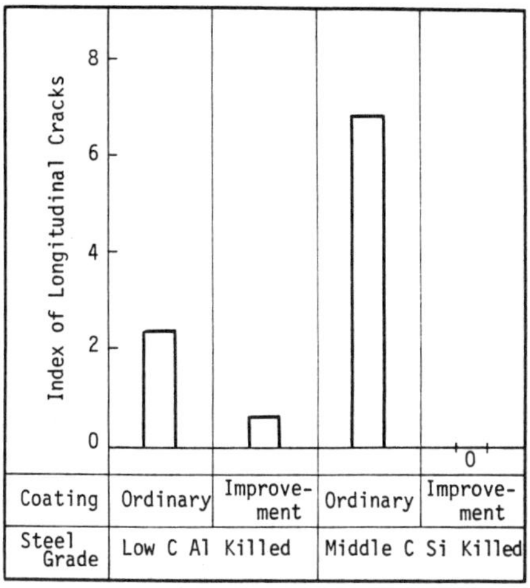

Figure 8 - Longitudinal cracks

deformation. In addition, for inclined type casting as this plant, the equalization of cooling rate of upper surface and lower surface at a transverse cross section is also important tehcniques to prevent longitudinal cracks.

(3) Hair cracks

Hair cracks are non-directional and can not be observed on the scaled surface, but are revealed after pickling. These cracks are less than 50 mm length and the depth is less than 1 mm. It is supposed that these cracks were caused at the early stage of solidification from the observation of the phosphorus segregation near by cracks.

As shown in Fig. 9, these cracks had related to the belt coating as same as longitudinal cracks, and were reduced extremely by the kind of coating materials and the thickness.

(4) Cold shut

At the early stage of casting test, cold shut had been appeared on the slab surface, however, could be prevented by the improvement of casting conditions, such as caster angle, belt coating materials and pool level control. The results are shown in Fig. 10.

(5) Ripple defects

These defects are observed on the upper surface. It is considered that these defects are caused by oxidation of the molten steel surface in the caster. Therefore, these defects are improved by adequate shrounding on the caster inlet. This result is shown in Fig. 11.

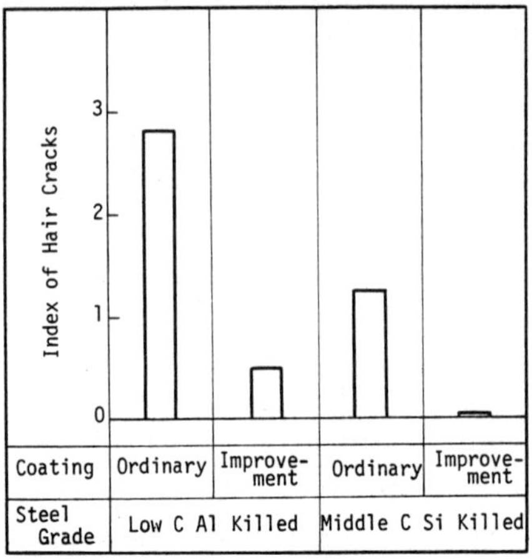

Figure 9 - Hair cracks

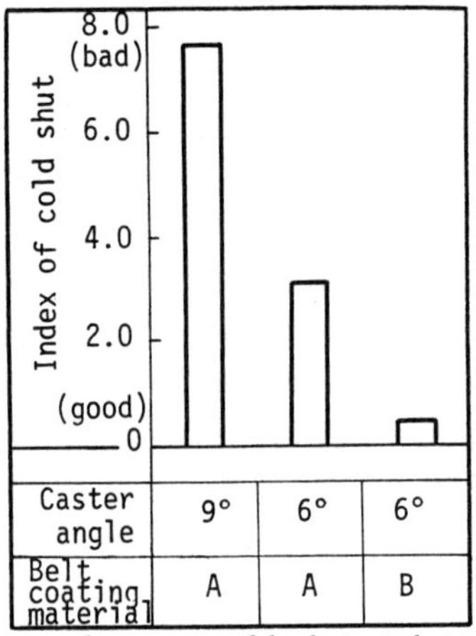

Figure 10 - Relation between cold shut and casting conditions

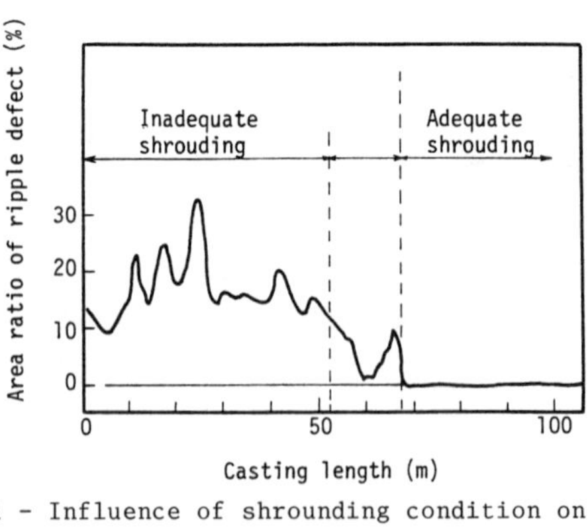

Figure 11 - Influence of shrounding condition on ripple defect

(5) Quality of narrow surface

On the narrow surface, 'dam mark' as shown in Photo. 1 has been appeared. These dam marks are caused by slight discrepancy of each dam block. But, this marks did not affect the qualities of rolled product.

At dam block stop, cold shut on the narrow surface as shown in Photo. 2 appeared.

From this results, it is considered that stationaly dam block is not suitable for this caster.

Photo. 1 Appearance of narrow surface

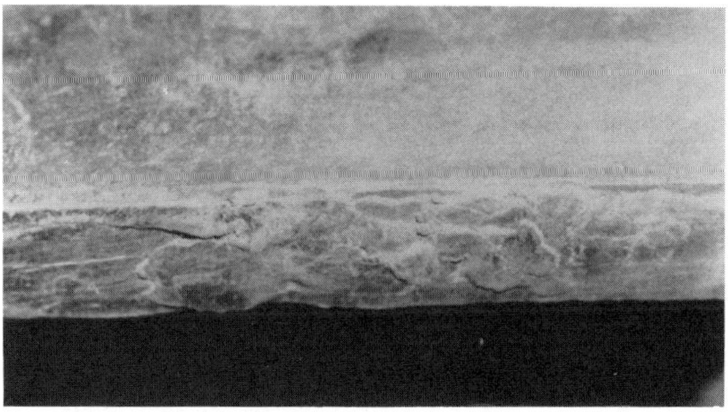

Photo. 2 Appearance of narrow surface at dam stop

(6) Non metallic inclusions

Fig. 12 shows the change of non metallic inclusions from ladle to slab. Non metallic inclusions could become the same amount as conventional CC slab by inert gas shrounding, especially at caster inlet, and by supplying the clean molten steel.

In Fig. 13, the distribution of non metallic inclusions across the slab thickness. Non metallic inclusions accumulated at about 5 to 10 mm from the upper surface of the slab. But, its amount was not detrimental to the qualities of final products.

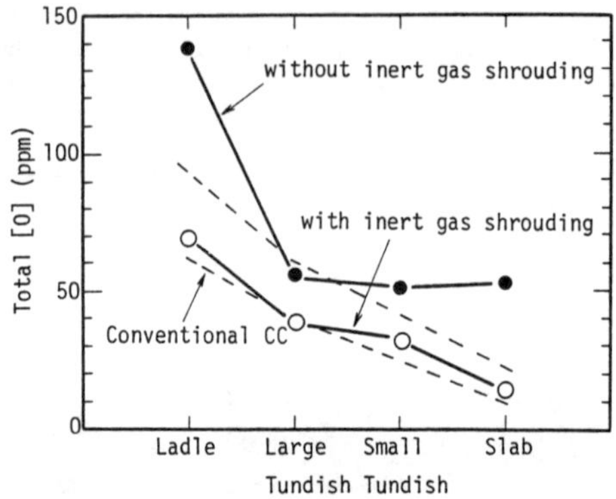

Figure 12 - Change of total oxygen content

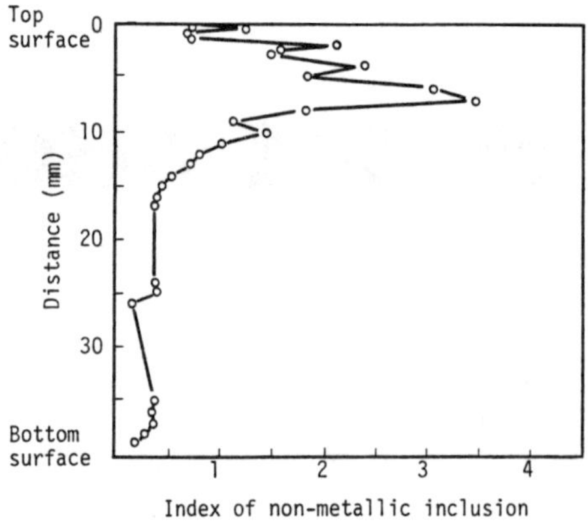

Figure 13 - Distribution of non-metallic inclusion across slab thickness

(7) Solidification structure

Photo. 3 shows the solidification structure of a cast slab. The structure is fine and columnar dendritic.

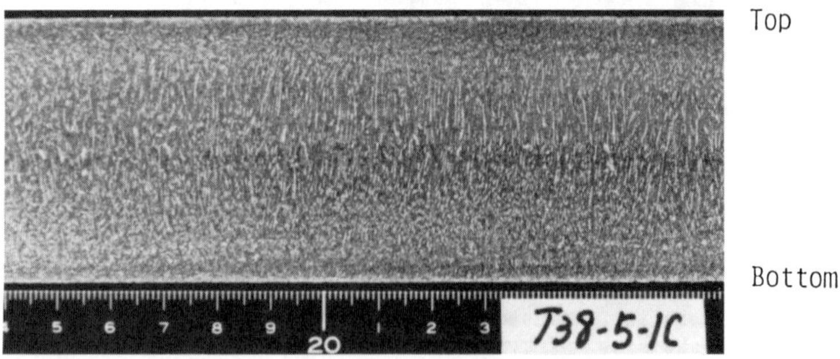

Photo. 3 Solidification structure (Longitudinal cross section)

(8) Centerline segregation

The centerline segregation of phosphorus and manganese, which are measured by X-ray macro-analyser, are shown in Fig. 14.

Centerline segregation were minimal. The reason is supposed that the dendrite structure became fine due to the rapid cooling rate and prevented the movement of interdentritic liquid steel enriched with solute.

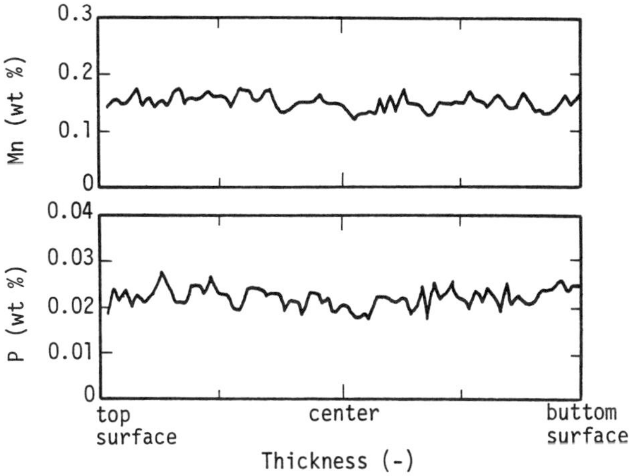

Figure 14 - P and Mn composition distribution across slab thickness by X-ray macro analyzer (Beam dia. 5 micrometer)

Result of Hot Rolling Test

The hot rolled plate and coil have been produced from cast slabs without conditioning in usual hot rolling process.

The properties of the hot rolled coil are shown in Table IV and Table V. The good results are obtained on the surface, internal qualities, and mechanical properties of middle carbon Si killed steel grade. The products of low carbon Al killed steel slab are not so good due to occurrence of slab defects, which is considered to be cause by slab surface defects such as hair cracks.

Table IV. Properties of hot rolled coil (middle carbopn Si killed steel)

	TSC sample	Standard
Plate thickness (mm)	3 - 4.5	16 ≥
Tensile test		
Y.P. (kg/mm^2)	37.6 - 43.1	29 ≤
T.S. (kg/mm^2)	54.7 - 56.2	50 - 62
El (%)	26.7 - 36.8	19 ≤
Bend test (N = 3) (180° - Bending)	Good	-
Surface defect	Good	-

Table V. Properties of hot rolled coil (low carbon Al killed Steel)

	TSC sample	Standard
Plate thickness (mm)	3 - 4.5	3 $\geqq$
Tensile test		
Y.P. (kg/mm^2)	24.4 - 28.4	
T.S. (kg/mm^2)	30.1 - 34.0	28 $\leqq$
El (%)	33.2 - 39.8	31 $\leqq$
Bend test (N = 3) (180° - Bending)	Good	
Surface defect	Bad (scab)	

Conclusion

The thin slab continuous casting process for steel using a twin belt type caster has been developed in Sumitomo Metal Industries, Ltd. since 1983.

Many techniques for stable casting have been developed, and as a result, it have been possible to cast in stable a 50 mm thick and 1320 mm width slab at casting speed 5 m/min.

Cast slab qualities are confirmed to be acceptable for commercial quality grade as light gauge steel for building material, gas pipe and water pipe, and so forth.

FUNDAMENTALS ASPECTS III: MODELING (continued)

Session Chairman: **P. Salmon-Cox**
U.S. Department of Energy
Washington, DC

MODELING AND DESIGN OF TWIN-ROLL CASTERS

J. W. Hlinka
J. A. Burgo
T. J. Conarty

Bethlehem Steel Corporation
Research Department
Bethlehem, PA 18016

Summary

At Bethlehem Steel Corporation, development of the twin-roll casting process was initiated in late 1980. The major incentives for the program were: (1) the operating cost reduction resulting from partial or total elimination of the hot-rolling process and (2) the capital cost reduction resulting from the installation of low-head machines and the elimination of the roughing stands in the hot strip mill. Simple mathematical models developed early in the program indicated that the direct strip casting of steel using the twin-roll process had potential, provided the rolls were large enough to meet the productivity requirements. Also identified was the need for various types of models in order to address important information gaps relative to the design of the machine components themselves. Described in this paper are the results in several areas including:

- Water modeling of the liquid steel feeding system to optimize the steel distribution to the pool formed by the casting rolls, while minimizing alumina plugging and freeze-offs.
- Thermal modeling and pouring trials in which the design requirements of the small transfer ladle and the required melting practices were identified.
- Thermal modeling in which critical design parameters such as the interfacial heat transfer coefficient, cooling water flowrate and the roll and drive size were estimated.
- Sensitivity analysis to define the significant process parameters as well as establish the specifications for the instrumentation and controls.

After over two years of work consisting of thermal modeling, physical modeling, liquid steel pouring trials and prototype operation, the design and construction of the 12-inch (305-mm) diameter twin-roll caster began in late 1982. Casting trials with Fe-Si steel began in August 1984 and

Casting of Near Net Shape Products
Edited by Y. Sahai, J.E. Battles,
R.S. Carbonara and C.E. Mobley
The Metallurgical Society, 1988

continued until late 1985. Experimentation since that time is being focused on the casting of low-carbon, aluminum-killed steels under a cooperative research program with Armco Inc., Inland Steel Co. and Weirton Steel Corp.

Discussed in this paper are the modeling results leading to the design and construction of Bethlehem's twin-roll caster. The results of the pilot program, including an evaluation of both the process and the product, will be described in a future publication.

Background

An overview of the past developments in the area of twin-roll casting of steel strip should begin with Sir Henry Bessemer. In his 1857 patent, Bessemer's arrangement had the rolls preset with a fixed gap which predetermined the strip thickness to be cast (except for the initial separation). In this arrangement and many others which followed, the rolls and roll drives are designed to deliver a force for hot rolling of the strip to this preset thickness. Several investigators have found difficulty with this approach as optimum product quality can only be achieved within a narrow range of a specific combination of thermomechanical casting parameters. If operation beyond that range is attempted, either tearing and cracking of the strip occurs due to excessive reduction or partially solidified strip leaves the rolls resulting in bleeding and distortion. Further impacting the requirement for uniqueness of the solidification point, the time dependent thermal variability of the liquid steel affects the rate of solidification. Over the past 125 years, the search for a direct strip casting process for carbon steels has continued and several alternative approaches to the twin-roll casting process have been developed. Recently a summary of these developments has been published[1].

Table I is an updated summary of the key historical dates in the development of modern steel strip casting and shows the various alternative processes and enhancements to Bessemer's ideas. From a review of the techniques employed, including twin-roll, single-roll, rheocasting, twin-belt, drum-wheel and electromagnetic casting, it was clear that the development of a direct strip casting process remained a process development challenge. Major concerns remained regarding both product quality and throughput.

Work on the twin-roll casting program at Bethlehem began in late 1980 with a study to investigate the feasibility of thin-section casting of various steel grades. For low-carbon, aluminum-killed steels, the major incentive for this work is the potential for a significant reduction ($20-40/annual ton) in capital cost due to the low-head requirements for the caster and the elimination of the roughing stands in the hot strip mill. Another area of potential savings is the area of operating costs. For the case of twin-roll casting to hot-band thicknesses of approximately 0.1 inch (2.5 mm), the reduction in operating costs could approach $20/ton. These potential capital and operating cost reductions would have to be verified by evaluation of strip produced from a pilot plant.

In order to design the pilot plant, a mathematical simulation of the thermal behavior of the twin-roll caster was developed. This approximate analytic model was very useful in understanding the fundamentals of the twin-roll caster -- a complex multivariable system. From the results of this simple model, it was concluded that, theoretically, the twin-roll process could provide the throughput to be competitive to the conventional casting machines. A major concern was the diameter of the rolls which would be required to meet the productivity requirements of approximately 30 TPH/ft of width (89.5 tph/m of width).

This preliminary modeling and analysis also revealed that the twin-roll process development did not lend itself to laboratory study. Scale-down to small rolls for research purposes was not useful because many of the anticipated process problems (e.g. thickness control, edge containment, liquid steel feeding) were machine size dependent. However, the model results did indicate that a relatively small, 12-inch (305-mm) diameter, twin-roll caster was sufficient to answer the throughput questions as well as produce the strip required for quality evaluations. With this size machine, the estimated thickness range of 0.02 inch to 0.08 inch (0.5 mm to 2 mm) could be cast at speeds of 300 to 80 ft/min (91 to 24 m/min), respectively. This thickness range could cover the needs of casting 0.02 inch (0.5 mm) Fe-Si steels as well as approach the thickness of hot-band for low-carbon, aluminum-killed sheet grades.

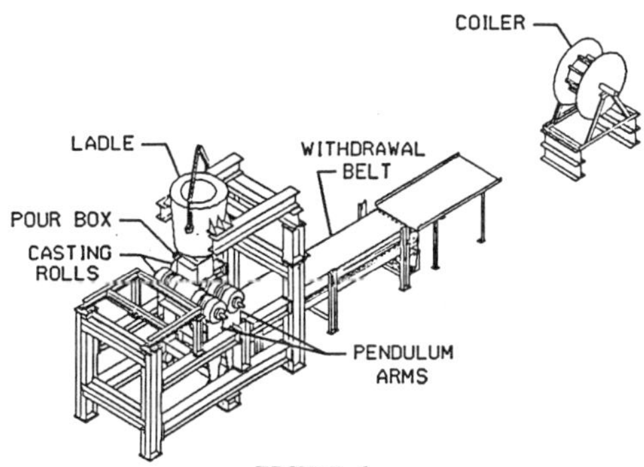

FIGURE 1
TWIN-ROLL CASTER

Once the roll size was established, the work was started on the detailed machine design. The general layout that was established for the process is shown in Figure 1. Liquid steel is supplied to the caster in 700 pound (320 kg) heat sizes via a ladle. Steel flows from the ladle via a manually controlled stopper-rod into a liquid steel pour box located between the separately driven twin-rolls (one roll is fixed; the other roll is spring loaded). Cast strip exits the rolls and is carried on a water-cooled conveyor to a coiler downstream. As the design proceeded, several information gaps relative to major machine subsystems were identified and closed using various modeling/testing techniques. The specific techniques used in the following subsystems of the twin-roll caster are the subject of this paper.

- Physical modeling of the liquid steel distribution system.
- Thermal modeling and pouring trials to ascertain ladle design and melting practices.
- Thermal modeling to estimate critical design parameters.
- Sensitivity analysis to define the experimental program and specify process instrumentation and controls.

Physical Modeling of Liquid Steel Distribution System

Early in the program, the method of introducing metal into the twin-roll caster was recognized as an important process consideration which is especially pertinent to product quality. Improper steel distribution will result in non-uniform physical properties, distortion and cracking, and uneven solidification across the strip. The design strategy was to define a distributor which would both control the flowrate of liquid steel and distribute the metal to the pool created by the rotating twin-rolls. Design problems are apparent in both of these areas. Control of steel flowrates at low tonnages by direct metering requires small orifices which may constrict due to alumina plugging or freezing. Distribution of the liquid steel without excessive turbulence over a large aspect ratio is also a difficult design challenge.

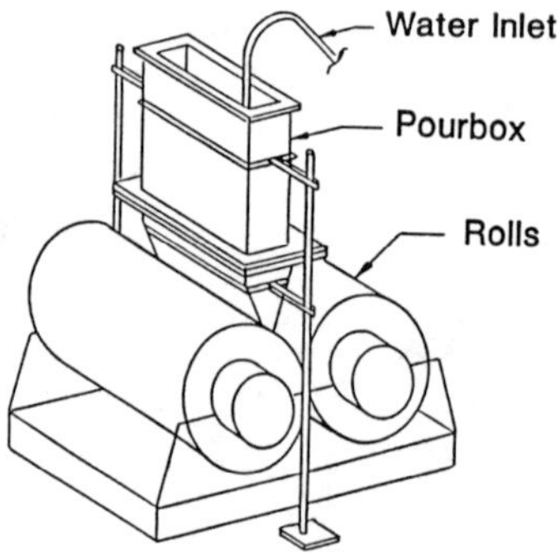

FIGURE 2
WATER MODEL OF POURBOX AND ROLLS

A full-scale water model was constructed to test various techniques for liquid steel flow control and distribution. Figure 2 is a schematic of this model in which two pour-box discharges were evaluated. Calculations had indicated that for 1 foot (30 cm) of ferrostatic head, a cross-sectional area of 0.12 square inches (77 square mm) was required to deliver 5 TPH/ft (15 tph/m) of width of steel to the pilot caster, assuming a discharge coefficient of 1.0. With this cross-sectional area requirement established, water modeling trials were conducted with pour box configurations consisting of: (a) a 0.010 inch (0.3 mm) slot across the entire 12 inch (304 mm) opening and (b) a series of 1/16 (1.6 mm), 3/32 (2.4 mm), or 1/8 inch (3.2 mm) diameter holes with a total cross-sectional area of approximately 0.12 square inches (77 square mm). The results of these trials were:

- The slot-type discharge box resulted in less agitation of the pool surface than any of the multi-hole distributors. Also, considerable air entrainment was observed from open stream pouring with the multi-hole distributor.

- With the multi-hole distributor (1/8 inch (3.2 mm) holes @ 1 foot (30 cm) of head), the streams penetrated the entire depth of the pool (approximately four inches) to the roll bite. Under this condition it was expected that undesirable hot spots across the width of the strip would be observed and lead to non-uniform physical properties.

- A submerged slot-type distributor is preferred for liquid steel distribution with minimal pool turbulence.

From these trials, two alternatives for flow control and liquid steel distribution were evaluated to the point where liquid steel trials were conducted to evaluate the start-up, freeze-off and alumina plugging concerns previously mentioned. Concurrent to these pouring trials for distributor box validation, the ladle design and melting practices were also established.

Liquid Steel Pouring Trials

Prior to conducting the liquid steel pouring trials, a one-dimensional mathematical model was used to simulate the conditions encountered for the ladle and pour box, including the level of preheating, transfer time, and casting duration with the pilot machine. The goals of this work were to estimate the temperature losses associated with the above events and to predict the tap temperature required.

The simulation procedure that was followed started with a preheat of the pouring box and the ladle by exposure to a 2800°F (1540°C) burner flame for one hour and three hours. Two time periods were chosen to provide a comparison. After the preheating and a two-minute transfer time the ladle was filled with 700 pounds (320 kg) of liquid steel at 3000°F (1650°C). A two-minute transfer time was used to move the ladle into teeming position over the pouring box. The steel was teemed into the pouring box which had the same preheat schedule as the ladle. The removal of metal from the ladle and the filling of approximately 100 pounds (45 kg) of metal into the pouring box were included in the simulation.

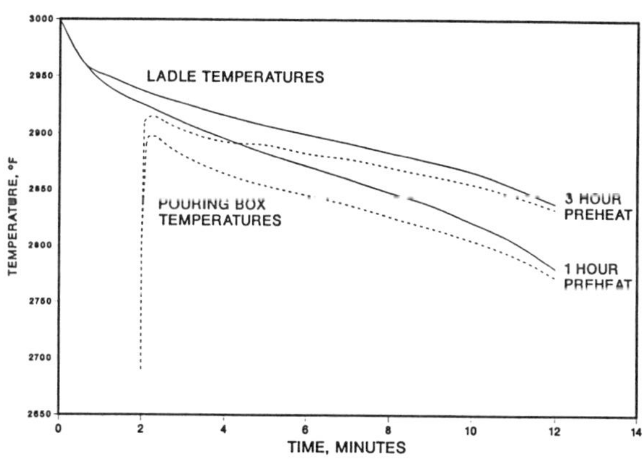

FIGURE 3
SIMULATED LADLE AND POURING BOX TEMPERATURE
LOSSES THROUGH REFRACTORY

Using the minimum one hour and maximum three hour preheat conditions, two sets of time vs. temperature curves shown in Figure 3 were developed. Each set corresponds to the temperature of the hot metal in the ladle and the metal in the pouring box after tap, transfer and cast while casting at a rate of 2 TPH (1.8 tph). This rate is the slowest anticipated casting rate and represents the worst case for temperature losses. Losses of approximately 8 to 12°F/min (4 to 6°C/min) were estimated for both the ladle and pouring box from this simulation.

Measurement of the ladle and pouring box temperature losses was then conducted in liquid steel pouring trials. In addition to the development of preheating and melting practices, testing of the ladle well-block and stopper rod designs was also completed. The major results of this testing were:

- The overall heat losses from the ladle were approximately 20°F/min (11°C/min). These consisted of 10°F/min (6°C/min) from the ladle refractory (in agreement with the thermal model) and 5°F/min (3°C/min) from both the graphite stopper rod and top surface.

- The heat losses from the pouring box were as expected from the modeling work, although a 3-hour preheat to 2200°F (1200°C) is required to assure flow through the distributor and provide superheated metal to the caster.

- The required preheating of the ladle was increased to 7 hours at a temperature of 2550°F (1400°C) to ensure a free opening of the stopper-rod.

- The aim tap temperatures for the pilot caster are 3050-3100°F (1675-1700°C).

Concurrent with these tests, the results from water modeling of the pouring box distributor were also verified. In this work, a series of hole and slot-type distributors were evaluated. In the limit, it was possible to pour the small, 700 pound (320 kg) Fe-Si heats through slots as small as 0.010 inch (0.3 mm) and holes as small as 1/16 inch (1.6 mm) in diameter. For these cases it is best to initiate flow by impacting the opening with the incoming stream (i.e. establishment of an impact pressure before a ferrostatic head develops in the pouring box).

For the pilot caster, the design shown schematically in Figure 4 was used in the early trials. Seven equally distributed 3/16 inch (4.8 mm) diameter holes were used for the casting trials. The basis for this choice was to insure a positive opening and continuous flow during the early part of the experimental program. Clearly, enhancements to the liquid feeding system would be made as the experimental program progressed.

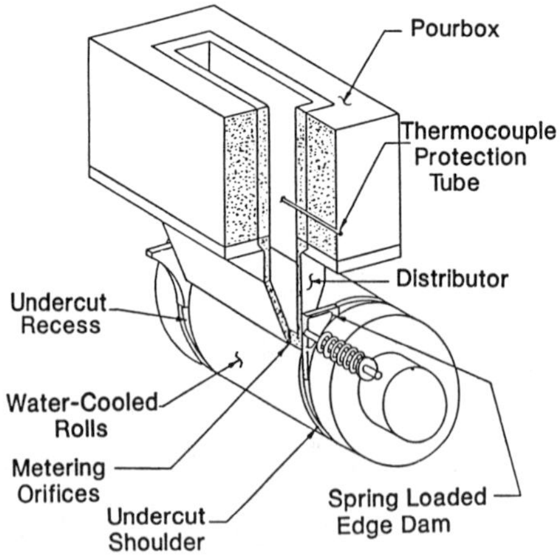

FIGURE 4
POURBOX ARRANGEMENT

Machine Design

The design of the 12-inch (305 mm) diameter twin-roll casting machine was aided by the development and utilization of a mathematical model. Before reviewing the results of this model relative to machine design, a description of the thermal system will be detailed.

Thermal Considerations

The twin-roll caster system is comprised of two rotating water-cooled rolls receiving a steady supply of liquid metal which solidifies onto the rolls and is discharged as a continuous solid strip. Figure 5 shows the distributor pouring box supplying the liquid metal to a wedge-shaped pool which is established between the rolls. As the roll passes through the pool a solidifying shell, moving with the roll, forms on the surface. The shell grows in thickness finally welding with its symmetric half at some point above the roll bite. Throughout this analysis the meeting of the shells is understood to be at the roll bite where the strip leaves the rolls as a single, completely solid, strip. In Figure 5 the solidified shells are shaded and are bounded by the solidus isotherm.

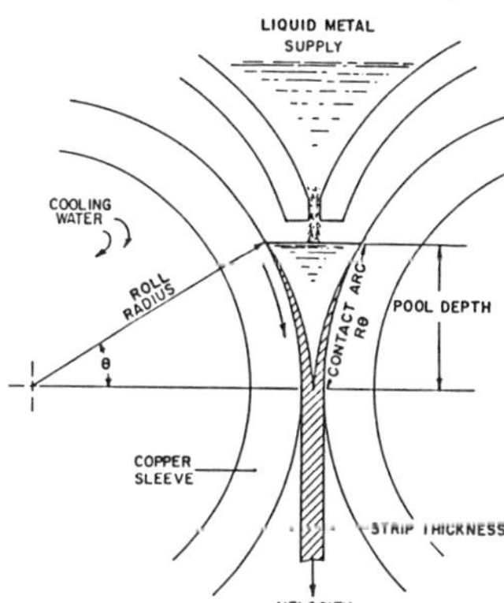

FIGURE 5
SCHEMATIC OF THERMAL SYSTEM

During the start-up transient a rising pool of liquid metal develops between the rotating rolls. As the process continues a quasi-steady state, sometimes referred to as cyclic equilibrium, ensues wherein the pool stabilizes and a balance in both the heat and mass flow is attained.

The thermal system involves two materials, the molten metal and the substrate roll, and two boundary conditions, a water-side heat transfer coefficient between the cooling water and the roll, and the interfacial heat transfer coefficient between the solidifying shell and the roll surface. An adiabatic condition is assumed at the top surface and along the centerline of the pool.

The casting rate (TPH/ft of width) and temperature of the incoming stream determine the amount of heat input to the system. There is heat flux from the liquid to the mushy zone, then through the solidifying shell and to the interface between the solid shell and the substrate roll. At the interface there is a contact conductance (characterized by an interfacial heat transfer coefficient) where the heat is transferred from the shell to the substrate roll.

The total amount of heat leaving the strip in one revolution may be equated to the amount of heat entering the roll which is also the amount of heat going to the cooling water. Thus the integral of the heat flux for the regimes can be equated as,

$$\int_0^{t^*} \ddot{q}_{STRIP} dt = \int_0^{t^*} \ddot{q}_{ROLL} dt = \int_0^{t_t} \ddot{q}_{WATER} dt \qquad (1)$$

where t* is the solidification time (i.e. contact arc/roll speed) and t_t is the time for one complete revolution.

Numerical models in various degrees of complexity have been used to analyze the thermal behavior of the twin-roll caster. However, a straight forward two-dimensional numerical approach requires a fair amount of computing time because: 1) the quasi-steady state condition usually requires many revolutions before the rolls reach cyclic equilibrium, and 2) the pool geometry is not known a priori. For example, to cast a specific strip thickness at a given roll speed one must iterate to equilibrium while changing the pool depth to satisfy the condition that solidification is complete precisely at a known point in the pool, e.g. at the roll bite.

The model can be simplified in a variety of ways and/or methods may be developed for more rapid convergence. Several models have been developed and used in the course of the present program. The most rapid and useful model developed is an approximate analytic solution to the above integral equations which converges to a final solution in less than three seconds on a desk-top computer. The model is described as approximate because liberties were taken with the formulation; however, when compared to results from a more rigorous numerical model, and later verified by experimental data, the approximate analytical model was found to be an excellent representation of the thermal behavior of the twin-roll caster.

Model Usage

Table II is a listing of the fixed and variable inputs to the analytic model. A total of 19 parameters are required, including:

- 12 parameters which define the thermal properties of the melt and the roll;
- 2 parameters which define the roll geometry (diameter and wall thickness);
- 2 parameters which express the casting dynamics (productivity and strip thickness);
- 2 parameters which relate to the cooling intensity (water-side heat transfer coefficient and water temperature);
- A value for the interfacial heat transfer coefficient.

An example of the use of the model is the effect of the water-side heat transfer coefficient on the productivity and thickness of strip cast from the 12-inch (305 mm) caster. As shown in Figure 6, the productivity is dramatically reduced with decreasing the water side heat transfer coefficient for a given strip thickness. Also, it should be noted that the relationship between productivity and cast thickness becomes flat as the heat transfer coefficient approaches 500 BTU/hr sq ft °F (0.28 w/sq cm°C). In this example the pool depth is maintained at a height of 5 inches (12.7 cm) or is lowered if the roll surface temperature (at the bite) exceeds 1000°F (550°C).

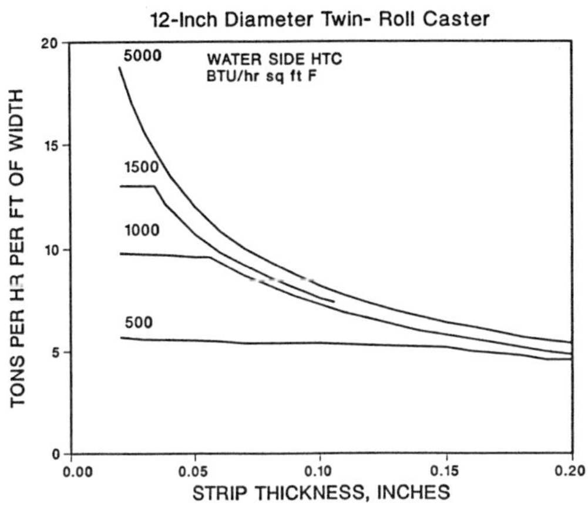

FIGURE 6
EFFECT OF WATER-SIDE HTC ON
PRODUCTIVITY AND THICKNESS

One advantage of an analytic model is the ability to exchange variables and/or parameters whereas inverse numerical models require a trial and error procedure. As a result, the model is rearranged so that the solidification time (contact arc/roll speed) is a known input while the interfacial heat transfer coefficient is a calculated output. Table III is a summary of the conditions and response characteristics for a typical inverse model run. The data generated on the experimental casting machine was limited to strip thicknesses of between 0.025 to 0.080 inches (0.6 mm to 2 mm) and solidification times of between 0.1 to 0.3 seconds; however, the literature[2-4] on twin-roll casting extended the range and permitted the development of a relation between the interfacial heat transfer coefficient and strip thickness.

Figure 7 is a plot of the solidification time vs. strip thickness. The points represent published data and data from casts made on the experimental machine at Bethlehem (open triangles). The line shown in the figure is the time-thickness relationship resulting from the approximate analytical model by using, at the interface, the relation,

$$HTC = 352/S^{0.547} \tag{2}$$

where HTC is the overall interfacial heat transfer coefficient in BTU/hr sq ft°F and S is the strip thickness in inches.

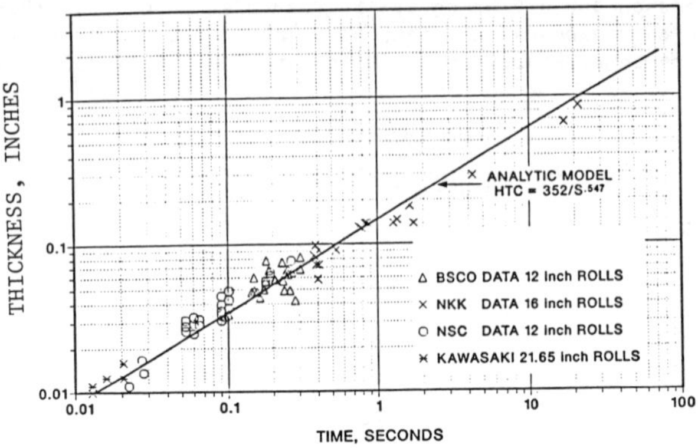

FIGURE 7
SOLIDIFICATION TIME VS STRIP THICKNESS

Another application of the thermal model is in the area of machine sizing. Figure 8 is a plot of maximum productivity as a function of cast thickness for various roll diameters. The area below the line characterized by roll diameter is the field of operation for the machine. There are practical constraints on the lower limit of machine productivity which disallow casting at low tonnages, especially with relatively large rolls. These are pouring box temperature control and/or attendant difficulties when casting with shallow pools. There is also a lower practical limit on strip thickness for certainly selecting a machine with 10 foot (3 m) diameter rolls to produce 0.02-inch (0.5 mm) strip would not be prudent.

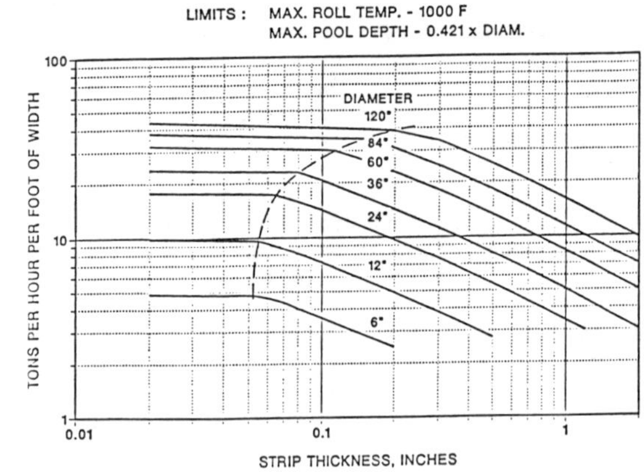

FIGURE 8
MAXIMUM TWIN-ROLL CASTER PRODUCTIVITY

Key Component Design

After the completion of the thermal analysis, a framework was established for the design of several key components of the twin-roll caster pilot machine. For the 12-inch (305 mm) caster, the results of the modeling showed that throughputs of 5-8 TPH/ft (15-24 tph/m) of width could be expected for as-cast thicknesses of 0.02 to 0.08 inches (0.5 to 2 mm). For roll cooling, a minimum water flow of 133 gpm (500 l/min) was established for each roll. This flowrate was determined from the expected heat flow to the water and a selected value (e.g. 10°C) for the overall water temperature rise, so that localized boiling would not be encountered. Shown in Figure 9 is the roll cooling water channel design which results in superficial water velocities of about 20 fps (6 m/sec). Although a theoretical water side heat transfer coefficient of about 2800 BTU/hr sq ft °F (1.6 w/sq cm°C) was computed for this set of conditions, practical experience on similar systems had shown that expected actual values would be 30-40% of theoretical.

Separate hydraulic motors were chosen for the pilot machine roll drives. Figure 10 is a schematic of the drive motor arrangement. Separate drives for each of the rolls were selected to provide a precise method for speed

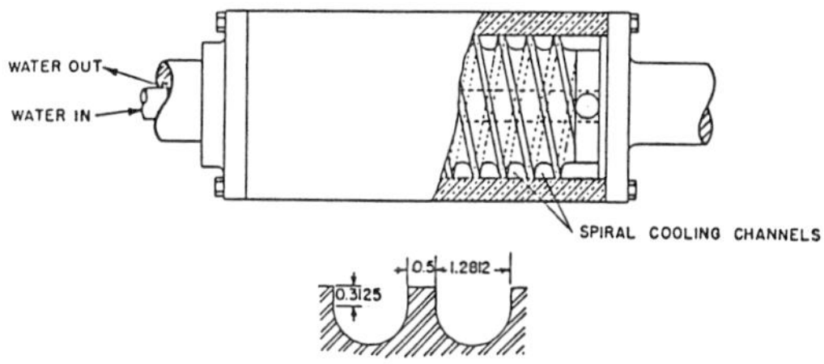

FIGURE 9
ROLL COOLING WATER CHANNEL DESIGN

matching. Hydraulic motors with appropriate speed controls were chosen based on local expertise. Electrical motors (d.c.) would also be suitable and may be preferred. In terms of motor sizing, a major factor which must be considered (in addition to machine frictional and inertial effects) is the level of hot rolling required. The level of hot rolling is directly related to the applied roll force. As the roll force increases, the point of solidification rises above the bite and the amount of hot reduction increases.

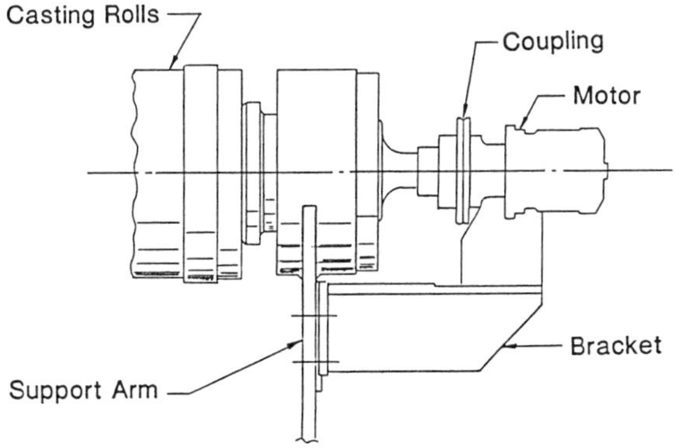

FIGURE 10
MOUNTED ROLL DRIVE MOTOR ARRANGEMENT

Power calculations using the method provided by Simms[5] were made to provide an upper limit for the demands anticipated for the pilot operation. The planned roll force for early operation of the pilot machine was approximately 2000 lbs/ft (29 kN/m) of cast width. As a result, the design specification for the motors was 2440 in-lbs of torque at 2000 psig (13.7 MPa) and 242 rpm.

Several alternatives were considered for roll mounting on the 12-inch pilot machine. Rather than placing the casting rolls on a sliding track, they were mounted as inverted pendulums to reduce extraneous forces. The patented idea of the inverted pendulum was a way to reduce the effects of friction and inertia on the moveable roll while utilizing gravitational forces to provide a variable loading on the strip. Figure 11 is a schematic of the system which shows the use of springs which passively change the force applied to the strip as a function of thickness. The system was designed to supply a roll force which could be varied from 0 to 10 tons/ft (0 to 290 kN/m) of width. The rolls can be hydraulically separated to a "full open" position in the event of emergency shut-down of the caster.

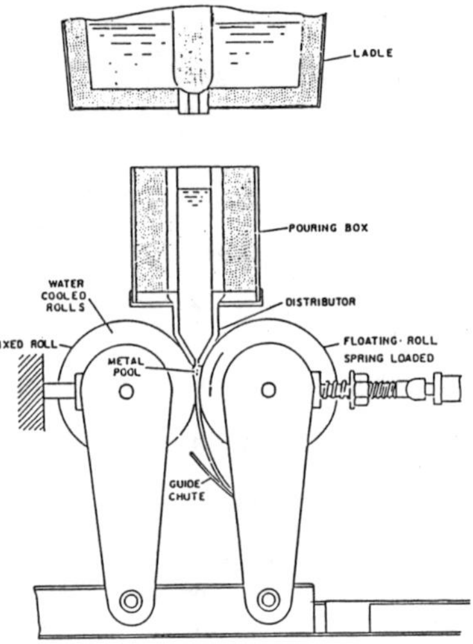

FIGURE 11
PENDULUM ROLL MOUNT AND
ROLL SPRING ARRANGEMENT

Instrumentation and Control Specifications

A final element of the design of the twin-roll caster pilot machine involves the specification of the instrumentation and control circuitry. The overall block diagram of this circuitry is shown in Figure 12. The major components of this system include:

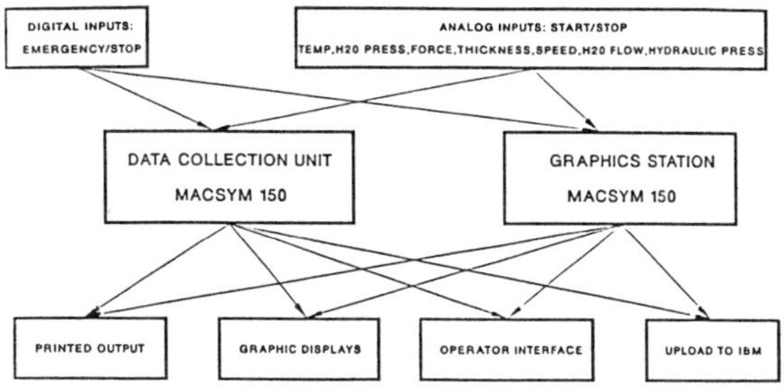

FIGURE 12
SCHEMATIC FOR DATA ACQUISITION
TWIN-ROLL CASTER

- Two Analog Devices, Macsym 150 microcomputers which operate in tandem for data acquisition, process control and display of key process parameters.

- Digital and analog inputs for major process parameters and machine control functions.

- Input/output devices for machine operation and generation of reports.

- Upload capability to a database on an IBM main frame.

During a cast, data are acquired from about 25 sensors on a five point averaging basis once per 0.25 seconds and stored in RAM. For typical casting speeds trials this frequency would be equivalent to a set of process data for each 4 inches (10 cm) of strip cast. Data tabulations are generated within minutes of the end of each cast. The capability for closed-loop control of strip thickness was also designed into the system.

After the completion of a sensitivity analysis to estimate the range of various process variables, the sensor list shown in Table IV was developed. Major components in this list include: 1) infrared pyrometers for strip temperature measurement, 2) resistance type devices (RTD's) for measurement of the cooling water temperature differentials, 3) load cells for roll force measurement, 4) LVDT's for strip thickness measurement, 5) optical encoders for speed sensing 6) transmitters for hydraulic pressures and water flows, 7) a type S thermocouple for continuous temperature measurement in the pouring box and 8) a sliding contact thermocouple for roll surface temperature measurement.

Conclusions

Physical and thermal models were used extensively in the design of Bethlehem's twin-roll caster. Also important were the liquid steel trials in which the liquid steel feeding system was tested. After almost two years of construction and shake-down operation, casting trials were started in August, 1984. Early trials were conducted with Fe-Si steel until late 1985. Subsequently, the focus was changed to low-carbon, aluminum-killed steels and continues to the present. Results of these trials will be presented in a future publication.

Acknowledgements

The authors wish to acknowledge with thanks the efforts of James H. Cox, now with Hatch Associates Ltd., who produced the relationship between time and solidification thickness for the twin-roll casting process.

REFERENCES

1. A. W. Cramb, Iron and Steelmaker, May 1988, Vol. 15, No. 5, pp. 65-68.
2. A. Kasama et al, Tetsu to Hagane, Vol. 71, #10, A225-228.
3. S. Nishiori et al, ibid, A241-244.
4. M. Ozawa, F. Kogiku, M. Yukumoto, S. Miyake and T. Kan, "Heat Transfer Analysis of Twin-Roll Casting", Conference on Modeling of Casting and Welding Processes, IV, April 17-22, 1988.
5. R. B. Simms, "The Calculation of Roll Force and Torque in Hot-Rolling Mills", Proc. Inst. Mech. Engr., 1954, V. 168, 6, 191-200.

TABLE I
KEY DATES IN MODERN STRIP CASTING

Date	Inventor	Process
1954	Hazelett	Armco attempts to cast a 0.25-inch strip using a Hazelett machine.
1957-69	Bessemer	Experimentation with Bessemer process at Waseda University,Japan. Steels cast were mainly high silicon sheets, cast iron and 18-8 stainless steels.
1965	J&L	Development of drum-in-drum process for steel strip begun.
1971-75	J&L	Further development of drum-in-drum process for strip casting.
1972	M.C. Flemmings	Discovery of rheocasting.
1975-79	British Steel	Attempts to develop a channel-type caster for strip abandoned due to cost and size mismatch with rolling mill.
1980		Rheocasting process proposed for steel strip
1983		Work ongoing at Japanese research labs on twin-roll process. Investigators include Nippon Steel, Kobe, NKK and Kawasaki.
1983?	M.C. Flemmings	DOE supports four preliminary studies of single-roll strip casting.
1984	Allegheny Ludlum	Single roll strip casting process developed for stainless steels.
1984	Armco-Westinghouse	DOE sponsors work on single-roll strip casting.
1984	Keyworth Initiative	Argonne Labs. designated as lead laboratory for study of direct strip casting. Decision made to study electromagnetic casting.
1987	AISI	Decision made to terminate Argonne study of electro-magnetic casting.
1987	Armco-Westinghouse	DOE continues single-roll strip casting support for low-carbon, aluminum-killed steel.

TABLE II
TWIN-ROLL CASTER APPROXIMATE ANALYTICAL MODEL

Type	Parameter	Value	Units
Fixed Steel Properties	Solid Thermal Conductivity	17.5	BTU/hr ft F
	Solid Density	475	lb/cu ft
	Solid Specific Heat	0.16	BTU/lb F
	Mush Thermal Conductivity	17	BTU/hr ft F
	Mush Density	460	lb/cu ft
	Mush Specific Heat	0.18	BTU/lb F
	Latent Heat	120	BTU/lb
	Liquidus Temperature	2770	F
	Solidus Temperature	2740	F
Fixed Copper Properties	Thermal Conductivity	180	BTU/hr ft F
	Density	550	lb/cu ft
	Specific Heat	0.0955	BTU/lb F
Fixed Conditions	Water-Side H.T.C.	1000	BTU/hr sq ft F
	Water Temperature	60	F
Variable Conditions	Interfacial H.T.C.		BTU/hr sq ft F
	Strip Thickness		inch
	Productivity		TPH/ft width
	Roll Diameter		inch
	Wall Thickness = 0.5 + Roll Diam./24		inch

TABLE III
TWIN-ROLL CASTER APPROXIMATE ANALYTICAL MODEL

Group	Parameter	Value	Units	
Operating Parameters	Strip Thickness	0.04	inch	Inputs
	Pool Depth	2.875	inch	
	Roll Speed	146	ft/min	
	Roll Force	5800	lb/ft	
Steel Properties	Solid Thermal Conductivity	17.5	BTU/hr ft F	
	Solid Density	475	lb/cu ft	
	Solid Specific Heat	0.16	BTU/lb F	
	Mush Thermal Conductivity	17	BTU/hr ft F	
	Mush Density	460	lb/cu ft	
	Mush Specific Heat	0.18	BTU/lb F	
	Latent Heat	120	BTU/lb	
	Solidus Temperature	2740	F	
	Freezing Temperature Range	30	F	
Casting Roll Characteristics	Roll Diameter	12	inch	
	Wall Thickness	1	inch	
	Conductivity	180	BTU/hr ft F	
	Density	550	lb/cu ft	
	Specific Heat	0.0955	BTU/lb F	
Water Cooling	Water-side H.T.C.	1100	BTU/hr sq ft F	
	Cooling Water Temperature	81	F	
Heat Transfer and Throughput	Interfacial H.T.C.	2374.68	BTU/hr sq ft F	Response
	Throughput	7.154	TPH/ft width	
	Roll Speed	46.732	rpm	
	Contact Arc	2.99823	inch	
Thermal Response	Max Roll Surface Temperature	838.866	F	
	Heat to Cooling Water	2.108×10^6	BTU/hr ft	
	Avg. Heat Flux	4.220×10^6	BTU/hr sq ft	
	Strip Temp at Bite	2440.34	F	
	Peak Strip Temp Downstream	2590.17	F	
Mechanical Factors	Hot Rolling Reduction	4.47648	%	
	Torque	586.334	inch-lb/ft width	
	Horsepower	0.4323	HP/ft width	
	Avg. Pressure applied to Strip	4477.43	psi	

TABLE IV
TWIN-ROLL CASTER SENSOR LIST

Group	Sensor	Input Range	Output Range
Thermal Data	Strip Temp (1)	1000 to 2500 F	0 to 10 volts
	Strip Temp (2)	1000 to 2500 F	0 to 10 volts
	Cooling Water to Rolls	-350 to 1400 F	-8.14 to 43 mv
	Roll A Cooling Water		
	Differential Temp	0 to 29 F	4 to 20 ma
	Flow Rate	0 to 160 gpm	4 to 20 ma
	Roll B Cooling Water		
	Differential Temp	0 to 29 F	4 to 20 ma
	Flow Rate	0 to 160 gpm	4 to 20 ma
	Roll Surface Temp	0 to 2500 F	0 to 2.5 volts
	Pour Box Temp	0 to 3200 F	-0.09 to 18.7 mv
Strip Data	North Side Roll Force	0 to 10000 lb	0 to 10 volts
	South Side Roll Force	0 to 10000 lb	0 to 10 volts
	North Strip Thick	0 to 0.100 in	-10 to 10 volts
	South Strip Thick	0 to 0.100 in	-10 to 10 volts
Roll Data	Roll A Speed	0 to 135 RPM	0 to 10 volts
	Roll B Speed	0 to 135 RPM	0 to 10 volts
	Withdrawal Belt Speed	0 to 270 RPM	0 to 10 volts
	Roll A Diff Press	0 to 50 psig	0 to 5 volts
	Roll B Diff Press	0 to 50 psig	0 to 5 volts
	Roll A Hydraulic Press	0 to 2500 psig	0 to 2.5 volts
	Roll B Hydraulic Press	0 to 2500 psig	0 to 2.5 volts
	Hydraulic Line Press	0 to 2500 psig	0 to 2.5 volts

MATHEMATICAL MODELLING OF THE SPRAY FORMING PROCESS

#E. Gutierrez-Miravete, +G. Trapaga and *J. Szekely

Assistant Professor, Metallurgy. Hartford Graduate Center, Hartford CT 06120.

+,* Research Assistant and Professor of Materials Engineering, respectively Massachusetts Institute of Technology, Cambridge, MA 02139.

ABSTRACT

Spray forming is simulated employing two mathematical models in order to describe thermal behavior both at the scale of the preform as well as at the scale of individual splats. Modelling has helped to identify the process parameters to which spray forming is most sensitive. It is found that the computed cooling rates during solidification in the preform can be made to agree with values estimated from the scale of the microstructure as long as the model is capable of resolving temperature differences at the scale of the splat size. The models developed are complementary and are helpful in determining the operating conditions appropriate to obtain preforms with desirable properties.

Casting of Near Net Shape Products
Edited by Y. Sahai, J.E. Battles,
R.S. Carbonara and C.E. Mobley
The Metallurgical Society, 1988

1. INTRODUCTION

The recently developed Spray Forming Processes (SFP) provide an attractive means for the production of bulk shapes of alloys directly from the melt. An added benefit of these processes is the possibility of producing fine, equiaxed , and practically segregation free microstructures which can not be obtained using more conventional casting techniques. Thick preforms (up to 0.5 tons. in weight) have been successfully produced using this technique. The materials processed include various aluminum alloys, steels and superalloys. In most cases the properties of the resulting products have been at least as good as those of their conventionally processed counterparts [1].

The spray forming process can be divided into three interrelated stages:

a) atomization of the melt,

b) cooling of the resulting droplets while in flight, and

c) collecting the partially solidified spray onto a suitably designed substrate (Figure 1).

Although the problems of atomization and of cooling the droplets in flight have received considerable attention [2]-[4], the deposition step has not been extensively studied. In this paper we address the issue of heat transfer in the preform during deposition, using mathematical models.

Prior modelling work in this area has produced a simplified analysis of the most appropriate conditions for producing quality preforms [5]. Somewhat more detailed analyses of heat transfer during deposition have been presented by El-Kaddah et al [6], for the case of plasma spraying, and by Mathur et al [7] for spray deposition. However, a satisfactory explanation of the mechanism producing the fine microstructure in the as-deposited preforms is not yet available. In our work, we have attempted to improve the quantitative understanding of this key aspect of the process. The final objective is to develop predictive relationships between the process parameters and the microstructure of the resulting products.

Probably the most important experimental result that must be explained by any reasonable model is that the microstructure of the spray forming (SF) deposit consists of fine, equiaxed, essentially segregation-free grains at high spray densities. According to Singer [5], the spray density, i.e. the deposition rate per unit area, is a critical process parameter in (SF). The physical picture presented in [5] was that when spraying at low rates the time interval between the impingement of two successive droplets at the same location is relatively large; in contrast when the spray density is

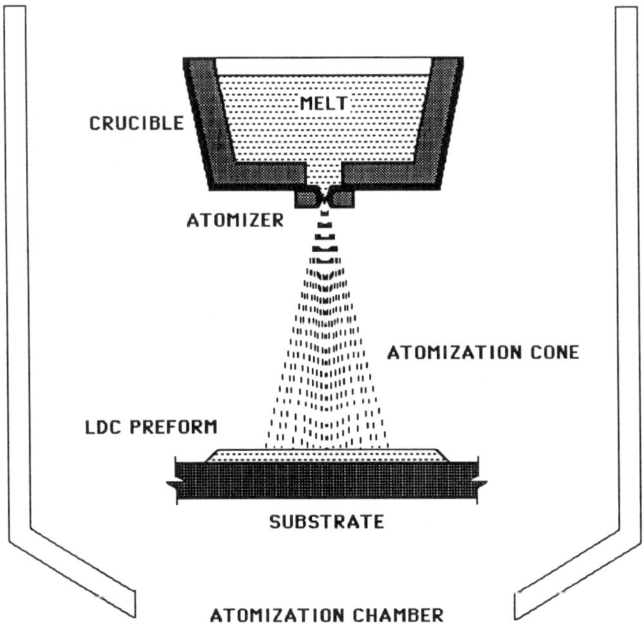

Figure 1.- Schematic representation of the spray forming process.

high, the frequency of droplet impingement is rather larger. So, when spraying at high densities, it is possible that a layer of liquid will form on the top surface of the deposit (spray casting), while when spraying at low densities the newly arrived droplets will tend to impinge onto already solidified splats (spray deposition). Although it would seem that these two extreme conditions would produce entirely different microstructures, in both cases fine grains are usually obtained. However, in the case of high spray densities the grains are generally equiaxed and no splat boundaries are observed. In contrast, at low spray densities, splat boundaries are frequently observed with the tiny grains inside individual splats elongated in the direction of heat flow [5]. This is a puzzling aspect of the process and one that requires explanation.

As mentioned before the ultimate objective of the modelling approach is to determine the microstructure of the preform from calculated values of transient temperature fields. Microstructural measurements have been frequently used to obtain estimates of the cooling rates during solidification [8],[9]. Expressions are available which relate the primary dendrite arm spacing d_p to the cooling rate, dT/dt during solidification for the specific case of nickel-base superalloys. A typical expression of this type is [10]

$$d_p = 141 \left(\frac{dT}{dt}\right)^{-0.4} \tag{1}$$

where d_p is in μm and dT/dt is in K/sec. It is interesting to estimate the cooling rates during solidification in spray forming from the observed scale of the microstructure in the as-deposited preform. Table 1 below is obtained by combining the results of various microstructural studies with Equation (1). Note that the estimated cooling rates span a wide range of values up to a maximum of about 4200 K/sec corresponding to the smallest grain size (5 μm). In contrast, it has been proposed that operation at high spray densities produces fine grains because conditions at the top surface of the deposit lead to the creation of a large number of nucleation sites [11]. It has also been suggested [4] that the solid-liquid region on the upper surface of the deposit consists of a more or less interconnected network of dendritic debris surrounded by small pockets of liquid and that the size of these pockets determines the final grain size of the deposit. As mentioned before, the fineness of the as-deposited microstructure is yet to be explained, because previous mathematical analyses tended to predict much smaller cooling rates which would have corresponded to a rather coarse structure.

TABLE 1. COOLING RATES DURING SOLIDIFICATION IN SPRAY FORMING ESTIMATED FROM THE SCALE OF THE MICROSTRUCTURE FOR VARIOUS NICKEL-BASE SUPERALLOYS.

MEASURED GRAIN SIZE (μm)	ESTIMATED COOLING RATE (K/sec)	REFERENCE/ALLOY
6-12	2600 - 470	Bricknell [11] / Rene 80
20-30	130 - 50	Ogata et al [12] / Merl 76
5-10	4200 - 750	Nell et al [13] / IN-100

2. MATHEMATICAL MODELLING OF THE SPRAY FORMING PROCESS

In a physical sense the SF process is a highly complex operation, with metallic droplets showering onto a substrate surface, spreading, cooling and solidifying after impingement. A comprehensive representation of the heat and fluid flow phenomena in these systems would be quite difficult. For this reason in the present case, we shall consider heat transfer and solidification only. More specifically we shall examine two modelling approaches, namely:

2.1 A continuum model, where the behavior of the individual droplets will be averaged. This is the approach that has been taken by most previous investigators.

2.2 An allowance for the behavior of the individual droplets, with temperature gradients in the individual splats.

We will see that the continuum model is easy to run but is not capable of resolving temperature differences at the scale of single splats. In contrast, the individual droplet model allows the calculation of the temperature fields across individual splats but the analysis of more than a few splats is rather time consuming.

2.1 THE CONTINUUM MODEL

Like most previous investigators, if we consider the growing preform as an expanding continuum bounded by the (cooled) substrate and the arriving molten droplets, as sketched in Figure 2, then the problem may be readily represented by a one-dimensional transient thermal energy balance. Thus, working in terms of the enthalpy of the system we have:

$$\frac{\partial H}{\partial t} = \frac{\partial}{\partial x}\left(k\left(\frac{\partial T}{\partial x}\right)\right) \tag{2}$$

where H is the enthalpy, T the temperature, k the thermal conductivity and x is the distance normal to the surface of the substrate. Of course, in order to incorporate the phase change, the enthalpy must undergo a step change at the melting point. It should also be noted that the deposit thickness increases with time and this has to be taken into account during the development of the solution algorithm.

Regarding the boundary conditions we have the following:

a) At the deposit-substrate interface the rate of heat extraction by the chill is given by a heat transfer coefficient, h_s, i.e.

$$-k\frac{dT}{dx} = h_s\,(T - T_{wall}) \tag{3}$$

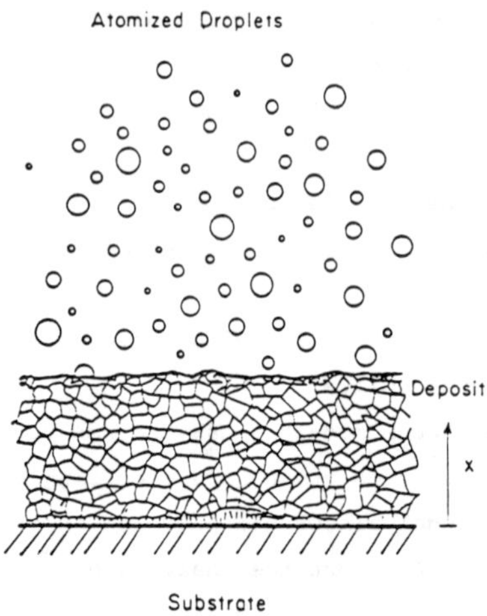

Figure 2.- Schematic representation of the growing preform

b) At the top surface of the growing deposit, the heat flux is equal to the sum of the heat input from the spray, and the heat output due to the combined effects of convection into the gas and thermal radiation, i.e.

$$- k \frac{dT}{dx} = - (\overline{H}_{top} - H_n) \frac{Q_v}{A_s} + h_{top} (T - T_{gas}) + \sigma T^4 \qquad (4)$$

where $\overline{H}_{top}$ is the average enthalpy of the spray, H_n is the enthalpy at the top surface of the deposit, Q_v the volumetric melt flow rate and A_s is the area of the collecting substrate. Note that the average enthalpy of the spray must be known when solving the equations (2)-(4). This quantity can be estimated from an analysis of heat transfer rates of the in-flight droplets prior to impact [3],[4]. Once deposition is completed, the heat input due to the impinging spray becomes zero.

Equations (2)-(4) are readily solved using standard numerical methods. For example, in [4] an explicit numerical scheme is used to solve on a fixed grid containing typically 100 grid points. In [7] the same problem is solved on a variable grid. The results obtained in both cases are fairly similar. Here, as in [4], Equation (2) is discretized using an explicit finite difference method. This explicit scheme is conditionally stable with the maximum allowed time step being given by $\Delta t = \Delta x^2/2\alpha$, where α is the thermal diffusivity of the melt. In a test problem consisting of deposition without heat extraction, the calculated temperature of the accumulated melt deviated by less than 0.3% from the exact value (the temperature of the impinging droplets) when a total of 100 grid points were being used. It should be stressed, however, that this model can only resolve the temperature fields over distances comparable to a splat thickness, if several grid points may be placed to cover this domain. While the resolution can be improved by increasing the number of grid points, a single run using 500 grid points (e.g. corresponding to a grid spacing of 50 μm for the cases considered here) would require over 5 hours of CPU time on a microVax II workstation. This would make the computation of cases with a thick deposit rather impractical.

2.2.- THE INDIVIDUAL DROPLET MODEL

The definition of the temperature profile in the growing preform can be improved by a model that accounts for the behavior of the individual impinging droplets. A simple heat transfer model of the deposition process can be constructed by assuming that a random dispersion of equally sized droplets of radius r_d impinges on the preform and each droplet becomes a splat of radius $r_s = \xi\, r_d$, where ξ typically has a value between 3 and 6, [14]. The value of ξ = 3.65 has been assumed in the calculations described here, which gave a 5 μm splat thickness when postulating 100 μm droplets. The time interval between the arrival of two successive droplets (i.e. time between impacts) τ, is given by [5]

$$\tau = \frac{4}{3} \frac{A_s\, r_d}{Q_v\, \xi^2} \tag{5}$$

The small scale phenomena that occur during the impingement of a single, partially solidified droplet onto a partially solidified deposit are as follows (Figure 3):

a) the droplet impacts the top surface of the deposit and flattens or spreads. This process is estimated to take place rapidly, taking only a few microseconds [14]. Therefore, the splatting time can be safely assumed to be much smaller than the interval between the arrival of successive droplets.

b) the resulting splat exchanges heat with the underlaying deposit and eventually solidifies. This sequence of events is repeated with a periodicity of τ. Of course, there are other cases that could be considered. For example, a solid droplet falling onto a solid deposit, and a liquid droplet falling onto a liquid top layer. Other intermediate cases are also possible.

Based on equations (2)-(5) a model can be constructed to represent the thermal history of individual splats from the time of impingement. Equation (2) can still be used to describe the transient temperature profile across the thickness of an individual splat; however, the initial conditions have to be modified appropriately. Simple modifications of equations (3) and (4) allow the calculation of the temperature fields of contiguous splats which impinge on top of each other, with the time interval between the arrival of successive splats being given by Equation (5). While the same numerical scheme may be used as for the continuum case, practical computational considerations limit one to the examination of just a few droplets or splats. As will be seen subsequently, interesting information may be deduced from these results.

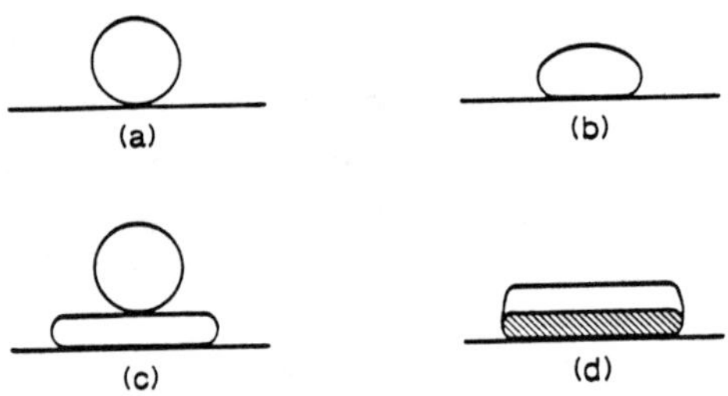

Figure 3.- Schematic representation of droplet impact and splatting for two droplets.

3. COMPUTED RESULTS

We have selected computed results derived from the models described above, for a system corresponding to the spray forming of a Nickel-base superalloy. The physical properties and the process variables are summarized in Table 2. The values of the operating variables in this table are typical of experiments on the deposition of superalloys carried out in the laboratory of Professor N. Grant at MIT. As mentioned in the introduction, the grain sizes in typical as-deposited preforms suggest that the cooling rates during solidification were in the range of 50 to 4200 K/sec.

TABLE 2. INPUT DATA FOR A NICKEL-BASE SUPERALLOY.

-	Deposition rate:	Q_D = 0.11 Kg/sec
-	Deposition time:	t_D = 28.45 sec
-	Substrate area:	A_s = 1.53 x 10^{-2} m^2
-	Final preform thickness:	X_f = 2.5 x 10^{-2} m
-	Heat transfer coefficient (bottom):	h_s = 10^3 W/m^2 K
-	Heat transfer coefficient (top):	h_{top} = 10 W/m^2 K
-	Temperature at the walls:	T_w = 298 K
-	Temperature of the gas:	T_{gas} = 358 K
-	Average enthalpy of the spray:	H_{top} = 155,000 J/Kg
-	Heat of fusion:	H_{fus} = 202,730 J/Kg
-	Specific heat:	Cp = 611 J/Kg
-	Thermal conductivity (solid):	k_S = 60 W/m K
-	Thermal conductivity (liquid):	k_L = 60 W/m K
-	Liquidus temperature:	T_L = 1,668 K
-	Solidus temperature:	T_S = 1,568 K
-	Density (solid):	ρ_S = 8,410 Kg/m^3
-	Density (liquid):	ρ_L = 7,990 Kg/m^3

3.1- COMPUTED RESULTS USING THE CONTINUUM MODEL

Figures 4, 5 and 6 show the behavior predicted by the continuum model. Figure 4 depicts the calculated temperatures as a function of time for the top, bottom, and middle of the preform, and shows that the temperature history markedly varies in the course of the deposition, especially for the upper surface of the preform. Although the temperature at the substrate-deposit interface decreases monotonically with time, the temperature of the top surface first decreases to below the solidus and then rises. Apparently the radically different thermal behavior exhibited by the top and bottom surfaces is the result of localized heat accumulation in the upper layers of the growing deposit. Figure 5 shows the corresponding profiles of the fraction solidified as a function of time calculated for the same conditions as in Figure 4. It is seen that a two-phase region, with a high solid fraction, is produced in the preform at about 12 seconds after the start of deposition. This mushy region solidifies almost immediately once deposition is completed.

Additional insight into the behavior shown in Figure 4 can be seen in Figure 6 where the cooling rate is plotted as a function of time. It is important to note that these computed cooling rates are quite low, less than

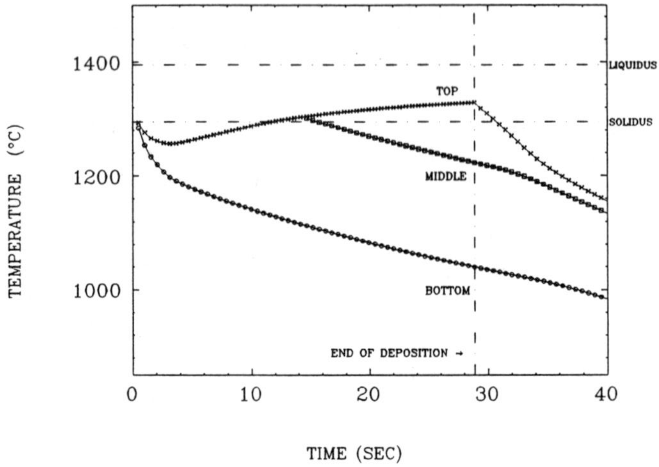

Figure 4.- Computed temperature history of the preform (continuum model).

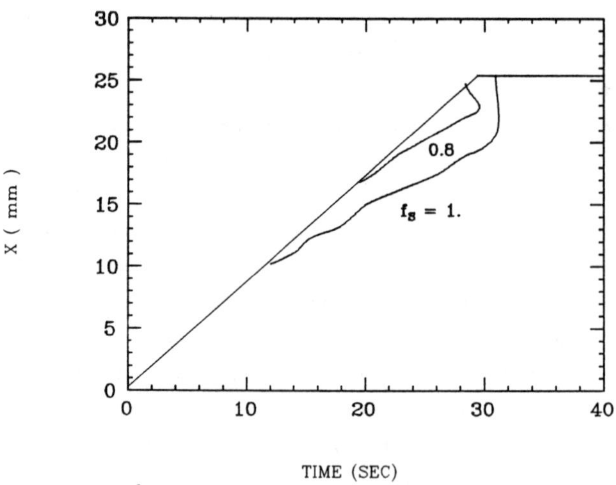

Figure 5.- Computed iso-fraction solid lines as a function of time (continuum model).

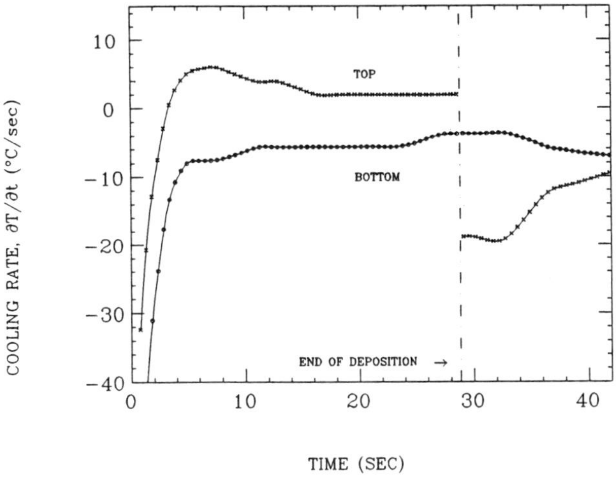

Figure 6.- Calculated cooling rates for the same locations in the preform as in Figure 4 (continuum model).

about 20 K/sec. This is in marked contrast to the cooling rates that may be estimated from the microstructure, e.g. through the use of Equation 1, which would indicate cooling rates that are at least two orders of magnitudo higher than this value. This discrepancy is an important drawback of the continuum model that needs an explanation. We shall return to this question subsequently, when discussing the discrete droplet or splat models.

Nevertheless, the continuum model does serve a useful function in predicting the overall behavior of the system. As an example, the experimentally measured cooling rates by Bewlay et al [15] using thermocouples were found to be in good agreement with predictions based in a continuum model; we should note that the linear scale of a typical thermocouple junction would correspond to several splat thicknesses, so that an averaging procedure must have been involved here.

Clearly the continuum model will have to satisfy the overall heat balance, so that the average temperatures (the averaging process taking place over a time scale that is large compared to the inverse droplet frequency) should be well predicted by this approach.

One major advantage of the continuum model is its speed of execution. For example, the complete thermal calculation for a 0.025 m thick preform using a net with 250 grid points takes about 1 hour of CPU time in a microVax II workstation. Thus, the continuum model can conveniently be used to perform useful parametric studies. In the following we give three selected examples of this type of analysis.

Figure 7 shows the computed top surface temperatures of the preform as function of time with the average enthalpy (where the solidus temperature is taken as the zero reference point) of the impinging droplets as a parameter. It should be stressed that the actual value of this enthalpy is directly related both, to the cooling conditions during atomization, and to the distance between the atomization nozzle and the substrate. The qualitative behavior of the different curves is very similar. However, it seems there is some threshold value of the average enthalpy (somewhere around 100 KJoules/Kg, which corresponds to a droplet temperature of 1617 K), for which the top surface temperature barely reaches the solidus temperature at any time during deposition. This implies that no liquid layer exists at the top surface of the preform. However, for larger values of the enthalpy, a liquid phase will eventually form at the top surface during deposition.

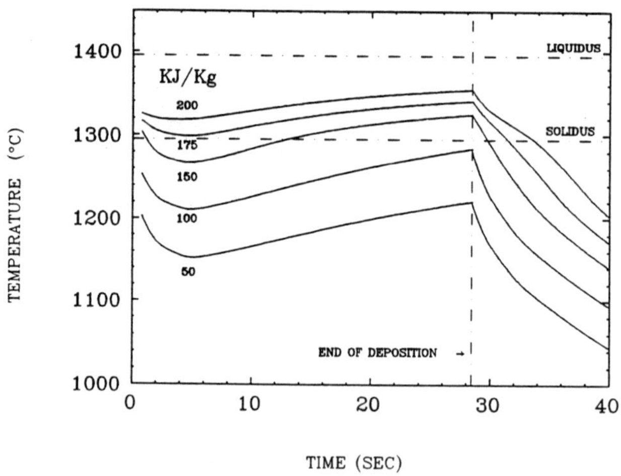

Figure 7.- Effect of the average enthalpy of the impinging droplets on the temperature at the top surface of the preform (continuum model).

The effect of the spraying rate on the top surface temperature, for a constant final preform thickness of 0.0254 m, is shown in Figure 8. As expected, at the higher spraying rates a high rate of energy transfer takes place into the growing deposit which can not be dissipated by the cold substrate. This excess energy results in a nearly constant top surface temperature throughout deposition (spray casting [5]). In contrast, for the lower spraying rates in the range shown, the top surface temperature decreases rapidly and can result in the layered microstructure typically observed in deposits produced at low spray rates - this condition has been termed spray deposition [5]. Figure 8 also shows that increasing the spray rate produces an approximately constant top surface temperature during deposition. The maximum possible top surface temperature is determined by the average enthalpy of the impinging droplets. However, the highest spray rates in the range considered result in the development of undesirable thick solid-liquid regions in the upper portions of the preform. This in turn results in the formation of coarse, heterogeneous microstructures, a condition akin to spray casting . It has been suggested that the enthalpy of

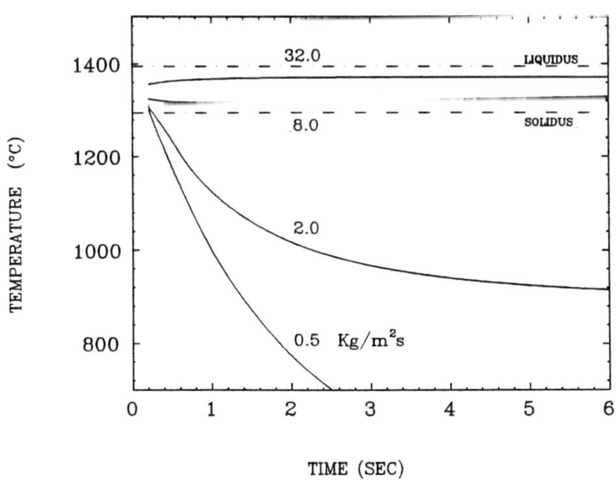

Figure 8.- Effect of the spraying rate on the temperature at the top surface of the preform (continuum model).

the melt at the point of impact must be maintained within a narrow range so there is, at all times, a definite liquid to solid ratio to ensure good consolidation of the growing preform [16]. A cold deposit leads to poor bonding and excessive porosity, while a hot deposit leads to splashing and excessive porosity. For satisfactory operation these extreme conditions have to be avoided.

Let us recall that Figure 4 showed that the top surface layers of the preform became mushy after about 12 seconds of deposition and that the proportion of liquid in the mushy region increased continuously until the end of the process. Since this two-phase region is exposed to the continuing impingement of sprayed droplets, one can expect it to be well agitated. The enhanced heat transfer due to this stirring may be modelled approximately by assuming an increased value of the thermal conductivity of both the liquid and the mushy phase. Figure 9 shows the computed temperature history in the preform for conditions similar to those in Figure 4, except for a higher average enthalpy of the impinging droplets (200 KJoules/Kg), and for three different values of the effective thermal conductivity of the liquid. As expected, the increased conductivity reduces the temperature gradients in

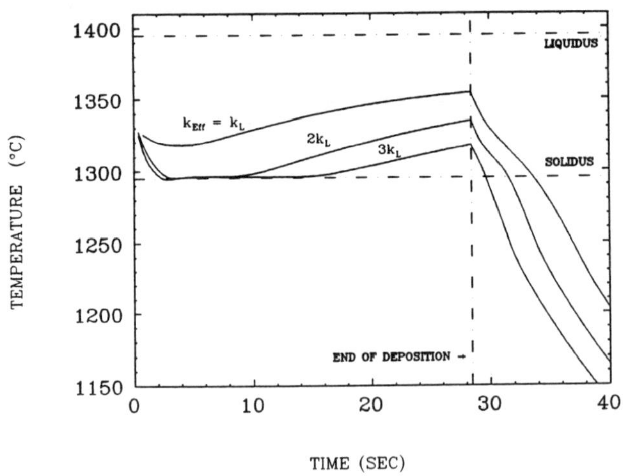

Figure 9.- Computed temperature histories at the top surface of the preform for different effective thermal conductivities of the liquid (continuum model).

the solid-liquid region and this suggests that deposition at relatively large melt flow rates without formation of an excessive liquid fraction is possible because of the stirring of the two-phase region by the impinging droplets.

3.2.- COMPUTED RESULTS USING THE INDIVIDUAL DROPLET MODEL

To better understand the thermal behavior of the splats during deposition, the individual droplet model described in Section 2.2 can now be used. Figure 10 shows the instantaneous average temperature for the first three droplets (initially at the liquidus temperature) that were impinging onto an initially cold (298 K) substrate, separated by an interval τ between impacts. Only the average temperature across the splat thickness is given since calculations have shown that the temperature differences in a given splat were less than 1 K. Curve 1 shows the behavior of the first splat, curve 2 depicts the mean temperature of the first two splats combined, while curve 3 shows the temperature history of the first three splats combined. Figure 10 indicates that the average temperature has

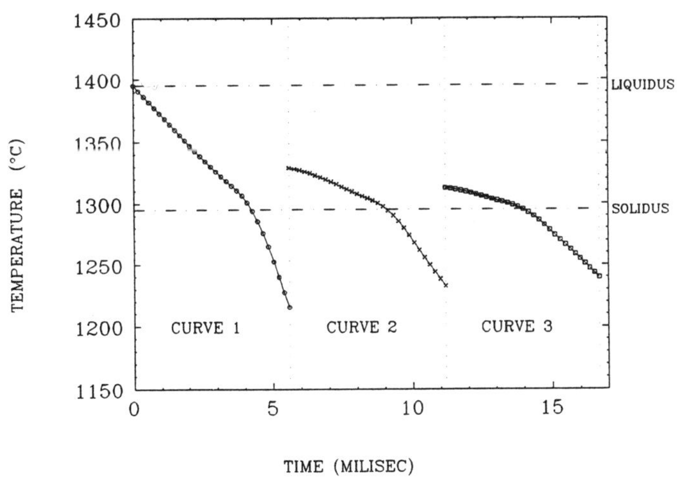

Figure 10.- Calculated average temperatures as function of time for, two and three splats at the beginning of deposition (individual droplet model).

a downward trend, a behavior similar to that observed during the first few seconds in Figure 4. One should note, however, that the temperature of the composite rises when the next splat impinges upon it. The net result is that, for the case considered here, the two-splat composite reaches a temperature corresponding to a two-phase region. This behavior is repeated when the third splat impinges. However, the maximum temperature reached by the composite on impact still exhibits a definite downward trend during these initial stages.

In Figure 11 we seek to represent the thermal history of three subsequent splats at a later stage in the deposition process, e.g. when the preform has reached a thickness of about 10 mm. Using estimates from the continuum mode, we can postulate a preform temperature of 1533 K, just below the solidus, and have assumed a high heat transfer coefficient of 10^5 W/m^2 K to represent a near perfect thermal contact. It is seen that now the temperature excursion is within the two phase region, but the cooling rates are still quite high, in the region of several thousand K/sec.

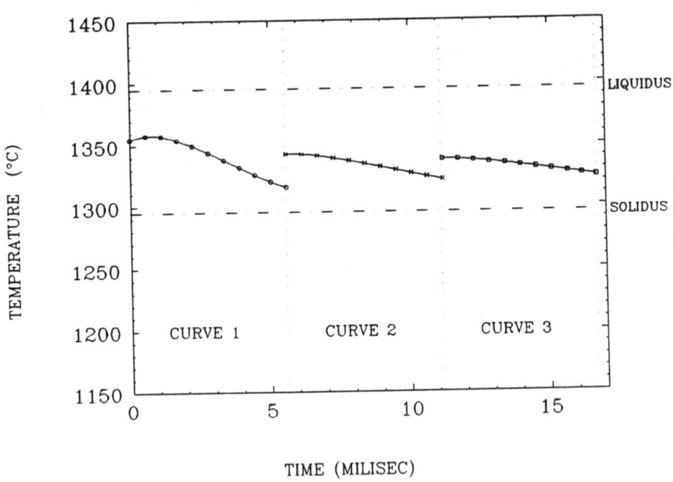

Figure 11.- Calculated average temperatures as function of time for one, two and three splats at some intermediate stage of deposition (individual droplet model).

It is important to stress that the values found for the cooling rates in Figures 10 and 11 are at least two orders of magnitude higher than those seen in Figure 6.

It is thus seen that by taking into account the behavior of the individual droplets, it is possible to predict quite high cooling rates, which in turn are in much better agreement with the observed microstructures than predictions based on continuum considerations. The implications of these findings will be discussed subsequently.

4. DISCUSSION

In the paper we presented two distinct mathematical models aimed at describing the thermal history of the deposit in spray forming processes.

One of these approaches involved the use of a continuum model, which is quite similar to the treatment that has been developed by previous investigators. In essence this continuum model relies on the solution of the transient heat conduction equation, with two moving boundaries. This continuum model is helpful in providing information on the global characteristics of the process, such as the thermal history of the top surface of the deposit and that of the substrate-deposit interface. This model also enables one to relate the temperature fields in the system to the key operating parameters, such as the spray rate and the enthalpy of the molten (or partially molten) metal spray.

However, this continuum approach has not really been successful in predicting the grain sizes found in the deposits; indeed judging by the significant discrepancy between the experimentally measured and the theoretically predicted grain sizes, it appears that the continuum model seriously underpredicts the cooling rates. In a philosophical sense this behavior is understandable, because when using the continuum model, both the length scale and the time scale are quite large, compared to the size of the individual splats and the time interval, between the arrival of the individual droplets.

In order to resolve this problem, an alternative model has been proposed, which describes the process in terms of the behavior of the individual droplets. A key feature of this model is that in allowing for the impact and the subsequent spreading and cooling of these droplets, we can represent the fact that during a significant part of the lifetime of a given splat, there will be a rapid rate of heat extraction, without a corresponding heat supply. This will occur during the time interval, between the arrival of the individual droplets. During this time period very rapid cooling rates are possible, which may account for the small grain sizes that have been observed in practice. Indeed, the predictions made for the grain size, when using this individual droplet model, were found to be in quite good agreement with the experimental data. An important drawback of this individual droplet model is that it requires a great deal of computational effort. Nevertheless, the preliminary results shown here indicate this to be a very promising approach in representing the small scale phenomena, the understanding of which is critical for the proper representation of the microstructure. Work is continuing, with the objective of improving the efficiency of the computational approach, so that this technique may be readily adopted to tackling process design problems.

ACKNOWLEDGEMENTS.

The authors gratefully acknowledge Professor D. Apelian of Drexel University for providing us with a copy of his paper [7] prior to publication. Thanks are also due to Dr. E. Lavernia (MIT, now at the University of California at Irvine) for providing experimental information used in the calculations.

5. REFERENCES

[1] E.J. Lavernia and N.J. Grant: Materials Science and Engineering, 1988, vol. 89, accepted for publication.

[2] J.B. See, J.C. Runkle and T.B. King: Metallurgical Transactions, 1973, vol. 4, pp. 2669-2673.

[3] J. Szekely and R.J Fisher, Transactions of the AIME, 1968, vol. 242, pp. 612-613.

[4] E. Gutierrez-Miravete, E.J. Lavernia, G. Trapaga, J. Szekely and N.J. Grant: Metallurgical Transactions B, accepted for publication.

[5] A.R.E. Singer and R.W. Evans: Metals Technology, 1983, vol. 10, pp. 61-68.

[6] N. El-Kaddah, J. McKelliget, and J. Szekely: Metallurgical Transactions B, 1984, vol. 15B, pp. 59-70.

[7] P. Mathur, D. Apelian, and A. Lawley: Acta Metallurgica, 1988, submitted for publication.

[8] H. Matyja, B.C. Giessen, and N.J. Grant, Journal of the Institute of Metals, 96(1968), 30-32.

[9] M.H. Burden and H. Jones, Journal of the Institute of Metals, 98(1970), 249-252.

[10] H. Jones: Rapid Solidification of Metals and alloys, (London, The Institution of Metallugists, 1982), 43

[11] R.H. Bricknell: Metallurgical Transactions A, 1986, vol. 17A, pp. 583-591.

[12] K. Ogata, E. Lavernia, G. Ray, and N.J. Grant: International Journal of Rapid Solidification, 1986, vol. 2, pp. 21-35.

[13] J.M. Nell, G. Rai, E.J. Lavernia, and N.J. Grant, Proceedings of the 1986 Annual Powder Metallurgy Conference and Exhibition, APMI, May 18-21, 1986, Boston, Ma.

[14] J. Madejski: International Journal of Heat and Mass Transfer, 1976, vol. 19, pp. 1009-1013.

[15] B.P. Bewlay and B. Cantor in: International Conference on Rapidly Solidified Materials, 1986, American Society for Metals, Ohio, pp. 15-21.

[16] A.R.E. Singer: Metals and Materials, 1970, vol. 4, pp. 246-257.

COUPLING OF SOLIDIFICATION KINETICS MODELS WITH

HEAT TRANSFER MODELS FOR CASTINGS

Doru M. Stefanescu, D. Bandyopadhyay, G. Upadhya

Solidification Laboratory
Department of Metallurgical and Materials Engineering
The University of Alabama
Tuscaloosa, AL 35487-9644

ABSTRACT

It is now generally accepted that a complete model for solidification of castings should include both macro- and micro-models. At the macro-level, it is required to describe the heat transfer from the metal to the mold, the fluid flow of liquid metal during mold filling, and stress analysis. At the micro-level, the model should include solidification kinetics and fluid flow during solidification. At the present time, the task of including all these models into a comprehensive solidification package is far from being complete, although significant progress has been achieved in each field.

The paper presents the state of the art on coupling two of the required models, i.e. macro heat transfer (HT) and micro solidification kinetics (SK). Various techniques of coupling these two models and of calculation of fraction solid are discussed. A new method of calculating the fraction of solid based on computer-aided cooling curve analysis is also discussed. Several models for nucleation of equiaxed eutectic grains are analyzed. A number of experimental and calculated results are shown for some typical casting alloys, to include cast iron and aluminum-silicon alloys. The paper heavily emphasizes the need and means of correlating modeling and experimental work.

Casting of Near Net Shape Products
Edited by Y. Sahai, J.E. Battles,
R.S. Carbonara and C.E. Mobley
The Metallurgical Society, 1988

INTRODUCTION

It is now generally accepted that a good solidification model for castings should include modeling of heat transfer (macro-level), solidification kinetics (micro-level), fluid flow during mold filling (macro-level), fluid flow during solidification (micro-level) and stress analysis (macro-level). Such a model will allow not only for accurate prediction of shrinkage, but also of cracking due to thermal stress as well as of microstructural features. This paper is concerned mainly with one aspect of solidification modeling, that is solidification kinetics (SK). Nevertheless, it is obvious that in order to validate any type of solidification kinetics model, it is important to incorporate such a model in a heat transfer (HT) model, or in other words, to couple the two models. A variety of computer programs (FDM or FFM) are available for heat transfer modeling, and any of them can be used to this effect. As it will be explained in this paper, the main problem in coupling solidification kinetics and heat transfer consists in incorporating the latent heat evolved during solidification, which is calculated by the SK model, in the HT model which calculates the overall cooling of the casting. To calculate latent heat evolution, it is necessary to calculate the evolution of fraction of solid, which in turn is calculated from nucleation and growth laws. Thus, to perform SK modeling, a good data base for nucleation and growth laws is required.

This paper will discuss some aspects involved in calculation of fraction of solid and number of nuclei required in SK modeling and will present some validation results on a number of commercial alloys.

BASIC PRINCIPLES INVOLVED IN COUPLED HEAT TRANSFER - SOLIDIFICATION KINETICS MODELS

As discussed in previous papers dealing with this subject (1,2,3), HT modeling requires solving the energy conservation equation for heat conduction:

$$\nabla.\nabla[K(T) \; . \; T(x,t)] + \dot{Q} = \rho \; C_p(T) \; \frac{\partial T(x,t)}{\partial t} \qquad (1)$$

where ρ is the density, C_p is the specific heat and K is the thermal conductivity. The source term $\dot{Q}$ describes the latent heat evolution during the liquid-solid transformation and may be written as:

$$\dot{Q} = L \; \frac{\partial f_s(x,t)}{\partial t} \qquad (2)$$

where L is the volumetric latent heat of fusion and f_s is the fraction of solid. In conventional HT modeling, f_s is unavailable, and the problem is circumvented, for example, by assuming that the enthalpy changes as a linear function of temperature within a fictitious solidification range ("enthalpy method").

HT - SK modeling has the advantage of a more direct approach to solving of the heat evolution problem, via the "latent heat method" (3). Indeed, it is quite obvious that if f_s in eq. (2) is known, solving of eq. (1) becomes trivial. A number of approaches in calculating f_s will be discussed later in the paper.

Another method of coupling HT and SK models is the "micro-enthalpy model" (4). Basically, in this approach a macroscopic enthalpy variation is calculated by the HT model, and within the time step of the HT model, microscopic changes are calculated at a smaller time step by the SK model. A comparison between the latent heat method, the micro-enthalpy method and the classic enthalpy method is shown in Fig. 1 in terms of enthalpy variation versus temperature.

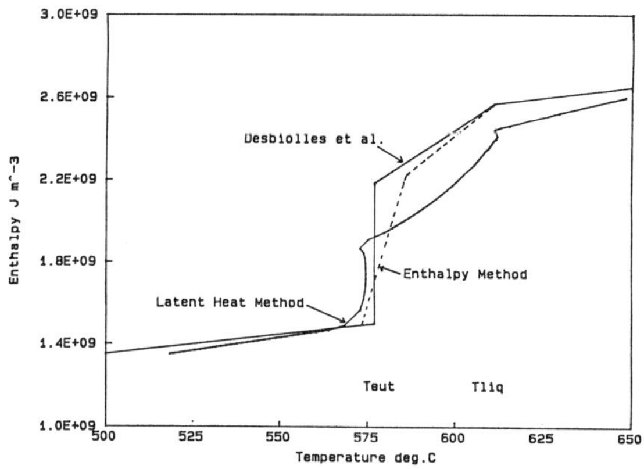

Figure 1
Calculated variations of enthalpy with temperature for different modeling methods for Al-7%Si alloys.

Both coupling methods require knowledge of fraction of solid. At least two avenues are possible in calculation of f_s, as follows:

- from cooling curve data using computer-aided cooling-curve-analysis
- from nucleation and growth laws.

CALCULATION OF FRACTION OF SOLID FROM COOLING CURVE DATA

Experimental cooling curves can be used to provide information on the evolution of fraction of solid during solidification. The technique which can be used to this effect is computer-aided cooling-curve analysis (CA-CCA) which consists, in principle, of using the first derivative, dT/dt, of the cooling curve and a simulated first derivative of a cooling curve with no transformation to calculate the heat evolution during solidification. A detailed description of CA-CCA is given in reference (5). Using this technique, $\dot{Q}$ in eq. (2) can be calculated at each time step for a given cooling curve corresponding to a given cooling rate. Eq. (2) can then be rewritten as:

$$\dot{Q}(x,t) = L \frac{\partial f_s(\dot{T},t)}{\partial t} \tag{3}$$

where T is the cooling rate.

It now becomes obvious that experimental values for f_s must be determined not only as function of T, but also as a function of $\partial T/\partial t$ (cooling rate), to allow for the calculation of heat evolution in different nodes. With $\partial f_s/\partial T$ known for experiments for different $\partial T/\partial t$, Q in eq. (1) can be calculated via eq. (3), thus coupling the SK model with the HT model. It must be noted that $\partial T/\partial t$ before the start of solidification is readily obtained from the HT model.

An example of application of CA-CCA technique to HT-SK modeling is given below for eutectic Al-Si alloys. Figure 2 shows two experimental cooling curves for the same eutectic Al-Si alloy poured in two different size cylindrical bars molded in sand. Using the CA-CCA technique, the evolution of fraction of solid with time was calculated and plotted in Fig. 3. The cooling rate at the beginning of the eutectic solidification corresponding to each cooling curve was obtained from the first derivative of the cooling curves.

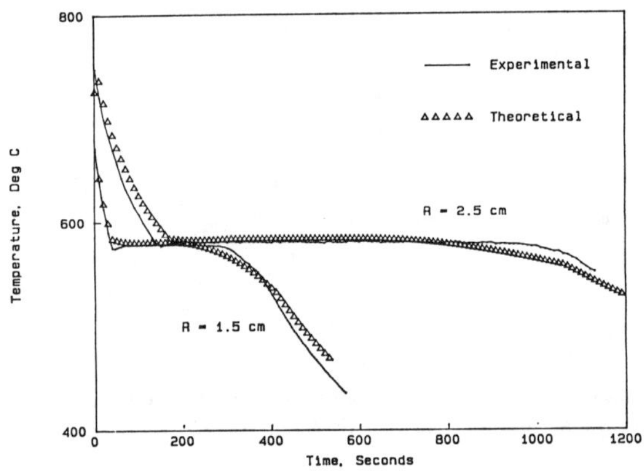

Figure 2
Experimental and calculated cooling curves for an Al-7% Si alloy for two different size molds (5 cm and 3 cm dia. bars).

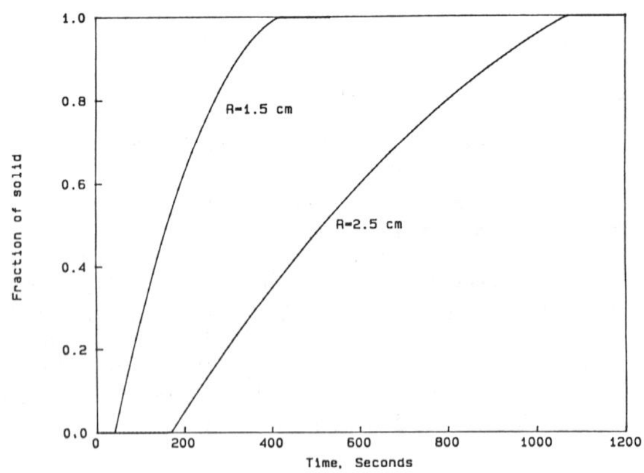

Figure 3
Evolution of fraction of solid versus time for an Al-Si eutectic alloy solidified for two different size molds (5 cm and 3 cm dia. bars).

It can be seen that the curves are second order polynomials. It follows that, for a given cooling rate, $\dot{T}$, $\partial f_s / \partial t$ is a first order polynomial that is:

$$\frac{\partial f_s}{\partial t} = At + B \tag{4}$$

where A and B are constants. In fact, A and B are only constant for constant $\dot{T}$. For various $\dot{T}$, eq. (4) becomes:

$$\frac{\partial f_s}{\partial t} = A(\dot{T})t + B(\dot{T}) \tag{5}$$

Then, by interpolating experimental data of the type shown in Fig. 3, a final equation for the rate of evolution of fraction of solid can be written as:

$$\frac{\partial f_s}{\partial t} = (a\dot{T}+b)t + c\dot{T}^2 + d\dot{T} + e \tag{6}$$

where a, b, c, d, e are constants.

Equation (6) is then used to calculate $\dot{Q}(x,t)$ via eq. (3), which allows for solving of equation (1).

The calculated cooling curves were superimposed on the experimental curves in Fig. 2. The various constants used in modeling are given in Table 1. It is seen that a good match was achieved for both undercooling and solidification time.

Table 1

Constants used for Al-Si eutectic solidification modeling

Latent heat of fusion = 473018 Joules/kg
Density of Aluminum = 2600 kg/m^3
Density of sand mold = 1500 kg/m^3
Thermal conductivity of mold = 1.5 watts/m.^{o}C
Thermal conductivity of metal = 221 watts/m.^{o}C
Temperature of eutectic = 557°C
Bar radii = 2.5 cm and 1.5 cm
Mold thickness = 2 cm
Heat transfer coefficient at mold-air interface = A * exp (B * Temp), where A & B are constants and Temp is the temperature at last node of metal.
A = 0.081002 watts/m^2 for 2.5 cm bar,
= 0.550000 watts/m^2 for 1.5 cm bar.
B = 0.012156 $(^{o}C)^{-1}$ for 2.5 cm bar,
= 0.009174 $(^{o}C)^{-1}$ for 1.5 cm bar.

CALCULATION OF FRACTION OF SOLID FROM NUCLEATION AND GROWTH LAWS

At a given time t and position x in the mold, the fraction of solid for an alloy solidifying with equiaxed structure can be calculated as:

$$f_s(x,t) = N(x,t) \cdot \frac{4}{3}\pi R^3(x,t) \tag{7}$$

where N(x,t) is the density of grains and R(x,t) is the average equiaxed grain radius. The rate of formation of fraction of solid is:

$$\frac{\partial f_s}{\partial t} = 4\pi R^2 \frac{\partial R}{\partial t} \frac{\partial N}{\partial t} \tag{8}$$

where $\partial R/\partial t$ is the growth rate and $\partial N/\partial t$ is the nucleation rate.

If the appropriate nucleation and growth laws are known, $\dot{Q}$ in eq. (1) can be calculated and then eq. (1) can be solved numerically. It is again evident that R and N will both be functions of t and dT/dt. Since the growth problem is somehow simpler and has been extensively discussed for both equiaxed eutectics (1) and equiaxed dendrites (6), only the nucleation kinetics problem will be addressed in this paper.

NUCLEATION RATE

Calculation of the nucleation rate $\partial N/\partial t$ is, apparently, the most complicated problem in SK modeling, since the formal theory of heterogeneous nucleation cannot be used, in its present form, for these calculations. To date, a number of approaches have been used:

- parabolic dependence of number of nuclei on undercooling (9):

$$N = A \cdot \Delta T^2 \tag{9}$$

which gives a nucleation rate of:

$$\frac{\partial N}{\partial t} = -2A \cdot \Delta T \frac{\partial T}{\partial t} \tag{10}$$

where A is a constant. It must be noted that according to eq. (10) nucleation rate is a function of both undercooling and cooling rate.

- instantaneous nucleation (1), where all nuclei are considered to generate at the same time at a given critical undercooling (nucleation temperature), ΔT_n.

- continuous nucleation between the equilibrium temperature, T_E, and maximum the undercooling temperature (2,11).

Only the last two approaches will be discussed here.

Instantaneous Nucleation

The rate of heterogeneous nucleation can be written as (7):

$$\frac{\partial N}{\partial t} = I_{het} = n'\gamma \exp\left(\frac{\Delta G_c^{het}}{kT}\right) \exp\left(-\frac{\Delta G_A}{kT}\right) \tag{11}$$

where n' is the number of surface atoms of the substrate per unit volume of liquid, γ is the vibrational frequency, ΔG_c^{het} is the energy of formation of the critical nucleus, ΔG_A is the diffusional activation energy for growth, k is Boltzman's constant, and T is the temperature.

According to J.D. Hunt (8), considering site saturation occurring in the system, eq. (11) may be rewritten as:

$$I_{het} = (N_s - N)\ K_1\ \exp\left(-\frac{\Delta G_c^{het}}{kT}\right) \tag{12}$$

where N_s is the number of heterogeneous substrates originally available per unit volume, N is the number of particles that have already nucleated per unit volume, and K_1 is a constant.
Since the liquid diffusivity can be approximated with:

$$D \simeq a^2 \gamma\ \exp\left(-\frac{\Delta G_A}{kT}\right) \tag{13}$$

where a is the atomic diameter, eq. (11) can be written as:

$$I_{het} = \frac{n'D}{a^2}\ \exp\left(-\frac{\Delta G_c^{het}}{kT}\right) \tag{14}$$

Then K_1 can be calculated from eq. (12) and (14) as:

$$K_1 = \frac{n'D}{a^2 (N_s - N)} \tag{15}$$

Calculations for the constant K_1 for Fe-C alloys nucleating on SiO_2 or Al-Si alloys nucleating on Si (see Appendix I), resulted in numbers ranging from 1.15×10^{18} to $5.57 \times 10^{22} s^{-1}$, depending on the type of alloys and the assumed radius of the substrates. An average number of $K_1 = 10^{22} s^{-1}$, which is similar to the one proposed in (2) will be used in this work.
Now:

$$\frac{\Delta G_c^{het}}{kT} = \frac{16}{3}\ \frac{\pi \sigma^3 T_M^{\ 2}}{\Delta H^2 \Delta T^2 kT}\ f(\theta) \tag{16}$$

where σ is the surface tension, T_M is the melting point, and ΔH is the latent heat. For metallic systems, T is very large compared with ΔT and can be considered constant. For a given nucleant $f(\theta)$ is a constant too, and thus one can write:

$$\frac{\Delta G_c^{het}}{kT} = \frac{K_2}{\Delta T^2} \tag{17}$$

Substituting eq. (17) in (12):

$$I_{het} = (N_s - N)\ K_1\ \exp\left(-\frac{K_2}{\Delta T^2}\right) \tag{18}$$

Following again J.D. Hunt's suggestion (8), K_2 can be calculated if ΔT is defined as the nucleation temperature ΔT_n, where the initial nucleation rate is $I_{het} = 1m^{-3}s^{-1}$. Since for this case $N = 0$, eq. (8) becomes:

$$K_2 = \Delta T_n^{\,2} \ln (N_s K_1) \qquad (19)$$

and K_2 can be calculated using the numbers for K_1 given in Appendix I.

Using this approach, the nucleation rate for different nucleation temperatures was calculated as an example for a cast iron having $N_s = 6 \times 10^{11} m^{-3}$. The results are shown in Fig. 4. It is apparent that nucleation starts below the nucleation temperature, at an undercooling $\Delta T_n' = \Delta T_n + \Delta T_{lag} = T - T_n$ and that once started it occurs very rapidly. For all practical purposes, the complex nucleation equation (18) can be substituted with a Dirac Delta function, $\delta (T-T_n)$ having the following definition:

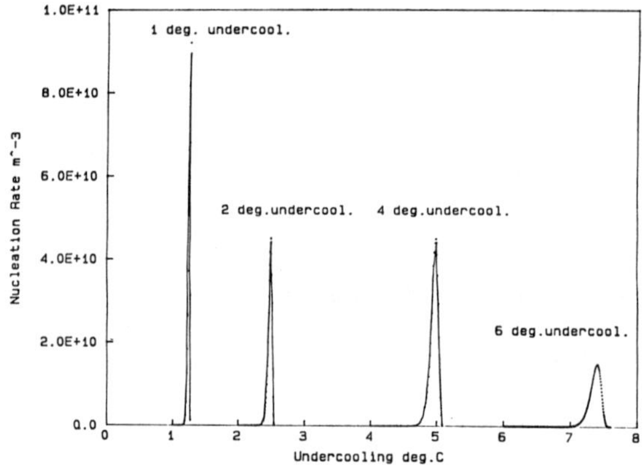

Figure 4
Dependency of nucleation rate on nucleation temperature.

$$\text{for } T \neq T_n \qquad \delta(T-T_n) = 0 \qquad (20)$$
$$\text{for } T = T_n \qquad \delta(T-T_n) = \infty$$

Also:

$$\int_{-\infty}^{+\infty} N(T)\,\delta(T-T_n)\,dT = N_s \qquad (21)$$

Thus, eq. (18) can be substituted by:

$$\frac{\partial N}{\partial T} = N_s \cdot \delta(T-T_n) \qquad (22)$$

which after integration, gives a total number of nuclei at the nucleation temperature T_n of $N = N_s$.

The nucleation law in eq. (22) predicts instantaneous nucleation, that is that all nuclei are generated at the nucleation temperature T_n. This approach was used in HT-SK modeling of a number of casting alloys to include gray and shperoidal cast iron and aluminum-silicon (1,3).

Comparison between experimental and HT-SK modeled data are shown in Fig. 5a for gray cast iron and in Fig. 5b for eutectic Al-Si alloy. The EUCAST and BAMACAST programs are FDM codes for HT-SK modeling described elsewhere (11). It is evident that for these particular alloy systems, the assumption of instantaneous nucleation works fairly well. Nevertheless, such an approach will predict uniform grain size across the casting, which is obviously not the case in real castings.

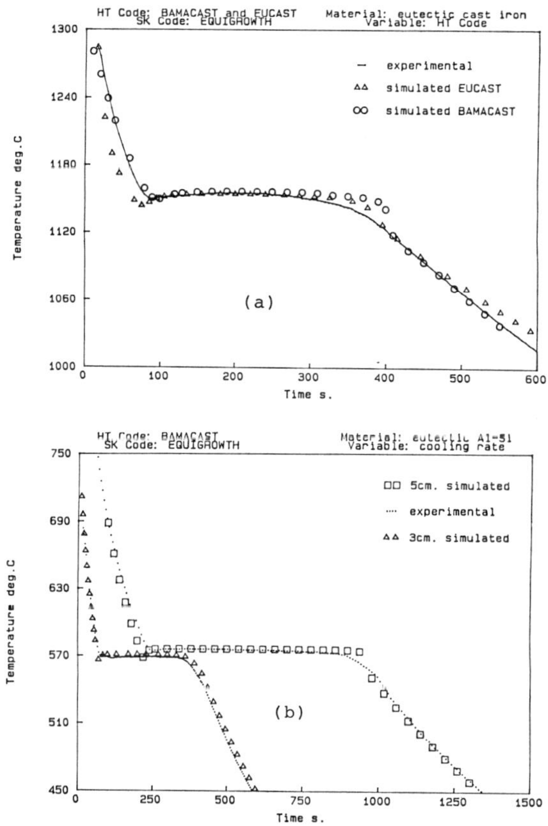

Figure 5
Experimental and simulated cooling curves for: a) gray cast iron of eutectic composition; b) eutectic Al-Si alloys, poured in sand molds of different sizes.

Solidification can occur on more than one type of nuclei, which means that more than one ΔT_n should be considered. In order to investigate the implications of multiple nucleation temperatures when instantaneous nucleation is assumed, the EUCAST code was used to simualte solidification of a eutectic gray cast iron. Three different cases were considered: a) all nuclei were formed at $\Delta T_n = 0°C$; b) half of the nuclei were formed at $\Delta T_{n1} = 5°C$ and the other half at $\Delta T_{n2} = 10°C$; c) half of the nuclei were formed at $\Delta T_{n1} = 0°C$ and the other half at $\Delta T_{n2} = 10°C$. The results are plotted in Fig. 6 for $N_s = 6x10^{11}$ m^{-3}. The cooling curves have been intentionally translated in the x direction to avoid superimposition. It

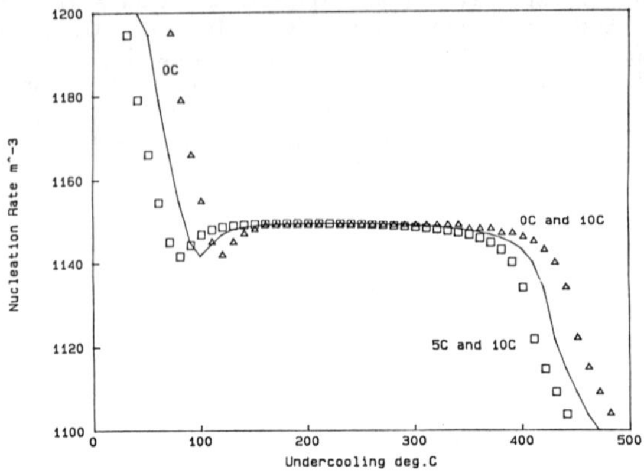

Figure 6
Simulated cooling curves for eutectic gray iron for singular and multiple nucleation temperatures.

can be seen that the differences are minimal, although slightly higher maximum undercooling occurs, as expected, when ΔT_n increases or when multiple nucleation temperatures are considered.

It is thus apparent that using a single nucleation temperature and the instantaneous nucleation assumption, HT-SK models can predict rather accurately the T-t cooling curves, i.e., the solidification path. Nevertheless, if microstructural features are to be modeled, a more complex model for nucleation is required.

Continuous Nucleation

Investigators using the continuous nucleation assumption consider that nucleation occurs continuously once the nucleation temperature is reached until the onset of recalescence. Nucleation laws such as those given by eq. (10) are used to calculate N in eq. (7) by integrating over time or temperature:

$$N = \int_{t_0}^{t} \frac{\partial N}{\partial t} dt = \int_{T_E}^{T} \frac{\partial T}{\partial t} \cdot \frac{\partial N}{\partial T} dT \tag{23}$$

where t_0 is the time at the beginning of solidification and T_E is the equilibrium temperature.

Another method is to produce a set of Dirac functions corresponding to nucleation sites becoming active at different nucleation temperatures and then replacing this discrete distribution by a continuous distribution (e.g., exponential) (2,12).

Regardless of the mathematical model, the basic concern when using the continuous nucleation approach is that it is not at all certain that nucleation ceases at the temperature of maximum undercooling. Indeed, directional solidification experiments on cerium treated cast iron have demonstrated that nucleation occurs continuously until the end of solidification is reached (13). The experimental number of nuclei, N(T) is given by a curve which can be interpolated by a third order polynomial, rather than by an exponential law as shown in Fig. 7, for a gray cast iron having a total number of nuclei $N_s = 6x10^{11}$ m^{-3}. It becomes obvious that

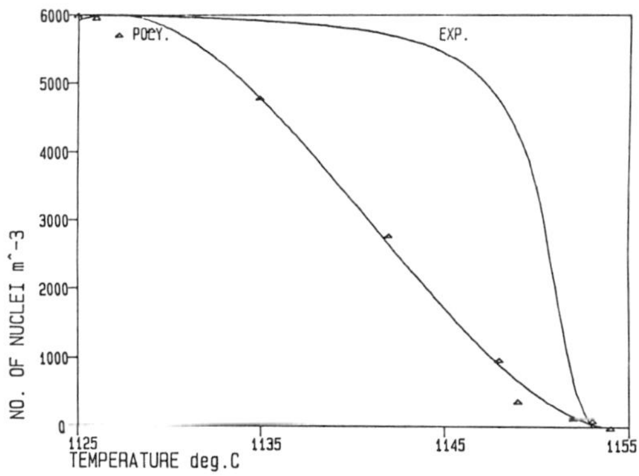

Figure 7
Variation of number of nuclei with temperature for a gray cast iron using polynomial and exponential laws.

nucleations laws used in HT-SK modeling should have a strong experimental support, if an accurate number of grains is to be predicted.

CONCLUSIONS

The basic principles of heat transfer - solidification kinetics modeling were discussed, and the latent heat method of solving the heat source term in the heat conduction equation was compared with other methods.

A new approach to calculating the source term, based on experimental data obtained by computer-aided cooling curve analysis, has been introduced, and its validity demonstrated. This method is straightforward, has the advantage of accurate calculation of fraction of solid, but also the disadvantage of limiting the amount of microstructural information which can be obtained.

It was also shown that the instantaneous nucleation approach allows for accurate modeling of solidification, but again does not provide enough microstructural data. The continuous nucleation approach allows for better modeling of microstructure, but requires substantial data base, which can be obtained experimentally, but is presently unavailable.

ACKNOWLEDGMENTS

This work was supported by NSF EPSCoR Alabama, Contract No. R11 8610669, NSF CADCAST, Contract No. DMC - 8612019, and by General Motors Technical Center, Contract No. PMS 11611. The authors are indebted to Profs. J. Berry and N. El-Kaddah for many helpful discussions.

APPENDIX I

CALCULATION OF K_1

Consider a Fe-C alloy nucleating on SiO_2 particles. Assuming that the substrates are spherical in shape with a radius r_s, the number of surface atoms of substrate per unit volume of liquid can be written as:

$$n' = 4\pi r_s^2 N_s n_s^{\frac{2}{3}} \tag{I.1}$$

where n_s is the number of atoms in the heterogeneous substrate per unit volume for SiO_2.

$$n_s = \frac{2.65\text{x}10^3 \text{ kg m}^{-3}}{(28.1+32)\text{x}10^{-3}\text{kg mole}^{-1}} \text{ x } 6\text{x}10^{23} \text{ at mole}^{-1} = 2.645 \text{ x } 10^{28} \text{ at/m}^3$$

Assuming $r_s = 1\ \mu m$:

$$n' = 1.11 \text{ x } 10^{20}\ N_s$$

At the beginning of nucleation N = 0 and eq. (15) becomes:

$$K_1 = \frac{n'D}{a^2 N_s} \tag{I.2}$$

Substituting $D = 10^{-9} m^2 s^{-1}$ and a = 0.25nm in eq. (I.2) yields $K_1 = 1.15 \text{ x } 10^{18}\ s^{-1}$.

For $r_s = 100\ \mu m$ it is calculated that $K_1 = 1.15 \text{ x } 10^{22} s^{-1}$.

Thus for $r_s = 1$ to $100\ \mu m$ and a = 0.25 nm, $K_1 = 1.15 \text{ x } 10^{18}$ to $1.18 \text{ x } 10^{22} s^{-1}$.

Consider now an Al-Si alloy nucleating on Si (slightly off-eutectic composition is assumed). Then:

$$n_s = \frac{2.33 \text{ x } 10^3}{28.09 \text{ x } 10^{-3}} \text{ x } 6\text{x}10^{23} = 4.97\text{x}10^{28} \text{ at/m}^3$$

Assuming again that $r_s = 1$ to $100\ \mu m$, $D = 10^{-9} m^2 s^{-1}$ and a = 0.14nm then: $K_1 = 5.57 \text{ x } 10^{18}$ to $5.57 \text{ x } 10^{22}\ s^{-1}$

REFERENCES

1. D.M. Stefanescu, C. S. Kanetkar, in State of the Art of Computer Simulation of Casting and Solidification Processes, H. Fredriksson, editor, Les Editions de Physique, Paris, 1986, p. 255.
2. M. Rappaz, D.M. Stefanescu, in Solidification Processing of Eutectics, D.M. Stefanescu et al., editors, TMS, Warrendale, PA, 1988, pp. 133-152.
3. C.S. Kanetkar, I.G. Chen, D.M. Stefanescu, N. El-Kaddah, J. of ISI of Japan, in print.
4. J.L. Desbiolles, M. Rappaz, J.J. Droux, J. Rappaz, ibid. 1, p. 49.
5. I.G. Chen, D.M. Stefanescu, AFS Transactions, 1984, vol. 92, p. 947.
6. M. Rappaz, P. Thevoz, Acta Metal, 1987, vol. 35, no. 7, p. 1487.
7. J.W. Christian, The Theory of Transformation in Metals and Alloys, Pergamon, Oxford, 1975, p. 450.
8. J.D. Hunt, Materials Science and Engineering, 1984, vol. 65, pp. 75-83.

9. W. Oldfield, ASM Trans., 1966, vol. 59, pp. 945-959.
10. K.C. Su, I. Ohnaka, I. Yaunauchi, T. Fukusako, in The Physical Metallurgy of Cast Iron, H. Fredriksson and M. Hillert, editors, North-Holland, New York, 1984, pp. 181-189.
11. C.S. Kanetkar, Ph.D. dissertation, The University of Alabama, Tuscaloosa, 1988.
12. Zou Jie, P. Thevoz, M. Rappaz, private communication, 1987.
13. D. Bandyopadhyay, D.M. Stefanescu, I.G. Chen, unpublished work at The University of Alabama, 1987.

CASTING TECHNIQUES I:

PROCESS CONTROL

Session Chairman: **W.A. Obenchain**
U.S. Department of Energy
Idaho Falls, Idaho

PROCESS CONTROL IN DIRECT STRIP CASTING

R.S. Carbonara*

BATTELLE
Columbus Division
505 King Avenue
Columbus, Ohio 43201-2693

INTRODUCTION

The technical and economic benefits that are possible with the Direct Strip Casting (DSC) process continue to push its development. Over the past decade, a wide variety of alloy systems have been cast into strip and sheet by the DSC process. Not surprisingly, one of the most important findings from all of this work is the wide range of casting conditions and parameters needed to successfully cast these alloys. Also, some understanding has been gained about the sensitivity of strip characteristics to the casting conditions and parameters. Clearly, control of the processing conditions and parameters is vital to the commercial success of DSC.

There are several types of DSC processes, viz., single roller, double roller, continuous twin belt, segmented twin belt, and horizontal strip casters. This paper will focus primarily on process control considerations for the single roller process. In single roller DSC, molten metal is metered through a tundish onto the surface of a single rotating drum or wheel where it solidifies into sheet or strip. This one step strip casting technique eliminates most of the rolling and reheating normally used in forming strip. This translates into vast savings of energy, labor, and production time which results in lower product cost. The high speed at which strip is produced (50-1000 m/min) by DSC, makes process control very important.

PROCESS OVERVIEW

There are four major elements in the single roller DSC process, viz., melting, molten metal transfer, strip casting, and post-casting treatment/ handling (see Fig. 1). Each of these elements has its own specific process control requirements, but the interaction of each of these elements with the others must also be considered.

The melting step requires more than rudimentary consideration. The melt stock must be clean and the melt must be kept clean in order to minimize

---

* Present address: SEA, Inc.
7349 Worthington-Galena Road
Worthington, Ohio 43085

Casting of Near Net Shape Products
Edited by Y. Sahai, J.E. Battles,
R.S. Carbonara and C.E. Mobley
The Metallurgical Society, 1988

and eliminate inclusions and inclusion formation. This requires monitoring and control of the atmosphere over the melt and careful selection of the crucible and tundish materials. Also, the materials that come in contact with the melt during transfer from the crucible to the tundish and materials that hold or protect sensors and probes in the melt must be selected with care so that they do not contaminate the melt. Transfer tubes can be used to protect the melt during transfer from the crucible to the tundish; even so, the transfer of the molten metal from the tundish onto the casting wheel requires a gas shroud to protect it from atmospheric contamination until it has solidified and cooled to the point where inclusions and other contamination induced defects cannot form. Of course, the entire process could be done in a closed and atmospheric controlled chamber, but this is not practical for a commercial operation, except for very special alloys.

The temperature of the melt must be controlled for two reasons. First, contamination of the melt is sensitive to the temperature of the melt. Second, melt temperature is a primary consideration in controlling strip thickness and cooling characteristics. Thus, even before strip is cast, the DSC process must be carefully controlled. Both melt chemistry and melt temperature are factors that affect the quality and characteristics of the strip.

The strip casting step is the focus of the entire process and is affected by the melting and molten metal transfer steps and, in turn, affects the post casting treatment and handling. The exact nature of the casting step is not yet sufficiently understood so that a model describing the interaction or relationship between the process parameters can be devised. This is a major barrier to the ultimate success of DSC and is currently being studied. The importance of understanding the direct casting step lies in the necessity to monitor and control the process on a real-time basis. This can be done only if the controlling parameters are known and their relationship to one another and to the strip quality and characteristics are understood. This indeed is a formidable task and will evolve as the process develops.

This post casting treatment and handling steps would be ideally set up to produce on-line a finished product ready for shipment to a customer. In order for this to be realized, the as-cast strip must be characterized to the extent that the appropriate post casting treatment can be defined and carried out as part of the continuous DSC process. Among the as-cast strip characteristics needed are thickness, width, surface finish, size and number of defects and grain structure. The last characteristic, grain structure, will most likely be inferred from the thermo-mechanical history of the strip unless on-line microstructural monitoring becomes possible.

With the current state-of-the-art, samples must be taken at the end of a coil run and subsequent treatment of that coil delayed until those results are analyzed and the appropriate post casting treatment determined. This is a less desirable situation from a production efficiency point of view, but is currently the most likely operational mode. However, work by Rosen (1), Winter and Green (2) and Reno et al (3) shows some promise for dynamic determination of the microstructure of the strip as it is produced by flash x-rays and ultrasonic methods. With this on-line analysis it would be possible to give the strip the appropriate thermomechanical treatment immediately after casting and to produce a "finished" mill product ready for shipment to the customer.

When viewing the DSC process from melting to post casting treatment, it is clear that there are numerous necessities for process control (see Fig. 1). the melting and metal transfer part of the process are reasonably well understood and some control is currently being used, although much more is needed. The casting process is perhaps the least understood and hardest to monitor since casting speeds are high (hundreds of feet per minute), space limitations for placing sensors is severe, and the environment is usually hostile to most sensors. Post casting treatment is the easiest area to deal with if delayed final processing can be tolerated.

Looking at the DSC process, two major factors are clearly evident. One is the need for sensors or sensing procedures that can reliably monitor the high speed DSC process and its hostile environment. The other factor is the need for a more complete understanding of the molten metal flow and the casting process. This second factor is critical to determining what parameters to monitor and then how to adjust those parameters to produce the desired strip characteristics.

Melting and Molten Metal Flow

Control of the melting step involves using very clean melt stock and/or a ladle chemistry control system that will remove any impurities or inclusions. In either case, atmospheric control over the melt must be maintained. Using an inert cover gas, such as argon, is the most reliable, since it will not react with any molten metal constituents and is heavier than air and will usually stay in place over a melt pool. The major drawback for using argon is cost. Using nitrogen instead of argon is possible in some instances, but nitrogen is <u>not</u> inert and can react with molten metals to form undesirable inclusions. The use of a mixture of gases or gases which are reducing to most metals must be assessed on an individual basis. Another aspect of melt chemistry that must be considered is the addition of alloying elements. This can best be done by having these elements in the master alloy. However, there may be instances, for example, in the cases of high vapor pressure or very reactive elements where adding these elements late in the melting cycle or even in the tundish is best. Once the "alloy" is made in the molten state, its chemistry must be maintained until it is cast. It is possible to monitor the melt chemistry by in-situ transient emission spectroscopy (4). However, great care must be taken to assure that a representative sample of the bulk melt is analyzed. Both slag formation and volatilization of high vapor pressure constituents must be dealt with here. It may also be possible to use immersible probes that are sensitive to a specific element or ratio of elements to monitor melt chemistry. However, the availability of such probes that can survive the molten metal environment is very limited.

If it is not possible to keep the melt inclusion-free, then a filtering system will be needed to remove those inclusions prior to the casting step. The nature and placement of a filtering system depends on specific circumstances of the alloy and the DSC apparatus. If the system is 100 percent effective and does not require adjustments, then there is no need for monitoring the inclusion content of the melt. However, such a system may not be feasible for DSC, which would then necessitate a melt-inclusion monitoring device and a feedback system to adjust the filtering system.

Temperature control is another very important part of the melting and molten metal flow system. Temperature controls nearly all the characteristics of the melt--including such things as melt chemistry and reactivity, viscosity, surface tension, density, etc. Although many of these are weakly dependent on temperature, they may have a major affect on the casting process so that small temperature changes are important. There is some evidence that melt temperature fluctuations in the tundish should be maintained within $\pm$ 30 degrees (5). Even tighter control must be maintained in the contact pool where the molten metal contacts the casting drum. The control of melt temperature in the contact pool is made more difficult because of the desire to have the molten metal flow onto the casting drum in a quiescent manner--perhaps in a laminar flow pattern. In this mode, there is very little mixing of the melt and temperature gradients will form if external heat is not provided. Adjusting this heat source requires feedback from sensors that are monitoring the temperatures in various parts of the melt pool. The concern here is to make temperature measurements without disturbing the flow pattern onto the casting drum. By understanding the fluid flow pattern in a given tundish, it may be possible to adjust external heaters at the contact pool or tundish heaters in such a way to minimize thermal gradients and satisfactorily control the temperature of the contact pool. Melt temperature can be monitored by thermocouples and infrared thermometry systems. Also, it may be possible to use penetrating radiation, such as ultrasonic, to measure internal temperatures in the melt--similar to the technique for measuring the internal temperature of a solid (6).

The melt level in the tundish is another critical control parameter in the DSC process. Melt level affects the flow pattern through the tundish and, depending on design, the contact length of the casting pool, i.e., the distance the molten metal is in contact with the casting drum. The other factors affecting pool length are drum diameter, tundish design and position, however, these are usually fixed parameters. Both contact and non-contact techniques can be used to control melt levels in the tundish. Non-contact methods use lasers or other types of electro-optic and electromagnetic devices. These can be placed away from the melt so they do not interfere with the flow pattern in the tundish. Contact techniques, such as floats and level probes are more intrusive but can be reliably used if they are positioned carefully. Weighing the molten metal in the tundish is another and fairly reliable approach to controlling the tundish melt level. No matter what sensing method is used, a key part of the level control system is the feedback loop and the gating method used to increase or decrease the flow from the crucible into the tundish. The "best" method depends strongly on the alloy system and the detailed design of the DSC equipment.

A great deal of care must be used in preparing the molten metal for the casting step in the DSC process. The reasons for such a degree of care are the very strong affects the molten metal characteristics have on strip characteristics, especially strip quality. Melt chemistry affects the solidification morphology and solidification temperature. In some cases, its improper control can lead to defects in the strip, viz., hot cracking and inclusions. Even though the melt is under a protective atmosphere throughout melting and molten metal transfer to the melt can come into contact with oxygen (air) that adheres to the casting drum as it rotates. This boundary layer is always present unless the drum is in a vacuum or in its own protective atmosphere. The presence of this oxygen (air) boundary layer can have an affect on the casting process in several ways. It will react with the molten metal as it exits from the tundish and flows onto the casting drum. In the case of casting steel strip, this can lead to an exothermic oxidation of the molten metal. There is usually more oxidation taking place near the strip edges--due to the greater amount of strip surface area exposed to the boundary layer--and this often leads to the presence of excessive molten metal on and around the edges of the solidifying steel strip. The heat from the excess molten metal must be absorbed by the strip and the drum. In many cases, the strip must do most of the heat absorption since it is beginning to release from the drum--this often leads to defects in the strip, such as edge curls and cracks. The entrained boundary layer has another affect on the casting process. As the molten metal comes in contact with the casting drum, some of the boundary layer which strongly adheres to the drum surface becomes entrapped between the molten metal and the drum and prevents direct contact of the molten metal and the drum. This reduces the surface area of contact between the solidifying strip and the drum and strongly affects the heat transfer. In fact, when making calculations using heat flow, it is necessary to use an "effective" heat transfer coefficient. Not only does the entrapped air reduce the contact area for heat transfer, but it can also decrease the surface quality, lead to casting defects and cause inhomogeneities in the solidification pattern of the strip. These are normally not a problem in casting thick strip (>1 mm). Attempts have been made to use skimmers and other such devices to reduce the boundary layer or to change its composition by adding an inert gas to it, but most are not very effective. Again, the degree of the problem of reaction with the boundary layer is dependent to a large extent on metal chemistry.

To control melting and molten metal flow in the DSC process, it is necessary to control the following:

- melt chemistry
- melt temperature
- inert gas flow rates
- molten metal flow rates
- casting flux additions (if used)
- boundary layer composition (if possible).

The Casting of Strip

Once the molten metal comes in contact with the casting drum, strip is formed. The width of the strip is controlled by the width of the tundish orifices that feeds the molten metal onto the drum. Strip thickness, however, is controlled by several factors. Some factors are intrinsic to the composition of the strip material being cast, such as viscosity, surface tension, heat capacity, thermal conductivity, heat of fusion, melting and freezing temperatures, and density. The other factors controlling the casting process are related to the design of the casting system, primarily the casting drum and the ancillary equipment associated with it.

Drum composition and size are key considerations which should be selected based on the alloy system being cast. Drum composition affects both wetting by the molten metal and heat conduction away from the solidifying strip. It is usually not advisable to select the drum composition based on a single criteria, such as maximum thermal conductivity. Although not all the factors regarding drum composition are known, in addition to wetting and thermal conductivity, thermal fatigue and surface maintenance are two more important ones to consider. Drum size should be viewed as affecting contact pool length and thermal cycle time (i.e., cycle time between hot metal contact at the tundish orifice). Other considerations are the cost, water cooling design pattern, drum drive motor size, support system size, and speed control system.

Contact pool length is a geometric factor that is determined by the radial position of the tundish on the drum (9 o'clock, 12 o'clock, etc.), tundish orifice design and drum size. Contact pool length is one factor affecting strip thickness and quality. However, the DSC process is not understood well enough at this time to establish a quantitative relationship between pool length and strip thickness. For example, other key factors affecting strip thickness and quality are the drum temperature, surface geometry, and surface chemistry. The drum temperature must not only be controlled, but it must be uniform over the entire drum surface--since it affects the wetting characteristics and thus, the heat transfer surface area and any variations would cause variations in strip characteristics, especially thickness.

Once the drum material and size have been selected, the operational control factors are those just mentioned, viz., drum temperature, surface geometry, and surface chemistry. The issue regarding these factors is that their general affect on the strip characteristics is understood, more or less. However, the qualitative relationship between these factors is not understood and the specific relationship between these factors and strip characteristics is not understood at all.

Control of the drum temperature involves controlling the heat input from the molten metal as it flows onto the drum and the heat withdrawal via cooling water passing through the drum. It is not difficult to monitor and control drum cooling water temperature. In principle, it is possible to establish a correlation between the exiting drum cooling water temperature and the drum surface temperature. In practice, however, it is not easy to control the drum surface temperature since there is a large time constant or time lag between variations of the surface temperature and the exiting water temperature. By the time a change in cooling water temperature is detected, indicating a change in drum surface temperature, the surface temepraturе has most likely changed again. The heat input to the drum is regulated by the temperature of the molten metal as it exits the tundish and contacts the drum and the amount of thermal contact between that molten metal and the drum. In practice, this contact (effective heat transfer) can be effected by changes in the drum surface chemistry which in turn affects the degree of wetting of the drum by the molten metal. The importance of drum surface chemistry is directly related to wetting, which subsequently affects several other operational parameters. For example, as the degree of wetting decreases, the strip temperature leaving the drum increases. Wetting also affects the cooling rate and the microstructure of the strip. These two factors, strip temperature and microstructure, affect the post casting treatment and handling.

Post Casting Treatment and Handling

Post casting treatment and handling is a part of the DSC process that has received very little attention. The work that has been done has not been done "on-line". One of the few examples is 304 stainless steel strip that has been thermomechanically processed after it was cast and coiled. The general "rule-of-thumb" is that 50 percent reduction of DSC produced 304 SS strip will give it the same properties as conventionally produced 304 SS (7). By controlling the DSC process, it should be possible to include the thermomechanical process on-line as part of the DSC process. On-line post casting treatment would increase the process efficiency and hopefully reduce production costs.

Post casting treatment initially involves controlling the temperature of the solidified strip as it leaves the casting drum and enters the post casting treatment system. As mentioned earlier, this is strongly controlled by the degree of wetting of the drum by the molten metal. However, there is still an opportunity to control the strip temperature after it leaves the casting drum. This can be done by heating or cooling the strip as it travels from the casting drum to the rollers. There are several heating schemes that can be used and a water spray system can be used for cooling. In order to make appropriate heating or cooling adjustments to the strip, the strip temperature and/or microstructure must be known. This task can be complicated by variations in the thickness and surface roughness or oxidation of the strip. Also, the thickness and any thickness variations, either longitudinal or transverse, and the width of the strip must be measured before the appropriate thermomechanical treatment can be applied.

CONCLUSION

Although this paper has suggested that there are substantial challenges to the commercialization of the DSC process, it should be mentioned that strip is being made in pilot operations even though some of the challenges remain unsolved at present. It may turn out that some present concerns will go away without a solution. It is also likely that new challenges will arise as the process gets closer to commercial operation and as new alloys and alloys systems are used.

No matter the specifics, the overall challenge of establishing process control technology for the DSC process remains.

EPILOGUE

The following is a brief list of some of the major process measurement and control requirements for the DSC process:

- Melt chemistry
- Melt temperature in the crucible and tundish
- Molten metal flow rate
- Melt level in the tundish
- Casting drum speed
- Casting drum temperature
- Casting drum surface geometry
- Casting drum surface chemistry
- Strip dimension
- Strip dimensions variation
- Strip temperature
- Strip microstructure
- After rolling strip dimensions.

REFERENCES

1. M. Rosen, "Ultrasonic Sensors for Characterization of Phase Transformations", Intelligent Processing of Materials and Advanced Sensors, H.N.G. Wadley, P.A. Parrish, B.B. Rath, and S.M. Wolf eds., p 29-45 (1987), The Metallurgical Soc., Inc.

2. J.M. Winter, Jr. and R.E. Green, Jr., "Real Time Monitoring of Microstructural Transformations Using Synchotron and Flash X-Ray Diffraction", Intelligent Processing of Materials and Advanced Sensors, H.N.G. Wadley, P.A. Parrish, B.B. Rath, and S.M. Wolf eds., p 61-75, (1987), The Metallurgical Soc., Inc.

3. R.C. Reno, A.V. Clark, G.V. Blessing, R.J. Fields, A. Govada, R.B. Thompson, P.P. DelSanto, R.B., Mignogna, and J.F. Smith, "Use of Neutron Pole Figures to Calibrate Ultrasonic Techniques for On-Line Texture Control of Al Plate", Intelligent Processing of Materials and Advanced Sensors, H.N.G. Wadley, P.A. Parrish, B.B. Rath, and S.M. Wolf eds., p 77-83, (1987), The Metallurgical Soc., Inc.

4. H.N.G. Wadley, "Sensors for Intelligent Processing of Materials", J. of Metals, October 1986, p 49-53.

5. R.A. Gleixner, "Process Control Considerations for Single Roller Direct Strip Casting of Steel Sheet", Presented at 1988 AISE Spring Conference, May 2-4, 1988, Baltimore, Maryland.

6. H.N.G. Wadley, "Sensors for Intelligent Processing of Materials", J. of Metals, October 1986, p 49-53.

7. D.B. Love and J.D. Nauman, "Controlling the Physical Mechanical Properties of Cast Stainless Steel Band", presented at the Near Net Shape Processing Symposium sponsored by the Pittsburgh Chapter ASM International on March 31, 1988 at Carnegie-Mellon University, Pittsburgh, Pennsylvania.

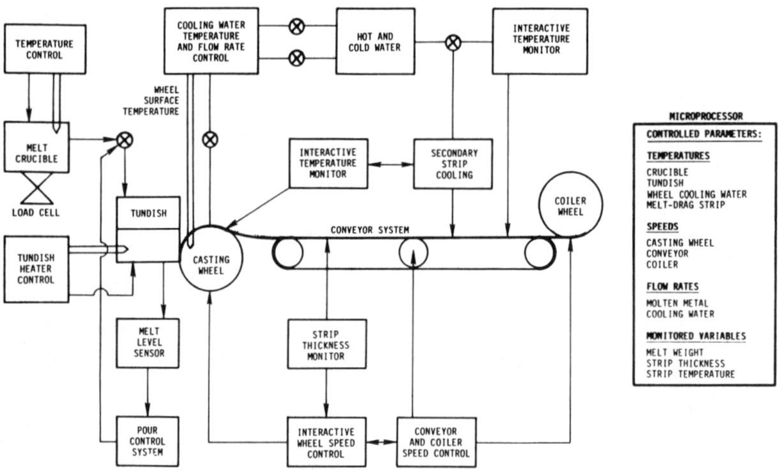

FIGURE 1. MELT-DRAG PROCESS CONTROL SCHEME

THE OHNO CONTINUOUS CASTING PROCESS AND ITS

APPLICATION TO NEAR NET SHAPE CASTING

A. Ohno

Department of Metallurgical Engineering, Chiba Institute
of Technology, Narashino, Chiba-ken, Japan
and
Department of Metallurgy and Materials Science, University
of Toronto, Toronto, Ontario, Canada

Abstract

This is a new process by which the molten metal is fed into a heated mold. With the heated mold, the formation of crystals on the mold surface is prevented and the solidification occurs without contacting the mold surface resulting in ingots with smooth, mirror-like surface and unidirectional grain structures. The products produced by the OCC process are free from cavity, porosity and segregation.

Any crystals formed at the beginning will decrease in number as solidification progresses and eventually the single crystal material is obtained.

The OCC process is well suited for production of wire, plate, tube and products having other cross sectional shapes of small dimensions which are very close to the final product size. It is a near net shape continuous casting process.

Casting of Near Net Shape Products
Edited by Y. Sahai, J.E. Battles,
R.S. Carbonara and C.E. Mobley
The Metallurgical Society, 1988

Introduction

Due to its distinct qualities, the Ohno continuous casting (O.C.C.) process drew the attention of several metal producers, and considerable interest from related industries. Since then, the OCC process has been applied commercially to cast single crystal wires and plates of metals and alloys, normally impossible to produce by the conventional continuous casting process (1,2).

In theory, the principle of the OCC process is simple. It uses a heated mold instead of a water-cooled mold, which in turn, gives several unique advantages over the conventional continuous process.

The aim of this paper is to review the fundamentals of the OCC process, and in particular to emphasize the unique features which are applicable to near net shape casting.

The Principles of the OCC Process

In 1971, the author investigated the origin of the equiaxed grain in castings by directly observing the solidification phenomena under the microscope. The results of this investigation provided sustantial evidence for the theory that the formation and subsequent separation of equiaxed grains occurs on the mold wall during the initial stage of solidification and prior to the formation of solid skin and columnar grains (3,4).

An attempt was made to eliminate equiaxed grains from castings by heating and maintaining the mold temperature above the solidification temperature of the metals. This idea was applied to the continuous casting process and subsequently led to the development of the OCC process.

The comparison between the conventional and the OCC process is shown in Figure 1.

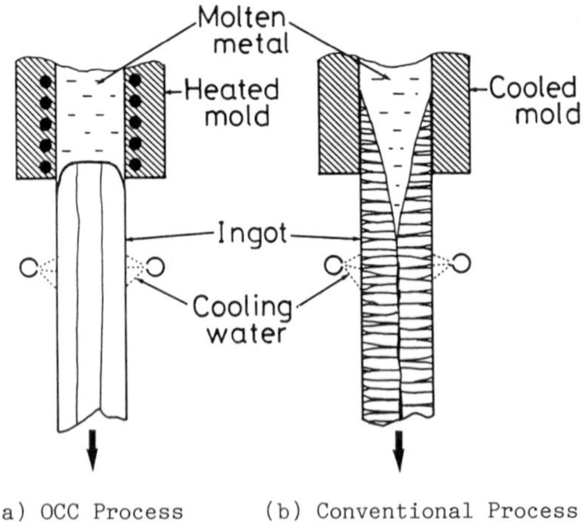

Figure 1 - Comparison of conventional casting process and the OCC process

The process which differentiates the OCC method from conventional continuous casting methods is the delivery of molten metal into a heated mold, the temperature of which is held just above the solidification temperature of the metal being cast. In this situation, the solid skin is formed only at the tip of the dummy bar at the start of casting, and, as the dummy bar travels away from the mold outlet, solidification progreses without contacting the mold wall, producing a unidirectionally solidified structure. Furthermore, the configuration of the solid-liquid interface is convexed into the liquid, as shown in Figure 1-a, leaving a thin peripheral layer of liquid film between the billet being cast and the mold outlet. The result is a very clean cast product with a mirror-like surface.

The Integrated Melting and Casting Facility of the OCC Process

The principle of the OCC process can be applied to horizontal, vertical and rotor-type casting methods. A detailed diagram of a horizontal OCC apparatus for experimental use is shown in Figure 2.

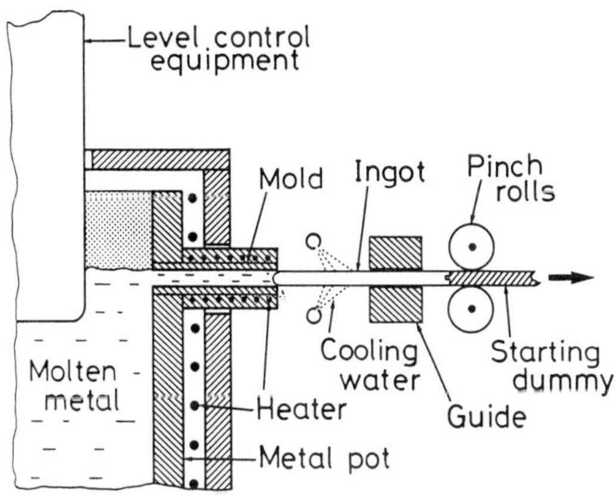

Figure 2 - A schematic diagram of a horizontal-type OCC apparatus

This method is well suited for the casting of very fine wires and plates. The apparatus consists of a melting crucible placed in an electric furnace, a heated mold, a cooling device, pinch rolls and a level control device to supply liquid metal into the mold.

To start up casting, the dummy bar is set at the mold outlet and the level control ramrod is lowered into the molten metal to raise the metal level to the desired height, so that a constant supply of liquid metal is delivered into the mold. As soon as the molten metal comes in contact with the dummy bar, water is introduced into the cooling device and the pinch rolls initiate and maintain movement of the dummy bar. The withdrawal and ramrod controls are set to the desired speeds taking into consideration such parameters as cooling capacity, mold temperature and melt temperature.

The horizontal-type OCC apparatus is shown in Figures 3, 4 and 5.

Figure 3 - Photograph of the horizontal-type OCC apparatus used for the casting of wire and having a capacity of 10Kg of copper.

Figure 4 - Horizontal casting unit used for plate production.

Figure 5 - Close-up view of horizontal casting unit used for plate production.

Unique Features of the OCC Process

The first of several unique features of the process is: that it is a near net shape casting process. It is extremely well suited for continuous production of wire, sheet, tube and products having other cross sectional shapes of very small dimensions, thus eliminating the need for costly high energy-consuming operations such as break down, rolling, drawing and subsequent intermediate annealing.

Since the mold is heated above the solidification temperature of the metal being cast, the molten metal flows into the mold despite the narrow dimensions of the mold cavity. As well, the metal solidifies out of contact with the mold wall, thus eliminating friction between the mold and the product being cast. The combination of these factors makes it possible to produce small castings.

Using the OCC method, it is possible to cast wire materials having dimensions as fine as 0.5mm, which can be used without any further drawing operation and with very little reduction in size (2). A few of the OCC products are shown in Figure 6.

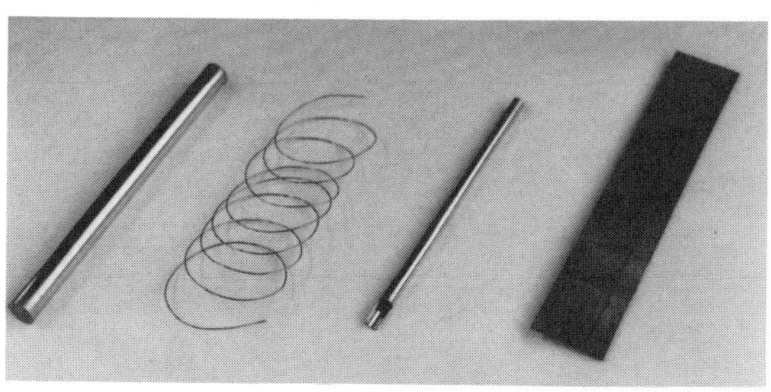

(a) (b) (c) (d)

Figure 6 - Examples of OCC Products
(a) 15 mm ø Cu billet, (b) 1.5 mm ø Cu wire
(c) 8 mm OD, 6 mm ID Cu tubing
(d) 1 mm thick sheet

Since this process has an applicability to a wide range of metals and alloys, it is also possible to directly cast wires, plates and tubes of metals and alloys which are very troublesome to roll or draw down to the desired size. For example, it can be used for the production of certain solder and brazing alloys which are normally very difficult to form using conventional casting methods.

The second unique feature of the OCC process is: that it enables the manufacture of products with a clean mirror-like surface. The casting conditions are arranged in such a way that a thin peripheral layer of liquid film solidifies just outside the mold outlet, so that there is no friction involved between the mold and the product being cast. This results in a cast product with an extremely smooth and shiny surface. Cast pure aluminum wire, having a diameter of 15mm, is shown in Figure 7. The wire exhibits an extremely clean, shiny surface which can be used in subsequent products as cast, eliminating the need for any chemical surface treatment or scalping operation.

Figure 7 - Pure aluminum wire having a diameter of 15 mm produced by the OCC process.

The third unique feature is that: the OCC process produces a single crystal. Since no grain growth occurs on the mold wall surface, any crystals formed at the tip of the dummy bar will decrease in number as solidification progresses, and will eventually become a single crystal material.

Figure 8 shows the macrostructure of a Sn plate indicating the reduction in the number of crystals and the subsequent formation of a single crystal.

Figure 8 - The formation of a single crystal

The continuous casting of a single crystal by the OCC process not only makes a significant contribution to the improvement of anti-corrosion properties of the product, but also enables crack-free production of very fine wires and thin foils (6), due to the elimination of grain boundaries which are susceptible to cracking and corrosion.

The fourth unique feature of the OCC process is that: it produces materials with excellent workability. For example, Figure 9 shows a sample of a 25μm Al-1%Si alloy wire, which was drawn down to its very fine size from a 20 mm ø billet without intermediate annealing.

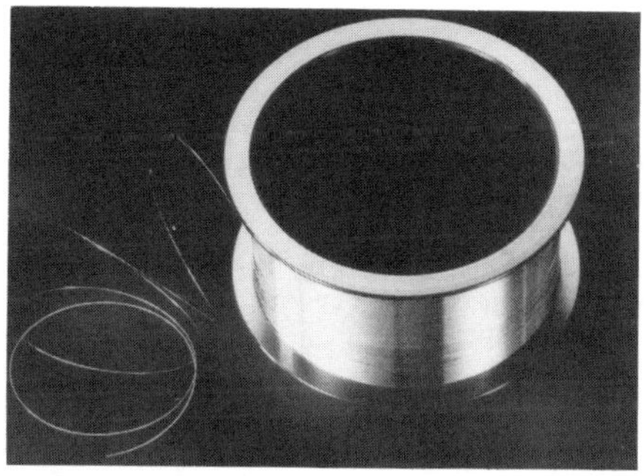

Figure 9 - 25μm Al-1%Si alloy wire drawn down from 20 mm ø billet without intermediate annealing.

Other OCC materials such as phosphorus bronze, Sn-30%Zn eutectic alloy and pure magnesium also show remarkably improved workability. For instance, the Sn-30%Zn alloy wire, was easily drawn down from 4mm to 100μm, and the phosphorus bronze plate was rolled down from 5mm to 100μm, both without intermediate annealing (7). The workability comparison between the OCC and conventionally cast materials of pure magnesium is shown in Figure 10.

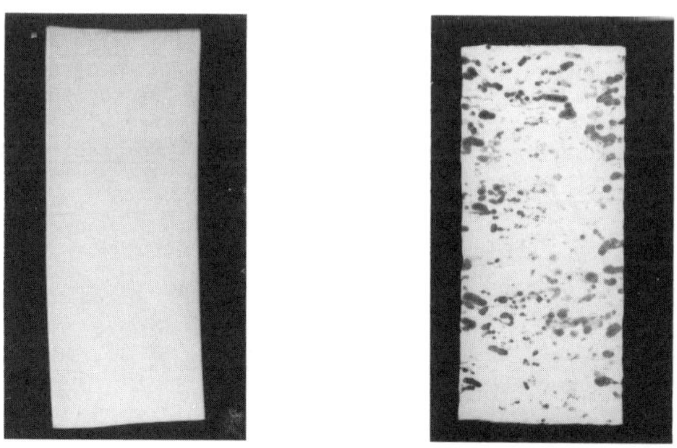

OCC Material Conventional Material

Figure 10 - Example of workability comparison of pure magnesium plates (30% reduction).

Both materials were reduced by 30% in thickness. The OCC material shows no sign of cracks, however the conventional material exhibits many surface cracks. The reason for this excellent workability may lie in the fact that the materials produced by the OCC process possess superbly clean surfaces, impossible to obtain by conventional casting methods, and are composed of either a single crystal or of several crystals of unidirectionally solidified structure which reduce the chance of intergranular fracture when being rolled or drawn down to size.

Summary and Conclusions

The principles of the OCC process and its unique features have been presented in this paper. The process is extremely well suited for the continuous casting of small-sized materials which are very close to the final product size, and elminates high energy consuming operations. As well, the materials produced by this process exhibit very desirable properties such as:

- a remarkably clean cast surface
- a composition of either a single crystal or several crystals of unidirectionally solidified structure
- excellent workability
- fewer casting defects
- reduced segregation

Based on these combined features, OCC process can be used for the production of fine wires and thin foils from "hard to form" materials, by casting them to small size first and then rolling or drawing to the desired size.

References

1. M. Shinohara, K. Nakano, A. Yamasaki and M. Shibata, "Copper Wire Produced by OCC Process," Furukawadenko Jiho, 79 (1986) 105-112.

2. K. Sawada, Y. Nakai, H. Shiraishi, T. Nakano, Y. Matsushita and M. Oshima, "Development of a Continuous Casting Single Crystal Copper Conductor," Sumitomodenki 131 (1987) 110-113.

3. A. Ohno, T. Motegi and H. Soda, "Origin of the Equiaxed Crystals in Castings," Trans. ISIJ 11 (1971) 18-23.

4. A. Ohno, T. MOtegi, "Formation Mechanism of Equiaxed Zones in Cast Metals," AFS International Cast Metal Journal, March (1977) 28-36.

5. A. Ohno, Japanese patent 1049146 U.S.A. patent 4,515,204.

6. A. Ohno, "Continuous Casting of Single Crystal Ingots by the OCC Process," Journal of Metals, Jan. (1986) 14-16.

7. A. Ohno, "Production of Single Crystal Materials by OCC Process," Sokeizai 12 (1986) 21-26.

THE EFFECTS OF MACHINE DESIGN ON POOL FEEDING,

STRAINS AND SHRINKAGE IN THE TWO-SIDED STRIP CASTING PROCESS

C. C. Gerding and E. Aukrust

LTV Steel Company, Technology Center
6801 Brecksville Road
Independence, Ohio 44131

Abstract

The strains incurred in forming a two-sided strip with a converging gap strip casting machine of the twin-roll type (e.g., the Bessemer) are analyzed. A new machine configuration is identified that minimizes these strains.

The feeding gap geometry and feeding dynamics resulting from the new arrangement is much more favorable, having the advantage of parallel, vertical mold sidewalls at the casting pool surface.

Across-the-width shrinkage of wide strip and the potential problem of longitudinal cracks has been studied and a critical experiment has been performed. The results show that this problem is resolved in the new process.

Casting of Near Net Shape Products
Edited by Y. Sahai, J.E. Battles,
R.S. Carbonara and C.E. Mobley
The Metallurgical Society, 1988

Introduction

LTV Steel over the last 20 years has made a substantial development effort in strip casting. At the AIME Spring Meeting in Toronto, an overview of this process development was presented with details on a 12-inch wide by 1/4-inch thick pilot plant operation using an inverted twin roll machine[1,2], a design in which the curvature of one of the rolls was concave instead of convex[3]. Conventional and inverted twin roll (ITR) configurations are shown in Figure 1.

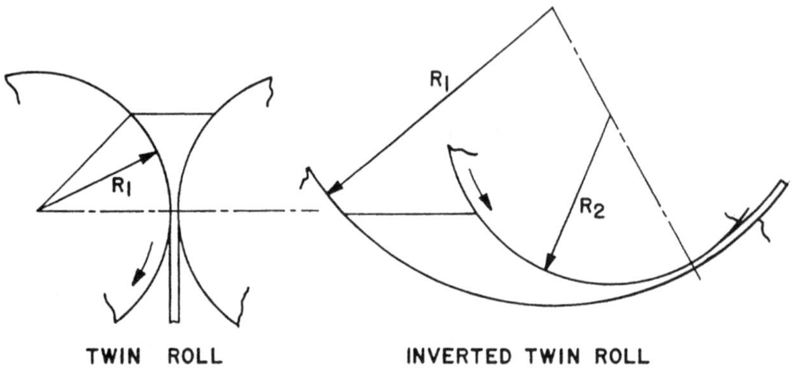

Figure 1 - The Bessemer configuration and its inversion.

This paper presents an analysis of some of the variables involved in any such two roll process and will show the principal advantages of the inverted twin roll in regard to minimizing strains in the strip at the critical "nip" and certain aspects of the feeding dynamics as well as providing some new experience and data in regard to the scale-up of the process.

Strains Incurred in the Nip

Consider the situation at the nip of a Bessemer twin roll machine shown in Figure 2. As shown in the vector diagram, the linear velocity at a point "B" on the lefthand sheet just before the nip is greater than that at a point "C" by an amount depending on the thickness of the sheet. However, when the two sheets are welded together and exit the machine, the linear velocities of "B" and "C" are the same. In welding the two sheets, the drum side is stretched and the free side is compressed so that the velocity of stretching adds to the outside of the composite strip and the velocity of compression subtracts from the velocity of the center. A bending deformation has to occur to accommodate the curvature change. These strains are much less in the inverted twin roll machine (see Figure 3). In this case there is no curvature change and thus no bending strain imposed on the bottom sheet at the nip. However, due to the minor unbending of the top sheet as point "C" must speed up, and "B" must slow down when passing through the nip while the velocities of "F" and "D" remain unchanged. The strains however are much less than in the conventional Bessemer design.

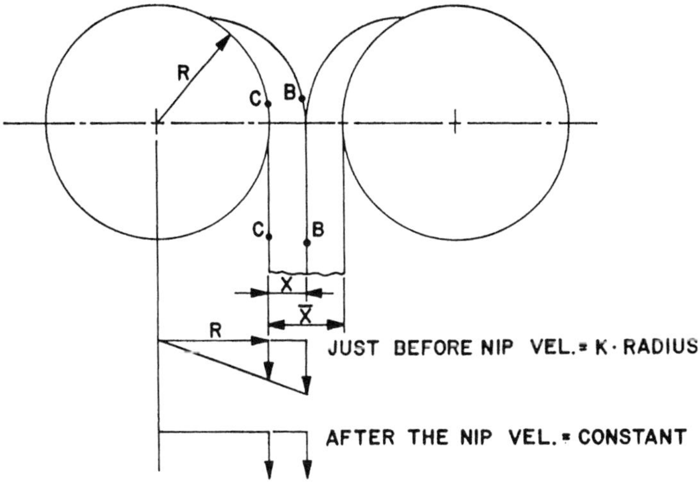

Figure 2 - Velocity changes at the nip due to unbending of the strip.

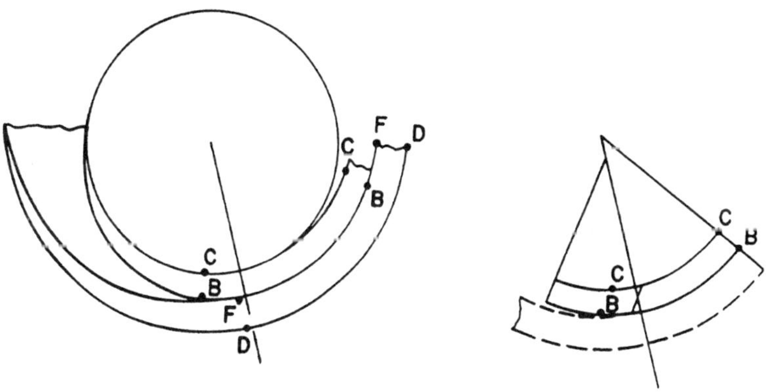

Figure 3 - Velocity changes at the nip in the top sheet of the inverted twin roll.

In Figure 4 we isolate a typical cross section of the Bessemer twin roll and note that plane strain conditions exist.

The strip may be thought to unbend about a neutral axis somewhere in the body of the strip -- that point of balance located such that the moment due to plastic stretching of material about this point on one side equals the moment from the plastic compression of the material from the other side. Since the material at the nip has a large center to surface thermal gradient, the material in the center is a lot hotter and weaker and the

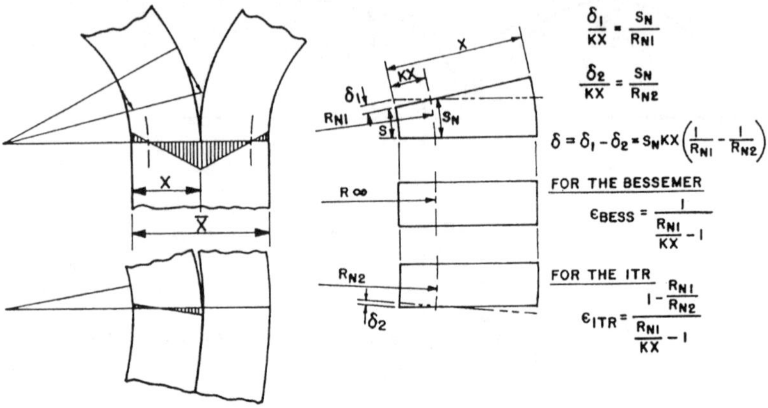

Figure 4 - Elongation and strain at the strip surface.

neutral axis is thought to be close to the cooler surface where the material is relatively strong.

The curvature of this neutral axis in the case of the Bessemer design changes from $1/R_{N1}$ to zero at the nip. In the ITR machine the unbending is only to the somewhat lesser curvature $1/R_{N2}$ and consequently the tensile strain is lessened considerably. The derivation of the strains for both the Bessemer and the ITR design are given in Appendix I and the resulting expressions are shown in Figure 4.

As the strip half thickness is small compared to the roll diameter in either machine, the radius to the neutral axis can be replaced by the roll radius in the calculations as shown in Figure 5. It is evident that the strain for the ITR becomes small as R_1 and R_2 become more nearly equal.

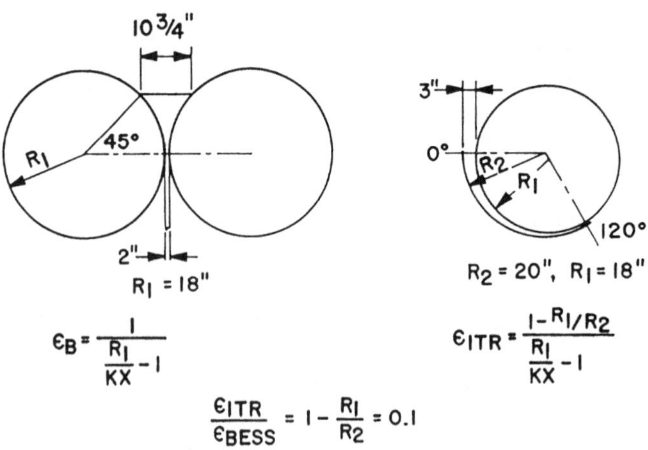

Figure 5 - Relative strain of the two machines.

For example, with the machine dimensions shown in Figure 5, the ratio of the strains of the two machines is 0.1, that is, the elongation of the surface of the Bessemer strip is 10 times larger than that of the ITR. Note that the ratio of the strains is independent of both the thickness of the sheet X and the location of the neutral axis K. (If "K" were 1/4th of a 0.1-inch sheet the twin roll strain would be 0.14% and the inverted twin roll strain would be 0.014%.) Decreasing the roll diameters would, of course, increase the strain.

Strain Rates in the Nip

We have so far considered how much the outside of the strip is stretched in passing through the nip, but not how fast this elongation occurs, so with this comparison of the relative strains in mind, we now look at the strain rates incurred in the two machines[4]. If we consider the extension of the top surface as in Figure 6 where the sheet is bent suddenly along a-a from R_1 to any larger radius R_2, it is seen that our simple picture predicts an infinite strain rate for any two roll machine.

Although the length along the neutral axis is unchanged by going through the nip, the top surface is stretched from S_1 to S_2 in time Δt. The apparent strain rate is as shown in the first equation of Figure 6.

The elongation of the material approaching the nip is not spread out uniformly but rather occurs locally at the nip. To illustrate this concept arcs θ_1 and θ_2 in Figure 6, are divided into 10 pieces and it is seen that only the piece just before a-a stretches into that just following a-a. That is, 1/10 of S_1 stretched to 1/10 of S_2 in 1/10 of the time Δt. This stretch was over the original length 1/10 of S_1, and the strain rate has increased tenfold.

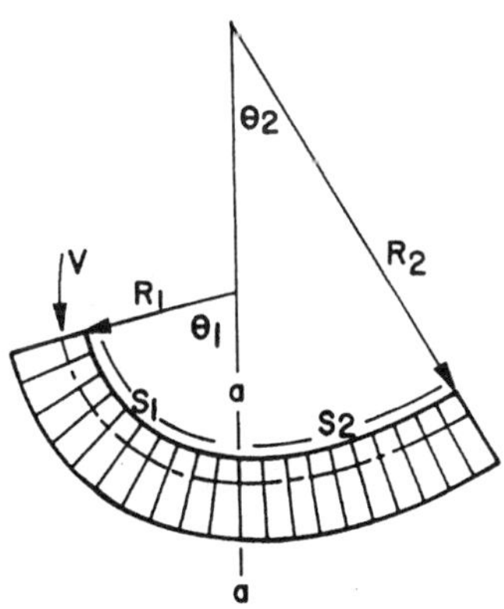

$$\frac{\Delta\epsilon}{\Delta t} = \frac{S_2 - S_1}{S_1 \Delta t} = \frac{R_2\theta_2 - R_1\theta_1}{R_1\theta_1 \Delta t}$$

$$\frac{\Delta\epsilon}{\Delta t} = \frac{R_2 \frac{\theta_2}{n} - R_1 \frac{\theta_1}{n}}{R_1 \frac{\theta_1}{n} \cdot \frac{\Delta t}{n}} = n\,\frac{R_2\theta_2 - R_1\theta_1}{R_1\theta_1 \cdot \Delta t}$$

$$\lim_{n \to \infty} n\,\frac{R_2\theta_2 - R_1\theta_1}{R_1\theta_1 \Delta t} \Rightarrow \infty$$

Figure 6 - Apparent infinite strain rate at the nip.

According to this model, the total strain occurs in the last segment. Since this can be arbitrarily small, the strain rate (and consequently the stress) is evidently infinite. An instantaneous change in velocity however small implies an infinite acceleration, so, all material properties aside, the forces and hence the stresses on a particle would be infinite. As this cannot happen, we suggest an alternate hypothesis for the behavior of the material at the nip.

Referring again to Figure 4, the notion of a neutral axis off center of the 0.1-inch sheet thickness does not conform to the conservation of matter, for if the strip is straightened about the neutral axis shown, a lot more material is being compressed than is being stretched, and no palpable compression of this material is possible since it is essentially plastic at these temperatures. In other words, Poisson's ratio is very nearly 1/2 and the volume is conserved. Also, the stretching of the outer fibers and the compression of the inner fibers in the direction of strip travel are accompanied by a strong tendency to contract and expand respectively in the orthogonal directions. This does not happen under plane strain conditions as long as the material remains continuous, which means that a plane element however defined may change in shape but not in area when going through the nip. In order that the velocity in the center of the strip may be slowed down, the wedge shaped elements entering the nip must be transformed into elements of constant thickness.

The surplus material which would otherwise be developed in the center of the strip just below the nip leads us to speculate that the strip thickness after the nip is slightly greater than the roll gap, so that the strip does not immediately straighten, but bulges slightly just after the nip and the curvature of the surface, although changing rapidly, varies continuously. Such a scheme would provide room for the extra material in the center, as well as softening the otherwise infinite strain rates.

However, the machine with inherently less strain should exhibit proportionally less strain rate, whatever the true mechanism is for mollifying the infinite rate suggested by any theory.

Reduction of the Strain Rate

The inverted twin roll design provides another degree of freedom for controlling the straining of the outer surface. Since the bottom strip is not bent as it comes through the nip, it is possible to overspeed the upper roll slightly so that a slight longitudinal compression is added to the top of the top sheet -- just enough to negate the tension due to the unbending, the greater top speed putting a little more top material into the nip so that the stretching velocity is not required. We speculate that a post-nip bulge would occur here also since there is still extra material that must go somewhere, although with any reasonably sized roll this bulge would be very small.

We have been following this practice of overspeeding the top roll at J&L and at LTV for some time and have never seen a regular pattern of longitudinal cracks on the upper surface due to unbending nor have we ever seen a pattern of transverse cracks due to the Poisson's ratio (lateral contraction) effect*.

---

* To operate a Bessemer machine in the same way would require that the sheet on the (say) right hand roll remained in contact with the underside of that roll for some distance after the nip and the left hand roll would be speeded up to compensate for the sheet from the left roll being bent twice as much (i.e., to twice the strain) to follow the right hand roll and be welded continuously to the other sheet.

Machine Productivity

Using the familiar freezing formula $X = a\sqrt{t} - b$ as a basis it is seen in Figure 7 that for equal casting lengths and therefore equal productivity the inverted twin roll is a much smaller machine.

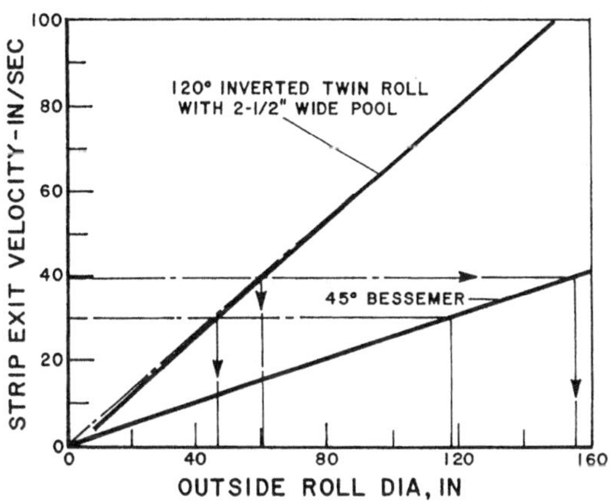

0.2-Inch Thick Strip Width
A = 0.1 and B = 0.025
in $X = A\sqrt{S/v} - B$

Figure 7 - Strip velocity as a function of roll diameter.

For example, the inverted twin roll machine with a 5-foot outside roll diameter produces 0.2-inch strip at 1 m/sec, while to achieve the same casting speed, a Bessemer requires a roll diameter of 13 feet*. For a strip speed of 30 in./sec an ~ 4-foot inverted roll or a 10-foot Bessemer roll is required. Formulae for the productivity of the several machines appear in Appendix II.

In the previous section we compared the strains in a Bessemer twin roll and an inverted twin roll with equal drum diameters. If we now compare the two on the basis of equal productivities, say 1 m/s, we find that the Bessemer strains are still 5-1/2 times those of the inverted twin roll.

Aspects of Machine Shape on Pool Feeding

Having considered the strains and strain rates of the two-roll machines in general, the geometry as it affects feeding of hot metal into the apparatus will be examined.

---

* The curve for the inverted twin roll does not start at the origin, but is asymptotic to one that does. This is because of non-symmetries and the fact that the thickness of strip being cast changes the casting length of the inner roll slightly.

Effect of the Sloped Surface

The sensitivity of the casting surfaces at the meniscus to wave action in the pool is an inverse function of the sine of the beach angle theta as shown in Figure 8, where a 45-degree slope amplifies local disturbances in pool height against a vertical wall by an additional 40 percent. Surface disturbances due to liquid metal turbulence, machine vibration, etc. will cause local casting length variations which will be reflected as slight variations in casting thickness. This is an unwanted condition and is best mollified by pool sides that are essentially vertical.

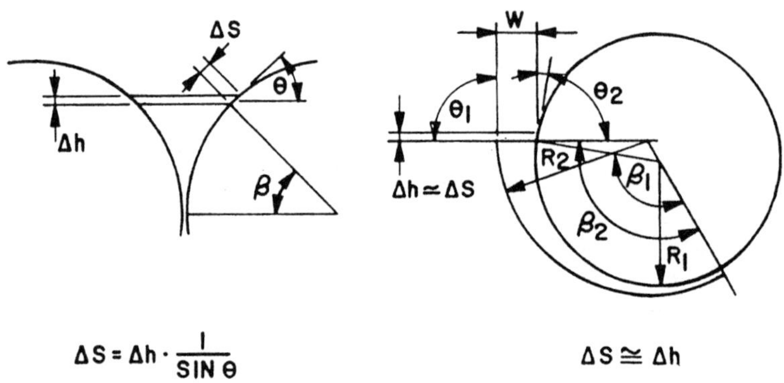

Figure 8 - Amplification of pool surface disturbances by beach angle.

In a Bessemer machine of a given roll diameter, casting length can only be gained by increasing pool depth which decreases the beach angle. Also, as pool depth is increased, pool cross-sectional shape is degraded as the edges of the pool become shallower, and the pool surface area is rapidly increased.

Generalized Twin Roll Design

The twin roll machine in its most general embodiment need not have rolls of equal diameter[5)]. In Figure 9 it is seen that this generalized machine and the generalized inverted twin roll have the same number of degrees of design freedom, i.e.:

r/R - The ratio of the two roll diameters,

θ - The angle of the line of roll centers with the horizontal (i.e., the direction of gravity), and

H - The elevation of the pool surface.

Still, the best twin-roll machine design appears to be the Bessemer where the two rolls are of equal diameter -- this gives the greatest beach angle for a given pool height, and the deepest pool under the metal stream to avoid metal impingement on the partly formed casting. Also more of the available casting surface is used in this arrangement.

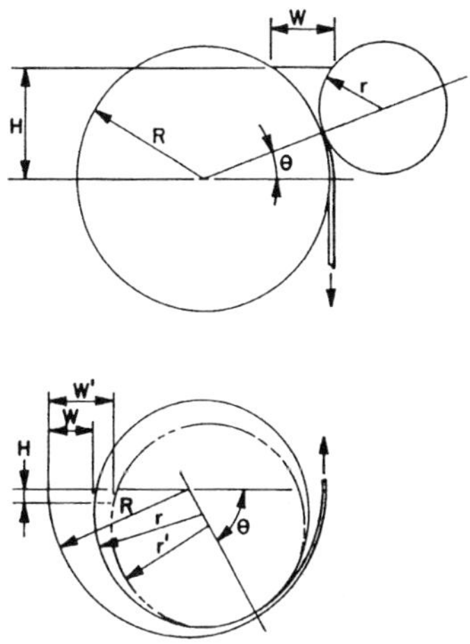

Figure 9 - Comparison of generalized twin roll and inverted twin roll casting machines.

In the inverted machine the two radii cannot be equal (because if they are, W goes to zero), but this ratio and the angle theta can be varied in such a way that the pool width W can be selected for best effect. Normally H is set to near zero for the best beach angle, θ is chosen for a reasonable ratio of cooling-to-heating time (say, 2 to 1) for the drum surface, and r/R is then selected for the desired value of W. By the inversion of one of the rolls in a twin-roll machine, the bottom of the strip can stay in contact with the lower roll for some distance after the nip. This gives additional cooling and strengthening to the strip.

Scale-Up of Width -- Lateral Shrinkage

So far we have been considering the cross section from a machine of indefinite width. The question naturally arises as to how wide a caster can be made without encountering shrinkage problems in the width dimension, in particular, longitudinal shrinkage cracks. In previous experimental work we have built inverted twin roll machines in one form or another in successively bigger sizes with widths up to 13-1/2 inches. However, we had seen no such cracks even though we might expect thermal shrinkages of up to 0.003 inch per inch of width during the casting process.

With a strip of essentially constant thickness, there are only two things left about the strip itself that can be scaled up; the speed of the strip (length per unit time) and the width. It was the width that presented some concern since the only wider strips we had ever cast were from a 24-inch dip stick where the edges were locked in and these strips had hanger cracks.

So, a critical experiment was designed to show whether frictional locking between the rolls or indeed <u>any</u> restraint across the width would cause cracking before the strip was out of the machine.

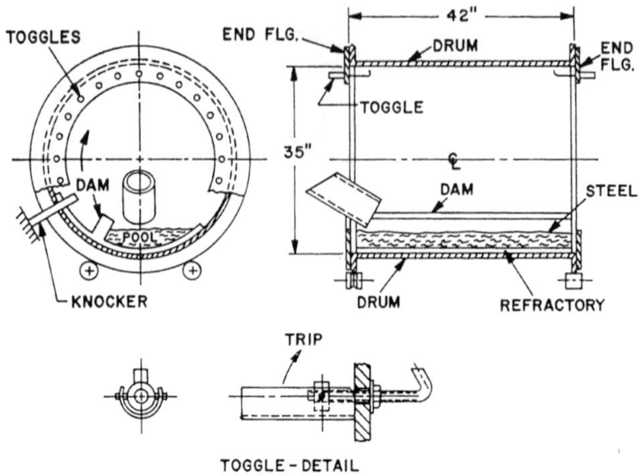

Figure 10 - One-sided casting drum with removable end restraints.

The experimental apparatus as shown in Figure 10 consisted of a hollow drum 36 inches OD x 35 inches ID x 42 inches long fitted with end flanges and with a pouring tube on one end. The drum was suspended on rollers so that it could rotate about its horizontal centerline. A portion of the inner circumference was covered with a layer of refractory which had a longitudinal dam in the center.

In operation liquid steel was poured into the apparatus from a 300-pound induction melting furnace to form the starting pool as shown. The drum was then turned through about a 300-degree arc at about 30-degrees per second during which time the pool rolled off of the starting refractory and along the bottom of the drum and cast a 42-inch wide strip. What was left of the pool then overran the other half of the refractory pad and the drum was stopped. The drum was spray-cooled from the outside and after cooling the 1/8-inch thick one-sided casting was removed from inside the drum.

The purpose of this experiment was to determine if the one-sided castings could be made with the ends of the casting being completely restrained from any across the width contraction during the time that the casting was being formed. To this end, a number of restraining hooks as shown in the figure were mounted in the end flanges of the apparatus. Each hook was fitted with a special toggle mounting so that the hook was held rigidly in place until such time as the toggle might be tripped by passing a carefully placed tripping bar. Once tripped the hook was perfectly loose.

As the drum turned during the casting the hooks dipped into the pool one-by-one and were frozen into the end of the casting. The tripping bar was so positioned as to trip the toggles at some short time after the hooks (now covered with rigid casting) emerged from the pool -- usually 1 second.

A number of runs were made with this device both with and without the tripping bar. Invariably the casting showed a wide longitudinal crack over its whole restrained length when the toggles were not tripped, but when they were, no such crack appeared. This was true over a range of casting speeds.

Photos of the apparatus are shown in Figures 11 and 12.

Our conclusion from these experiments was that complete end restraint of a casting however wide during its formation would not cause hanger cracks in the casting, and that cracks of this nature only appear if restrained during post nip cool-down of the solidified strip.

Figure 11 - Photo of the apparatus.

Figure 12 - Photo of the apparatus.

Summary

With this look at the gamut of possible twin roll machinery we conclude that the inverted twin roll shows strong advantages over any configuration in which both rolls are convex as indicated in the following table comparing machines of equal roll size:

	Symmetrical Twin Roll (Bessemer)			Inverted Twin Roll (ITR)		
Solidification Angle (ß, see Figure 8)	30°	45°	60°	90°	120°	150°
Beach Angle (Θ, see Figure 8)	60°	45°	30°	Over 80° All Cases		
Relative Strip Speed for Equal Roll Diameter	1.0	1.5	2.0	3.0	4.0	5.0
Relative Strain (see Figure 4)	$(1/R_1)$*			$(1/R_2 - 1/R_1)$		
	0.033			0.002		

* R_1 = 30 inches, R_2 = 32 inches.

The advantages of the ITR summarized above speak for themselves and we believe that castings in the 1/4-inch thickness range produced in an inverted twin roll embodiment of any desirable width represent the most viable path to the attainment of strip casting as an efficient and relatively inexpensive tool for the production of direct rolled hot band.

References

1. C. C. Gerding, E. A. Mizikar, and E. Aukrust, "Progress Report on LTV's Unique Strip Casting Process," 71st Steelmaking Conference, Iron and Steel Society of the AIME, Toronto, Ontario, Canada, April 18, 1988.

2. E. Aukrust, E. A. Mizikar, and C. C. Gerding, "Recent Progress in a Unique Strip Casting Process," AIME 4th Advanced Technology Symposium on "Near Net Shape Casting," Myrtle Beach, South Carolina, May 10, 1987.

3. C. C. Gerding, "Direct Casting of Channel-Shaped Strip," U.S. Patent No. 3,773,102, November 20, 1973.

4. Eisuke Niyama, et al., "Belt-Wheel Type Continuous Caster for Steel Billets," Iron and Steel Engineer, August 1981, pp. 39-40.

5. C. W. Hazelett, "Strip Forming Apparatus," U.S. Patent No. 2,450,428, October 5, 1948.

Appendix I

In Figure 4 if the wedge shaped element is flattened we see by a simple proportion that

$$\frac{\delta_1}{KX} = \frac{S_N}{R_{N1}} \tag{1}$$

So that for the Bessemer the elongation at the surface is

$$\delta_1 = \frac{KX\ S_N}{R_{N1}} \tag{2}$$

If the flat element is bent back into a less acute wedge

$$\frac{-\delta_2}{KX} = \frac{S_N}{R_{N2}} \qquad \text{or} \qquad -\delta_2 = \frac{KX\ S_N}{R_{N2}} \tag{3}$$

The net elongation at the surface is

$$\delta = \delta_1 - \delta_2 = KX\ S_N \left(\frac{1}{R_{N1}} - \frac{1}{R_{N2}}\right) \tag{4}$$

Since the strain is δ/S, and since again by simple proportion

$$\frac{S}{R_{N1} - KX} = \frac{S_N}{R_{N1}} \tag{5}$$

It is easily shown that

$$\varepsilon = \frac{1 - \dfrac{R_{N1}}{R_{N2}}}{\dfrac{R_{N1}}{KX} - 1} \qquad \text{for the inverted twin roll} \tag{6}$$

or, if as for the Bessemer $R_{N2} \to \infty$

$$\varepsilon = \frac{1}{\frac{R_{N1}}{KX} - 1} \tag{7}$$

Therefore if RN1 is the same for Bessemer and ITR,

$$\frac{\varepsilon_{ITR}}{\varepsilon_{BESS}} = 1 - \frac{R_{N1}}{R_{N2}} \tag{8}$$

Appendix II

Given the usual form of freezing equation with X = thickness, t = time

$$X = a\sqrt{t} - b \tag{1}$$

where a and b are constants, the thickness of a two-sided casting for a pure metal may be described by:

$$\bar{X} = X_1 + X_2 = a\left(\sqrt{\frac{S_1}{v}} + \sqrt{\frac{S_2}{v}}\right) - 2b \tag{2}$$

where S_1 and S_2 are casting lengths and v is casting velocity. If $S_2 = \eta S_1$,

$$v = \left(\frac{a}{\bar{X} + 2b}\right)^2 \cdot S_1 \cdot (1 + \sqrt{\eta})^2 \tag{3}$$

In the Bessemer machine $\eta = 1$, in the ITR $0.9 < \eta < 0.95$ usually.

In the Bessemer of roll radius R and casting angle ß,

$S = R\beta$	casting length
$H = R \sin \beta$	pool height above nip
$W = 2R(1 - \cos \beta) + \bar{X}$	width of pool

$$v = \left(\frac{a}{\bar{X} + 2b}\right)^2 \cdot 4 \cdot R\beta \qquad \text{(see Figure 8).} \tag{4}$$

The ITR is somewhat more involved. If the radii and angles are R_1 and β_1 (inner roll) and R_2 and β_2 (outer roll) (see Figure 8) then, if W, β_2 and R_2 are known it can be shown that

$$R_1 = \frac{2\, R_2\, (1 - \cos\beta_2)\, (R_2 - \bar{X} - W) - 2\, W\, \bar{X}\, (\cos\beta_2) + \bar{X}^2 + W^2}{2\, [W\, (\cos\beta_2) - \bar{X} + R_2\, (1 - \cos\beta_2)]} \tag{5}$$

$$\beta_1 = \pi - \sin^{-1}\left[\frac{R_2 - W}{R_1}\,(\sin\beta_2)\right] \tag{6}$$

then

$$v = \left(\frac{a}{\bar{X} + 2b}\right)^2 \cdot \left[\sqrt{R_2\,\beta_2} + \sqrt{R_1\,\beta_1}\right]^2 \tag{7}$$

LUNCHEON LECTURE

Introduction by: **R.S. Carbonara**
SEA Inc.
Worthington, Ohio

NEAR NET SHAPE FORMING OF RAPIDLY

SOLIDIFIED METALS AND ALLOYS

Nicholas J. Grant

Department of Materials Science and Engineering
Massachusetts Institute of Technology
Cambridge, MA 02139

Abstract

The need to conserve energy at every step of the steelmaking process, while maintaining quality, has focused heavily, in particular, on the continuous casting processes which are becomming a bigger and bigger fraction of total steel output. But many of the emerging processes offer significant improvements in quality as well as in cost and energy utilization. The interest and activity in thin slab casting (50mm: 2 inches thick), and in thin strip casting (1.5 to 3mm thick) are prime indicators of such interest and activity. Avoidance of the casting step by substituting spray forming (atomization and deposition against a metallic substrate for more rapid cooling), as demonstrated by the Osprey and Liquid Dynamic Compaction Processes, offers highly improved initial and final structures, much improved properties, and also faster, cheaper, energy saving processes. These developments are examined and discussed in this presentation.

Casting of Near Net Shape Products
Edited by Y. Sahai, J.E. Battles,
R.S. Carbonara and C.E. Mobley
The Metallurgical Society, 1988

Continuous Casting

A look at international steel production trends confirms major moves to thinner cast sections in contrast to current general thicknesses of 7-9 inches. Current emphasis has been on 50mm (2 inches thick) thin slabs with an aim of direct hot rolling, and on thin cast strip (thickness values from about 1.5 to 3mm), intended for cold rolling. In spite of resultant dendritic crystallization patterns with attendant preferred orientation and related problems with porosity and dendritic segregation, the move to thinner cast sections has merit and offers considerable improvements in overall costs; in simplicity of rolling equipment; and in improved structures and properties in the final products.

Both for the thin slab casting as well as the thin strip casting, liquid metal is delivered to the "freezer" unit; accordingly, all aspects of processing, conversion and utilization are based in liquid metal technology.

In an effort to further define and classify conventional continuous casting and the two current new developments, clearly current continuous casting processes cannot be classified as rapid solidification processes. Thin slab casting at 50mm thickness can not meet an arbitrary minimum designation of 10^2 °K per sec. (K/s) frequently used as the minimum rate to separate rapid solidification processes from other (slower) solidification processes. Thin strip casting, on the other hand, may, at best, approach the above solidification rate, but current solidification rates are not well reported. Further, 3mm thickness need not be the upper limit for the thickness of thin continuously cast strip.

In question at this early stage of development is whether hot direct rolling of thin slabs is possible. Depending on the width of the slab and the transfer distance to the rolling stand, it seems unlikely that a uniform temperature can be achieved with natural cooling (for example, without added heating to keep the slab edges from cooling too rapidly).

Concerns are also raised regarding the amount of hot reduction necessary to break down the as-cast oriented dendritic structure. With low carbon, plain carbon steels this may not be too severe a problem, but with increasing alloy content and alloy complexity, refinement and dispersion of phases will present major problems.

As regards thin strip casting intended for cold rolling, the above problems are bound to be magnified. Starting with 1.5 to 3mm (0.06 to 0.12 inch) thick cast strip, using the preferred (for now) twin roller technique, the resultant dendritic solidification patterns and necessary dendritic breakdown and grain size refinement and homogenizations through cold work, with limited reductions, seem formidable tasks. For alloyed steels, the problems are significantly more serious.

The twin roller strip casting technique offers among the fastest solidification processes but considerably slower solid state cooling rates, depending on the selected roll speed and desired thickness of strip. A significant number of critical parameters must be taken into account; they are:

(a) Roll diameter
(b) Roll speed
(c) Quench capacity of the rolls
(d) Depth of the poured liquid metal puddle (metal delivery rate)
(e) Stability of the melt puddle: no splash, no surging, and a sharply controlled melt puddle height
(f) Temperature of the melt puddle and ability to control closely
(g) Roll gap and its precise control over a significant time period.

As indicated in Figure 1, the interplay of the above variables will determine when the strip is fully solidified. If the center of the strip is molten on entering the minimum roll gap, roll pressure will result in a smaller thickness. If the strip is fully solidified on entering the roll gap, there will be spring back and a thicker strip will emerge. The variation in strip thickness will of course depend on the continuous reproducibility of the solidification rate in line with a number of the above variables. If condition B prevails (premature through thickness solidification), soft rolls such as copper or ferritic, low carbon alloy steels will undergo deformation in view of the very high temperatures of the strip and the resultant roll pressures.

Control of the twin roller system is not a simple task and becomes more difficult to control with increasing strip width and thickness. In part this is due to the dendritic, columnar solidification, wherein surface crystallization is strongly influenced by several uncontrolled variables and the rate of dendritic columnar growth is influenced by the surface crystallization.

Other potential strip solidification processes based on liquid metal delivery are the melt overflow process and the melt-drag process, both of which have been studied and tried a number of times for steel strip production.

To its advantage, the twin roller technique offers among the fastest solidification processes, but somewhat slower solid state cooling rates because the strip loses contact with the cooled twin rolls, as seen in Figure 1. Edge definition has been a problem, as has end leakage for the wider strip widths.

It is not the purpose here to throw negatives at either the thin slab or thin strip processes; instead it is intended to point up areas of expected (experienced) processing problems so that comparisons can be made against other developing continuous strip, sheet and plate forming processes.

Whereas the twin roller thin strip continuous casting process is currently being tested for production potential, single wheel melt spinning has already received significant attention over many years (2-4). The aim was to produce net shape thin strip in plain carbon, low carbon steels. The effort was to parallel the successful production of glassy alloy ribbon and strip in very thin sections, typically less than about 50μm (0.002 inch) (2-4). Thin crystalline strip was melt spun with varying degrees of success (5). Top surfaces tended to be good to excellent but bottom surface (substrate contact side) showed very fine, abundant defects as small, shallow gas pockets. Not unexpectedly the crystalline ribbon and strip is dendritic in structure, with a finer grained chill zone at the copper wheel surface.

Efforts to melt spin ribbon and strip up to 0.5 to 1mm in thickness simply lead to longer dendrites, generally encompassing the thickness of the ribbon. The combination of the oriented grain structure with long dendrites and fine dendritic porosity did not promise useful properties or applications. Figures 2 and 3 show typical dendritic melt spun structures. Figure 2 also shows the chill (substrate) zone; Figure 3 shows breakdown of the columnar dendritic structure into a cellular dendritic structure due to the imposition of an electrical field during pouring (5). The structural transition sharply varies from the very fine chill (substrate) zone of very fine dendrites to the much coarser cellular structure over a distance of about 500μm.

Rapidly Solidified Powder Metallurgy Processes (RSPM)

The technologies for producing fine, rapidly solidified powders in an extremely broad range of alloy types (Al, Mg, Cu, Fe, Ti, stainless steels, super alloys, etc.) and of compositions varying from low alloy content to extremely complex, high alloy content materials, is well documented (6-12). Solidification rates from about 10^1 to 10^5 K/s, or even slightly higher rates are well documented (13). Grain sizes in the submicron range are readily achieveable in many alloys after appropriate hot working (14). Consolidation to fully dense bar, plate, shapes, sheet, strip, etc. has been demonstrated, often with truly outstanding structures and properties (8-12). Hot extrusion, hot forging, hot isostatic pressing (HIP) to full density, to be followed by other forms of hot and cold work have all been used successfully.

Unfortunately, at least two important problems must be addressed in the RS-PM area. First, the relatively fine powders (1 to 250μm), even when produced in an inert atmosphere, for economic reasons are subsequently screened and canned in air to prepare the powders for production of suitable forms for HIP or hot working processes. These are slow expensive steps. Further, extrusion presses of suitable size for large tonnage operations are not available, and HIP units able to handle single canned compacts in excess of several thousand pounds have only recently appeared on the industrial scene. Single canned compacts up to 20,000 lbs. have now been HIP processed, and are suitable for hot forging or pressing or extrusion to full density. The largest HIP unit now in operation for such large canned compacts has a working I.D. of 60 inches (1.5 meters) (15). However, the cost of HIP processing qualifies, preferably, the higher alloy materials which in general are poorly hot workable in cast form (16). Strip, sheet and plate of more common grades of alloys are not processed today via the RSPM route because of cost considerations.

Spray Deposition Processes

In sharp contrast to the continuous casting processes is the newer emerging technology of spray atomization and collection, which is also referred to as Spray Deposition. Such processes are identified as the Osprey Process (17,18), the CSD Process (Consolidated Spray Deposition) (19,20) and LDC (Liquid Dynamic Compaction) (21-25). These processes are not labeled as continuous casting processes, but as continuous (spray) forming processes, resulting in totally different solidification phenomena, as-deposited densities, microstructures and initial and final properties. The first efforts to utilize spray deposition processes for sheet production were reported by Singer in 1960 (26), with fair to good results considering the early stage of rapid solidification technology and atomization techniques in general. Porosity, perhaps one of the worst problems, ranged from 15 to 20% and rolling had to be done very soon thereafter, or the strip had to be maintained in a suitable protective atmosphere to avoid internal oxidation. A 50% reduction of thickness appeared necessary to increase the density to about 95%, which resulted in sealing of the pores against oxidation.

Pilot scale operations were reported for the CSD Process with suggestions of success, however with an acknowledgement as well that the atomized spray was quite coarse, ranging up to 0.5 to 2mm diameter droplets. This frequently led to excess liquid metal accumulation within the strip with the formation of coarse grained zones and dendritic crystalline growth through normal solidification.

The Osprey Process, which initially concentrated on producing preforms suitable for forging (17,18), was more recently applied to sheet and plate forming, a development currently under study on a pilot scale.

In the steel production world, continuous casting is concerned with production rates of 100 or more tons per hour. In spite of major time and energy savings over ingot processes, there are still problems of slab reheating, breakdown of coarse, oriented dendritic structures, segregation, porosity, etc., in continuous casting. It is no surprise to find major interest in thin (50mm, 2 inch thick) slab casting, using liquid metal delivery to a freezing unit as a means of further refining the structure, improving the hot working properties and further improving mechanical properties, while still working at high tonnage rates. Thin slab continuous casting has as a primary aim the ability to directly hot roll, without intermediate heating between the caster and the roll stand.

A still more ambitious plan is the continuous production of thin strip in the 1.5 to 3mm thick range, intended for cold rolling. As noted earlier, the emphasis currently is on the twin-roller process to produce the thin strip.

For wide or relatively wide thin slabs and thin strip, using liquid metal delivery, production rates can be maintained at the 100 tons per hour, plus or minus, level. For the Singer (26), CSD (19,20), and Osprey (17,18) spray deposition processes noted above, circular atomization nozzles were used with metal stream diameters of about 5mm. This typically leads to metal delivery rates of about 1 to 2 tons per hour, far short of the 100 tons, plus or minus, per hour prevalent in current continuous casting of steel.

Accordingly, research was undertaken to develop a linear atomization nozzle (27) using geometries prevailing in the Kohlsva ultrasonic gas atomization (USGA) nozzle (28), combining atomization variables developed in the author's laboratory over the past 10 to 15 years (29-32). The linear atomization nozzle is extendable to permit metal delivery rates for atomization from 5 to 100 tons per hour. Calculations and laboratory trials indicate that a 50mm long slit, 2mm wide can provide atomization rates of 5 to 10 tons per hour. Multiples of this length indicate 50 to 100 tons per hour for nozzle lengths of about 500mm.

The other advantages of the linear nozzle over circular nozzles are an adjustable included angle between the opposite halves of the atomizer pair; more freedom to position the gas jets (number of rows, spacing of rows, etc.); and easier manufacture of the nozzle with its large number of jets.

An obvious advantage of a linear nozzle is the even, transverse profile of strip, sheet or plate produced during deposition. A circular nozzle characteristically delivers more metal (and somewhat coarser than the average) at the center of the as-deposited product, as shown in Figure 4. To minimize this uneven distribution of the atomized spray, experiments have been run with the atomizer in an "oscillating mode," that is, operation as a "scanning" atomizer which atomizes the stream and deflects the spray in a wider arc to try to achieve greater lateral uniformity (26). For wide strip or sheet more than one scanning atomizer can be used while striving to obtain uniform overlaps of the separately sprayed nozzles.

A possible linear atomizer arrangement is shown in Figure 5. Deposition will continue to be heavier along the length at the center of the spray, but on a moving belt, wheel, or a moving linear substrate, the deposition is averaged out and a uniform linear strip or sheet results.

Thickness of strip, sheet or plate depends on the rate of metal spray delivery and substrate movement, however one can increase thickness by combining several spray deposition stations. Alternatively, cladding or surface coating can also be accomplished by using different alloy compositions.

Current Continuous Thin Strip and Sheet Forming Developments

In a laboratory scale study using a moving substrate, atomizing with both the circular USGA nozzle and with a 50mm long by 2mm wide nozzle, plates measuring about 9 inches (230mm) in width by 14 to 16 inches (350-400mm) in length and of plate thicknesses varying between about 3 and 12mm in thickness are being produced of complex aluminum alloys, 1020 steel, 4130 steel and a 316L stainless steel. Melt size is only about 10-12 lbs. in the case of steel processing.

Processing variables include metal superheat temperature, gas pressure (therefore atomized droplet size and velocity), gas to metal flow rates, distance of spray travel to the substrate, and substrate velocity. The aim is to have a narrow droplet size range, a minimum of minus 10μm droplets (which solidify in flight for our conditions), and a minimum of +250μm droplets. With the USGA nozzle configuration, staying in this range of droplet sizes appears quite practical.

A number of substrate materials and substrate surface finishes have been evaluated. The extremes of choices included low, plain carbon ferritic steels and copper, with and without water cooling. A spectrum of shot peening and grit blasting finishes have been studied. A positive aim is to achieve rapid solidification, in which case water cooling was a desired variable. Copper, an excellent conductor, is a high expansion metal, is soft and easily deformable, and, relatively, an expensive material. Simple ferritic steels are low expansion materials, good thermal conductors, quite inexpensive and much more wear and deformation resistant than copper.

To serve as effective substrates, the as-deposited strip, sheet or plate must remain in contact through the solidification interval to guarantee a fine, uniform grain size; and must remain in contact during solid state cooling to obtain uniform precipitation, or uniform structure if phase transformations occur (austenite to ferrite), and to avoid peeling or warpage of the as-deposited product. In this respect, grit blasted, water cooled simple ferritic steels have performed well.

As-Deposited Structures

An as-deposited LDC plate of 1020 steel, bottom and top surfaces, is seen in Figure 6, deposited against a grit blasted ferritic steel (water cooled) substrate. Figure 7 shows the substrate side of another LDC plate with an oustandingly excellent surface.

Figure 8 shows typical microporosity in 1020 steel, as-deposited, at 82X, with a measured density of about 95%. Figure 9 shows the as-deposited grain structure: non-dendritic, non-directional, fully recrystallized ferrite plus pearlite. The ferritic grain size varies from about 10 to 40μm and is relatively uniform throughout the LDC deposit.

Figure 10 shows the as-deposited LDC structures of an austenitic 316L stainless steel formed against an uncooled ferritic steel substrate, with a grain size which varies between 10 and 50μm, uniformly throughout the plate.

Hot and Cold Rolling

Rolling of the LDC as-deposited 1020 steel plates was done primarily at 700°C (ferrite plus carbide) and 1050°C (austenitic); some of the higher density deposited plates were also rolled at room temperature.

When as-deposited densities were about 94-96% or more, rolling at 20°, 700° or 1050°C produced strip which was completely free of edge cracking. When as-deposited densities were as low as 88-92% rolling at 20°C was not practical; however rolling at 700° and 1050°C could be achieved with some edge cracking in the first 20 to 30% reduction. Removal of the cracked edges then permitted unlimited further reductions without significant edge cracking. Table I lists densification trends for rolling at 700° and 1050°C. Full densification and totally refined structures are achieved after hot reductions of about 50% (2:1), in contrast to reductions for ingot derived materials with 35 to 50:1, and somewhat less for continuously cast thick slabs.

Rolling at 1050°C resulted in a fully recrystallized fine grain size of about 5 to 10μm, whereas rolling at 700°C resulted in a cold worked, elongated grain size which in cross section was 2 to 4μm diameter and 7 to 10μm in length. Typical microstructures are shown in Figure 11.

The 1020 LDC hot and cold rolled steels responded to quench and temper treatments, normalizing treatments and annealing, yielding a broad range of strength and ductility values at very attractive levels.

Mechanical Properties

Table II lists tensile properties for several rolling and heat treating conditions for the 1020 LDC produced steels. In examining Tables I to V, it should be remembered that hot rolling at 1050°C in the austenitic field results in a fully annealed (soft) condition, whereas "hot" rolling at 700°C, in the ferrite plus carbide two phase field, leaves the alloy in a partially cold worked state, the degree of cold work depending on the final reduction and the final thickness of the strip or sheet. This can be seen in Figure 11. On this basis one can expect a fairly wide range of yield and tensile strengths, as well as ductility, for alloys rolled at 700°C. This is quite apparent from the results listed in Table III.

Table II shows that subsequent annealing (at 1050°C) or normalizing (at 700°C) leads to highly reproducible yield, tensile, and ductility values in a series of LDC rolled plates. Annealing or normalizing of the 1050°C rolled product (which is in an annealed state) results in significantly lower strength values than similar heat treatments of the 700°C rolled product. Further, normalizing either hot rolled material (1050° or 700°C) results in superior strength properties over annealed material, with no penalty in tensile elongation values. The best combinations of strength and ductility, at excellent levels of yield and ultimate, are achieved by 700°C normalizing of the 700°C rolled LDC plates. The elongation values are for a gage length of one inch or less.

Table III compares tensile values for product rolled at 700°C and 1050°C, and normalized in each instance at 890°C, just into the austenitic field. Note the variability of yield, tensile and elongation values for the 700°C rolled product, which is due to the residual cold work retained by the strip on cooling to room temperature. Normalizing at 890°C results in excellent combinations of yield strength (54 to 64 ksi), ultimate (81 to 84 ksi) and ductility values (gage length of 1 inch or less) of 24 to 28%. The normalized, austenitic rolled product shows values only slightly lower.

Table IV illustrates the softening of the 700°C rolled product, with resultant decreases in strength and increases in ductility. The first sharp decrease in yield strength occurs on annealing at 600°C with further large

Table I. Densification of LDC As-Deposited 1020 Steel Rolled at 700° and 1050°C

Material	Final Amount of Reduction, %	Total Reduction, Percent	Density g/cc	Percent of Full Density
Run 21F		0	6.98	88.8
700°C		50	7.80	99.0
	52	76	7.87	100
Run 22F		0	7.16	91.0
700°C		29	7.48	95.0
		50	7.84	99.6
	55	78	7.87	100
Run 23F				
(top), 700°C		0	7.60	96.5
		33	7.79	99.0
		53	7.85	99.8
	50	77	7.87	100
Run 23F		0	7.24	92.0
(bottom),		33	7.81	99.2
700°C		59	7.84	99.6
	43	77	7.87	100
Run 22F		0	--	--
700°C		52	7.78	98.9
	44	75	--	--
Run 22A		0	--	--
1050°C		27	--	--
		46	--	--
	56	76	7.85	99.8
	32.5	84	7.87	100
Run 22A		0	7.06	90
1050°C		34	7.84	99.7
		52.6	--	--
	39	71	7.87	100
Run 23A		0	7.20	91
1050°C		37	7.84	99.7
		58	--	--
	12.7	68	7.87	100

Table II. Representative Tensile Data for Hot Rolled, Heat Treated 1020 LDC Steel Strip

Alloy	Condition	Y.S., ksi Upper	Y.S., ksi Lower	UTS ksi	Elong.* %
11	Rolled 1050°C: 55% Reduction	53+		78	22
18	Rolled 700°C : 60% Reduction	61	60	89	10
18	Rolled 1050°C: Annealed 1050°C	41	39	55	20
19	Rolled 1050°C: Annealed 1050°C	34	36	55	--
20	Rolled 1050°C: Annealed 1050°C	55	49	70	20
18	Rolled 1050°C: Normalized 700°C	47	44	65	20
19	Rolled 1050°C: Normalized 700°C	49	46	64	21
20	Rolled 1050°C: Normalized 700°C	55+		80	21
18	Rolled 700°C : Annealed 1050°C	39	40	59	21
19	Rolled 700°C : Annealed 1050°C	39	40	58	21
20	Rolled 700°C : Annealed 1050°C	--	--	58	21
18	Rolled 700°C : Normalized 700°C	54+		77	22
19	Rolled 700°C : Normalized 700°C	48+		69	21
20	Rolled 700°C : Normalized 700°C	56+		84	23

* Elong. in one inch or shorter gage length
+ No upper-lower yield

Table III. Tensile Properties of As Hot Rolled and Hot Rolled and Normalized 1020 LDC Steel Strip

Alloy	Condition	Y.S. ksi	UTS ksi	Elong.* %
21	As Hot Rolled at 700°C	95	126	7.3
21	Hot Rolled 700°C, Normalized 890°C	58	84	25.
22	As Hot Rolled at 700°C	117	123	4.6
22	Hot Rolled 700°C, Normalized 890°C	59	81	28.
23	As Hot Rolled at 700°C	121	124	5.8
23	Hot Rolled 700°C, Normalized 890°C	54	83	24.
23	As Hot Rolled at 700°C	112	125	3.5
23	Hot Rolled 700°C, Normalized 890°C	64	83	24.
22	As Hot Rolled at 1050°C	68	87	16.5
22	Hot Rolled 1050°C, Normalized 890°C	56	80	25.

Table IV. Tensile Properties of Annealed 1020 LDC Steel Strip After Hot Rolling at 700°C: Alloy 23

Condition	Y.S.* ksi	UTS ksi	Elong. %
As Hot Rolled at 700°C	121	124	5.8
Hot Rolled 700°C, Annealed 500°C, 1 hr.	95	97	10.
Hot Rolled 700°C, Annealed 550°C, 1 hr.	91	93	15.
Hot Rolled 700°C, Annealed 600°C, 1 hr.	71	90	13.
Hot Rolled 700°C, Annealed 650°C, 1 hr.	53	77	16.
Hot Rolled 700°C, Annealed 700°C, 1 hr.	35	58	28.

* Upper yield strength for all 5 annealed tests

Table V. Tensile Properties of LDC 1020 Hot Rolled Strip : Quenched and Tempered*

Condition	Y.S. (0.2%) ksi	UTS ksi	Elong. %
Alloy 18, Hot Rolled 700°C, Q & T*	90	104	9
Alloy 18, Hot Rolled 1050°C, Q & T	99	104	9
Alloy 20, Hot Rolled 1050°C, Q & T	96	102	7

* Q and T: Quench from 890°C, Temper 550°C, 1 hr.

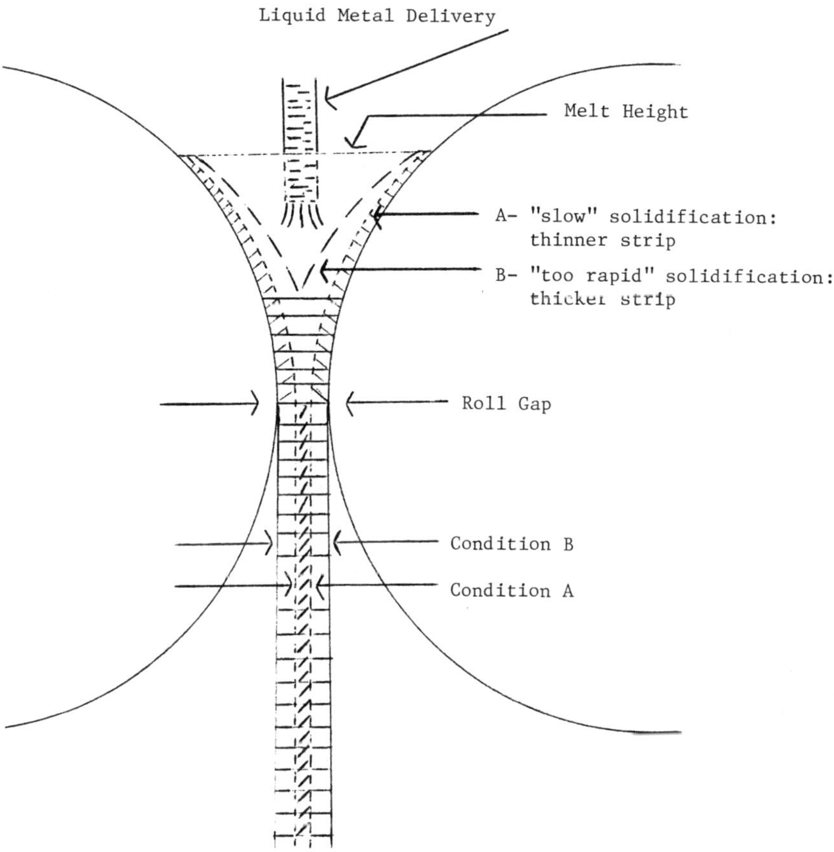

Figure 1. For a given roll diameter, roll material, roll speed, metal height and metal temperature, solidification may be too slow or too fast relative to the roll gap, and will result in different thicknesses of strip.

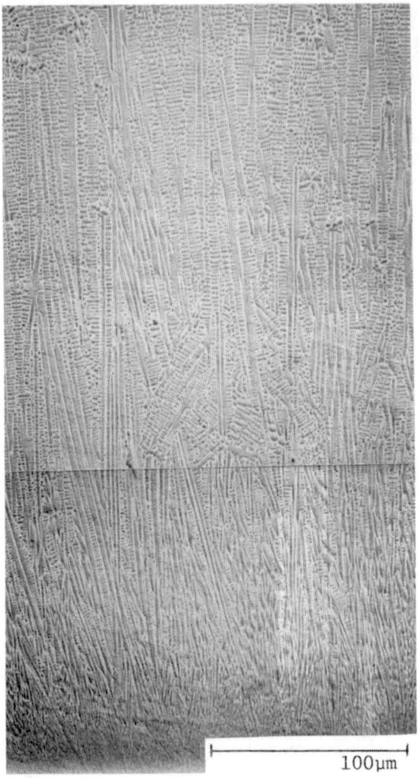

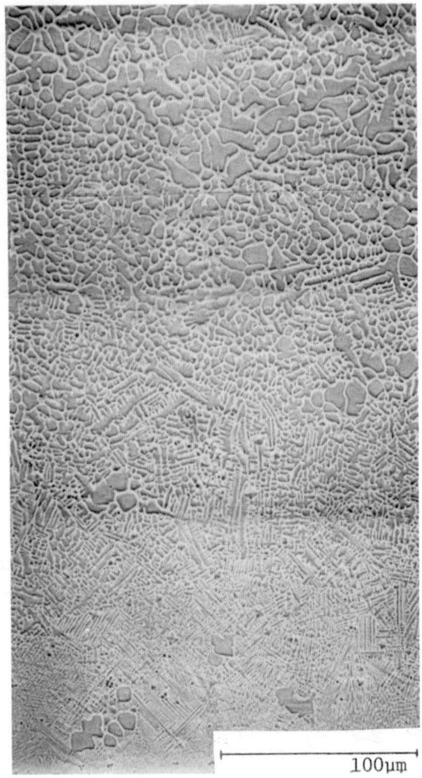

Figure 2 - (left) Shows fine columnar dendritic microstructure in a 304 stainless steel for melt spun ribbon 400μm thick.

Figure 3 - (right) Shows influence of imposed electric field during melt spinning; fine dentritic structure at substrate surface and cellular structure at top. Ribbon was 500μm thick.

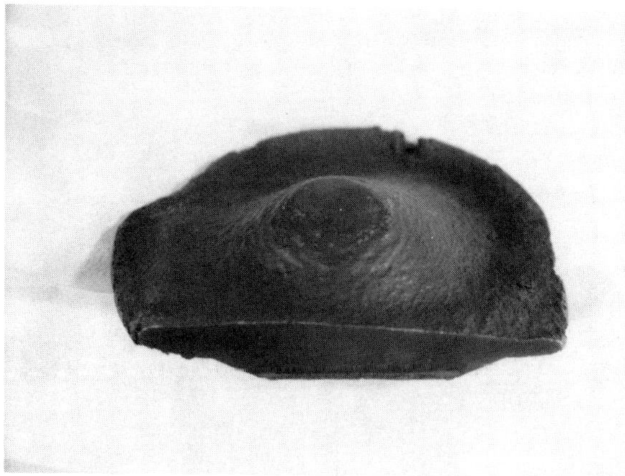

Figure 4. As-deposited preform of small turbine disc using circular atomizer. Uneven as-deposited profile is characteristic of circular nozzle spray deposition. 9Cr - 1 Mo steel, 96% as-deposited density; 100% density after hot isostatic pressing at 15,000 psi and 1100°C. (0.5X)

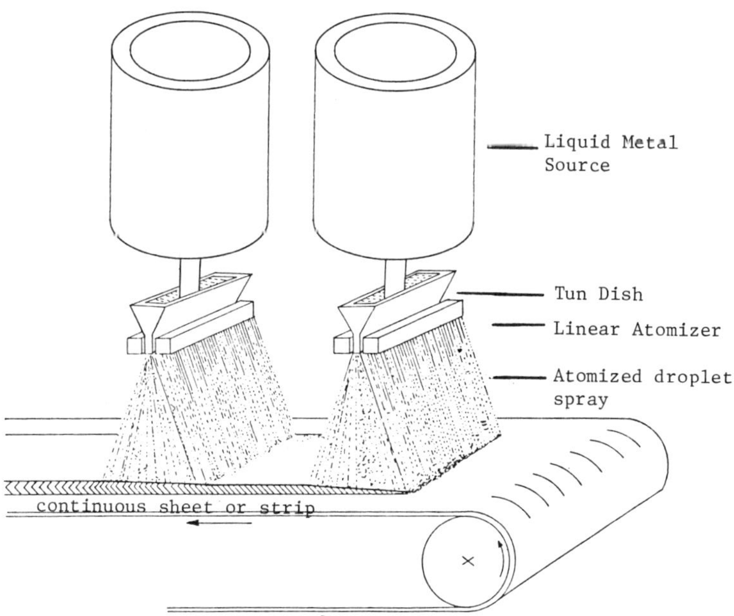

Figure 5. Continuous LDC spray forming of thin strip or sheet on moving belt using linear atomizer (singly or in series).

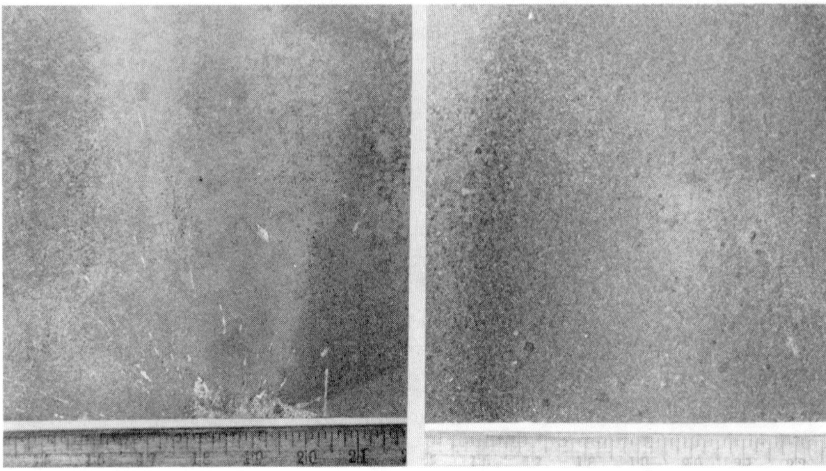

Figure 6 - LDC as-deposited 1020 steel plate using grit blasted, low carbon steel substrate, water cooled. Left: substrate surface; Right: top surface of deposited plate.

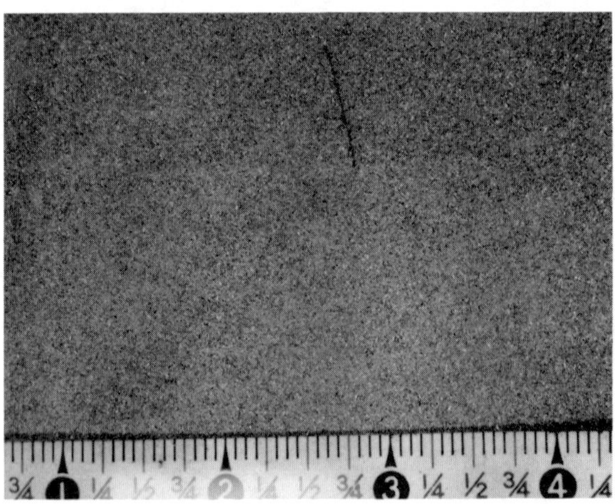

Figure 7 - LDC as-deposited 316L stainless steel plate against low carbon steel substrate showing substrate side of plate, uniform over a length of 28cm.

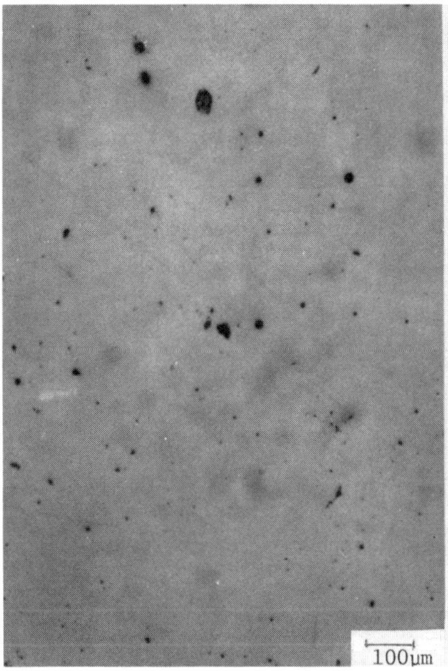

Figure 8 - LDC as-deposited 1020 steel plate showing normal pore size and distribution.

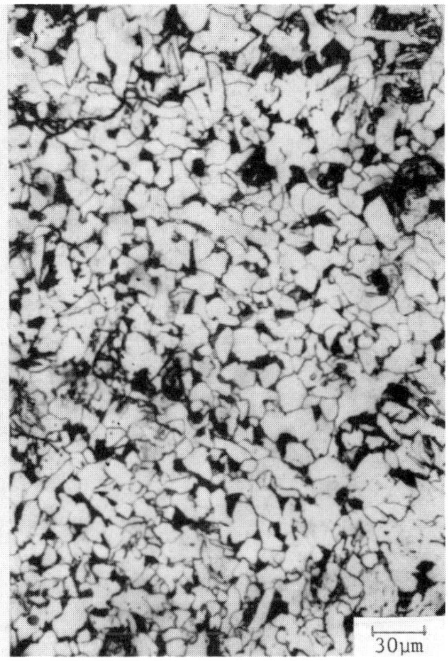

Figure 9 - LDC as-deposited 1020 steel plate showing ferrite plus pearlite microstructure.

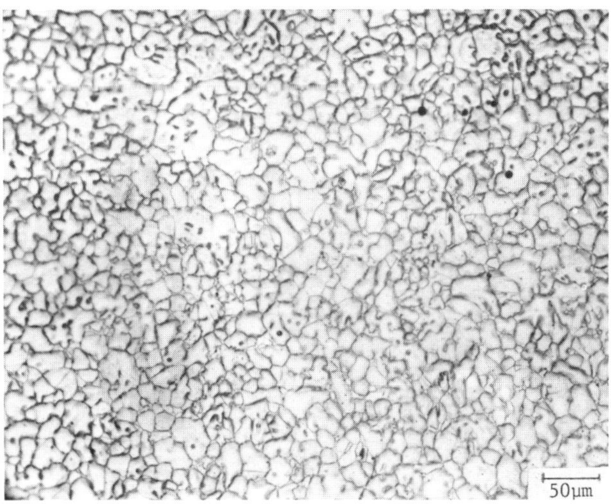

Figure 10 - LDC as-deposited 316L stainless steel microstructure, with grain size range of 15-40µm.

(a)

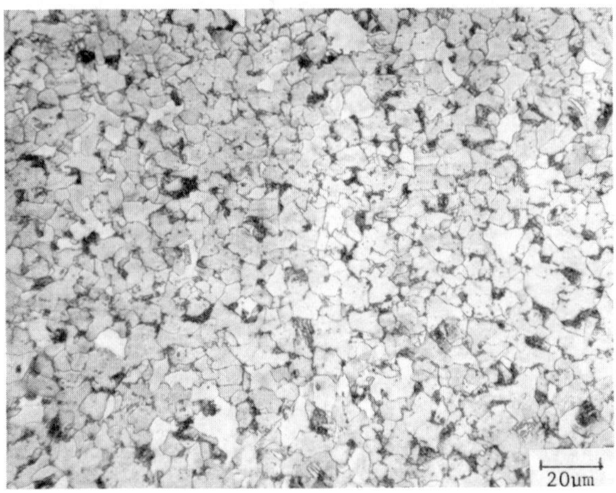

(b)

Figure 11 - As hot rolled microstructures of LDC 1020 steel plate. (a) Hot rolled at 700°C showing unrecrystallized microstructure (longitudinal view); (b) Hot rolled at 1050°C showing fine grained, fully recrystallized microstructure.

decreases at 650° and 700°C. Recrystallization in this fine grained, homogenous alloy starts between 550 and 600°C, depending on the residual cold work levels.

Conclusions

In laboratory studies, the LDC process was able to produce high density as-deposited plate (and strip) which has a fully recrystallized, fine grain size (10 to 50μm) with well scattered fine porosity.

Water cooled, plain carbon, ferritic steel substrates, appropriately grit blasted, provide adherence long enough to achieve uniform, reproducible, rapidly solidified microstructures, as well as adequate solid state cooling to avoid grain growth or excess phase coarsening.

As deposited densities of 94-96% permit hot rolling to full density at 700°C (ferrite plus carbide), at 1050°C (austenitic) or room temperature without observable edge cracking. Lower as-deposited densities (down to 88-90%), rolled at 700° or 1050°C, will develop edge cracking initially; however removal of the cracked edges permits further rolling free of further edge cracking. As hot rolled and heat treated grain sizes from 5 to 10μm are readily produced.

The hot rolled strip responds to annealing, normalizing and quench and temper treatments to provide a wide range of yield, tensile and ductility values. Yield strength values from 50 to 60 ksi, tensile values from 80 to 85 ksi and elongation values (for gage lengths of 1 inch or less) of 20 to 28% are reported. Quench and temper yield strengths of 90 to 100 ksi were obtained however at lower ductility values and are not as high as values attainable for the 700°C as rolled condition.

Acknowledgements

The author is pleased to acknowledge the excellent work of his research group comprising I. Ucok, A. Ibrahim, Chi Tsao, R. Mason and Ms. X Duan. The support of DOE and of Dr. J. Key of INEL are also much appreciated.

References

1. N.J. Grant, "Rapid Solidification of Metallic Particulates," Journal of Metals, 35(1)(1983) 20.

2. H. Jones, "Effects of Experimental Variables in Rapid Solidification from the Melt," Rapidly Quenched Metals: Second International Conf., N.J. Grant and B.C. Giessen, eds., MIT Press, Cambridge, MA., 1976, 1.

3. H.A. Davis, B.G. Lewis, and I.W. Donald, "Comparison of Roller Quenching and Melt Spinning for Continuous Casting of Strip," Rapid Solidification Processing, Principals and Technologies, Proc., R. Mehrabian, B. Kear and M. Cohen, eds., Claitor's Publ. Div., Baton Rouge, LA. (1978) 78.

4. E. Vogt and G. Frommeyer, "The Influence of Process Parameters on the Cooling Rate of the Meltspinning Process," Rapidly Quenched Metals, Proc. Fifth International Conf., S. Steeb and H. Warliamont, eds., Elsevier Sci. Publ., No. Holland Physics Publ., The Netherlands, (1985) 63.

5. H.A. Davis, Ibid., (1985) 101.

6. J.V. Wood and K. Boulby, "Rapid Solidification of Steel Particulates," Rapidly Solidified Materials, Conf. Proc., P.W. Lee and R.S. Carbonara, eds., ASM, (1985) 77.

7. W. Wang and N.J. Grant, "Structure and Properties of an R.S. Al-Mg-Li Alloy," International Jour. on Rapid Solidification, (1)(2)(1985) 157.

8. J.M Sater and T.H. Sanders, Jr., "A Quantitative Microscopy Investigation of Melt Spun Al-10 Wt % Co.", Rapidly Solidified Materials, Conf. Proc., P.W Lee and R.S. Carbonara, eds., ASM (1985) 357.

9. S. Kang, P.K. Domalavage and N.J. Grant, "Mechanical Properties of Rapidly Solidified Modified Cupro-Nickel Alloys:, Mats. Science and Engineering, (72)(1)(1986) 33-44.

10. C.H. Smith and N.J. Grant, "Rapidly Solidified O.D. IN-100: Structure and Properties," Rapid Solidification Processing: Princp. and Tech. III., R. Mehrabian, ed., National Bureau of Stds. (1983) 215.

11. R.G. Rowe and E.F. Koch, "Rapidly Solidified Titanium Alloys Containing Cerium Sulfide and Oxysulfide Dispersions," Rapidly Solidified Materials, Conf. Proc., P.W Lee and R.S. Carbonara, eds., ASM (1985) 115.

2. P.K Domalavage and N.J. Grant, "Rapid Solidification of Cobalt-Base Superalloys Alloyed with Hf. and Zr.", Metal Powder Report, (38)(10) (1983) 55.

13. N.J. Grant, "A Review of Various Atomization Processes," Rapid Solidification Processing: Proc., R. Mehrabian, B. Kear and M. Cohen, eds., Claitor's Publ. Div., Baton Rouge, LA (1977).

14. C. Ashdown, Y. Zhang, J. Megusar and N.J. Grant, "Grain Size and Superplasticity of Crystallized Fe-Cr-Ni-B Glasses," Proc. Fifth Conf. on Rapid Quenching and Solidification of Metals, S. Steeb and H. Warliamont, eds., No. Holland Physics Publ., Holland, (1985) 1763.

15. R. Widmer, "The Current Status and Future of Castings Densification via HIP," Proc. International Conf. on Hot Isostatic Pressing of Materials, The Royal Flemish Soc. of Engineers, Metallurgical Section, Antwerp, (1988) 3.1.

16. K. Ogata, E. Lavernia, G. Rai and N.J. Grant, "Structure and Properties of a Rapidly Solidified Superalloy Produced by Liquid Dynamic Compaction," International Jour. of Rapid Solidification, (2)(1)(1986)21.

17. U.S. Patent 3,826,301, Issued to R.G. Brooks, July 30, 1974. "Method and Apparatus for Manufacturing Precision Articles from Molten Articles."

18. U.S. Patent 3,909,921, Issued to R.G. Brooks, Oct. 7, 1975. "Method and Apparatus for Making Shaped Articles from Sprayed Molten Metal of Metal Alloys."

19. B.A. Rickinson, F.A. Kirk and D.R.G. Davies, "CSD: A Novel Process for Particle Metallurgy Products," Powder Metallurgy No. 1 (1981) 1.

20. "CSD Splatting Process Lifts Steel Quality:, Reprinted from Feb. 1981 British Steelmaker.

21. T.S. Chin, et.al., "(Fe,Co)-Nd-B Permanent Magnets by Liquid Dynamic Compaction," Jour. Appl. Phys. (59)(4) (1986) 1297.

22. E.J Lavernia and N.J. Grant, "Structure and Properties of a Modified 7075 Aluminum Alloy Produced by Liquid Dynamic Compaction," International Jour. of Rapid Solidification, (2)(2)(1986) 93.

23. J.M. Nell, G. Rai, E.J. Lavernia and N.J. Grant, "Microstructure and Properties of an IN-100 Tube Produced by LDC." Progress in Powder Met., (42) 1986 Annual P.M. Conf. Proc., E.A. Carlson and G. Gaines, eds., MPIF, Princeton, N.J., (1986) 67.

24. T. Ando, E.J. Lavernia and N.J. Grant, "A TEM Study of the Microstructures of Modified 7075 and X2020 Aluminum Alloys Produced by Liquid Dynamic Compaction," Materials Letters, (5)(1,2)(Dec. 1986) 17.

25. E.J. Lavernia, J.C. Baram and N.J. Grant, "Developments in Gas Atomization and Spray Deposition," Metal Powder Report, (42)(10)(1987) 688.

26. ARE Singer, Met. Materials, (4)(1970) 246.

27. U.S. Patent No. 4,485,834. Issued to N.J. Grant, Dec. 4, 1984, "Atomization Die and Method for Atomizing Molten Material."

28. U.S. Patent No. 2,997,245. Issued to E.O.F. Nilsson et.al., Aug. 22, 1961. "Method and Device for Pulverizing and/or Decomposing Solid Materials."

29. K. Sankaran and N.J. Grant, "The Structure and Properties of Splat Quenched Aluminum Alloy 2024 Containing Lithium Additions," Materials Science and Engineering, (44)(1980) 213.

30. J. Megusar, O.K. Harling and N.J. Grant, "Potential for Using Rapid Solidification for Improved Irradiation Performance in the Fusion Environment," Jour. of Nucl. Mats., (103 & 104)(1981) 961-966.

31. N.J Grant, "Powder and Particulate Production of Metallic Alloys," Advances in Powder Metallurgy: 1981 ASM Materials Science Seminar, G. Chen, ed. ASM (1982).

32. N.J. Grant, "Rapid Solidification of Metallic Particulates," Jour. Metals, (35)(1)(1983) 20.

CASTING TECHNIQUES II: SPRAY CASTING

Session Chairman: **G. Gillen**
Drexel University
Philadelphia, Pennsylvania

Near Net Shape Casting via Droplet Consolidation

Diran Apelian and Gerry Gillen
Drexel University
Solidification Processing Laboratory
Department of Materials Engineering
Philadelphia, Pennsylvania 19104

ABSTRACT

Near net shape casting of metals via droplet consolidation has advanced so quickly in recent years that it is fast becoming a commercial processing reality. Spray casting technologies such as the Osprey process and the Liquid Dynamic Compaction (LDC) process offer tomorrow's metallurgists unique alternatives to conventional ingot processing routes. Droplet consolidation technologies combine the advantages of near net shape casting and the beneficial characteristics of P/M processing without the problems associated with powder storage and handling. Exotic or complex alloy systems which cannot be cast by conventional methods can be fabricated more readily by these technologies. Novel rapidly solidified fine grained microstructures with reduced levels of segregation are obtainable. Incorporation of secondary phases or reactants during processing can result in novel composite structures. Economic savings arising from this one step processing do not preclude its use with the less expensive ferrous metals. Research at Drexel University is directed towards a fundamental science-based understanding of these droplet consolidation technologies. This keynote paper reviews current process understanding and will address the challenges which lie ahead; in particular, the development of "intelligent processing" by on-line sensors and controllers is a necessity to optimize both shape control and microstructural control in the finished product.

Casting of Near Net Shape Products
Edited by Y. Sahai, J.E. Battles,
R.S. Carbonara and C.E. Mobley
The Metallurgical Society, 1988

1. INTRODUCTION

Droplet consolidation technologies such as the Osprey spray deposition process [1,2] and the Liquid Dynamic Compaction (LDC) process [3] have developed so quickly in recent years that they will soon become commercial manufacturing processes. They have significant advantages over ingot metallurgy routes in that a rapidly solidified near net shaped component can be fabricated directly from the melt in a single operation. Complex and exotic alloys, which cannot be fabricated by conventional casting methods, can be produced more readily. Novel rapidly solidified fine grained microstructures containing lower levels of segregation are possible.

At present the major obstacle to the widespread acceptance of these processes is related to process control, product uniformity and reproducibility. If these processes are to continue developing and realise their full potential, complete process optimization will be critical. The most intelligent and efficient way of achieving this goal will be to develop on-line monitoring and sensing of the critical processing parameters, combined with integrated process control. In this manner, a more systematic approach, known as *"Intelligent Processing"* can be adopted for the near net shape casting of metal components via droplet consolidation. A necessary precursor to intelligent process control is a fundamental understanding of the critical process parameters. This keynote paper illustrates the requirements necessary for intelligent process control and will describe ongoing research at Drexel University which is directed towards a fundamental understanding of the critical process parameters.

1.1 Near Net Shape Casting

Since the dawn of civilization man has been able to fashion liquid metals into near net shaped components by various casting routes. The formation of intricate and ornate metal castings by the "lost wax" process dates back to a time before the discovery of bronze around 3000 B.C. In those times, the lost wax process was the most economical and direct method of producing an intricate shape from molten metal with the minimum amount of waste. Today the same principle still holds true; the foundry which produces a quality near net shaped component by the most direct casting route will have a distinct advantage over its competitors in today's global market. The key word is quality.

The near net shape casting of thin section ferrous products such as sheet and thin slab has been a problem which some of the great metallurgical minds have attempted to overcome. As early as 1856, Sir Henry Bessemer experimented with a twin roller method for casting malleable iron into sheet. After reporting some initial success he surprisingly gave up on the idea thinking it was "before its time." In the 1920's Hazelett developed several novel single roll and twin roll casters as well as a twin belt caster for the production of thin aluminum and copper slab. More recently the twin belt caster has been used for the production of steel slab. Problems arise due to the variable solidification structure that is produced in the as-cast slab. A chill zone is initially produced at the outside edges of the slab which are in contact with the moving belt substrate. This is followed by the columnar growth of grains inward and finally equiaxed grains in the middle of the slab. Invariably, all the twin roll and twin belt processes have been quite successful for producing aluminum and copper alloys in strip form but have always fallen short when it came to ferrous materials [4]. The major limitations of all these methods for the production of thin steel strip seems to be control of the feeding rate of liquid metal to the nip or gap between the rolls and the inability to conduct heat rapidly from the solidifying metal strip to the substrate or mold surfaces. The direct contact between the source of liquid metal and the moving substrate also leads to processing problems. If the metal begins to solidify in the feeding nozzle, the flow of metal is restricted and eventually further casting is prohibited. Any process which prevents direct contact between the source of molten metal and the substrate or roller will have a significant competitive advantage. Recently Kawasaki Steel has reported success with their twin roller method wherein the metal is "splashed" into the gap or nip between two rollers[5].

1.2 Droplet Consolidation Technologies

One group of processes that prevent the direct contact between the retaining crucible and the moving mold or substrate are generically known as droplet consolidation technologies. These technologies include Liquid Dynamic Compaction (LDC), electrohydrodynamic atomization, the Osprey Process and Low Pressure Plasma Deposition (LPPD). These processes all result in the formation of molten metal droplets, which are then consolidated on a suitable substrate. In the LDC process the metal droplets are created by the impingement of ultrasonic pulses of gas jets onto a stream of molten metal. In electrohydrodynamic atomization a stream of molten metal is destabilized by the application of electric and magnetic fields. The Osprey process is perhaps commercially the most viable of these technologies at the present time. In this process a stream of molten metal is atomized by inert gas jets creating a spray of fine droplets, which are accelerated towards a substrate where they consolidate to form a fully dense deposit (Figure 1). By maneuvering the substrate a number of near net shaped components can be fabricated (Figure 2). Spraying onto a rotating refractory disc enables a cylindrical billet to be fabricated. A coating can be applied to a base metal by spraying onto a rotating mandrel. With longer deposition times a tube of a required wall thickness can be fabricated. Spraying onto a conveyor belt, roller or flat substrate will produce thin strip or thicker slab, the thickness being determined by the deposition rate and the relative motion of the substrate. Deposition rates of up to three pounds per second have been achieved by the Osprey process and, therefore, it can truly be called a spray casting process. In contrast, low pressure plasma deposition (LPPD) has much lower deposition rates, of the order of grams per second. As a consequence it may not be thought of as a pure casting process but can be considered as more of a technology for the application of coatings under a controlled atmosphere. The Department of Materials Engineering at Drexel University is unique in having three droplet consolidation technologies currently in operation. These are a pilot plant Osprey spray casting research facility (presently the only one anywhere in an academic institute), a low pressure plasma processing unit and a Jet Kote flame spraying unit.

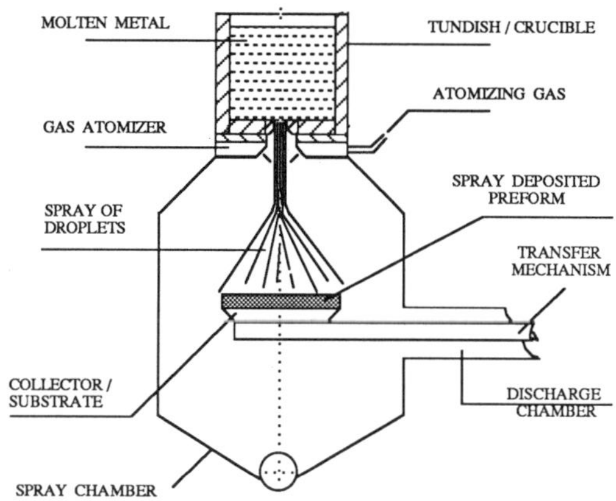

Figure 1. Schematic of the Osprey Process

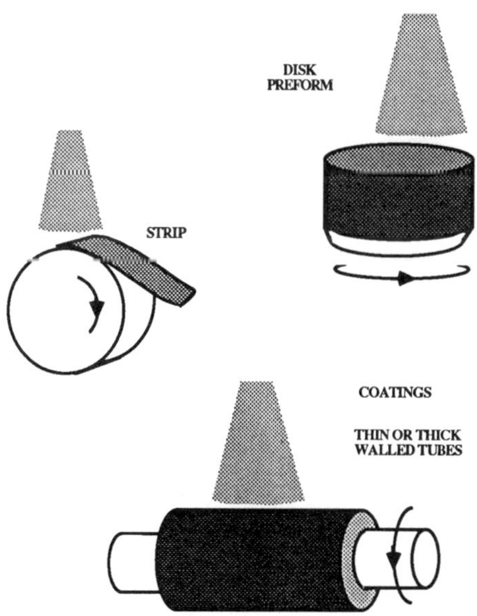

Figure 2. Schematic of different shapes produced by spray deposition

2. INTELLIGENT PROCESSING OF DROPLET CONSOLIDATION

To date very few metallurgical industries have embraced on-line process control. As a consequence there is often a high probability of component rejection during quality control inspection. For example, in the aluminum industry up to 10% of all castings are rejected because of inherent shrinkage porosity. An intense technological revolution is currently in progress worldwide. If metal processing industries are to remain competitive, they must turn to "intelligent" manufacturing. Intelligent processing of materials can be achieved by on-line monitoring and sensing of the process, combined with closed-loop controllers (Figure 3a). This will increase productivity and efficiency, and yield a more reliable defect-free product, which is manufactured in an easily reproducible fashion. Metal casting operations are particularly suitable for an intelligent processing approach (Figure 3b).

To control a process it is necessary to understand the significant parameters which determine the integrity of the final product. Initially, this requires the development of process understanding via computational modeling and experimental verification. This process understanding indicates which process parameters need to be controlled, where control needs to be applied and how it should be most effectively implemented. The three principal components of an automated process control system are:

- Process modeling and knowledge-based understanding.
- On-line monitoring and sensing with closed-loop feedback.
- Efficient implementation of process control.

The first step in the intelligent processing of metals is to understand the process parameters particular to a given manufacturing technology. In Drexel University's Solidification Processing Laboratory research has been directed towards a fundamental science-based understanding of droplet consolidation via the Osprey spray deposition process. During the course of this research a knowledge data base has been established which will serve

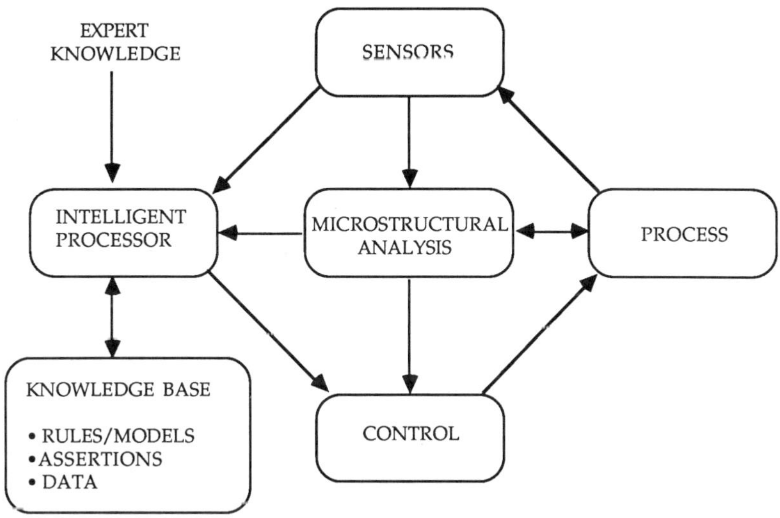

Figure 3(a). Intelligent Processing of Materials

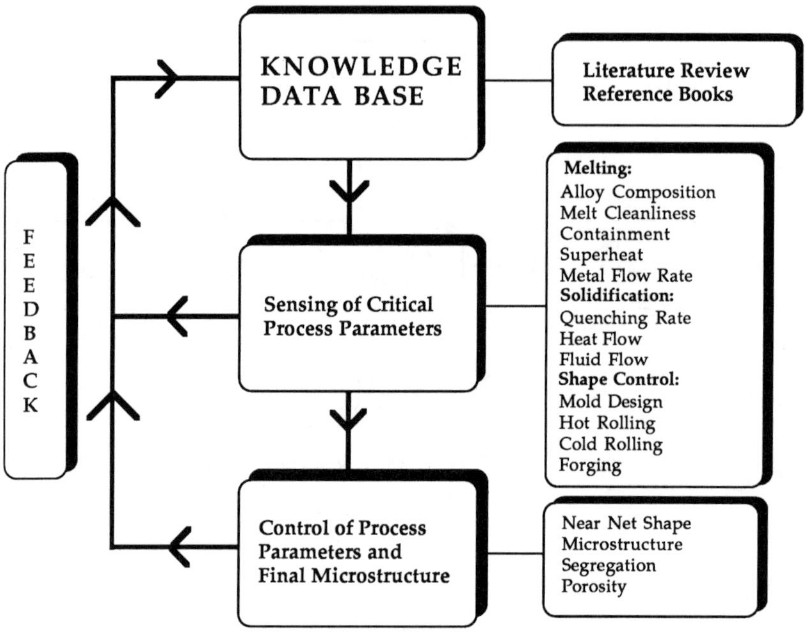

Figure 3(b). Intelligent Processing of Metal Casting by on-line Sensors and Process Control.

in the future as the operational "brain" for a more intelligent approach to manufacturing. New information and data produced by computational modeling and experimental results will be data "INPUT" to continuously update this knowledge base. A schematic representation of intelligent process control for droplet consolidation via Osprey is shown in Figure 4.

3. DROPLET CONSOLIDATION VIA THE OSPREY PROCESS

In droplet consolidation via the Osprey process, Drexel University has conducted research in two specific areas. Firstly, the critical process parameters have been studied to gain a fundamental scientific understanding of droplet consolidation via Osprey [6-8]. Secondly, an investigation of strip casting of plain carbon steels via Osprey has been undertaken [9]. These research investigations have been funded by the Office of Naval Research (ONR) and the National Science Foundation (NSF), respectively.

3.1 Osprey Process Fundamentals

Research at Drexel University has established the critical spray casting parameters which must be monitored and controlled to optimize the process. In their basic processing terms these are the casting temperature, metal flow rate, atomizing gas pressure and the flight distance of the droplets prior to deposition on the substrate. Optimization of the process is complete when a critical balance of these parameters is achieved. Starting with optimal processing conditions and then increasing the casting temperature necessitates decreasing the metal flow rate or increasing the atomizing gas pressure or flight distance to return to the original optimal conditions. On a more refined scale the process parameters can be characterized in terms of melt superheat, metal to gas mass flow rates, droplet size distributions, enthalpy of the spray, the fraction of a droplet solid or number of droplets solidified at the point of impingement, droplet-substrate interactions and droplet-deposit interactions. All of these macroscale and microscale process parameters contribute to the morphology and integrity of the overall deposit and the evolution of the final microstructure. The most critical process parameters will be addressed in the following sections.

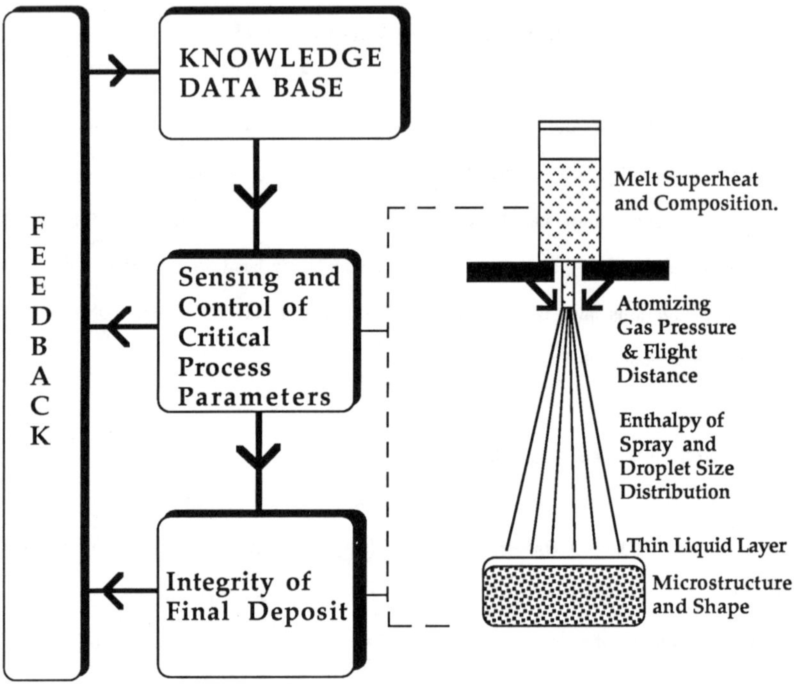

Figure 4. Intelligent Process Control of Droplet Consolidation via Osprey.

3.1.1 Droplet Size Distribution

Patternation studies have been conducted which indicate a bimodal population distribution of droplet sizes (Figure 5). The first and largest population peak occurs at ~ 10 μm and the second smaller peak occurs at ~ 60 μm. The large number of very fine droplets is formed as a consequence of the destabilization mechanism of the metal stream during atomization. It is these very fine droplets arriving at the substrate, partially or fully solid, which then act as the refining nuclei for the evolving microstructure (Figure 6). Thus the resulting microstructure is highly refined and similar in scale to that obtained during P/M processing. There are a significant number of still larger droplets, some as large as 1-2 mm. This results in a much higher mean droplet size of ~ 80 μm when a weighted mass average is calculated. These droplets when they impact the substrate are partially or fully liquid and form a thin liquid layer on the surface of the deposit. This thin liquid layer helps to remelt some of the partially solidified droplets and, therefore, there are no prior particle boundaries visible in the final deposit.

3.1.2 Droplet Velocity

Subsequent to their formation during atomization the droplets are accelerated by the gas until they attain the velocity of the gas (Figure 7). The gas itself decays exponentially after it leaves the atomizing nozzle. Smaller droplets (20μm) are accelerated to higher velocities more quickly than larger droplets. Droplet velocities in excess of 120 m/s have been measured by photographic experiments, thus confirming the velocities predicted by modeling studies. In terms of the final deposit morphology and microstructure, droplet velocity is thought to be a less critical parameter than droplet temperature.

3.1.3 Droplet Temperature

The temperature of a given droplet depends on the size of the droplet, and its degree of solidification at a given time. In turn, its degree of solidification is determined by the amount of undercooling the droplet may have undergone. Two extreme scenarios are possible; either the droplet experiences no undercooling and heterogeneous nucleation results; or the droplet undergoes maximum undercooling and homogeneous nucleation results. In reality, the actual scenario will be somewhere between these two extremes. Experimental studies indicate

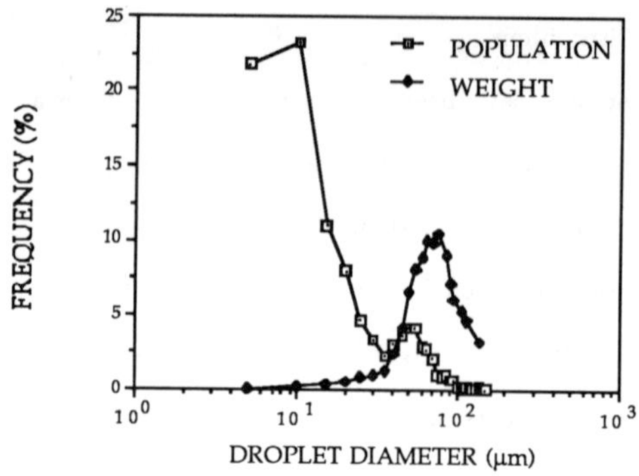

Figure 5. Droplet size distribution by population (bimodal) and weight.

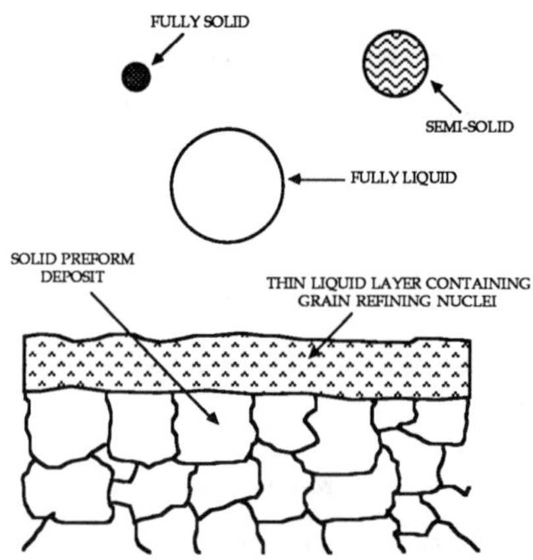

Figure 6. Droplets in different stages of solidification arriving at substrate.

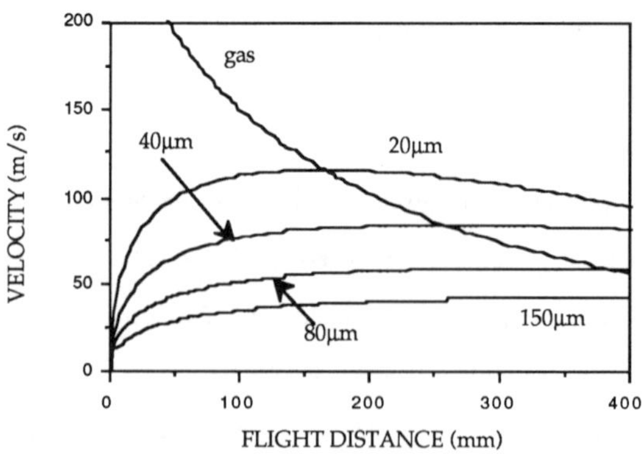

Figure 7. Predicted variation of droplet velocity with flight distance for four droplet sizes (Fe-5 w/o Ti).

that very small droplets (< 20 μm) nucleate homogeneously while larger droplets, which have a larger probability of containing nucleii, will solidify heterogeneously. Computer modeling of the heat transfer within a solidifying droplet has led to temperature profiles for both heterogeneous nucleation (Figure 8) and homogeneous nucleation (Figure 9). In homogeneous nucleation a droplet undercools and as solidification begins, the latent heat of fusion is liberated and recalescence begins to occur. The time it takes for this recalescence to occur and the temperature rise that it brings about within the droplet are critical to the spray casting process. If recalescence occurs quickly and the droplet does not attain the solidus temperature, the droplet will solidify instantaneously. If this happens to a significant number of droplets the final deposit may contain a large number of pre-solidified particles. If, however, during recalescence the droplet is heated above the solidus temperature, complete solidification will be achieved over a longer period of time. This results in a partially solidified droplet arriving at the substrate in the ideal "mushy" state. Remelting of partially solidified droplets in the thin liquid layer is possible and as a consequence no prior particle boundaries are evident in the microstructure.

3.1.4 Spray Enthalpy

The enthalpy (or heat content) of the spray upon arrival at the substrate is one of the most significant parameters governing the final microstructure of the deposit. If the enthalpy is too high, a microstructure similar to a conventional casting will be obtained. Conversely, if the enthalpy is too low, a large number of pre-solidified droplets with prior particle boundaries will be evident in the microstructure. The average enthalpy of the spray can be calculated from a summation of the enthalpies of individual droplets,which is given by the equation:

$$H_{\text{spray (average)}} = \Sigma\{H_d(d_i)\, d_i^3\, f(d_i)\} / \Sigma\{d_i^3\, f(d_i)\} \quad , \qquad ......(1)$$

where $H_d(d_i)$ is the enthalpy of a droplet of size d_i. The enthalpy of the spray can also be calculated as the heat content of the alloy at the solidus temperature plus the heat content associated with the freezing range:

$$H_{\text{spray}} = H_{\text{solidus}} + \Phi H_{\text{fr}} \quad , \qquad ......(2)$$

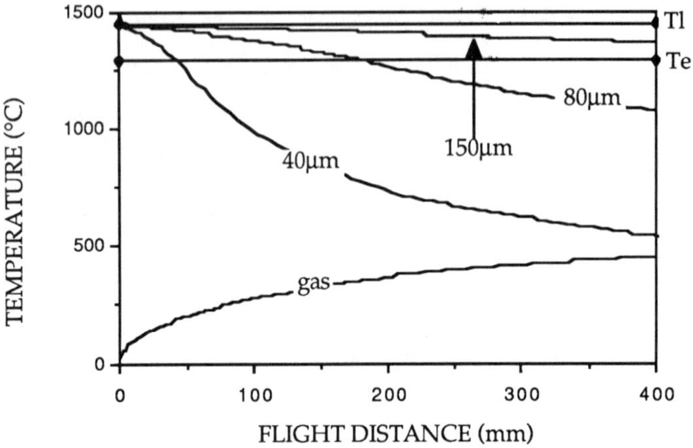

Figure 8. Predicted variation of droplet temperature for heterogeneous solidification

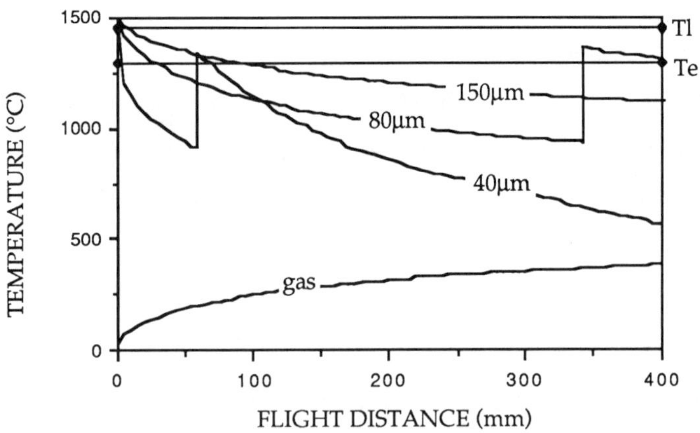

Figure 9. Predicted variation of droplet temperature for homogeneous solidification.

where Φ is a dimensionless constant associated with the heat content in the freezing range. Φ = 0 if the spray is totally solid and Φ = 1 if the spray is completely liquid. The predicted variation of Φ with flight distance for three alloy systems is given in Figure 10. At the point of atomization everything is liquid and Φ = 1. Φ drops rapidly during the first 100 mm of flight and then at greater flight distances levels off. This plot can be used to predict for a given flight distance, the % liquid of the spray arriving at the substrate. Since the fraction liquid arriving at the substrate largely determines the final microstructure this knowledge is of paramount importance.

3.1.5 Deposition Rate

Due to the high density of droplets along the axis of the spray the deposition profile has a Gaussian distribution. This has been confirmed by patternation studies and by the shape of deposits made on stationary flat substrates. Modeling studies predict similar results as shown in Figure 11. Complex shapes can be predicted and controlled by an extension of the modeling studies to incorporate rotational and translational movement of the substrate.

3.1.6 Isotherms within Deposit

The predicted temperature isoterms within a solidifying deposit are shown in Figure 12. Deposition occurs only during the initial 40 seconds. Thereafter the deposit thickness is fixed. Cooling occurs due to heat conduction through the substrate and by the forced convective cooling of the gas on top of the deposit. Radiative cooling also occurs on the outside surfaces of the deposit. The model predicts that a large area in the center of the deposit cools very slowly and depending on the freezing range of the alloy, may be partially liquid for some time. This variation in local solidification time can yield differences in microstructure and possibly regions of porosity.

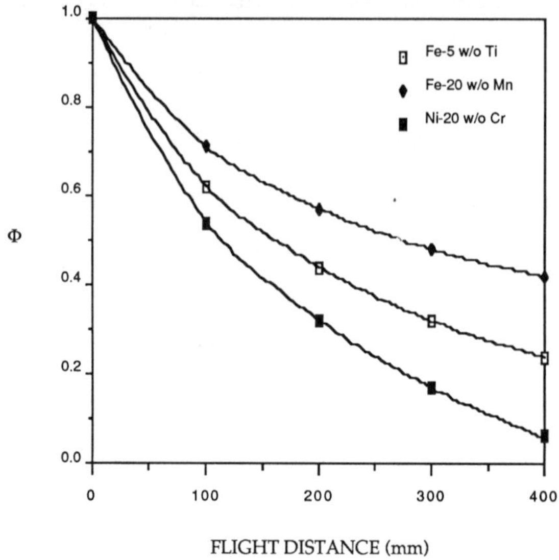

Figure 10. Variation of calculated spray enthalpy with flight distance for three alloy systems.

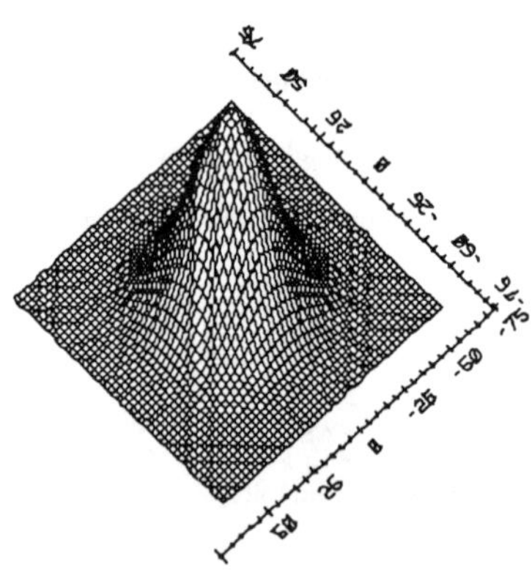

Figure 11. Geometry of a deposit produced on a stationary substrate; spraying time = 40s.

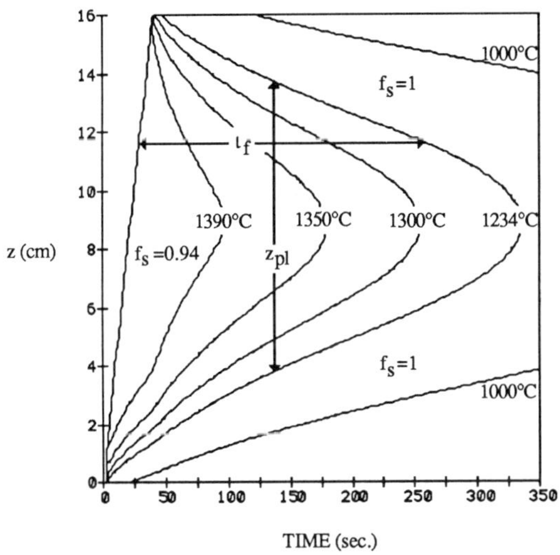

Figure 12. Predicted temperature profiles within a deposit.
Deposition rate = 4 mm/s; T spray = 1410° C.

3.2 Strip Casting via Droplet Consolidation

The fundamental understanding of the Osprey process, which has been developed at Drexel University has been instrumental in studying strip casting of plain carbon steel by spray deposition. Plain carbon steel (AISI 1026) has been cast into thin strip in a semi-continuous fashion by spraying onto various single roller substrates.

3.2.1 Strip Dimensions

For substrate velocities ranging from 1 m/s to 0.1 m/s the strip thickness ranged from 0.6 mm to 5.5 mm and the width of the strip produced varied from 0.08 m to 0.12 m (~ 3" to ~ 5"). The strip exhibited maximum thickness along the center-line since the droplet flux is highest along the spray axis. Wider strip widths are possible by scanning the atomized spray perpendicular to the motion of the substrate. This also yields strip which has a more uniform thickness. Mannesmann Demag in West Germany are currently developing thin slab casting via spray deposition. Thin slab, which is ~ 10 mm (thickness) x 1 m (width) x 2-3 m (length), has been produced by spraying onto a moving flat substrate.

3.2.2 Strip Microstructure

The microstructure at the edge of the strip, which was in direct contact with the substrate, consists of a chill zone with fine grains of ferrite and pearlite. The remaining thickness consists of equiaxed grains containing fine Widmanstatten plates. A small degree of porosity is noticeable near the edges of the strip. This is due to the morphology of the spray and the low deposition rate at the periphery of the spray. The initial and final material which is deposited often derives from this region of low droplet flux density. Due to the low deposition rate and rapid cooling on the substrate or by the gas, a region containing some porosity is obtained. Once again a scanning atomized spray with a more uniform spray morphology would overcome this problem.

3.2.3 Near Net Shape Casting of Thin Strip

Uniformity of strip thickness and strip microstructure can be improved by hot rolling shortly after the strip has solidified on the substrate. This would produce strip of the required thickness, remove the small amount of porosity present and refine the as-cast microstructure. The metal pouring nozzle is not in direct contact with the substrate so the feeding problems associated with other thin slab casters are eliminated. The semi-continuous production of steel strip or thin slab is therefore feasible by spray casting.

SUMMARY

To remain competitive in today's global market, primary and secondary metal manufacturers must turn to new innovative near net shape casting processes. If they are to use these processes in the most effective and cost efficient manner they must implement "Intelligent Processing Controls." Droplet Consolidation Technologies, such as the Osprey process, offer considerable advantages over conventional ingot metallurgy processes for the production of improved near net shape components. These technologies lend themselves well to intelligent processing approaches. In order to apply an intelligent processing approach, the fundamental process parameters must be investigated and characterized. Knowledge of these parameters will indicate how process control will be most effectively applied. To date, research at Drexel has addressed the critical process parameters of spray casting via the Osprey process. Research is currently underway to apply intelligent process controls to this droplet consolidation technology.

ACKNOWLEDGMENTS

The authors wish to thank Dr. Pravin Mathur and Suresh Annavarapu for their research on process fundamentals and strip casting, respectively. This work was sponsored by the Office of Naval Research and the National Science Foundation.

REFERENCES

1. "Modern Developments in Powder Metallurgy", A.G. Leatham, R.G. Brooks and M. Yaman, Metal Powder Industries Federation, Princeton, NJ, 15, p.157 (1985).
2. "Near Net Shape Manufacturing via the Osprey Process", D. Apelian, G. Gillen and A. Leatham, ASM Meeting, Orlando, October 1986, Conf. Proc. p.107.
3. "Liquid Dynamic Compaction of a Rapidly Solidified High Strength Aluminum Alloy", E.J. Lavernia, G. Rai and N.J. Grant, International Journal of Powder Metallurgy, V22, No. 1, p.9.
4. "New Developments in the Continuous Casting of Steel", A.W. Cramb, Iron and Steelmaker, February 1988 p. 42.
5. M. Ozawa, Kawasaki Steel Corporation. (Private Communication).
6. "Spray Deposition: The Interaction of Material and Process Parameters", G. Gillen, P. Mathur, D. Apelian and A. Lawley, Progress in Powder Metallurgy, MPIF, Princeton, NJ, 42, p.753 (1986).
7. P. Mathur, Ph.D. Thesis, Drexel University (1988).
8. "Analysis of the Spray Deposition Process", P. Mathur, D. Apelian and A. Lawley, Acta Met., (in press).
9. "Processing Effects on the Structure of Plain Carbon Steel Strip via Spray Deposition", S. Annavarapu, A. Lawley and D. Apelian, Met. Trans A, (in press).

SPRAYFORMING OF SHEET AND STRIP METALS

A. R. E. Singer

Chairman and Managing Director
Sprayforming Developments Limited
Innovation Centre
University College of Swansea
Swansea, UK

Abstract

The paper deals briefly with the principal methods of sprayforming sheet and strip and then proceeds to consider some of the ways in which a suitable profile across the width of the strip can be attained.

Other factors dealt with are the porosity at the substrate interface and the condition of the top spray-deposited surface.

In addition to near-zero segregation, sprayforming also offers the possibility of rapid solidification with all its attendant advantages. Examples of both of these benefits are described in the paper.

Casting of Near Net Shape Products
Edited by Y. Sahai, J.E. Battles,
R.S. Carbonara and C.E. Mobley
The Metallurgical Society, 1988

Introduction

During the last few years sprayforming has been shown to be a credible route from liquid metal to a wide variety of solid semi-finished metal products. Sprayforming constitutes a radical departure from conventional ingot routes, including continuous casting, from direct strip casting and from rapid solidification by melt spinning and similar routes. Because of the radical nature of the concepts an entirely new technology has had to be created to enable simple shapes to be produced for evaluation. In the circumstances it is therefore not surprising that early development was relatively slow because of the credibility barrier. The first sprayformed materials were produced and tested 20 years ago by the author and his team, but it is only now that the various sprayforming processes have become a credible alternative to conventional processing.

A further restriction on the application of sprayforming has been the fact that many metals and alloys produced on an immense scale in industry by conventional techniques have benefited from many years of optimisation and also enjoy the economic benefits of scale. Clearly any new process employing an entirely different technology will be applied first in those areas of premium products where manufacture is carried out on a modest scale but where enhanced properties of the product are recognised by a higher market price. This overcomes the formidable benefits of scale and allows the technology to be developed to the stage, some years hence, where it can be applied with confidence in the tonnage market.

Sprayforming comprises a new generation of metal-manufacturing processes producing a wide range of semi-finished products, each having certain distinguishing features. In all the cases at present under discussion the starting point is a stream of molten metal, usually from a melt, which is atomised either by high-pressure inert gas or by centrifugal means to form a spray of molten particles. The spray is directed towards a cool substrate onto which the particles splat to form a coherent deposit. In most cases the deposit is stripped from the substrate to be further hot or cold worked although in other cases it can be allowed to remain bonded to the substrate to produce a coated or laminated composite product. The sprayforming operation can take place at a considerable throughput rate and at present deposition rates are as high as 200 kg/min.

Such a radical technology will only be developed if it promises to make semi-finished products having very special properties, or having low processing costs, or both. In this respect sprayforming seems to have the hallmark of a winner for the following five reasons:

1. It enables many products to be made by a very short route from the melt to the final shape.

2. It has the capability of producing materials with near-to-zero segregation. In this respect it has strong resemblances to powder metallurgy.

3. It produces very fine grain sizes and structures leading, together with (2), to enhanced mechanical properties.

4. The products are easier to fabricate subsequently by hot or cold working.

5. Many special products in the field of laminates and composites can be made, which are difficult or impossible to make by conventional methods.

At this point it must be emphasised that sprayforming is entirely different from conventional thermal or arc-spraying for corrosion protection. In the case of sprayforming the final product has high density and excellent mechanical properties, whereas conventional thermal and arc-spraying produce a deposit that is porous and brittle. These differences are caused by the fact that conventional thermal spraying is not carried out in an inert atmosphere, with the consequence that the deposits have a high oxide content. Moreover, they are not consolidated and therefore remain relatively weak and brittle. The fact that metal has to be supplied in the form of wire or powder and that throughput is very limited means that massive products cannot be produced economically by means of conventional metal spraying.

Sprayforming applied to sheet and strip

When strip is to be made continuously or semi-continuously, the spray is usually directed onto a cooled rotating drum or onto a cooled moving band or linked belt. A suitable substrate for most purposes is a low-carbon steel, a low-alloy steel or cast iron, which is water-cooled to dissipate the large quantities of heat extracted. The spray deposit so produced will usually have some porosity, much of it at the substrate interface but some distributed throughout the thickness. As a consequence the mechanical properties of the 'as-deposited' strip are lower than those produced conventionally. This deficiency can be rectified by subsequently hot or cold rolling the strip. The most convenient and effective method is to hot roll in-line using the sensible heat of the deposit to obtain the required hot-rolling temperature. An alternative procedure is to direct the spray into the nip of a rolling mill, although for reasons given below this has not yet been developed to any extent. Other ways are to use the Centrifugal Spray Deposition (CSD) technique or the Simultaneous Spray Peening (SSP) procedure, but these are entirely different in concept and will be discussed towards the end of the paper.

Although in principle sprayforming applied to sheet and strip - or spray rolling as it may conveniently be called - is relatively simple, in practice it is one of the most difficult operations to accomplish satisfactorily. The difficulties can be classified in three categories:

1. Thickness control across the width of the strip and reduction of overspray.

2. Separation from the substrate and maintenance of substrate.

3. Structural considerations including both porosity and metallurgical structure.

Each of these three categories will now be examined in more detail.

1. Thickness control and overspray

For the spray-deposited strip to be successfully hot or cold rolled, either in-line or subsequently, it is essential to maintain a uniform strip thickness across the width. If this is not done rolling will produce a badly shaped strip that will be difficult or impossible to rectify.

Opinions differ as to thickness tolerances, and indeed a thick strip with a given percentage thickness variation across the width will be easier to correct than a thin one. During the work at Swansea it has been assumed that tight thickness tolerances will be needed and a target of $\pm 2\frac{1}{2}\%$ has been set.

Such a tight specification demands that the gas atomising process used for the majority of such work is very closely controlled to ensure that equal weights of metal are deposited at each point across the width. Many ways of accomplishing this objective have been considered and several have been tried in practice in Swansea and elsewhere.

Obvious methods for spreading the deposit, such as deflection of the spray by polished solid surfaces, or by jets of gas operating on the 'bat's wing' principle, were found either to be impracticable or inadequate when considered against the above target specification.

The first promising attempt to obtain uniformity across a significant width used three matched nozzles on an experimental plant designed to produce 6mm thickness Al alloy spray-cast strip 0.5m wide up to 6m in length, at a throughput rate of 30 kg/min. This work was described in a paper[1] in 1972. The best tolerance achieved was $\pm 7\%$ across the width, which proved to be not good enough for satisfactory rolling. Moreover, the unreliability of metal flow from small matched nozzles was a persistent problem. Nevertheless, multiple-nozzle systems were not abandoned but shelved until the potential for maximising uniform deposit width from a single source had been more fully explored. It must be said at this juncture that in 1972 methods of controlling molten metal delivery from matched nozzles was very unsophisticated, and unequal erosion of nozzles rapidly caused disturbances to the pattern of the deposit. With modern methods of controlling metal flow by electromagnetic means the possibility of using matched nozzles is undoubtedly extended.

By 1974 the potential for deflecting a single atomised spray of particles by a pair of alternately activated gas jets was being explored,[2] and early success led to a practical device being used in production.[3] The technique was so promising that a large concentrated effort was made, and is still being made, to develop a device which meets the target specification referred to above.

Pneumatic scanning was further enhanced by a technique in which a series of secondary jets of gas was allowed to impinge on the spray of atomised particles produced by a primary atomiser near its source in repetitive sequential bursts. Each nozzle was aimed at a different point across the width of the strip forming a fan-shaped array. By timing the bursts and direction of the secondary gas using a rotary valve it was possible to scan the strip at speeds up to 100 Hz. A controlled overlap of the gas ports on the rotary valve allowed strip contours close to the target tolerances to be reached with 20% overspray.

The techniques of scanning are being pursued by several organisations and without doubt great improvements will occur in the future, which adds to the attraction of single-nozzle systems, although clearly other single nozzles may be added downstream to increase throughput and improve the properties of the product.

The limits to which a nozzle can effectively scan are not yet determined. At Swansea we do not scan beyond 0.5m with an included angle of 90° because of experimental limitations, but there is no a priori reason why scanning width should not be increased by increasing the atomising distance and reducing the chilling effects on the particles in flight by using a heat-retentive spray chamber. Alternatively a second or third nozzle could be added laterally and matched to the others.

An important feature of any potential production process is the yield of usable product that can be attained. It is particularly important in sprayforming because there is necessarily a certain amount of overspray, i.e. particles either missing their target or bouncing off their target. To reduce overspray it is desirable to generate a relatively narrow cone of atomised particles so that, when scanned, only a small proportion of the particles will become overspray at the edges of the strip. In this respect scanning has an advantage over matched sprays because matched sprays will generally have a wider spread of each individual cone of atomised particles.

Generally speaking the wider the strip the smaller will be the percentage of overspray. For a 1m-wide strip an overspray of 20-25% would be practicable. A further decrease in the amount of overspray can be obtained by the use of horizontal subsidiary gas jets to blow part of the outer cone towards the centre of the strip. However this device must be used with caution because of its effect on strip shape and the fact that the amount of gas used is greater.

2. Separation from Substrate

Spray deposition on a substrate surface always involves a compromise of conflicting tendencies. On the one hand a very smooth polished substrate surface is desirable because, on replication, it produces a smooth deposit surface. On the other hand, such a smooth or polished surface will not retain the first layer of splatted particles. The particles will immediately detach themselves from the substrate and either be blown away by impinging atomising gas or slide sidways until halted by an adherent splat when further movement will be prevented. In this case a 'sand dune' effect occurs where buildup of deposit takes place in patches leading to an orange peel-type surface.

It must always be remembered that splat particles are only castings in miniature and show all the characteristics of castings. On solidification shrinkage occurs, and on cooling below the MP still further shrinkage takes place leading to internal tensile stresses in the splat and corresponding compressive stresses in the substrate. If the adhesion to the substrate is poor because of the presence of non-wetting oxide films on its surface and because of the absence of angular asperities that could result in mechanical interlocking, then the internal stresses will cause separation at the interface.

During spray deposition it is, of course, essential to obtain temporary adhesion of the first splats. The subsequent splats cause no problem because the surfaces presented by the top surfaces of the first splats are either rough, if fully solidified, or still partially liquid.

To achieve essential temporary adhesion it becomes necessary to use a slightly roughened substrate surface. Heating the substrate also helps because it decreases the rate of shrinkage of the first solidified splats, while annealing effects within the splats may reduce internal stresses.

At the end of spray deposition it is necessary to detach the deposit from the substrate, and in this respect good adhesion presents a problem. This is specially true with thin strip where the forces available for detachment are limited by the cross-sectional area of the strip, whereas the adhesive forces are directly related to the surface area at the interface. Fortunately, as strip thickness increases and the temperature falls, the internal stresses become larger and detachment is easier.

In this way the successful sprayforming of strip depends on working out a compromise between adhesion and detachment. It is usually found that a slightly roughened cool substrate having an adherent oxide film provides a satisfactory compromise and that heating the substrate tends to aid adhesion.

The forces needed for detachment are usually quite small. Provided the deposited strip is of adequate thickness - say 1 mm or greater - the pull of a coiler using a starter strip is adequate. However, it frequently happens that small pieces of the deposit stick tightly to the substrate. Clearly these must be removed before the substrate is re-used.

One advantage of a link-belt system such as that proposed by Mannesmann is that the links can be arranged to cause mechanical detachment of the deposit as they pass over the forward pulley.

The system in which the spray is directed into the nip of two contra-rotating rolls has the advantage that it provides an automatic stripping mechanism. This is caused partly by an extension of the deposit as it passes through the rolls and by the geometry of the system. However, major problems of gas flow in an axial direction occur with such equipment and at present it is not a favoured procedure.

A similar benefit is conferred by SSP, but this will be dealt with later in the paper.

3. Structure and Properties

Porosity is one of the problems with sprayformed materials in the 'as-deposited' condition. Although it can be reduced to values as low as 0.2% it is virtually impossible to get rid of it completely without further hot or cold working. The difficulty derives mainly from the statistical distribution of the position, size and temperature of particles in the spray. Although porosity can be reduced to very low levels by ensuring that incremental solidification takes place, it is usually necessary to carry out further working to obtain the density and properties required in a finished product.

In the case of strip, porosity is a special problem at the substrate interface. This is caused by two main factors. Firstly, atomised sprays consist of a random distribution of molten particles with the spray density highest in the centre and lowest at the outer fringes. When a substrate traverses beneath a spray the low spray density outer fringe is the first part to be deposited. In these circumstances of low spray density, splats formed on the cool substrate have time to freeze completely before the subsequent splats impact them. The effect of subsequent depositing on top of an earlier fully frozen splat is to leave small voids at the interface causing localised porosity.

Another factor leading to the same problem is that the greatest amount of chilling occurs when the first layer of splats is deposited on the substrate. After a few layers of splats have been build up the rate of heat extraction is reduced and more splats retain a liquid top surface on which subsequent particles impinge.

The total effect is that porosity is highest at the substrate interface and decreases rapidly as the deposit grows in thickness.

The metallurgical structure varies at the same time. The first layers of deposit often show a splat structure in which the original splat boundaries can be recognised. If the spray density is high compared with the chilling effect of the substrate, subsequent layers will form a cellular structure in which new grain boundaries completely obliterate the original splat boundaries. The new structure is caused by incremental solidification[4] in which later spray particles splat on earlier splats that are not fully solidified. In this way splat boundaries are obliterated and porosity is greatly reduced. Examples of the two types of structure are shown in Figs.1 and 2 depicting cross-sections of the two types of deposit.

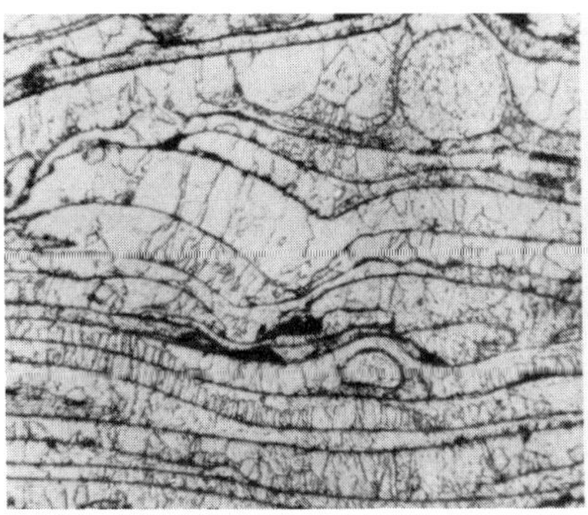

Fig.1. Gas atomised and spray deposited at low spray density.

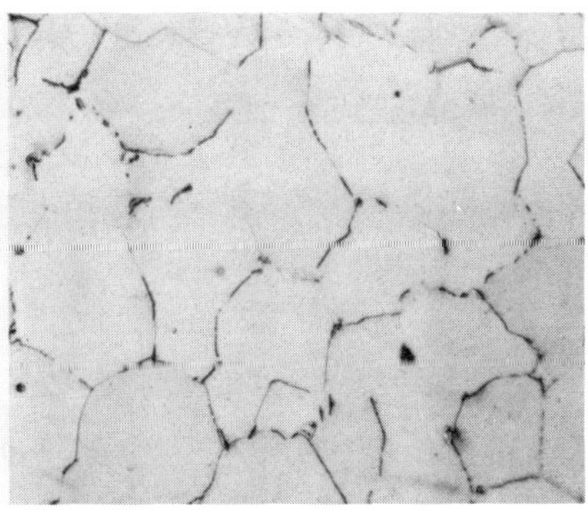

Fig.2. Gas atomised and spray deposited at high spray density.

Figure 1 shows characteristic splat boundaries and porosity which lies between 5% and 2%. Figure 2 shows complete obliteration of splat boundaries and replacement by a new equiaxed cellular structure together with very low porosity, typically between 1% and 0.2%. Generally speaking splat structures occur in rapidly cooled deposits and incrementally solidified structures with less rapid cooling.

In both the above cases segregation is extremely low. The reason quite clearly is that segregation can only occur within each splat as all particles will have the same composition as the original melt if atomised in an inert atmosphere. The effect is quite striking. For example, a 0.5 tonne ingot of highly alloyed tool steel showed the structure shown in Fig.3 (longitudinal section) after rolling to 25mm round rod where the carbides are in the form of highly undesirable stringers. By contrast, Fig.4 (also a longitudinal section and at the same magnification) was taken from a 1-tonne spray casting and rolled to 25mm round rod in precisely the same way. It is clear that the uniform distribution of carbides provides a much preferred product, which in fact turned out to have enhanced impact properties and edge retention.

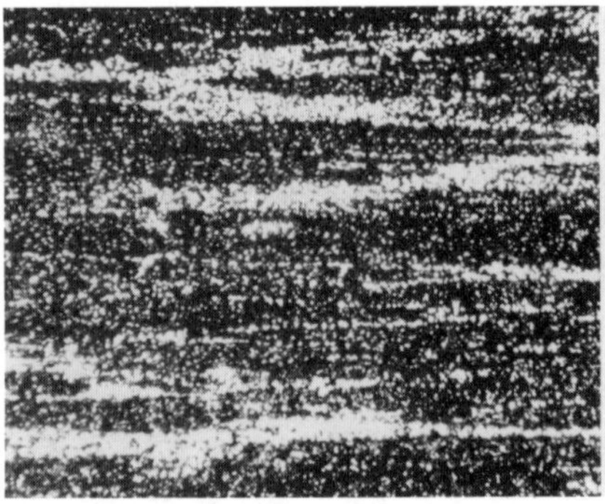

Fig.3. Heavy stringers of carbide in tool-steel product made by conventional casting and hot rolling.
[Longitudinal section x 100]

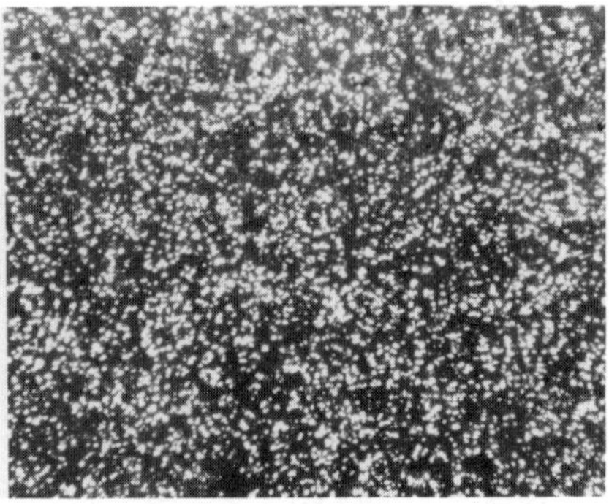

Fig.4. Fine distribution of carbide in tool-steel product made by CSD and hot rolled.

[Longitudinal section x 100]

Special Benefits of Sprayformed Products

The combination of near-zero segregation and fine metallurgical structure confers on sprayformed products enhanced mechanical properties and in particular much greater ease of hot and cold working. Two examples of this characteristic in the form of strip can be given.

The aluminium alloy 7091 is a complex, highly alloyed material that is needed in strip form for the development of advanced aircraft. It is so highly alloyed that it cannot be made in strip form by rolling from conventional continuous castings. As a consequence, it is manufactured by a powder-metallurgy (PM) route, which involves making the powder, canning, HIPing, extruding and rolling. This route is very long and expensive; moreover, the oxide content of the final product is high because of its powder origin.

When the alloy was sprayformed on a rotating drum at Swansea[5] and subsequently hot and cold rolled at the Lockheed laboratories in Palo Alto, California, it was found that the strip produced had properties at least as good as those obtained by the PM route and the oxide content was only one-tenth of the PM product. In other words, the sprayforming route enabled rolling to take place easily and at the same time the route was greatly shortened.

Similar behaviour was shown by a 6.5% silicon steel. This steel is likely to be required in the form of thin (0.5mm) strip for transformer laminations as it is the optimum composition with regard to magnetic and electrical properties. Unfortunately, as the silicon content rises the difficulties of rolling increase enormously. As a result it is normal practice to limit the silicon content to a maximum of 3.5% despite inferior electrical properties.

In contrast, a 6.5% Si steel can be sprayformed as a thin strip and then can easily be hot and cold rolled to 0.5mm thickness. Once more the ease of fabrication is a direct consequence of the low segregation and fine structure.

Other special benefits of the sprayforming process are that laminates can be made which are perfectly bonded and will not delaminate under the most severe conditions. An example is 18/8 stainless steel deposited on a low-carbon-steel base. In this case the tundish was first filled with low-carbon steel to deposit the base, after which it was filled with 18/8 when atomising continued. The result was a perfectly bonded product produced in a single operation. If necessary composition can, of course, be graded between one alloy and the next.

Other benefits are the ease with which particle composites can be produced by spray co-deposition. In this case the matrix alloy is spray-deposited in the usual way but the particles entrained in gas are introduced into the spray so that co-deposition takes place. The net result is that a very uniform distribution of the particle phase can be obtained without the segregation that often occurs when using the alternative PM route.

A further benefit is that it is possible by spray co-deposition to make composites containing particles that are inherently soluble in the matrix. Such products are virtually impossible to make by PM means. Examples of this are the use of chilled iron grit particles in an aluminium matrix, graphite in mild steel[6] and diamonds in stainless steel.

Special Sprayforming Procedures for Strip

Centrifugal spray deposition (CSD) has been described in the literature[7] where it has been shown that excellent products can be made without the use of large quantities of gas, but with a limitation on the geometric shape to large rings or tubes.

An alternative route to sheet is offered by the CSD process where the throughput is relatively small and where batch operation is acceptable. It has been shown that sheet can be made by first producing a large-diameter thick-walled tube, as shown schematically in Fig.5, and then slitting the tube vertically and opening it out to form a flat blank, which can be rolled into a sheet.

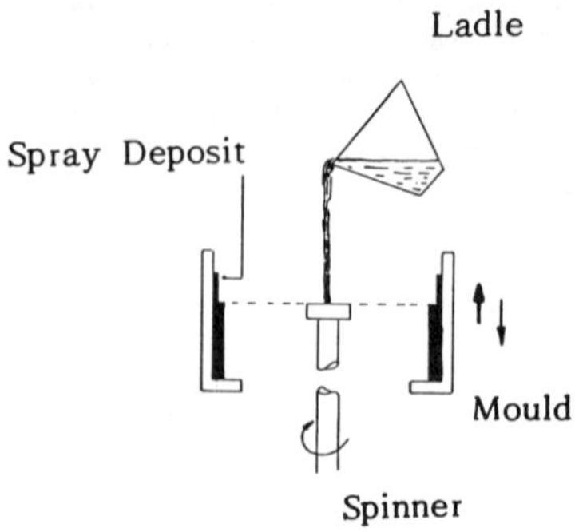

Fig.5. Schematic diagram of CSD equipment for producing a short, thick, large-diameter tube.

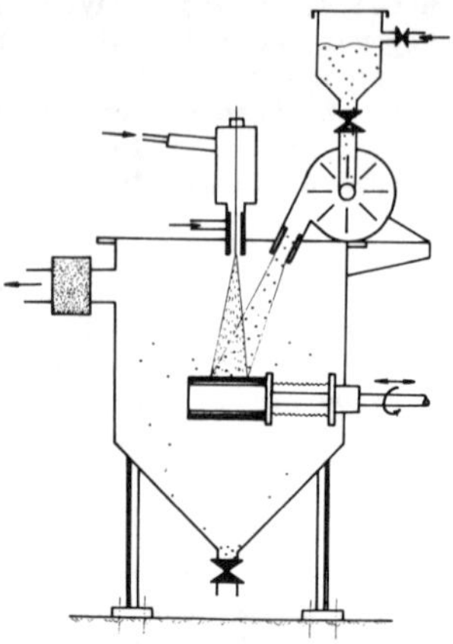

Fig.6. Schematic diagram of SSP using plasma.

Typically a tube, 1.5m diam. x 1m length x 50mm thickness, can be centrifugally spraycast and then slit into 3 parts and flattened to give blanks 1.5m long x 1m wide x 50mm thickness. This can be hot and cold rolled to plate or sheet of considerable size. The process is particularly attractive for titanium where preformed electrodes of titanium can be fed onto a spinning water-cooled copper disc and a current passed between the disc and the electrode to produce an arc. Metal is melted by the arc, forms a solidified skull on the spinning disc over which the remainder of the molten metal passes to be spun out centrifugally and atomised to form a spray of molten particles. The spray takes the form of a very narrow, slightly diverging horizontal disc. By moving the mould vertically a tube can be built up which has all the characteristics of the usual sprayformed products. In the case of titanium alloys, the special advantages are (i) maintenance of purity of the deposited titanium because of skull formation, (ii) very low usage of protective gas, (iii) very low segregation, (iv) a process that can be closely matched to the tonnage requirements of titanium.

The relatively new process of SSP can also contribute to the technology of sprayforming thin strip. SSP has been described in the literature[8] as a process particularly suited to making coated or laminated products and tools and dies. However, it seems likely that very thin strip or foil may be made economically by the process. The principle of operation in the case of the coating of a metal product is shown in Fig.6, from which it will be seen that the spray-deposited metal is peened as soon as it is splatted onto the substrate. The diagram shows an SSP unit in which molten metal is supplied from a plasma gun, but it can equally easily be adapted for a molten-metal feed. The process has the advantage of giving a hot-worked product that is fully dense and in which internal stresses can be controlled at will.

Applied to the manufacture of thin strip or foil the process would seem to offer the possibility of producing hot-worked and fully dense, very thin strip in a single operation. Moreover, the simultaneous peening operation is known to facilitate removal from the substrate roll. It might well be possible to make foil from very difficult alloys by this means.

Problems with the process are firstly that the foil has a top surface that is covered with very shallow peening marks giving a kind of 'hammered' effect. Perhaps more seriously, the concept is still at a very early stage of development and much work needs to be done before an assessment can be made.

The Present and the Future

Sprayforming is rapidly rising in popularity and more and more industrial applications are appearing on the horizon. The various applications differ greatly in the ease with which they can be adapted to industrial production. The making of sprayformed hollow billets is one of the easiest operations and CSD has been carried out on a fairly large (2 tonne) scale while large-diameter solid billets are also at an advanced technological stage.

Making strip and sheet by sprayforming (i.e. spray-rolling) is one of the most difficult processes even though the benefits of success would be enormous. The difficulties have been mentioned above and need not be repeated here. Suffice it to say that resources are now being funnelled into the spray-rolling of strip, and apart from the well-known Mannesmann initiative at least two other major strip pilot-plant developments are at present taking place. However, no strip has yet been made available on the market, and it will be 2 or 3 years at least before this is possible.

The prospect for the future is excellent. The enhanced properties and the short route make spray-rolling a very attractive goal. The fact that scanning technology is now able to offer a way through the difficult area of thickness and shape control is most encouraging, and we can confidently look forward to major developments in the future.

REFERENCES

1. A R E Singer, "Aluminium and Aluminium Alloy Strip Produced by Spray Deposition and Rolling", J.Inst.Metals, 100 (1972), 185.

2. British Patent No. 1455862.

3. "Melt Spray Process in Operation at GKN" - Report in Powder Metallurgy 25(4)(1982).

4. A R E Singer and R W Evans, "Incremental Solidification and Forming", Metals Technology, 10 (1983), 61-68.

5. A R E Singer et al., "A Novel Approach for Rapid Solidification Processing of High Strength PM Aluminium Alloys", SAE Technical Papers Series 841514, 1984, 27-33.

6. S Ozbek and A R E Singer, "Some Special Metal Matrix Composites Produced by Spray Co-Deposition" (paper given at The Institute of Metals Conference, London, UK, November 1987).

7. A R E Singer et al., "Centrifugal Spray Forming of Large-Diameter Tubes", Metals Technology 10 (1983), 105.

8. A R E Singer, "An Introduction to the Process of Simultaneous Spray Deposition Peening of Metals", Metals Technology, 11 (1984), 99-104.

"Recent Development in Producing
Near Net Shape Products by Spray Deposition
(Osprey Process)"

Wolgang Reichelt and Otto H.Metelmann

Mannesmann Demag

Duisburg, W. Germany and Pittsburgh, PA.

Summary

In our opinion, spray deposition is one of the few genuine innovations in the field of basic shaping. Its advantages comprise not only simple and direct obtainment of near net shape products from liquid metal, dispensing with the casting stage and with comparatively complex powder metallurgy; it also delivers a high-quality product.

Furthermore, the process makes possible the production of completely new composite and multi-layer materials, creating new potentials for the designer, the ultimate user of the product.

As with all innovations, certain questions do, of course, remain open with regard to spray deposition. However, a large number of companies around the world are at present conducting intensive development work; for this reason, more knowledge and more certainty with regard to the risks of industrial application of this new technology is achieved with the passing of time

Introduction

The following paper details the present-day development status of the Osprey-Process, which makes it possible to obtain near net shape products while eliminating the casting process or the somewhat more involved powder metallurgy approach.

Casting of Near Net Shape Products
Edited by Y. Sahai, J.E. Battles,
R.S. Carbonara and C.E. Mobley
The Metallurgical Society, 1988

The process is the result of development work performed for more than 15 years by OSPREY METALS LTD., Neath, Wales, and its licensees; Mannesmann Demag has been working in this field for some three years, concentrating on the development of plant for the process.

It is now possible to demonstrate that the process has reached the stage of development to fully permit industrial application.

First should be presented the basis of the process followed by status of development for various applications, and industrial plant concepts.

Process Principle and Development Status

If one examines the diagram (Figure 1) for the spray deposition process, it can be seen that it involves a large number of elements already familiar (particularly from powder metallurgy):

- tundish containing liquid metal,
- an outlet nozzle,
- a gas atomization system.

The substrate for the atomized liquid metal shown in the figure is practically the only point of difference as compared to a powder production system.

This "minor" difference, however, leads to a completely new process with completely new possibilities for the production of intermediate products or so-called preforms.

The processing principle:

Spray deposition consists of three processing stages:

- <u>Atomization</u> of a jet of liquid metal using an inert gas.

- Cooling of the droplets generated and withdrawal of the heat of solidification by the gas during flight of the droplets.

- Compaction of the particles, exploiting their high kinetic energy and high temperatures, on a substrate.

The liquid metal flows through a simple ceramic nozzle out of a tundish and is atomized by a gas atomization system.

Nitrogen is generally used as the atomization gas; however, other gases, for example argon, can also be used.

The droplets produced cool during flight - before striking the substrate - to a temperature between their solidus and liquidus temperature. They then compact at high energy to a dense product on the substrate.

The material produced in this way can then be cooled under defined conditions.

The process is thus characterized by rapid solidification followed by defined cooling.

Theoretical investigations - model calculations - indicate that, in accordance with the droplet size distribution of the particles generated, completely solidified, completely liquid and undercooled droplets must be present at the point of collision with the substrate. The correct proportion of the various states is the precondition for production of a dense product (Figure 2).

Optimum processing parameters depend on the material and the product required; delineation of them is one of the objectives of the present development project.

The process is also typified by a great flexibility with regard to capacity and the product geometrics which can be generated.

Spraying rates of some 10 to 50 kg/min. per nozzle are used for production of special alloys - superalloys, and aluminum composite materials; rates of 150 kg/min. per nozzle, equating to 9 t/h per nozzle, can be achieved without difficulty in the production of such materials as steel plate.

With regard to the geometries possible, it should be noted that rotationally geometrical small components, round bars, blooms and pipes, in addition, strip and sheet can be produced, according to the geometry of the substrate.

The figure 3 provides a summary of near net shape products, spray deposited presently, with details of the materials used. They range

from tubes through round bar and blooms, and flat product, to coated product.

This list very nearly encompasses the current status of development.

Production of Pipes

Development work for the production of pipe has been performed mainly by SANDVIK Steel AB, Sandviken, Sweden.

Intensive development work on the production of thin-walled and heavy-walled tubes using the OSPREY process was carried out for several years at Sandviken.

Thick-walled tubes and hollow bars are appropriate preforms for the production of rolling mill roll of special alloys or as extrusion slugs.

A pilot production plant for manufacture of tubes of up to 8 m length and 400 mm outside diameter and up to 40 mm wall thickness has been in operation at Sandviken, Sweden since last year.

Sandvik Steel has achieved spray deposition of tubes with such a good quality that the spray-deposited tube blanks can be directly further processed by means of cold pilger rolling after machining of the top and bottom surfaces, (i.e. the spray-deposited tube blanks are of optimum density).

Figure 4 provides a general impression of a typical spray deposition plant.

The most important part of the system is the spray chamber shown in the center of the figure, and a calidus ladle mounted on it, in which temperature controlling of the material for spraying is performed, and also facilitates regulation of delivery pressure .

The material is sprayed onto a thin-walled welded-sheet tube, which is drawn while rotating through the spray chamber using a drawing machine. The installation of two such machines makes continuous production possible.

After completion of spraying, the substrate tube is mechanically removed. Following machining of the surface, the blank can be further worked on a cold pilgering unit to produce a finished tube.

As far as pipe production is concerned, it can be stated that at present the industrial application of the spray deposition process is entirely feasible. Research and development work in this field is devoted at present to the optimization of plant and processing parameters, and is no longer concerned with the clarification of the fundamental questions of principle.

Production of small components, billets and round bar

The second broad range of application, for which pilot production plants have already been commissioned, is the field of rotationally symmetrical solid bodies:

- disks, e.g. as preforms for forges and presses,
- round bar and billets, also as feed material for forges and presses

As mentioned initially, such bodies can be generated by means of spray deposition onto a rotating table. Long bars or round billets can then be produced by lowering the substrate table (Figure 5).

For the sake of completeness, it should be noted that the production of round billets is also possible on horizontal machines; OSPREY has such a machine.

The initial interest in the production of round bars and billets was in the aluminun industry. Several major companies already have pilot plants in operation.

There is great interest in these sectors in the production of genuine composite materials using OSPREY technology. For instance, the incorporation of silicon carbide into aluminum, which is relatively simple. For this purpose, SiC powder is added to the atomization gas and thus incorporated homogenously into the metal. Incorporating of other powder materials has also been successfully demonstrated. Tests aimed at the incorporation of fibres are under preparation.

Production of Flat Products

Spray deposition is without doubt an ideal process for the production of rotationally symmetrical components. The effects on product geometry of the irregularities in material distribution in the spray jet, are all but eliminated through the rotation of the collecting surface.

The situation is different in the case of the production of a flat product. The passing of a flat substrate surface through a spray jet would lead to the formation of a strip with a thickness profile matching the material distribution in the spray jet.

It therefore appears an obvious step to use a rectangular nozzle for production of flat products in spray deposition, instead of the simple round nozzle used for all other products.

However, after tests on the flow conditions, we decided to abandon the concept of a rectangular nozzle. Instead, we began work on the attainment of a uniform sheet or strip geometry using a spray jet moving at right angles,in a scanning fashion,to the direction of travel of the substrate.This mode of operation produces the same kind of benefits which are obtained in the rotational production of symetrical components.

A unit was constructed at the Mannesmann Research Institute for the performance of such tests (Figures 6).

This unit makes possible spray deposition of sheet of a maximum of 2 m length and 1 m width. Since the principle problem in the obtainment of flat product is uniform thickness distribution across width, it was decided not to incorporate a facility for spray deposition of longer-continuous-sheet.

Up to the end of 1987, all the tests discussed here - up to sheet widths of 1 m - have been performed using a single nozzle system; the unit has been equipped for two-nozzle operation since March, 1988.

The actual spraying tower can be seen in the middle. To the left of it is an area in which a substrate can be preheated on a transport car, and to the right of this is an area in which the deposited material can cool down under an inert gas atmosphere after the spraying process.

The test procedure is as follows:

- The substrate - we generally use a steel sheet - is preheated on the car in the left-hand station.

  Preheating temperature influences the quality of the spray-deposited product.

- The substrate is then moved on the transport car at a prescribed speed through the spraying tower into the right-hand section of the unit.

The speed of the transport car determines deposition thickness at a given spraying rate; wide plate, at thicknesses of 8 to 40 mm, can be produced.

The quality and geometry of the plate produced in this way depends primarily on:

- substrate preheating temperature,
- the gas: metal mass flow ratio at atomization,
- superheat temperature of the liquid steel.

The figure 7 provides an impression of the internal structure of a spray-deposited plate:

A certain degree of porosity can be seen in a zone approximately 2 mm thick in the lower section of the plate, i.e. in the zone of contact with the substrate, whereas the material immediately above is completely dense.

The thickness of the porous base zone depends, inter alia, on the temperature of the substrate plate; the colder the substrate, the thicker this porous zone will be.

In the case of the plate shown in the figure, substrate temperature was approx. 800 degrees C.

The pores in this zone are closed, i.e. they do not communicate with the surface.

It was established in rolling tests that hot deformation of approx. 50% suffices to obtain the desired absolutely dense material (Figure 8).

Samples were heated under normal atmosphere for test rolling; no defects, as would be caused, for instance, by scaling of the formerly porous contact zone area, were established.

It can be stated that the quality of the spray-deposited plate completely fulfilled all expectations.

Physical characteristics are superior to those of cast material; the material is isotropic, and free of macro-segregation.

Thickness distribution is also excellant. Mean deviations from nominal thickness fluctuates between 0.5 and 0.8 mm, irrespective of total plate thickness (Figures 9).

In designing production plants (Figures 10) for the manufacture of strip and plate, we believe that hot compaction should be done in line, under inert atmosphere. Both the porous contact zone and surface roughness can be eliminated, without difficulty, in this way.

Our experience up to now indicates that multi-nozzle systems - series and parallel nozzle arrays - will make it possible to construct plants with highly flexible output and product dimensions for the production of flat product using the OSPREY process.

Furthermore, it is perfectly apparent that the process is also suitable for the production of multi-layer products and coated products.

Production of Multi-layer Materials and Coated Product

Series installation of nozzles makes it possible, without difficulty, to spray deposit differing materials in sequence, e.g. stainless steel onto plain carbon steel, or non ferrous-materials onto steel substrates.

It is equally possible to coat an intermediate product with a metallic material. For this purpose, the surface of the intermediate product, e.g. strip or tube, must be heated, in order to ensure good fusion between the basic and the deposited material.

A pilot plant is already in operation for coating of tubes with wear-resistant layers; coating of strip, in some cases with non-ferrous metals, has already been successfully tested.

Production Plant Concepts

Having explored the potentials of the process, we shall now turn our attention to plant engineering.

The process makes it possible to produce intermediate products of differing geometrical forms at a relatively low investment cost.

A spraying chamber with a movement and transportation system specific to the product to be spray-deposited is required (Figure 11).

All the other systems required to operate the process, for instance:

- gas supply,
- gas regulation system,
- possible gas recycling system,

are not specific to any particular product, and can therefore be used in conjunction with several, variously equipped, spraying chambers.

Figure 1

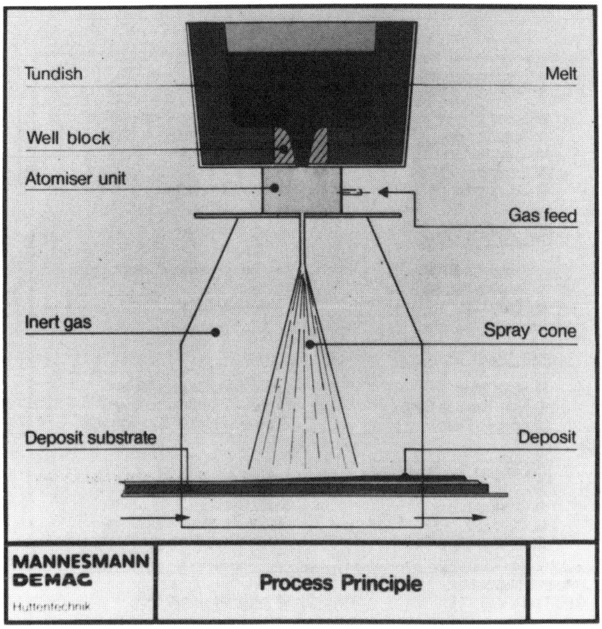

Figure 2

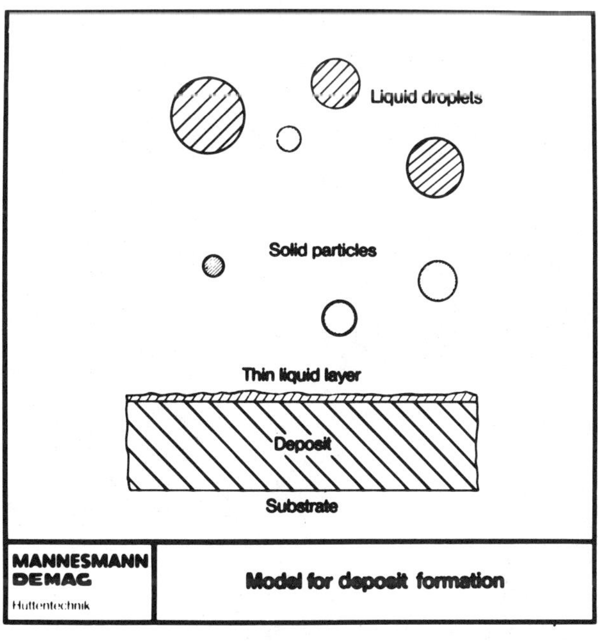

Figure 3

Tubular Products

● Tubes	– Stainless Steels/Superalloys
● Rolls/Rollers	– Cast Irons/High Alloy Steels
● Cylinder Liners	– Aluminium Alloys
● Turbine Rings	– Superalloys
● Bearing Rings	– High Speed Steels

Bars/Billets/Ingots

● Extrusion Billets	– Aluminium Alloys/Composites
● Forging Blanks	– Superalloys
● Round Bar	– High Alloy Steels

Flat Products

● Steel Strip	– Low Alloys/Silicon Steels
● Non-Ferrous Strip	– Copper/Aluminium Alloys
● Armour Plate	– Steel/Aluminium/Composites

Coatings/Clad Products

● Coated Tubes	– Boiler Tubes
● Clad Bar	– Stainless Steel/Tool Steel
● Coated Strip/Plate	– Wear Plates/Bearing Alloys

MANNESMANN DEMAG Huttentechnik

Near-net shape products

Figure 4

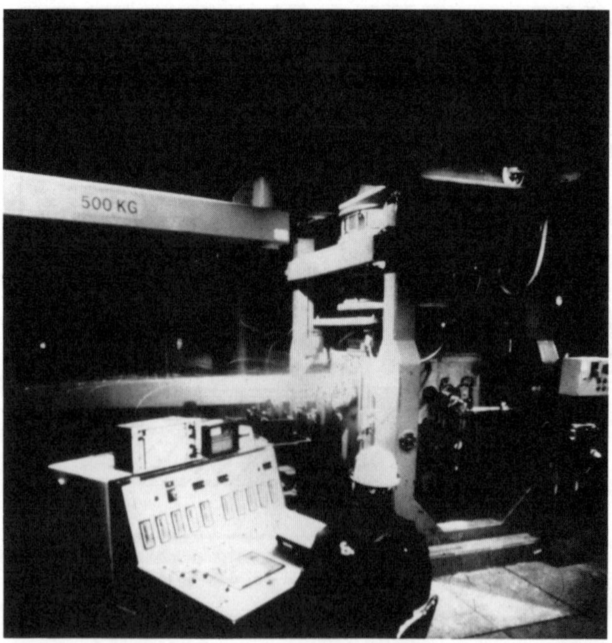

Figure 5

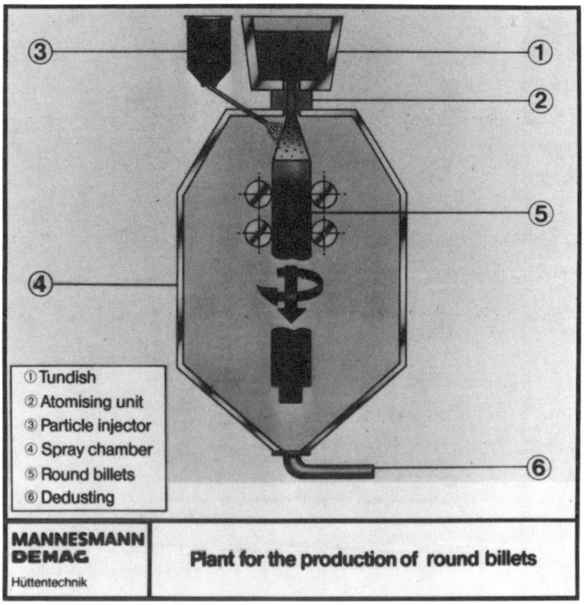

Figure 6

Figure 7

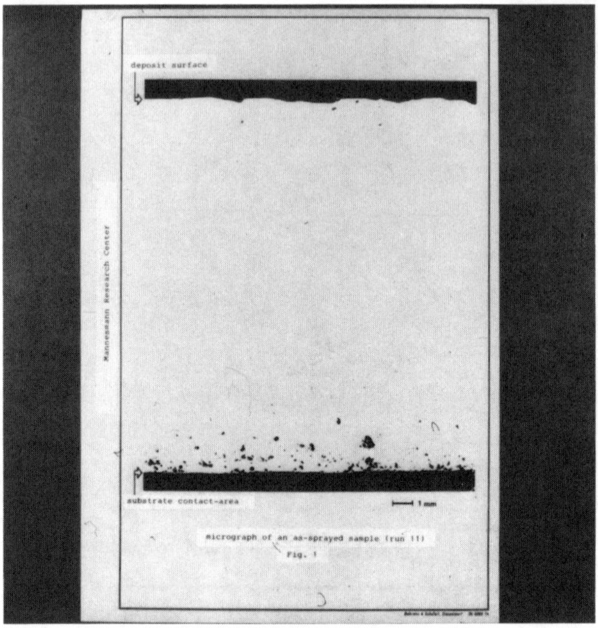

Figure 8

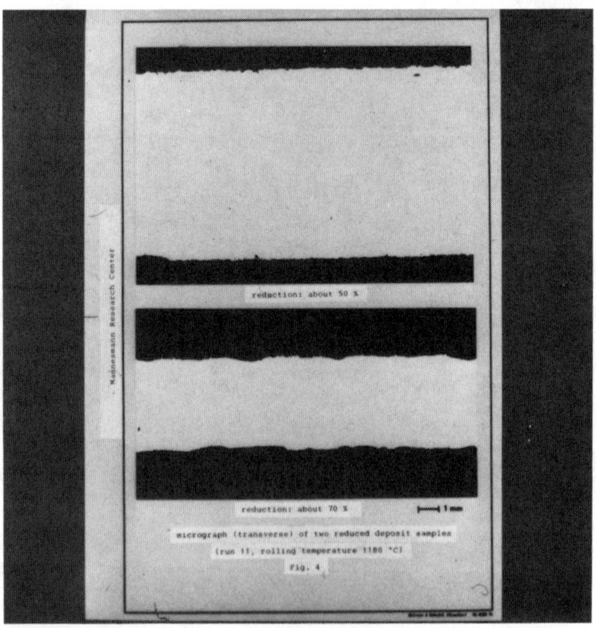

Figure 9

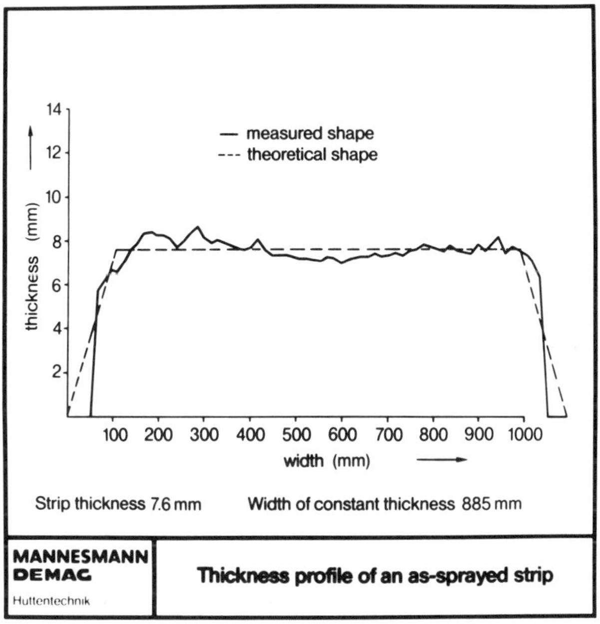

Figure 10

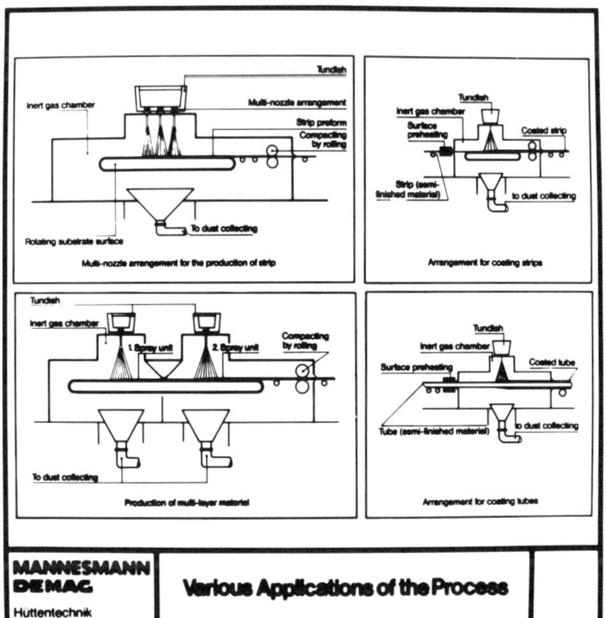

Figure 11

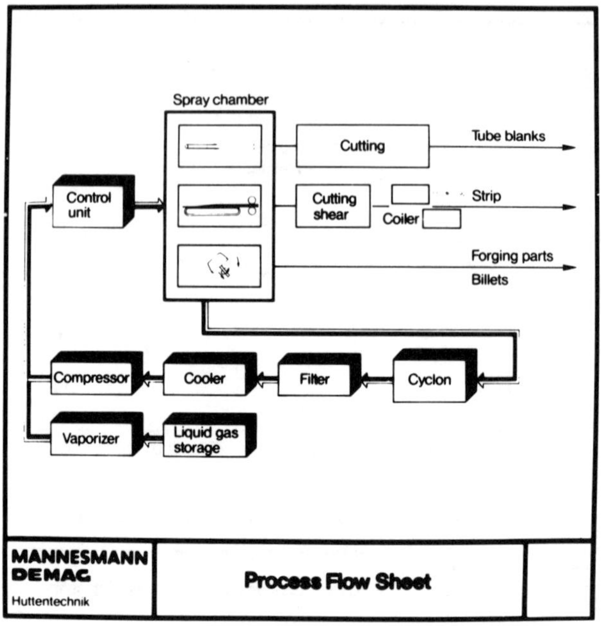
Spray chamber
Cutting
Tube blanks
Control unit
Cutting shear
Strip
Coiler
Forging parts
Billets
Compressor
Cooler
Filter
Cyclon
Vaporizer
Liquid gas storage
MANNESMANN DEMAG
Huttentechnik
Process Flow Sheet

NOZZLE-ASPIRATED METAL FORMING

L. Watson, S. Ploger, J. Fincke and J. Key

EG&G Idaho, Inc. P. O. Box 1625
Idaho Falls, ID 83415

Abstract

The Idaho National Engineering Laboratory is developing a novel approach to spray forming molten metals—using converging/diverging nozzles to aspirate liquid metal into an atomizing gas stream. Besides important low pressure, self-governing advantages in process control, such devices spray spatially uniform plumes and substrate deposits, with dynamic feedback on product characteristics from advanced instrumentation. Rapid incremental solidification dictates small grain sizes, while porosity observed from gas entrapment is minimal (~1%), even in complex molds. The uniform droplet distribution and empirical scaling relationships have enabled extending a bench-scale circular design to a true linear nebulizer with multiple liquid orifices. This prototype produces large aspect ratio, near net shape strips (4 in. wide) with negligible thickness variations (~1%) and minimal overspray losses (~5%). For convenience, this work has been performed primarily in tin, but recent efforts included aluminum and copper, and low carbon steel is the principal goal.

Casting of Near Net Shape Products
Edited by Y. Sahai, J.E. Battles,
R.S. Carbonara and C.E. Mobley
The Metallurgical Society, 1988

Near Net Shape Processing

By Spray Atomization and Deposition

+E. Lavernia and #E. Gutierrez-Miravete

+Assistant Professor, Department of Mechanical Engineering, University of California, Irvine, CA 92717. Formerly Research Associate, M.I.T.
#Curriculum Chair and Assistant Professor of Metallurgy, School of Engineering and Science, The Hartford Graduate Center, Hartford, CT 06120. Also Research Affiliate, M.I.T.

Abstract

A near net shape processing approach currently being utilized to overcome the drawbacks associated with traditional powder technologies is spray atomization and deposition (i.e., Osprey, and Liquid Dynamic Compaction). In addition, spray atomization and deposition technology is presently being studied to assess its potential as an alternate route for the fabrication of metal matrix composites.

In order to derive maximum benefit from the resulting structure and properties of spray deposited materials, it will be necessary to develop an understanding of the principles governing the evolution of the microstructure. The microstructural evolution during spray atomization and deposition can be separated into two distinct but closely related stages. The first stage consists of those phenomena that are primarily active in the atomized spray prior to impact. The second stage begins after the droplets have impacted the deposition surface and alteration of the microstructure from impact and annealing must be considered. An important aspect of the microstructural evolution is the resulting equiaxed grain morphology; this has been documented for numerous spray deposited materials, regardless of their composition.

In this paper, the solidification heat transfer phenomena taking place during spray deposition of a carbon steel and an aluminum alloy are analyzed and compared using a finite difference method based on the enthalpy formulation.

Casting of Near Net Shape Products
Edited by Y. Sahai, J.E. Battles,
R.S. Carbonara and C.E. Mobley
The Metallurgical Society, 1988

Introduction

Near net shape processing has become an attractive alternative for the commercial manufacture of metals and alloys as a result of economic pressures from abroad, and the anticipation of limited improvements in conventional technologies. The development of near net shape processes has been accelerated by the increasing pressures on the metal forming industry to reduce its costs, improve structures and properties, and bolster its efficiency. In a broad sense, near net shape processing encompasses a wide variety of technologies; these include powder metallurgy (PM), continuous casting, and more recently, spray atomization and deposition. An underlying objective, common to all three technologies, is to manufacture a product which is near to the ultimate size and structural requirements.

Because of the numerous advantages offered by PM technologies, Powder Metallurgy (PM) continues to achieve widespread commercial acceptance by the automotive, aerospace, tool steel, appliance and other industries. The use of PM technologies decreases machining, joining, and finishing requirements; it can also result in improved materials yields [1] by eliminating scrap losses of 30-80% through net and near-net shape production methods. In addition, PM components often have longer and more reliable service lives [2]. Powders are produced industrially by a variety of techniques, including atomization, chemical reduction, mechanical milling, and electrolytic deposition [3]. Of these techniques, atomization has become the most popular in industry [2,4] mainly because of its production rates, flexibility, and relatively low costs.

Continuous casting processes can be classified into two broad categories: thin strip casting and thin slab casting. The technology of direct net shape casting of very thin strip is at a very advanced stage of development as evidenced by the numerous processes developed to date. These include: melt extrusion, chill block melt spinning, planar flow casting melt drag processes, and melt extraction [5,7]. Unfortunately, thin strip casting is severely limited by the resultant throughput rates. The University of Basle and Concast Standard, for example, developed a one wheel caster capable of reaching speeds of 25 m/sec [8]. The resultant throughput, however, amounts to just a few kilograms per second. Also, since in thin strip casting the solidification events depend on conduction of heat into a substrate, and convection to the surrounding environment, there is usually a trade off between microstructural refinement, (as measured by secondary dendrite arm spacings) and thickness (see Table I); and between strip thickness and production rates (see Figure 1).

Table I: Cooling Rate of As-Cast Strips and Slabs Estimated from Secondary Dendrite Arm Spacing [8]

	Thickness (mm)	Distance From Surface (mm)	Secondary Arm Spacing (μm)	Cooling Rate ($^{\circ}C/sec$)
	0.65	0.1	2 ~ 4	$10 \sim 10^4$
	0.65	0.3	2 ~ 4	$10 \sim 10^4$
Twin	1.10	0.2	3 ~ 5	$\sim 10^3$
Roll	1.10	0.6	4 ~ 7	$10^2 \sim 10^3$
	1.70	0.2	3 ~ 5	$\sim 10^3$
	1.70	0.8	4 ~ 8	$10^2 \sim 10^3$
	150	<10	~ 40	$\sim 10^1$
Conven-	150	~70	~ 250	$\sim 10^{-1}$
tional				
CC	200 ~ 300	<10	~ 70	$\sim 10^0$
	200 ~ 300	~100	~450	$10^{-1} \sim 10^{-2}$

Finally, the as-cast structure resulting from thin strip casting is usually dendritic, making it necessary to include an annealing step in order to transform it to a non-directional structure [9].

In view of the limitations faced by thin strip technology, a sizable share of the current research and development activities are being devoted to thin slab casting technology.

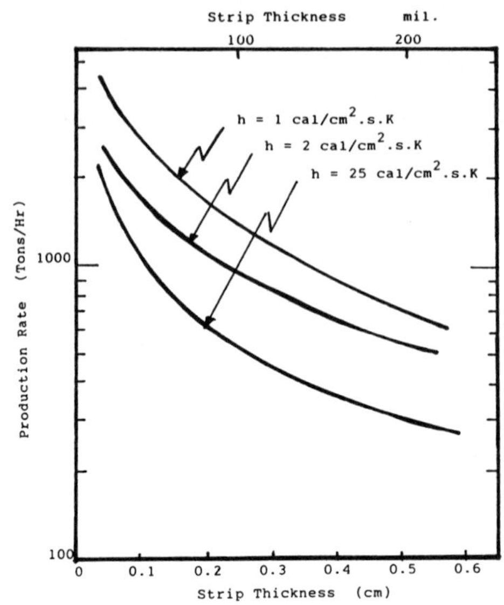

Figure 1 - Calculated production rates for thin strip casting as a function of heat transfer coefficient and strip thickness [9].

The technology of thin slab casting is based on the traveling mold concept which dates back to the ideas of R.W. Hazelett in the 1940's [8]. Current applications of this technology include: a two belt caster (Nucor 1984 [8]), a two wheel caster (Pechiney [10]), and a single wheel or belt caster (British Steel [10]). Although continuous casting of slab and strip is being done commercially for aluminum and copper, there are issues that need to be addressed; these include side containment, surface quality, and centerline segregation as a function of thickness. Moreover, continuous casting of higher temperature materials (i.e., steel) is complicated as a result of severe erosion of crucibles and nozzles, aggressive chemical environments and high mechanical stresses.

Spray atomization and deposition technology is an attractive alternative for the manufacture of reactive (i.e., Al, Mg) and high temperature (i.e., Fe) materials into near net shapes because of the relatively low temperatures involved, and its potential for improved structural control during high volume manufacturing.

In spray atomization and deposition, a stream of molten metal is disintegrated into a fine dispersion of droplets using high energy inert gases (i.e., Ar, He, N_2). The resultant distribution of metallic droplets is directed towards a substrate where it impacts and builds up into a deposit of predetermined thickness. The first high velocity droplets impact directly against the substrate surface, and then themselves act as quenching substrate for their successors. High rates of solidification are achieved as a result of the thickness of the splats, and the rapid heat extraction during flight. A number of property improvements have been reported for spray deposited materials; particularly interesting are those discussed for aluminum alloys [11-13], ferritic steels [14,15], nickel based alloys [16-21] magnesium alloys [22-24] and magnetic alloys [25-26]. A review on spray deposition of metals and alloys is presented in [27]. One application of considerable interest currently being studied [28] is using spray atomization and deposition to fabricate metal matrix composites.

Recently, the potential of using spray atomization and deposition technology for the manufacture of low carbon steels into near net shapes has attracted considerable interest [28,29]. In the less performance oriented, larger volume markets such as the steel market, spray deposition technology is appealing because of its potential to improve materials utilization, decrease energy consumption, and improve structure and properties. Materials substitution in these markets will depend on achieving competitive cost levels through efficient processing. Therefore in order to achieve optimum control over the resulting structure and properties of spray deposited materials, it will be necessary to develop an understanding of the various heat mass, solidification, and momentum transfer phenomena that take place during spray atomization and deposition.

In spray atomization and collection processes, the microstructural characteristics of the spray deposited material depend to a great extent on the condition of the droplets prior to impact, i.e., the relative proportions of liquid and solid present, degree of undercooling, temperature, velocity, and the size of the microstructure in both the partially and fully solidified droplets, etc. It is also important to know the size and distribution of the droplets, because the size is strongly related to the amount of heat that must be dissipated from the droplets in flight prior to impact.

The objective of this work is to use previously developed mathematical models of spray deposition [31-34] to study the effects of crucial processing variables on the microstructural evolution of two materials with different thermal properties: aluminum and steel.

Formulation

In order to study spray atomization and deposition, it is useful to separate the process into two distinct but closely related stages. The first stage consists of those phenomena which are primarily active in the atomized spray prior to impact. The second stage begins after the droplets have impacted the deposition surface, and alteration of the microstructure from impact and annealing must be considered. The results of previous studies [15,32,34] show that the variables which are relevant to the atomization stage, prior to impact with the substrate, are: droplet sizes, cooling rates, secondary dendrite arm spacings, temperatures, degree of undercooling, and volume fraction solidified. The present paper addresses the problem of collection and coalescence of the droplets after impact with the deposition surface.

A schematic representation of spray atomization and deposition is shown in Figure 2. A spray atomization and deposition facility is shown in Figure 3. The present paper addresses the deposition of an atomized droplet spray onto a stationary surface, however, variations of this concept, which are of interest, include using deposition surfaces which translate, (horizontally and vertically), and rotate [28]. Also, the use of parallel rows of gas jets (linear atomizer) to atomize a sheet of molten metal, is actively being studied for the production of sheet and strip [28,35].

In this study, three major simplifying assumptions must be stated; they are as follows:

a) The transfer of heat through the deposit takes place principally in the direction of the deposit thickness. Heat flow in the other two spatial directions are inconsequential and can be neglected.

b) The deposition rate is assumed spatially uniform and constant in time.

c) The thermal boundary condition at the upper boundary of the growing deposit is calculated as an average of the enthalpies of the falling droplets at the moment of impact.

In order to solve the solidification heat transfer problem, as the impinging droplets coalesce and the deposit increases in thickness, the time dependent location of the upper boundary of the deposit must be specified.

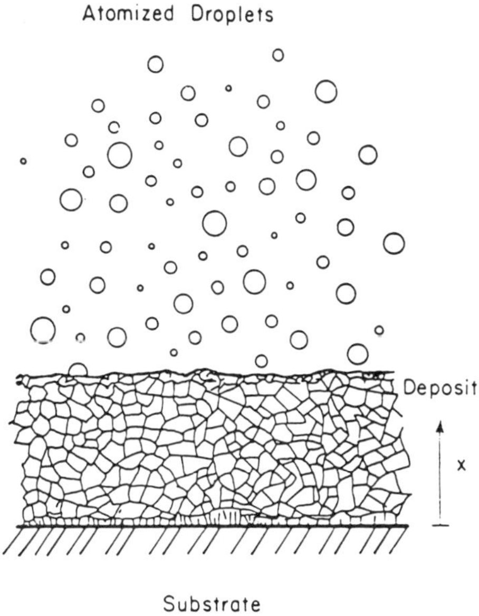

Figure 2 - Schematic representation of the deposition stage during spray atomization and deposition processing [28].

Figure 3 - High pressure atomization and spray deposition facility at the University of California at Irvine [28].

To calculate the deposit growth rate from the known melt flow rate one has to solve the equation of continuity; i.e.:

$$\frac{dH_d}{dt} = Q_m \,/\, \rho_m A_s \qquad (1)$$

Where H_d is the instantaneous deposit thickness; Q_m is the melt deposition rate, ρ_m is the melt density, and A_s is the area of the substrate. When the metal flow rate is assumed constant throughout deposition, Eq. (1) can be integrated to give:

$$H_d = H_{do} + (Q_m / \rho_m A_s)\, t \qquad (2)$$

It should be pointed out that at time t = 0.0 the initial deposit thickness, H_{do}, is in actuality always equal to zero. However, it should also be clear that in order to perform the heat transfer calculations there should be at minimum a few grid point present at the start of the calculation. Therefore, an artificial initial deposit thickness has been assumed in the model. This artificial initial thickness, however, has been selected in such a way that its effect on the results of the calculations is minimal. After a number of computer experiments, it was found that an artificial initial thickness less or equal than about 10% of the final deposit thickness was appropriate [31-34].

For the calculation of the thermal phenomena in the deposit we therefore have to solve the differential energy balance equation taking into account the liberation of the latent heat of fusion. This is most easily done by using the so called enthalpy formulation of the problem [36]. In the enthalpy formulation, the equation to be solved is [33]:

$$\frac{\partial H}{\partial t} = \frac{\partial}{\partial x} \left(k \frac{\partial T}{\partial x}\right) \qquad (3)$$

where H is the enthalpy, T the temperature, k the thermal conductivity and x the distance normal to the surface of the substrate. In order to incorporate the phase change, the enthalpy function must be discontinuous at the melting point. Also, it should be noted that the deposit thickness increases with time and this has to be incorporated in the solution algorithm. Next, the boundary conditions must be stipulated in order to address the solution to Equation (3). Regarding the boundary conditions we have the following:

a) At the deposit-substrate interface the rate of heat extraction by the water cooled deposition substrate is given by a heat transfer coefficient at this interface, i.e., h_{bot}:

$$-k \frac{dT}{dx} = h_{bot} \, (T - T_{wall}) \qquad (4)$$

b) At the top surface of the growing deposit, the heat flux is equal to the sum of the heat input due to the spray and the heat output due to the combined effects of convection into the gas and to radiation into the atomization chamber walls, i.e.:

$$-k \frac{dT}{dx} = -\bar{H}_{top} \left(\frac{Q_m}{A_s}\right) + h_{top} \, (T - T_{gas}) + \sigma T^4 \qquad (5)$$

where H_{top} is the average enthalpy of the spray and Q_m/A_s is the spraying rate of the metal. Once deposition is over, the heat input due to the impinging spray becomes equal to zero. The incorporation of a T^4 term into Equation (5) is necessary in order to account for the radiative heat losses resulting from the high temperatures encountered during processing of ferrous based alloys.

Results and Discussions

The set of equations (1) - (5) constitute a mathematical statement of the spray deposition process, after impact of the droplet distribution with the deposition surface; these can be solved numerically. An explicit finite difference scheme can be shown to be both consistent and conditionally stable [31-34]. The time step must be selected so as to satisfy the well known stability restriction of explicit methods. However, the algorithm is easily implemented for calculation in a computer. Some of the results obtained by numerical solution of Equations (1) - (5) using the input data shown in Tables II and III are described in the following section.

Table II - Input Data for Steel

Deposition rate:	Q_D=0.1 Kg/sec
Deposition time:	t_D=30 sec
Substrate area:	A_s=1.53x$10^{-2}$$m^2$
Final thickness:	X_f=2.5x10^{-2}
Heat transfer coeff. (bottom)	h_{bot}=10^3 W/m^2K
Heat transfer coeff. (top)	h_{top}=10^2 W/m^2K
Temperature at the walls:	T_x=25 C
Temperature of the gas:	T_{gas}=60 C
Average enthalpy of the spray:	H_{top}=174,500 J/Kg
Heat of fusion:	H_{fus}=184,305 J/Kg
Specific heat:	Cp=725 J/Kg
Thermal conductivity:	k=30 W/mK
Liquidus temperature:	T_L=1530 C
Solidus temperature:	T_S=1490 C
Density:	p=7,860 Kg/m^3

Table III - Input Data for Aluminum

Deposition rate:	Q_D=0.03 Kg/sec
Deposition time:	t_D=30 sec
Substrate area:	A_s=1.53X$10^{-2}$$m^2$
Final thickness:	X_f=2.5x10^{-2}m
Heat transfer coeff. (bottom)	h_{bot}=500 W/m^2K
Heat transfer coeff. (top)	h_{top}=10^2 W/m^2K
Temperature at the walls:	T_w=25 C
Temperature of the gas:	T_{gas}=60 C
Average enthalpy of the spray:	H_{top}=132,580 J/Kg
Heat of fusion:	H_{fus}=397,746 J/Kg
Specific heat:	Cp=862.5 J/Kg
Thermal conductivity:	k=209.2 W/mK
Liquidus temperature:	T_L=740 C
Solidus temperature:	T_S=660 C
Density:	p=2,700 Kg/m^3

In a previous section a schematic representation of the deposition stage during spray atomization and deposition was shown (see Figure 2). It is seen that the underlying deposit cools by transferring heat into the substrate while the atomized particles arrive on the top surface.

As an illustration of the capabilities of the mathematical model described in the previous section two cases will be discussed. The deposition of an aluminum-base alloy is analyzed first and then compared with the results obtained for the deposition of a low carbon steel. The two materials considered possess markedly different physical properties and it is thought that a comparison of their computed behavior during spray deposition could be of some interest.

Figure [4] shows the calculated temperature history in spray deposited aluminum. It is seen that there is a temperature difference between the top and bottom surfaces of the preform and that this difference increases with time. Also, it is worthwhile noting that the temperature of the top surface begins to increase approximately 5 seconds after the beginning of deposition. This behavior can be explained in terms of a heat imbalance in the top surface of the deposit; in this case the heat input exceeds the heat dissipation by conduction and convection. The temperature difference between top and bottom regions reaches a maximum value when deposition terminates and decreases rapidly thereafter. The time rate of change of temperature can be readily computed and Figure [5] shows the result. The computed cooling/heating rates during most of the duration of the deposition process are between - 10 and + 2 ^{o}C/s. Higher cooling rates occur both at the beginning and at the end of deposition, and also when the last liquid solidifies at the top region of the deposit. Cooling rates of this order of magnitude have been reported in other, more conventional casting processes. However, the microstructural size scale observed in numerous spray-deposited materials is considerably finer than that obtained in conventional processes. These results suggest that the microstructural evolution during spray deposition is strongly dependent not only on the extraction of heat and the solidification events that occur after deposition, but also on the heat flux and solidification history of the droplets prior to impact with the deposition surface; this matter is the subject of continuing studies.

Figure 6 shows the calculated iso-fraction solid lines in the deposit for various distances as a function of time. The results from the figure indicate that a large fraction of the preform thickness consists of a solid-liquid mushy zone throughout deposition. The large number of crystalline nuclei present in this mushy zone, combined with dynamic nucleation events, could well account for the grain refinement effect observed in the as-spray deposited materials.

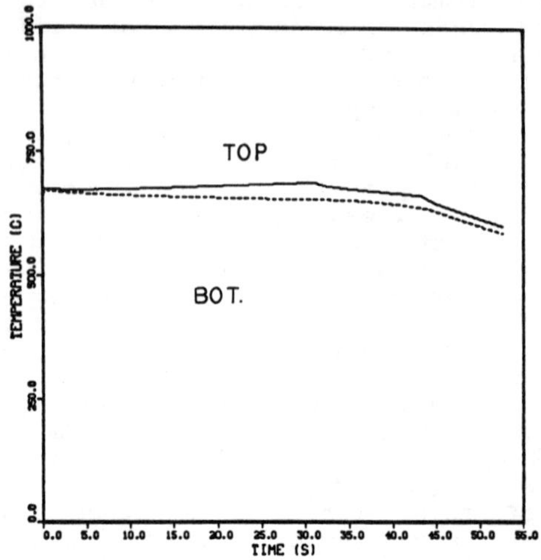

Figure 4 - Calculated temperatures for two positions in spray deposited aluminum.

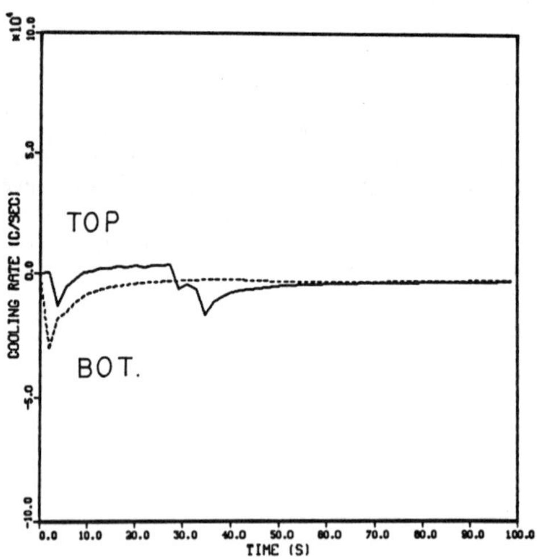

Figure 5 - Calculated cooling/heating rates for two positions in spray deposited aluminum.

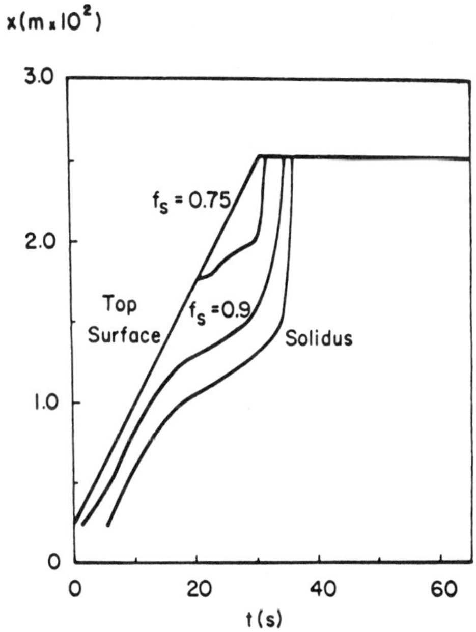

Figure 6 - Calculated iso-fraction solid lines for various distances as a function of time.

The model can now be used to calculate the thermal behavior of low carbon steel preforms during deposition. Figure [7] shows the computed temperature histories for two regions (top and bottom) of the preform. The behavior is somewhat similar to that obtained for the Al-alloy except that the temperature difference between top and bottom surfaces is more pronounced in the case of the steel. This is to be expected since the thermal diffusivity of steel is almost one order of magnitude smaller than that of the aluminum alloy. Even after deposition is over, the temperature difference in the case of steel is much larger than the maximum difference calculated for the aluminum alloy.

Figure [8] shows the computed cooling/heating rate for the steel preform. The results from the figure indicate that except for some relatively large values at the beginning and at the end of deposition, the time rate of change of the temperature fluctuates close to zero. Again, the computed cooling rates are similar to those encountered in the case of more conventional castings. However, the experimentally reported grain sizes are significantly smaller than those found in conventional castings. Figure [9] shows the calculated iso-fraction solid lines for the steel preform. In contrast with the aluminum alloy, here the mushy zone only appears close to the end of the deposition process. Also, it is worthwhile noting from the figure that the top surface of the preform becomes 100% liquid during the last few seconds of deposition.

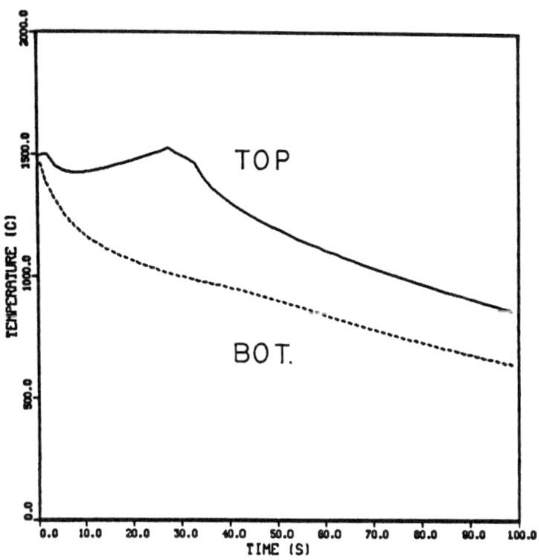

Figure 7 - Calculated temperatures for two positions in spray deposited steel.

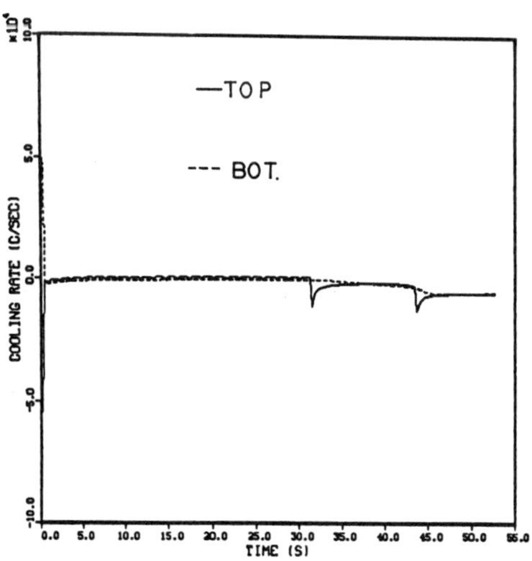

Figure 8 - Calculated cooling/heating rates for two positions in spray deposited steel.

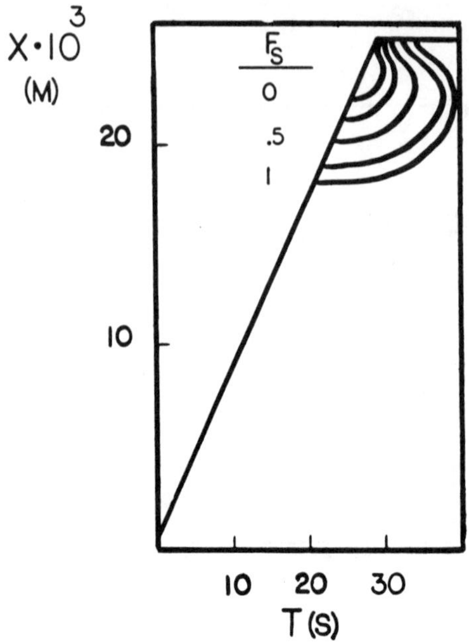

Figure 9 - Calculated iso-fraction solid lines for various distances as a function of time.

If the top surface of the deposit indeed becomes mushy and then fully liquid towards the end of deposition, it is quite conceivable that such liquid-mushy layer will be in state of intense agitation due to the continued impact of the falling droplets. The presence of a highly agitated layer at the top of the deposit means that convection in the melt might have a significant effect on the heat transfer and solidification events. A crude estimate of the effect of melt convection can be obtained by increasing the thermal conductivity of the liquid phase. Figure [10] shows the computed temperature histories for three points in the deposit but for the case in which the value of the thermal conductivity of the liquid has been doubled. Although the same two regions found in Figure [7] can still be distinguished in this plot, it is readily apparent that the effect of the increased thermal conductivity of the liquid phase is to reduce the rate of increase of the temperature when crossing the mushy zone. Indeed, the maximum temperature reached in this case is the liquidus temperature of the alloy.

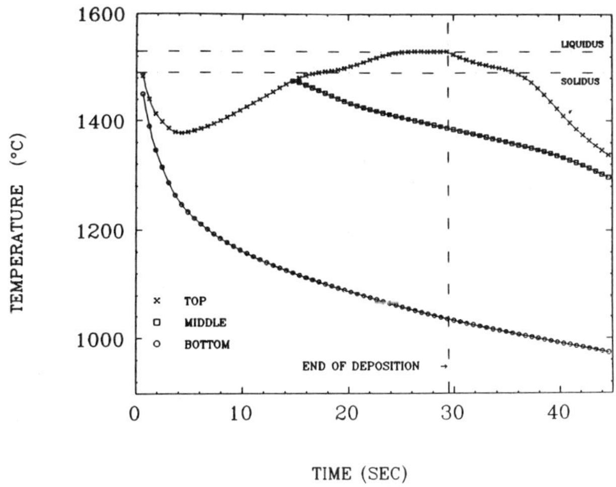

Figure 10 - Thermal history for three regions in spray deposited steel assuming $k_{effective} = 2\ k$ (see Figure 7).

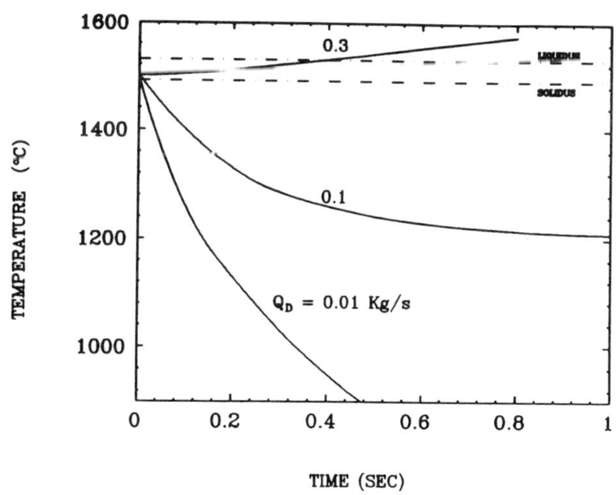

Figure 11 - Thermal history of the top region of the deposit as a function of time various deposition rates (steel).

Summary

Summarizing, a mathematical model of the spray deposition process has been proposed and calculations have been carried out for two systems of engineering interest. The results of the calculations indicate that the model is capable of describing the expectedly different behavior. Further work is required, particularly on the actual mechanisms leading to the formation of the as-deposited microstructure.

Additional information can be obtained from carefully planned implementation of the program. For example, one could study the initial stages of formation of the deposit as a function of processing parameters. For this purpose a relatively fine numerical grid (grid spacing = 10 microns) can be established and the thermal history of the top surface of the deposit can be calculated as a function of time for at least the first second of deposition. Figure [11] shows the results of such a calculation for various values of the melt flow rate while holding all other variables constant. The calculation is done for the Fe-based material. It is readily seen that the initial temperature history of the top surface of the deposit can be used to asses the feasibility of carrying out an experiment under specified metal flow rate conditions. For example, when operating at the low end of the flow rate range, it is possible that the temperature of the growing preform is too cold to promote diffusion and hence adequate adherence to the incoming spray particles. Another extreme can occur at the high end of the range, when the heat input is so large that a liquid layer forms very early during deposition making hard to maintain steady growth of the preform. Hence it is anticipated that a melt flow rate in the middle of this range would be more appropriate.

Acknowledgments

E. Lavernia would like to acknowledge financial support of this work by the Academic Senate Committee on Research of the University of California at Irvine. E. Gutierrez would like to thank the National University of Mexico for continued support.

References

1. D.W. Hall and S. Mocarski, Int. Jour. of Powder Metall. and Powder Technology, 21, 79 (1985).

2. Powder Metallurgy, Metals Handbook, 7, Ninth Edition, ASM, 1984.

3. F.V. Lenel, Powder Metallurgy, Principles and Applications, MPIF, Princeton, New Jersey, 1984.

4. P.U. Gummeson, Powder, Met., 15, 67 (1972).

5. M.J. Fleetwood, A.G. Law, J.E. Whitlle, A.G. Todd, and J.H. Vincent, "Control of Thin Strip Casting", 2nd ASM International Conference on Rapidly Solidified Materials, 7-9 March 1988, San Diego, California.

6. S.J. Savage and F.H. Froes, J. of Metals, 36 (4) (1984) 20-33.

7. W.A. Heinemann, in Rapidly Quenched Metals, ed. by S. Steeb and H. Warlimont, Elsevier Service Publishers, Amsterdam, (1984) 27-34.

8. M. Cygler and M. Wolf, IS&M Journal, (August, 1986) 27-34.

9. M.C. Flemings and N.J. Grant, "Thin Strip Metal Casting", DOE - Idaho Operations Office, Final Technical Report, Contract No. DE-AS07-83IDI2444, (April 26, 1983 - January 26, 1984).

10. G. McManus, Iron Age, August 1, 1986, 31-38.

11. E.J. Lavernia, G. Rai, N.J. Grant, International Journal of Powder Metallurgy, (1986) (22) 1, 9-16.

12. E.J. Lavernia, T. Ando, and N.J. Grant, Proceedings of ASM's International Conference on Rapidly Solidified Materials, San Diego, Ca., 3-5 Feb., 1986, P. Lee and R. Carbonara, eds., 29-44.

13. P.J. Meschter, et al., TMS Annual Meeting, New Orleans, LA., March 2-6, 1986.

14. J. Megusar, et al., Journal of Nuclear Materials (122) and (123), (1984), 789-793.

15. B.P. Bewlay and B. Cantor, Proceedings of ASM's International Conference on Rapidly Solidified Materials, San Diego, CA., 3-5 Feb., 1986, P. Lee and R. Carbonara, eds., pp. 15-20.

16. R.H. Bricknell, Met. Trans. A., (April 1986), Vol 17A, 583-590.

17. A. Leimkuhler Moran and W.A. Palko, Proceedings of the 1987 Annual Powder Metallurgy Conference and Exhibition, Progress in Powder Metallurgy,(43) 711-722.

18. A.I. Taub, K.M. Chang, and S.C. Huang, Proceedings of ASM's International Conference of Rapidly Solidified Materials, San Diego, CA., 3-5 Feb., 1986, P. Lee and R. Carbonara, eds., 297-302.

19. H.C. Fiedler, et al., Journal of Metals, (August 1987), (39), No. 8, 28-35.

20. J.M. Nell et al., Proceedings of the 1986 Annual Powder Metallurgy Conference and Exhibition, May 18-21, 1986, Boston, MA.

21. K. Ogata, et al. International Journal of Rapid Solidification, 1986 (2) No. 2, 21-35.

22. E.J. Lavernia, E. Gomez and N.J. Grant, Materials Science and Engineering, (95), 1987, 225-236.

23. S. Savage, L. Arnberg, and J. Liu, Swedish Institute for Metals Research, Stockolm, Sweden, current research.

24. J. Baram, Ben Gurion University of the Negev, Beer Sheva, Israel, current research.

25. M. Veistinen, Y. Hara, E.J. Lavernia, R.C. O'Handley and N.J. Grant: Proceedings of the 1987 Materials Research Society Meeting, Anaheim, CA., 93-101.

26. T.S. Chin et al., Journal of Applied Physics, 1986, (59) No. 4, 1297-1300.

27. E.J. Lavernia and N.J. Grant, Proceedings of the Sixth International Conference on Rapidly Quenched Metals, August 3-7, 1987, Montreal, Canada, to be published in Journal of Materials Science and Engineering.

28. E.J. Lavernia, University of California at Irvine, current research.

29. J. Key, J. Alvarez, Idaho National Engineering Laboratories, Idaho Falls, Id., current research.

30. E.J. Lavernia, N.J. Grant and J.P. Clark, "Economic Feasibility of Liquid Dynamic Compaction Powder Processing Operations", Conference on the Mathematical Modeling of Metals Processing Operations, sponsored by AIME and CIM, Nov. 29 - Dec. 2, Palm Dessert, CA., 1987.

31. E.J. Lavernia et al., Proceedings of the 1987 Annual Powder Metallurgy Conference and Exhibition, Progress in Powder Metalurgy, (43) 683-710.

32. E.J. Lavernia et al., "A Mathematical Model of the Liquid Dynamic Compaction Process, Part 1: Heat Flow in Gas Atomizaiton", accepted for publication, International Journal of Rapid Solidification.

33. E. Gutierrez, et al., "Part 2: Formation of the Deposit", accepted for publication, International Journal of Rapid Solidification.

34. E. Gutierrez, et al., accepted for publication, Metallurgical Transactions B.

35. C.Y. Tsao, E.J. Lavernia, and N.J. Grant, "Spray Forming 2024 Aluminum Alloys Using Linear Spray Atomization and Deposition", presented at Second ASM International Conference on Rapidly Solidified Materials, San Diego, CA., 3-5 Feb., 1986.

36. J. Crank, "Free and Moving Boundary Problems", Clarendon Press, Oxford, 1984.

CASTING TECHNIQUES III: GENERAL

Session Chairman: **H.H. Liebermann**
Allied Signal Inc.
Parsippany, New Jersey

CONTINUOUS CASTERS FOR ALUMINUM MINI-SHEET MILLS -

AN ALCOA PERSPECTIVE

Ronald Bachowski
Alcoa Laboratories
Alcoa Center, PA 15069

R. J. O'Malley
Alcoa Laboratories
Alcoa Center, PA 15069

M. M. Mohajery
1501 Alcoa Building
Pittsburgh, PA 15219

Abstract

The effect of mini-mills on the steel industry has been widely discussed. These small independent facilities generally focus on niche products, like reinforcing bar, but may pose a threat to sheet products previously dominated by the large fully integrated steel companies.

Does a similar threat exist for the aluminum industry? Evidence concerning installed capacity of mini-mills will show increasing penetration in some products. Business reasons for these successes will be presented.

The major types of continuous casters used in these facilities for producing slab and strip will be discussed. This will include process description, metallurgical characteristics, product application and technical constraints. The potential of producing more sophisticated products will be evaluated.

Process characteristics of new emerging technologies will be presented.

Casting of Near Net Shape Products
Edited by Y. Sahai, J.E. Battles,
R.S. Carbonara and C.E. Mobley
The Metallurgical Society, 1988

Introduction

No doubt most of you are familiar with the effect that mini-mills have had on the steel industry. These small independent facilities have generally focused on niche products, like reinforcing bar and structural shapes. Most use scrap as the melting stock, electric arc melting furnaces and some form of continuous casting to near-net shape in order to reduce the highly capital intensive equipment required for traditional fabricating methods. Clearly there are many other business reasons which contribute to the success of mini-mills--including such factors as lower labor rates, extensive use of scrap for feedstock and strategic geographic location. However, one common thread is the ability to avoid high capital costs by using continuous casters. Also, several steel companies are pursuing new technologies for continuous casting thin wide strip (1,2). These low capital processes may pose a threat for flat rolled sheet products previously dominated by large fully integrated steel companies.

So much for steel. The question now becomes whether mini-mills similarly equipped with continuous casters are a significant factor to the aluminum industry. Of course, we in Alcoa have particular interest in flat rolled sheet products due to the large proportion of our business in those markets. This is not an easy question to answer because many factors such as volume, quality, cost and alloy flexibility come into play. However, it is one we continue to debate. These are our findings to date, first from a business perspective.

As with steel, the aluminum mini-mill concept is based on lower capital, lower volume, special markets, scrap metal feedstock and typically incorporates continuous casters (3).

We find that many of the early development stages of mini-mills experienced by the steel industry are being repeated in the aluminum industry. Table I below is offered as a qualitative measure of the inroads made by continuous caster-based mini-mills in the aluminum industry. Exact data are difficult to come by because (a) a significant number of mini-mills can flex between low levels of operation and larger quantities based on market conditions and (b) "world" capacities and production numbers are necessarily "soft" due to lack of data regarding the planned economics.

Table I

		U.S.	World
Year	Caster Related Prod., $\overline{M}$ lbs.	% of Total Flat Rolled Production	% of Total Flat Rolled Production
1977	Small	Negligible	Small
1987	~ 1000	~ 15%	~ 30%
1990	~ 1300	~ 20%	>30%

These facilities typically rely on recycled (scrap) metal and, therefore, have developed sophisticated scrap acquisition and processing capabilities. Their operating costs are dominated by the cost of metal, energy and capital, with capital being significantly lower than traditional fabricating mills. Some of these caster-based facilities may be underutilized, but because of extreme low invested capital these operations are characterized by relatively low break-even points. As a result, they are capable of surviving economic downturns and are poised to take advantage of new opportunities by adding incremental capacity at relatively low capital cost.

Before discussing the types of products produced by these facilities, it is appropriate to briefly describe the most common types of continuous casters and how they differ from traditional flat rolled sheet plants.

Conventional vs. Continuous

For comparison purposes, Figure 1 shows a process diagram of the conventional method for manufacturing flat rolled sheet. Typically, the sequence starts by semicontinuous casting of DC ingots 16-30 in. thick followed by hot rolling to 0.100-0.250 in. thick coiled reroll stock. These "hot lines" are the starting point for many products ranging from common household foil to advanced aerospace applications. They are versatile, sophisticated and capable of producing up to 1200 $\overline{M}$ lbs./yr. One drawback is that they cost around \$200 $\overline{M}$. Most companies have chosen to incrementally add capacity to existing hot lines rather than commit to new green field installations.

Manufacturing Steps for Traditional Hot Line Method of Producing Reroll Stock. Each Box Denotes a Separate Step
Figure 1

The two basic types of continuous casters which have achieved wide commercial success are briefly described as follows.

Slab Casters

Continuous slab, typically from 1/2-1 in. thick, is cast by introducing molten aluminum between parallel traveling mold surfaces. As shown by Figure 2, low speed in-line mills reduce the hot slab to coils of reroll stock. The continuous nature of this method allows the opportunity to eliminate several fabricating operations (Figure 3).

There are two approaches to slab casters, as shown by Figures 4 and 5 (4,5). Hazelett has chosen to use water cooled steel belts for the molds, whereas the Alusuisse Caster II employs massive metal chill blocks. In both cases, the slab exits the caster at approximately 1000°F, so that little if any reheating is required prior to hot rolling. The combination of high casting rate and long mold length (typically 20 fpm and 6 ft., respectively) promotes significant solidification shrinkage, but this can

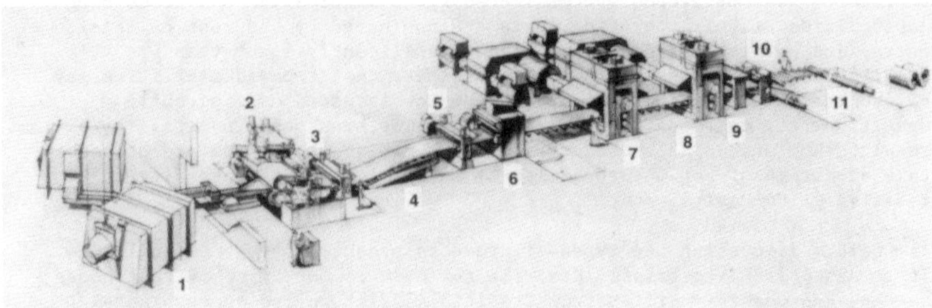

Hazelett Aluminum Strip Casting Line

1 Holding Furnaces
2 Hazelett Caster
3 Pinch Roll
4 Looping Table
5 Pinch Roll
6 Shear
7 Four High Rolling Mill
8 Four High Rolling Mill
9 Shear
10 No. 1 Coiler
11 No. 2 Coiler

Typical Arrangement for a Slab Casting Facility with In-Line Hot Mills
Figure 2

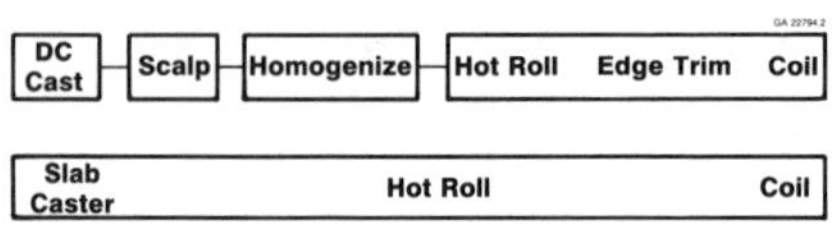

Compares Slab Caster Approach with Hot Line
Figure 3

easily be healed during the in-line hot rolling step. Since the lineal rate of continuous casters is width insensitive, the term generally used to define productivity is lbs./hr. per inch of width. On this basis the productivity of slab casters is 1000-1300 lbs./hr.-in.

At first glance the necessity for having an expensive multistand hot mill in line with a slab caster is in conflict with the low capital strategy of mini-mills. No doubt this could adversely affect economics if the hot mills were brand new. However, the utilization of low cost reconditioned mills from idled steel facilities makes the slab caster/hot mill combination economically viable.

Strip Casters

The most common form of strip casting is the well-known roll casters principally manufactured by Hunter Engineering, Lauener, Pechiney (SCAL), Davey-McKee and Krupp. Figure 6 illustrates that nominal 1/4 in. thick

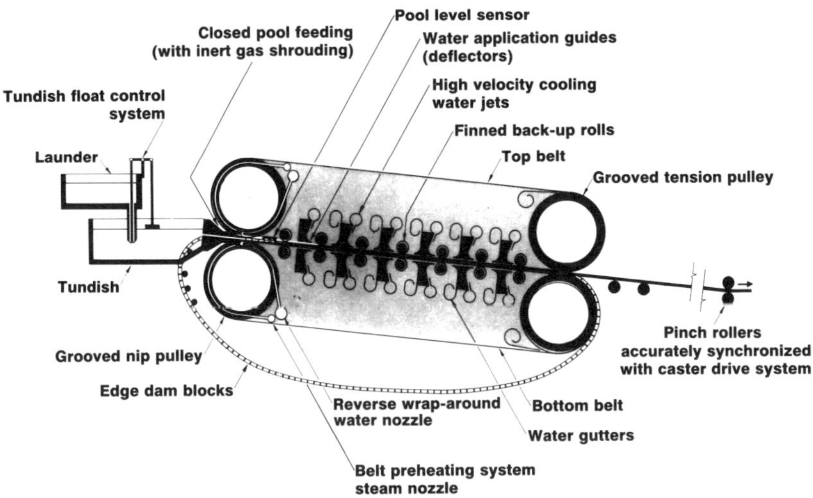

Hazelett Twin Belt Slab Caster
Figure 4

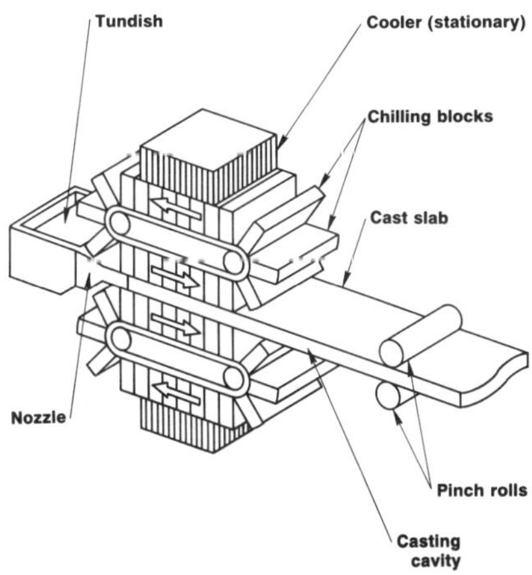

Alusuisse Caster II
Figure 5

coiled reroll stock is produced by introducing molten metal between water cooled rolls (6). A closer look (Figure 7) reveals that significant hot deformation also takes place between the rolls. Linear speeds are 6-10 fpm and productivity ranges from 60-100 lbs./hr./in. depending on alloy.

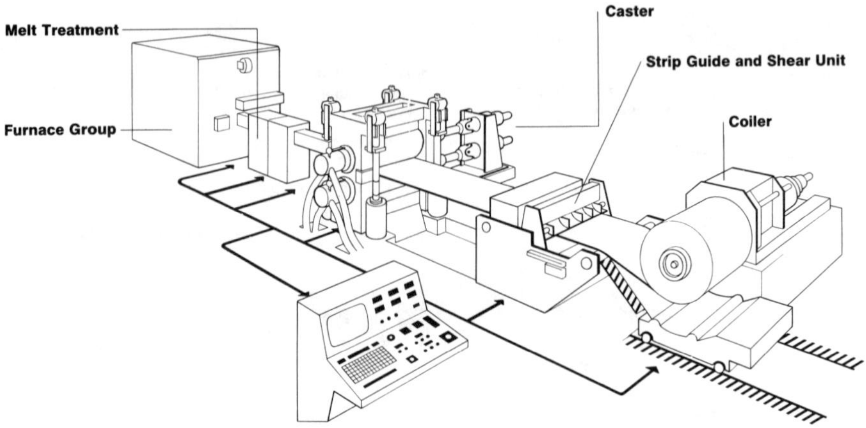

Typical Arrangement of a Roll Caster Facility
Figure 6

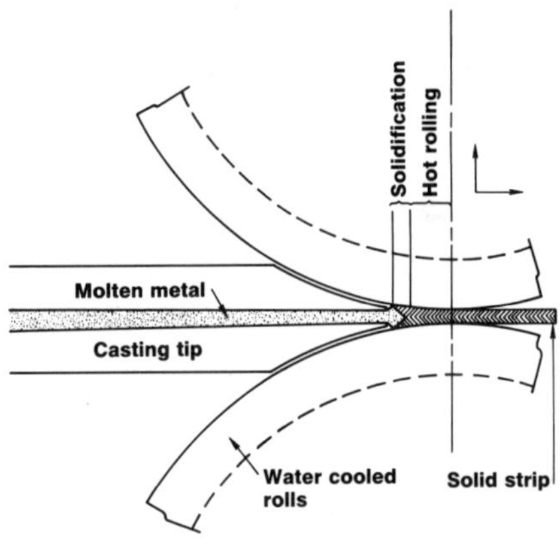

Schematic Representation of Solidification/Deformation Sequence in Roll Casting Process
Figure 7

Figure 8 compares the process flow of continuous casters with that of the hot line.

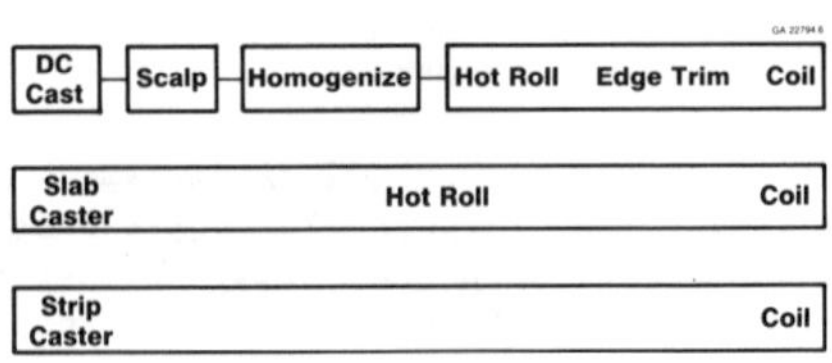

Compares the Three Most Common Methods for Manufacturing Aluminum Reroll Stock
Figure 8

Additional productivity and metallurgical data concerning slab and strip casters are listed in Table II:

Table II

	Slab	Strip
Typical Capacity	30 $\overline{M}$ lbs./mo.	3 $\overline{M}$ lbs./mo.
Width Range	10-80 in.	30-60 in.
Solidification Rate	1°-10°F/sec.	100°-1000°F/sec.

Commercial Application of Continuous Casters

Previous data showed that installed casters had the capacity to produce a significant share of U.S. and world flat rolled aluminum sheet. Where are these facilities and what do they make?

Table III describes the location of nine operational slab casters along with the company, type and width. Five are in the U.S., two in Japan and one each in Canada and Venezuela. A tenth caster, believed to be operated by Alcan in Canada, is not listed because information is not available. Roll casters, on the other hand, are much more numerous and widely distributed. As shown by Table IV, a previous survey revealed approximately 160 roll casters scattered throughout the world (7).

Table III. Summary of Aluminum Slab Casting Operations

Date Installed	Company and Location	Type	Cast Width, In.
1963	Algoods (Alcan) - Canada	Hazelett	20
1970	Nihon Atsuen Kogyo KK - Japan	Hazelett	12
1979	Barmet-Carson (Alflex) - USA	Hazelett	28
1983	Barmet-Livia - USA	Hazelett	14
1984	Nihon Atsuen Kogyo KK - Japan	Hazelett	18
1984	Coors (Golden Recycling) - USA	Caster II	28
1985	Advanced Aluminum Products - USA	Hazelett	30
1986	Newport Rolling Mills (Barmet) - USA	Hazelett	52
1987	Pivensa - Venezuela	Hazelett	41

Table IV. Location and Distribution of World Strip Casting Operations

Location	Number	Percent
North America	58	37
Western Europe	37	23
Eastern Block	29	18
Asia	18	11
Africa	9	6
South America	7	4
Australia	2	1

Thus far, caster based mini-mills have concentrated on less sophisticated product lines. An example would be painted sheet used for building products such as siding and gutters. When compared to hot line stock, there have been several quality issues that inhibited caster stock from being used for more demanding products. These issues included anodizing quality, formability, segregation and gauge/profile control. In recent years, however, there is evidence that the pace of technical developments and improvements has quickened. Perhaps having consolidated a firm footing in less demanding products, the mini-mills can afford to investigate possibilities for extending product lines. Some examples of this are:

- Coors is successfully producing end stock for two-piece aluminum cans with an Alusuisse Caster II. The alloy is Al-2 Mg-0.7 Mn.

- Barmet-Carson (Alflex) has produced fully qualified quantities of 5042 (Al-0.35 Mn-3.5 Mg) tab stock for easy-open ends on its Hazelett caster. It also produces 3004 (Al-1.25 Mn-1.0 Mg) stock for anodized furniture tubing.

- Reports indicate that computer memory discs are being produced from roll cast stock. Also, Hunter Engineering has patented a novel metal treatment/roll casting process for producing memory discs of Al-4 Mg (8). The significance here is that roll cast material has the potential of fewer reading errors because the more rapid solidification reduces constituent size.

- Hunter Engineering reports that roll casting is the preferred method for producing high purity aluminum sheet used in sophisticated lightweight batteries. Battery performance is enhanced by the uniform distribution of small constituents provided by the rapidly solidified roll cast strip.

- Roll cast stock is being widely used for products such as fin stock and formed container stock, which require relatively good formability.

Thus, there seems to be a trend to expand the product line both in areas where continuous cast material may have inherent advantages (memory discs and battery sheet) or into more sophisticated alloys traditionally produced by hot lines (end and tab stock). However, it is likely that the quality of caster-produced stock will differ from hot line material.

These differences may be insignificant or, as indicated above, advantageous for a particular application. However, where quality standards have been optimized for hot line material, there could be major barriers to inroads by caster stock.

Alcoa Experience

Continuous casters are not a new phenomenon. The original Hazelett twin belt development dates back over 40 years and commercial models have existed for at least 30 years. Strip casters have also been commercially utilized for over 30 years.

Likewise, Alcoa is not a newcomer to casters. A production size Hazelett caster complete with in-line hot mills was installed and operated on an experimental basis in two separate Alcoa plants in the period 1950-60. Our most recent experience was with a 51 in. wide Hazelett at the Alcoa Laboratories in 1969-72. Thus far, we have been unable to identify an appropriate fit for slab casters in our business plans. The opportunity still exists, however, providing issues concerning market size, quality requirements and economic return can be satisfied.

Roll casters are a growing part of our business tactics, a major reason being the lower volume/capital features provided by this approach. Alcoa currently operates eight casters in the U.S., Mexico and Brazil, producing strip for a variety of markets including building products, closures and formed containers. Four additional units are planned for the near future.

Productivity - Current vs. Theoretical

As indicated previously, the current production capabilities for slab casting and roll casting differ by about a factor of 10; slab casting being over 1000 lbs./hr.-in. width and roll casting being at about 100 lbs./hr.-in. width. Although the production capabilities of the two processes are somewhat alloy dependent, both processes have shown slow but steady productivity gains since their initial development, primarily through improvements in process understanding and the application of improved process control. How much further can these processes be pushed? We will show in this section that the windows for productivity improvements are still very wide relative to the theoretical limits for the two processes.

The maximum production rates for slab casting and roll casting can be calculated from heat transfer considerations by assuming that the only resistance to heat flow is within the solidifying sheet or slab. This calculation provides the absolute theoretical limitations in productivity for the casting processes. Additional heat flow restrictions may also be included in the calculations to develop more conservative productivity estimates. For example, the rate of heat extraction from the surface of the sheet can be restricted by imposing a convective heat transfer coefficient of a specified value, or the resistance to heat flow imposed by a thick mold (such as a steel roll shell or copper chill block) may be included. In either case, including the additional resistance to heat flow in the calculations further reduces the maximum productivity predictions.

The current production capabilities for slab casting and roll casting are compared to the theoretical limits for the two processes in Figures 9 and 10. The top line in each figure represents the absolute theoretical limit for each process for the conditions listed. The 0.5×10^6 W/m^2 K heat transfer coefficient used to generate the more conservative productivity estimates shown in the same figures is taken from heat extraction rates measured in the laboratory for the solidification of thin aluminum ribbons. This heat transfer coefficient represents the highest heat extraction rate measured to date in our labs for the solidification of aluminum sheet. The most conservative productivity predictions shown are for calculations that include the resistance to heat flow imposed by a thick copper or steel mold.

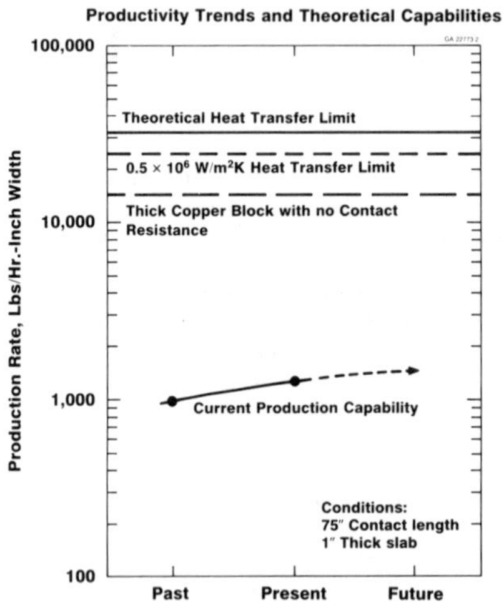

Slab Casting
Figure 9

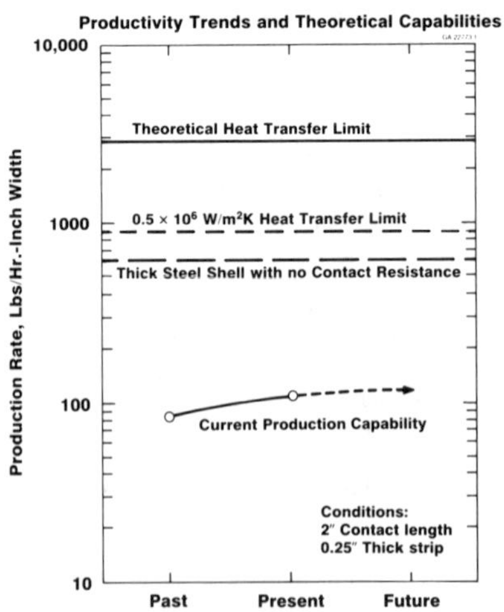

Roll Casting
Figure 10

It is interesting to note that the theoretical production capacity for slab casting is also about 10 times larger than for roll casting, and that the actual production capabilities are about an order of magnitude lower than their most conservative theoretical counterparts for both processes. The 10X gap that exists between today's production capabilities and the theoretical capabilities for the two processes is primarily a result of imperfect thermal contact at the mold/metal interface due to solidification shrinkage and thermal stresses that promote air gap formation, and the presence of lubricants, coatings and naturally formed oxides at the interface that moderate heat flow. Of course, other operating considerations that impact strip or slab quality are also contributing factors.

In slab casting, thermally insulating coatings or textures are applied to the mold surfaces to purposely moderate heat transfer to prevent thermal distortion of the belt or blocks and to improve the uniformity of heat extraction from the solidifying strip. Thermal management of the solidification sequence by intentionally moderating heat transfer is essential to improve the metallurgical quality of the slab. In those situations where this moderation has an adverse effect on productivity, the caster length can be increased to restore throughput to a reasonable level.

In roll casting, the development of a sticking condition between the roll and the strip is a constraint to productivity. The sticking problem forces producers to apply parting agents, such as graphite or magnesium hydroxide, to the rolls to avoid sticking at reasonable productivity levels. Even with the use of these parting agents, sticking occurs above a critical casting speed which is well below the maximum production capability of the process.

What steps can be taken to move closer to the theoretical capabilities for these casting processes? In roll casting, a solution to the sticking problem must be found before significant advances can be made in the productivity of the process. In slab casting, techniques for improving the thermal and mechanical stability of the process need to be found that do not restrict the rate of heat flow from the slab. Alternatively, new casting technologies for slab or strip may allow us to push closer to the theoretical limits by bypassing certain limitations inherent in today's casting technologies.

Emerging Technologies

Thus far, we have only discussed the continuous casting methods that have achieved wide commercialization. There are other technologies under development that may be of interest, as follows.

- Single wheel melt drag (also known as melt overflow) casting shown by Figure 11 is an emerging technology that produces light gauge sheet direct from the molten state at speeds of 50-400 ft./min. The principal developers of this technology for steel appear to be Battelle, Allegheny Ludlum and Voest Alpine. Battelle has reported the results of laboratory studies concerning wheel-related parameters including material, surface condition and temperature (9). Both Allegheny Ludlum and Voest Alpine are operating pilot plants to investigate the casting of stainless and low carbon steel, respectively (10,11). Armco-Westinghouse also have a DOE sponsored program for carbon steel. Nippon Metal Industry Company has reported laboratory casting trials with stainless steel using a modified method that uses a small upper roll in combination with the large primary casting roll (12).

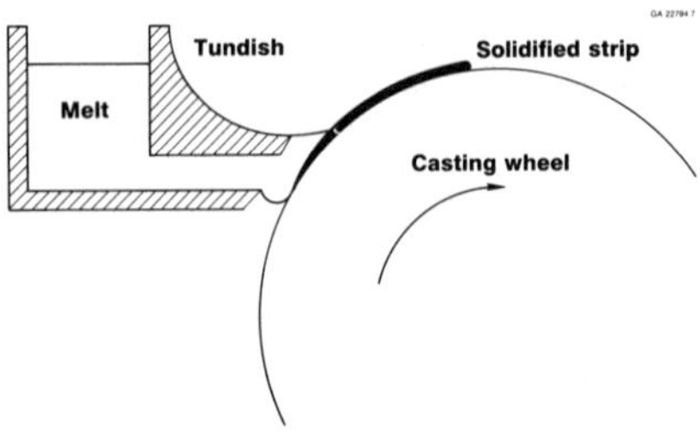

Schematic of Single-Roller Casting System (3)
Figure 11

There are some indications that the melt overflow process may be applicable to aluminum alloys. For example, Ribbon Technology Corporation reported the production of 0.002-0.003 in. thick x 12 in. wide strip of alloys A360 (Al-9.5 Si-0.5 Mg) and 7075 (Al-5.6 Zn-1.6 Cu-0.23 Cr) from 200 lb. melts. Strip was cast at rates up to 400 ft./min. depending on the desired thickness (13). A second piece of information about activity in aluminum was found in the patent literature, where National Aluminum Corporation described the casting of strip 0.0002-0.4 in. thick at rates of 50-5000 fpm. The alloys included A356.2 (Al-7.0 Si-0.35 Mg), 1100 (Al-0.5 Fe-0.25-0.15 Cu) and 3105 (Al-.55 Mn-0.5 Mg). The patent stated that strip quality was such that further rolling was possible if desired and as-cast strip was successfully used to produce building products (14).

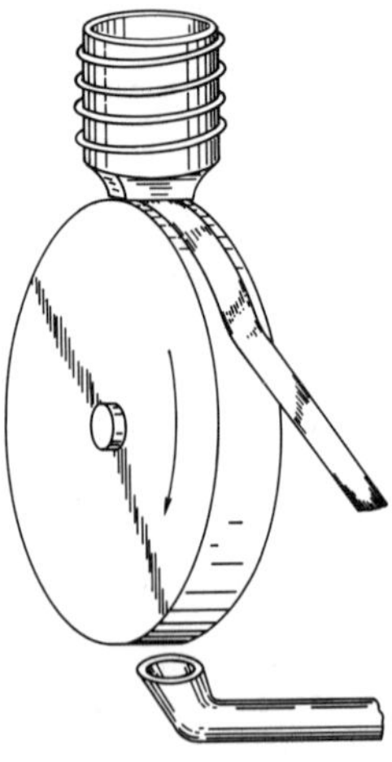

Schematic of Single Wheel Planar Flow Casting Process (6)
Figure 12

- Single wheel planar flow casting (Figure 12) is a commercial technology that produces 0.005-0.012 in. gauge sheet directly from the molten state at speeds of 1000-3000 ft./min. and production rates of 1000-1500 lb./hr./in. width (15,16). The principal developer of this technology is Allied Signal, which uses the process commercially to produce Fe based amorphous sheet, brazing foils and RSP aluminum powders. However, the production of crystalline aluminum alloy sheet suitable for commercial use has not yet been announced.

- SMS Concast slab casting (Figure 13) is a semicommercial steel casting technology that produces a 1.5-2 in. thick slab using an oscillating slab mold which is funnel shaped at the inlet to accommodate a feeding spout (17,18). Casting speeds of 12-20 ft./min. have been demonstrated with steel. The principal developer of this technology is SMS Schloemann-Siemag AG. They have an operating pilot plant in Buschhütten, Germany and are installing their first commercial unit for Nucor Corporation in the U.S.

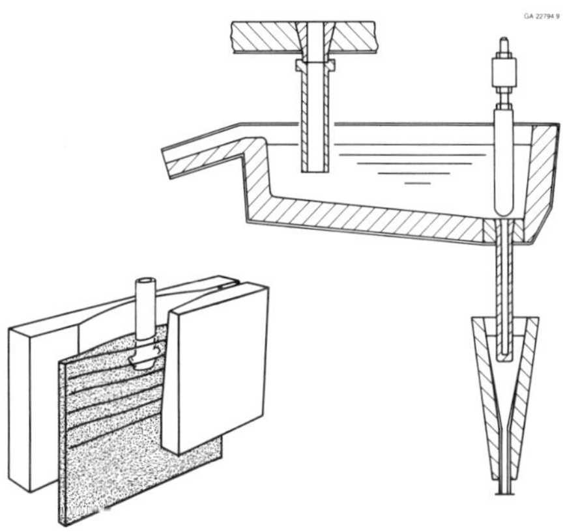

Metal Transfer and Mold Arrangement of SMS Concast Slab Casting System
Figure 13

- Electromagnetic strip/slab casting is a technology that appears to be in the early stages of development. The concept is similar to EMC casting of conventional ingot in that an electromagnetic field is used to contain the liquid metal to avoid any mold contact. Heat is extracted by direct gas and/or water impingement. The challenge is in the electromagnetic shaping and containment of a wide flat liquid sheet of 1 in. thickness or less that can be quenched to a solid sheet.

  Several approaches to this version of EM casting have been considered. As shown by Figure 14A, the Olin Corporation has patented an EM method of casting thin strip of a variety of metals and semiconductors (19). A unique feature is the use of a shaping inductor ahead of the solidification zone to contain the reservoir of molten metal. A similar system (Figure 14B) for casting aluminum up to 3 in. thick was described in patent literature by V. S. Shipilor et al. (20). In contrast to the vertical casting mode of these designs, Figure 15 illustrates patented means for horizontal strip casting (21,22). The U.S. Department of Energy method (Figure 14A) pertains to steel whereas the Sumitomo design (Figure 14B) appears to be directed at aluminum. The absence of additional information on these processes does not permit an accurate appraisal of their current status.

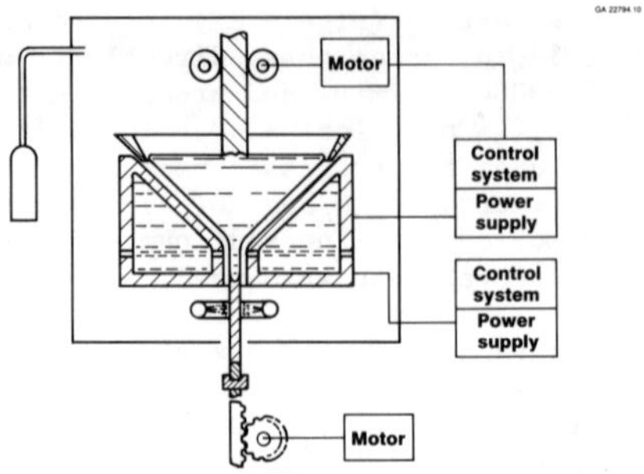

(A)

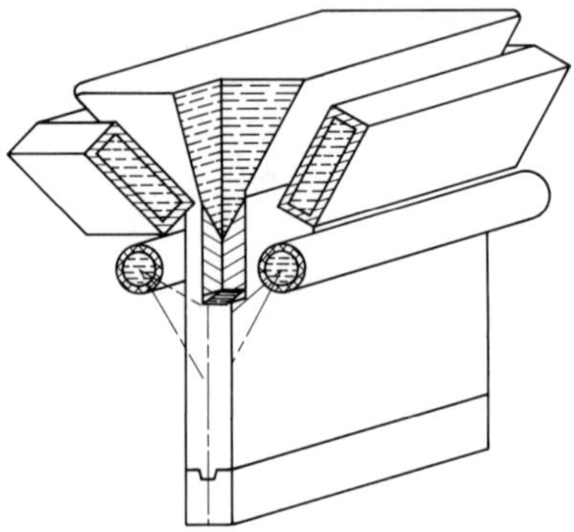

(B)

Vertical EM Casting of Strip/Slab
Figure 14

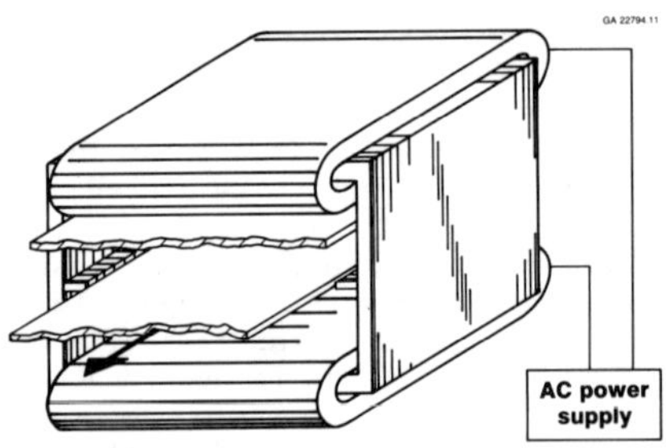

(A)

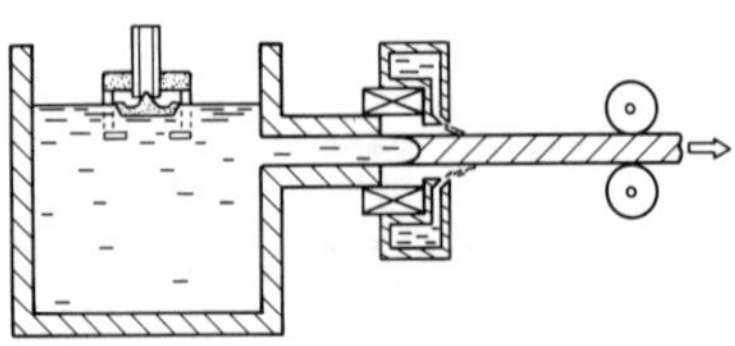

(B)

Horizontal EM Casting of Strip/Slab
Figure 15

Summary

There is no doubt that traditional hot lines provide the most versatile, proven way of producing flat rolled products suitable for a wide range of applications. However, capital cost is relatively high and the facilities must be highly utilized to achieve economic success.

On the other hand, caster equipped mini-mills skirt these economic disadvantages and have gained significantly in flat rolled aluminum products, with approximately 1.0 billion lbs. produced domestically in 1987. Thus far, these mini-mills have primarily penetrated niche markets having less demanding quality requirements, but they seem to be expanding their capabilities. Further expansion is possible if some key technical problems are solved. Therefore, it is conceivable that mini-mills can play a large role in both reducing the traditionally high capital intensity of the aluminum industry and reshaping future manufacturing methods.

Another reason that mini-mills may be more important is that the growth rate in consumption of aluminum in the industrialized economies of the U.S., Western Europe and Japan has slowed. Future growth in consumption is likely to occur in developing countries, where initial market size is probably limited but a full product line is desirable. Based on this premise, future manufacturing facilities need to be much smaller than today's, flexible in nature but offering a versatile range of products. Continuous casters, either existing or emerging, may provide the solution.

The authors express their gratitude to the following Alcoans for their input and support:

Mr. E. L. Rooy, Pittsburgh Office
Mr. A. J. Sartschev, Pittsburgh Office
Mr. F. P. Swigon, Lebanon Works

References

1. J. Isenberg-O'Laughlin and W. P. Jacob, "Thin-Slab Casting: Has Its Time Come?," <u>33 Metal Producing</u>, (1985 March).

2. G. McManus, "Will Strip Casting Reshape the Metals World," <u>Iron Age</u>, (1985 August).

3. M. Marley, "Mini-Producers are Invading Lower-Tier Aluminum Markets," <u>Iron Age</u>, (1987 September).

4. Private communication from Hazelett Strip Casting Corporation.

5. A. M. Odok and I. Gyöngyös, Alusuisse Caster I and Caster II, Aluminum Association, Continuous Casting Seminar, 1975.

6. Lauener Engineering sales brochure.

7. B. Frischknecht and K. P. Maiwald, "Roll Caster Applications and Developments," <u>Light Metals</u> (1987).

8. U.S. Patent 4,681,152.

9. R. A. Gleixner and R. S. Carbonara, "Considerations for Single Roller Direct Strip Casting," (Proceedings International Symposium on Near-Net-Shape Casting of Strip, 89th Annual General Meeting CIM - 1987).

10. J. D. Nauman and D. B. Love, "Controlling the Shape of Cast Stainless Steel Strip," (Proceedings International Symposium on Near-Net-Shape Casting of Strip, 89th Annual General Meeting CIM - 1987).

11. K. Schwaha et al., "Strip Casting of Low Carbon Steel at Voest Alpine," (Proceedings International Symposium on Near-Net-Shape Casting of Strip, 89th Annual General Meeting CIM - 1987).

12. N. Tayama et al., "Direct Strip Casting of 18-8 Stainless Steel by Unequal Diameter Twin-Roll Method," (Transactions 151 J, Vol. 26, 1986).

13. L. E. Hackman and T. Gaspar, "Producing Strip by Melt Overflow Process," Industrial Heating, (1986 January).

14. International Publication No. WO 87/02285, 1987 April 23.

15. U.S. Patent 4,221,257.

16. U.S. Patent 4,676,298.

17. H. Fastert and V. Lackinger, "Strip-Casting with In-Line Hot-Rolling," (Mechanical Working and Steel Processing Conference, AIME, 1987).

18. E. Hoffken et al., "The Production of Continuously Cast Strip on the Pilot Plant at Buschhütten," Metallurgical Plant and Technology, (1987 February).

19. U.S. Patent 4,375,234.

20. U.S.S.R. Patent 908,487.

21. U.S. Patent 4,678,024.

22. European Patent Application 0 191 586.

NEAR-NET-SHAPE CASTING OF SHEET AND BAR OF

ORDERED NICKEL ALUMINIDE ALLOYS*

V. K. Sikka

Metals and Ceramics Division
Bldg. 4508; MS 083
Oak Ridge National Laboratory
P. O. Box 2008
Oak Ridge, Tennessee 37831

Abstract

This paper presents the details of direct casting of sheet and bar of ordered nickel aluminide alloys from liquid metal. The intermetallic alloys (IC-50 and IC-218), based on the Ni_3Al compound, were continuously cast into both 1- to 1.5-mm-thick sheet and the IC-50 alloy continuously cast into 8- and 12.5-mm-diam rod. The important aspect of this study is to extend the use of sheet and bar casting processes developed for well-established alloy systems to newly developed ductile ordered intermetallics of nickel and aluminum systems. Thermomechanical processing of near-net-shape cast sheet and bar was carried out. In both cases, cold working with intermediate anneals produced equiaxed fine-grain structure and expected levels of mechanical properties were achieved. We believe that the near-net-shape processes for sheet and bar described in this paper are the most economical methods for producing these products as opposed to conventional techniques.

*Research sponsored by the Division of Materials Sciences, U.S. Department of Energy, under contract DE-AC05-84OR21400 with Martin Marietta Energy Systems, Inc.

Casting of Near Net Shape Products
Edited by Y. Sahai, J.E. Battles,
R.S. Carbonara and C.E. Mobley
The Metallurgical Society, 1988

Introduction

Near-net-shape manufacturing offers the best approach for reducing the cost of many complex-shape components. Numerous near-net-shaping processes are currently available. These can be divided simply into two classes: one where the component requires a minor repair and/or machining step prior to using, and the other one where the component requires additional thermomechanical processing to refine the structure followed by a possible repair and/or machining step prior to using. Some examples of the near-net-shape processes where only minor repair and/or machining is required include sand casting, precision casting, squeeze casting, rapid omnidirectional compaction (ROC), hot isostatic pressing (HIP), and isothermal or superplastic forming. Examples of a second set of near-net-shape processes include single or twin-roller casting of thin sheet, direct casting of bar for wire fabrication, spray casting, consolidation under atmospheric pressure (CAP), and injection molding. In both of the near-net-shape processes, cost is reduced by energy savings and high-production rates that are achievable in near-net-shape processing as compared to conventional processing.

A flowchart describing various near-net-shape processes that are possible from liquid metal, where liquid atomization is the starting step, is presented in Fig. 1. A similar flowchart where liquid metal is the starting step is presented in Fig. 2. In all of the processes, the number of steps required prior to achieving the desired final shape is shown. These shapes can vary from the simplest form of sheet to very complex disks for turbine and turbocharger rotor applications. Figures 1 and 2 show that the number of steps vary from one to four in most of the near-net-shape processes. It should be noted, however, that the near-net-shape process selected for a component will depend upon component design, tolerance requirements, process characteristics, and material behavior.

This paper describes the newly developed nickel aluminide intermetallic alloys, direct casting of sheet and bar from liquid metal, thermomechanical processing of sheet and bar, and their mechanical properties of the processed materials.

Description of Nickel Aluminide Intermetallic Alloys

Ordered intermetallic alloys (1) constitute a unique class of metallic materials which form long-range-ordered crystal structures below their melting points (T_m) or critical-ordering temperatures (T_c). The various species in these alloys tend to occupy specific sublattice sites and form superlattice structures. In ordered lattices, dislocations travel in pairs or groups, and their motion is thus subject to certain constraints, particularly at elevated temperatures (2). In general, the strength of ordered intermetallics does not degrade rapidly with increasing temperature. In many cases, these alloys exhibit yield stresses that increase with test temperature. Interest in ordered intermetallic alloys for structural applications has been limited because of their low ductility and brittle fracture. The Ni_3Al intermetallic system investigated in the present study is highly ductile when used as a single crystal but extremely brittle as polycrystalline (1,3). Recently the polycrystalline Ni_3Al intermetallic has been ductilized (4,5) by small additions of boron and control of aluminum to stoichiometrically deficient levels. The boron addition ductilized the alloy at all temperatures up to the melting point. However, the higher ductility is prevalent when tested in vacuum. In the intermediate temperature range from 600 to 900°C, ductility of Ni_3Al is very low when tested in air. However, recently the intermediate temperature ductility in air has also been enhanced by additions of up to 8% chromium in Ni_3Al-base alloys. Nominal compositions of Ni_3Al-base-ordered intermetallics under development at the Oak Ridge National Laboratory (ORNL) and commercialization at ORNL in conjunction with industry are given in Table I. The IC-50 (IC = intermetallic compound) alloy has the ordered structure (γ'-phase) at room temperature (RT) to its melting point. The IC-218 and IC-221 are primarily ordered structure (γ'-phase) with small amounts of the disordered structure (γ-phase) at room temperature. The fraction of disordered structure in these alloys increases at high temperatures (≥1000°C).

Intense efforts have been under way (6-9) at ORNL in conjunction with industry to develop the processing technology for intermetallic alloys listed in Table I. Among various processing technologies under development, the direct casting of sheet and bar have the best chance for near-term commercialization for nickel aluminide alloys.

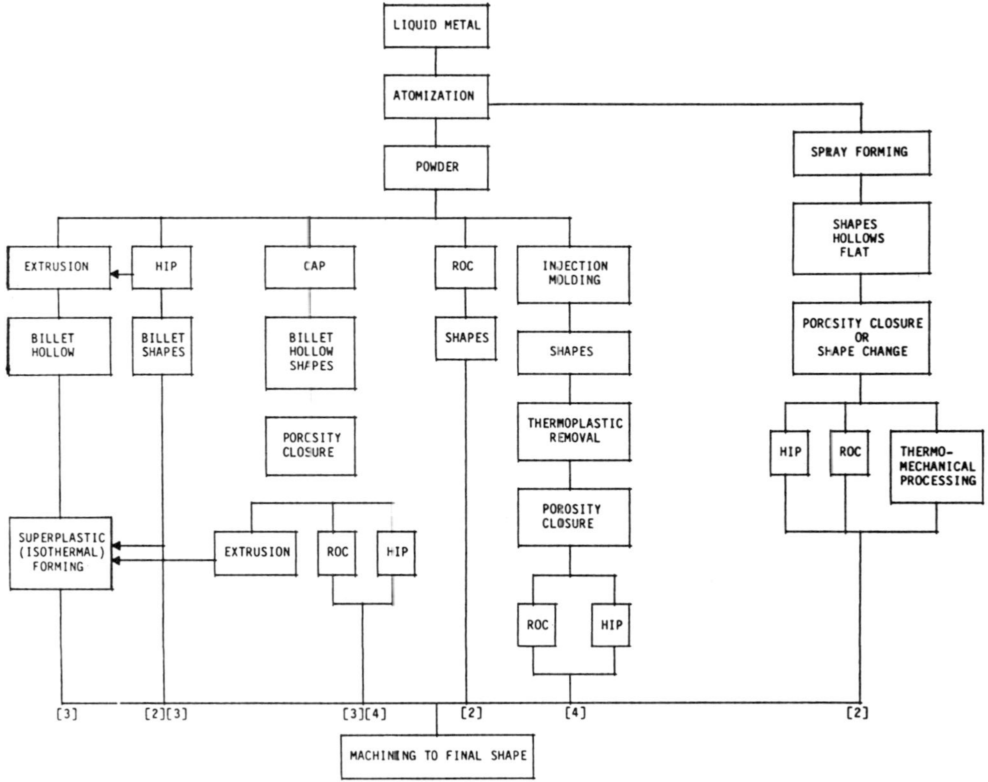

Figure 1 - Near-net-shape processes, where liquid atomization is the starting step. [ ] = Numbers represent the steps required in each near-net-shape process prior to the common minor machining step.

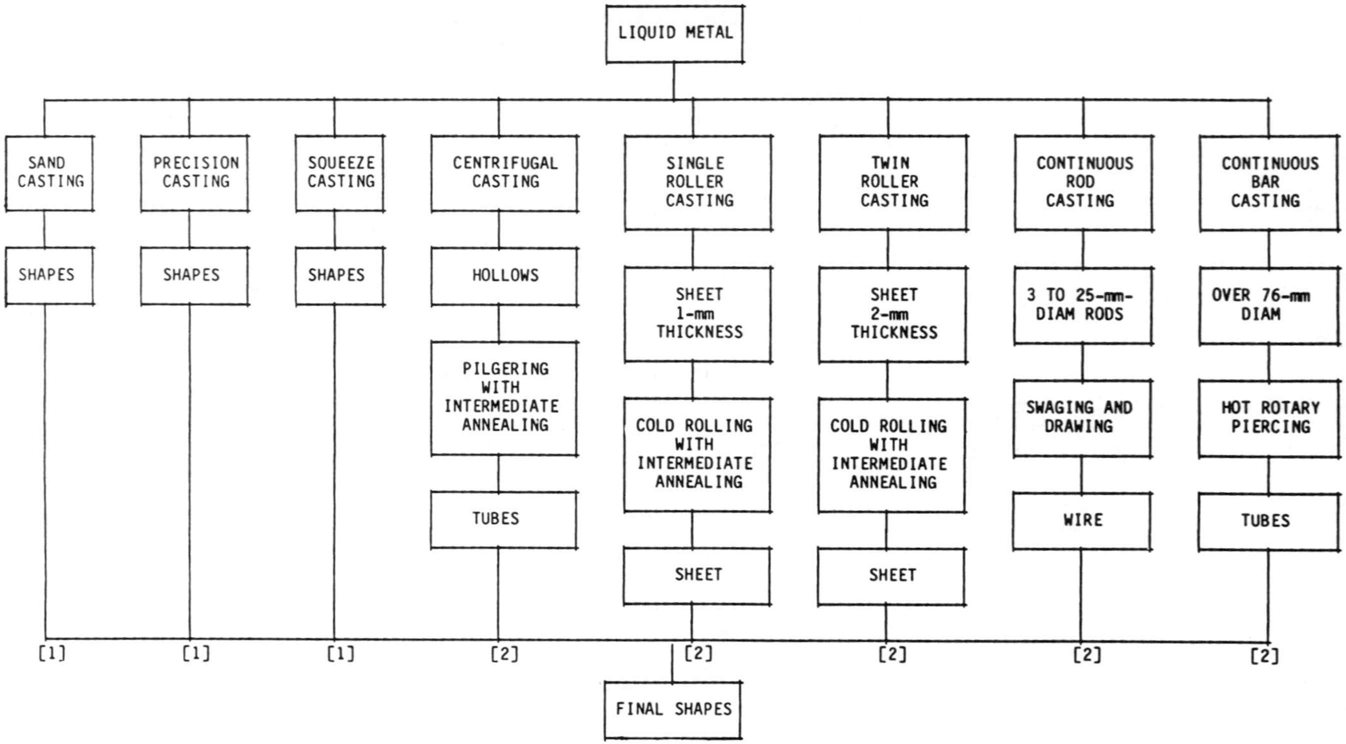

Figure 2 - Near-net-shape processes, where liquid metal is the starting step. [ ] = Numbers represent the steps required in each near-net-shape process prior to the common minor machining step.

Table I. Composition of Three Nickel Aluminide Intermetallics Being Fabricated at ORNL

Element	Weight Percent		
	IC-50	IC-218	IC-221
Al	11.3	8.5	8.5
Cr	--	7.8	7.8
Zr	0.6	0.8	1.7
B	0.02	0.02	0.02
Ni	Bal.	Bal.	Bal.

Direct Cast Sheet

The nickel aluminide alloys IC-50 and IC-218 have been air-induction melted successfully in heats of sizes approaching 900 kg. A 900-kg heat of IC-50 and one 900-kg heat of IC-218 have been direct cast into sheet, which was continuously coiled, Fig. 3. The sheet thickness ranged from 1.0 to 1.3 mm, and the width for the first two heats cast was 300 mm. However, the caster has a potential to go to widths of 600 mm. One side of the sheet was of excellent quality. However, the other surface contained rough spots, which have been suggested to result from localized crystallization. The rough spots were metallic in nature and could either be ground readily or rolled in. Additional heats are currently planned to eliminate the crystallization observed on the first two heats mentioned above. The thickness of the cast sheet was uniform within ±3% except 20-mm edges on both sides which were generally much thinner (approximately -10%).

Chemical analysis of the direct cast sheets of IC-50 and IC-218 are compared with the target in Tables II and III. Being first trials of 900-kg heats, there is a good general agreement between the target and actual analysis. Some difference in oxygen and nitrogen analysis between the vendor and check is possibly due to sampling. The sheet specimens that ORNL tested probably had surface oxide as opposed to vendor analysis, which was on a specimen taken from the melt. Our previous experience (10) has shown that air-melted heats of nickel aluminides generally contain 14 to 35 ppm of oxygen by weight, which is in good agreement with vendor analysis.

The as-cast sheet possesses excellent ductility as demonstrated by the fabrication of a tray with tight bends shown in Fig. 4. Optical microstructure of as-cast sheets of both IC-50 and IC-218 were similar. A typical set of micrographs for as-cast sheet of IC-50 alloy is shown in Fig. 5. The as-cast microstructure is fine-grain columnar, which is refined to equiaxed finer-grain structure, Fig. 6, by cold rolling ≥50% followed by annealing at 1100°C. The RT tensile properties of annealed IC-50 as a function of percent cold work prior to annealing is presented in Fig. 7. These figures show that isotropic properties in direct cast sheet of nickel aluminide alloy IC-50 can be obtained by cold rolling ≥50% followed by annealing at 1100°C. Furthermore, cold rolling more than 50% prior to annealing does not seem to further improve the RT tensile properties of the alloy studied here. Results observed for IC-50 were also observed for IC-218. Tensile properties of direct cast sheet of IC-50 as a function of temperature for the longitudinal and transverse orientation are plotted in Fig. 8. Once again, this figure shows that cold rolling the direct cast sheet by 60% with final anneal of 1100°C produces near isotropic material for the entire test temperature range.

Direct Cast Bar

The direct cast bar of nickel aluminide alloy, IC-50, was produced using the horizontal caster at Steel Casting Engineering in Orange, California. Two sizes of bar, 8- and 12.5-mm

Figure 3 - Photograph of direct cast sheet of IC-50 and IC-218 alloys. The large coil of IC-50 alloy weighs 740 kg.

Table II. Chemical Analysis of IC-50 Sheet Prepared by Direct Cast Process

Element	Analysis (wt %)			Analysis (at. %)
	Target	Vendor	Check	Vendor
C	--	0.044	0.011	---
S	--	0.0002	0.0006	---
Si	--	0.05	0.06	0.09
Al[a]	11.3	12.0	11.90	22.92
Zr	0.6	0.94	1.05	0.53
Cr	--	0.09	0.09	0.08
B	0.02	0.018	0.016	0.09
O	--	0.005	0.0142	---
N	--	0.0020	0.0034	---
Ni	Bal.	86.90	86.3	76.29

[a]Al (at. %) Equivalent = (Al + Zr + 1/2 Cr) = (22.92 + 0.53 + 0.04) = 23.49.

Table III. Chemical Analysis of IC-218 Sheet Prepared by Direct Cast Process

Element	Analysis (wt %)			Analysis (at. %)
	Target	Vendor	Check	Vendor
C	--	0.040	0.012	---
S	--	0.0003	0.0008	---
Si	--	0.04	0.05	0.07
Al[a]	8.5	11.4	10.95	21.87
Zr	0.8	1.14	1.33	0.65
Cr	7.8	6.5	6.63	6.02
B	0.02	0.017	0.020	0.08
O	--	0.004	0.0097	---
N	--	0.0013	0.0055	---
Ni	Bal.	80.90	80.3	71.31

[a]Al (at. %) Equivalent = (Al + Zr + 1/2 Cr) = (21.87 + 0.65 + 3.01) = 25.53.

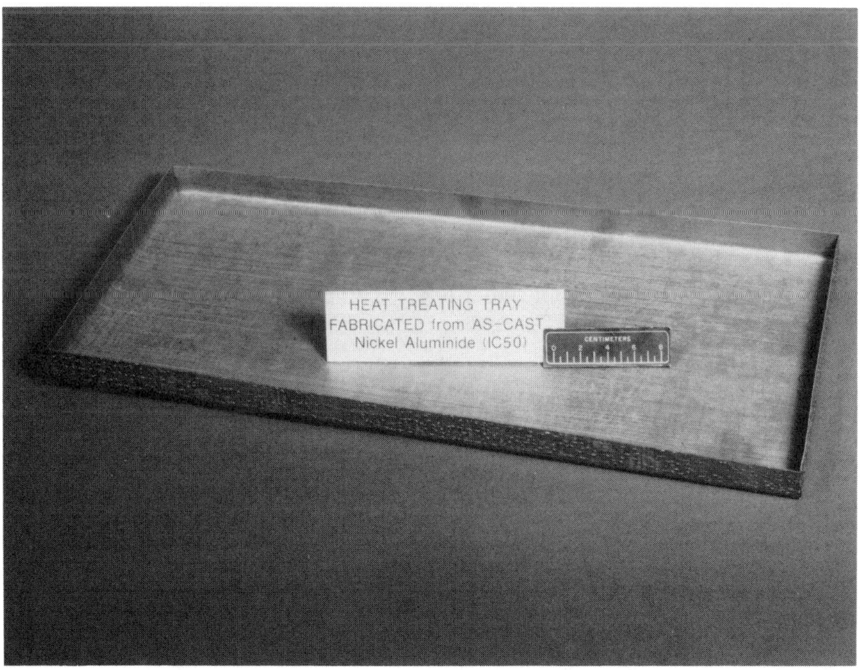

Figure 4 - Photograph of a heat-treating tray fabricated from direct cast sheet of IC-50. The tight bends in this tray demonstrate good ductility of direct cast IC-50 sheet.

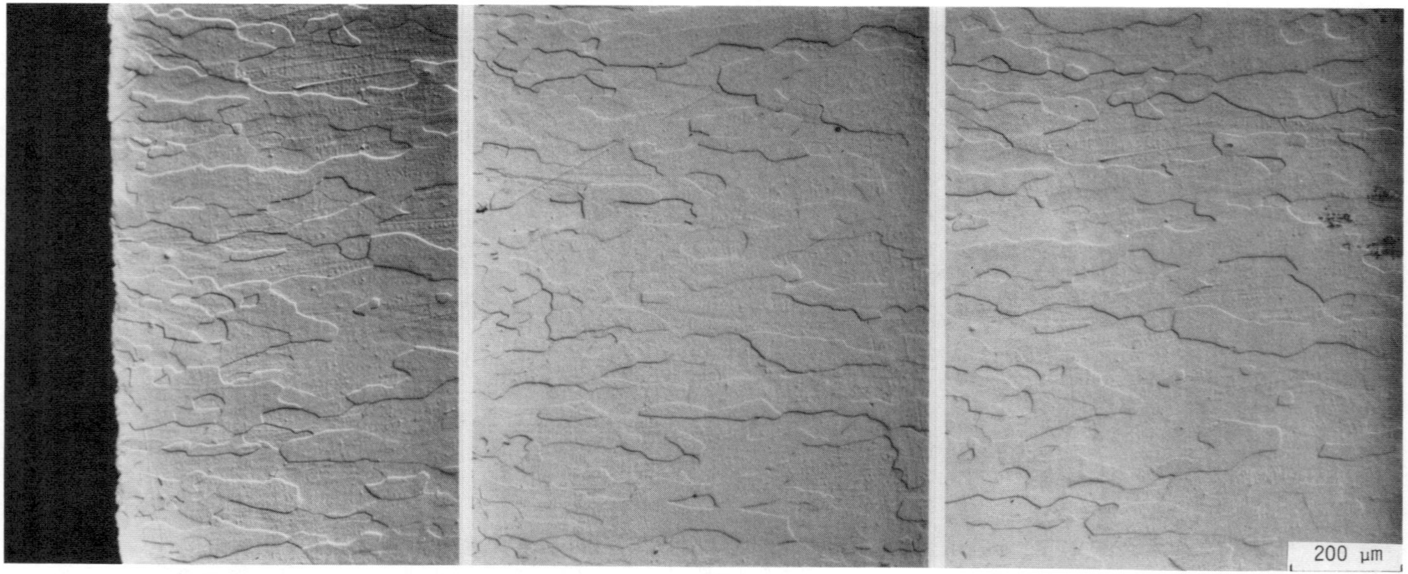

(a) (b) (c)

Figure 5 - Optical microstructure of IC-50 sheet in the as-cast condition. (a), (b), and (c) show various locations. Magnification: 100×.

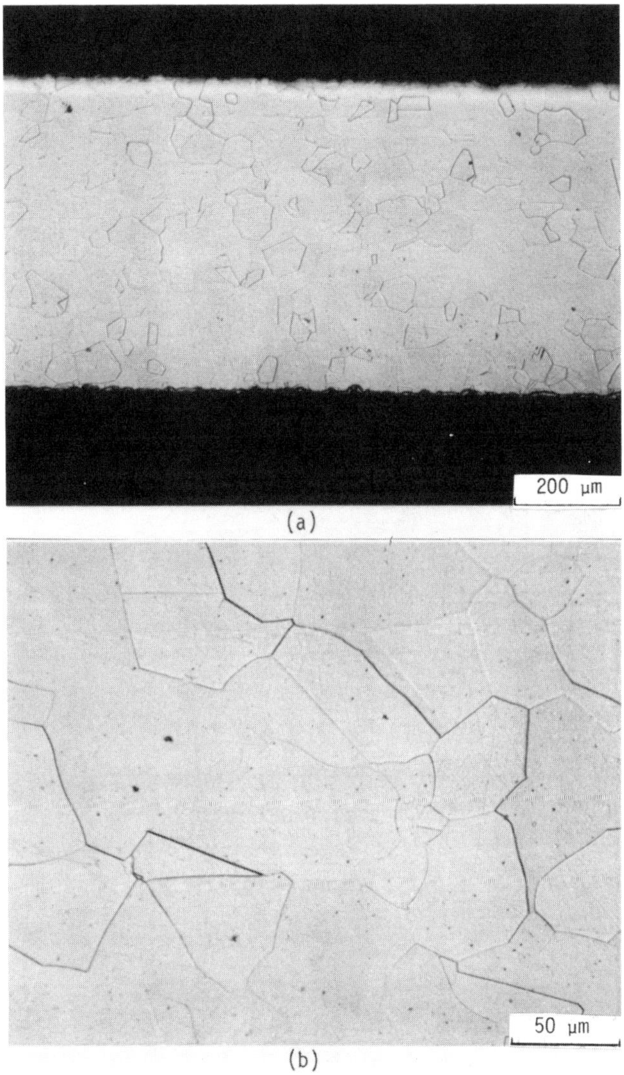

Figure 6 - Optical microstructure of IC-50 sheet produced and processed by cold rolling and annealing. Magnification: (a) 100× and (b) 400×.

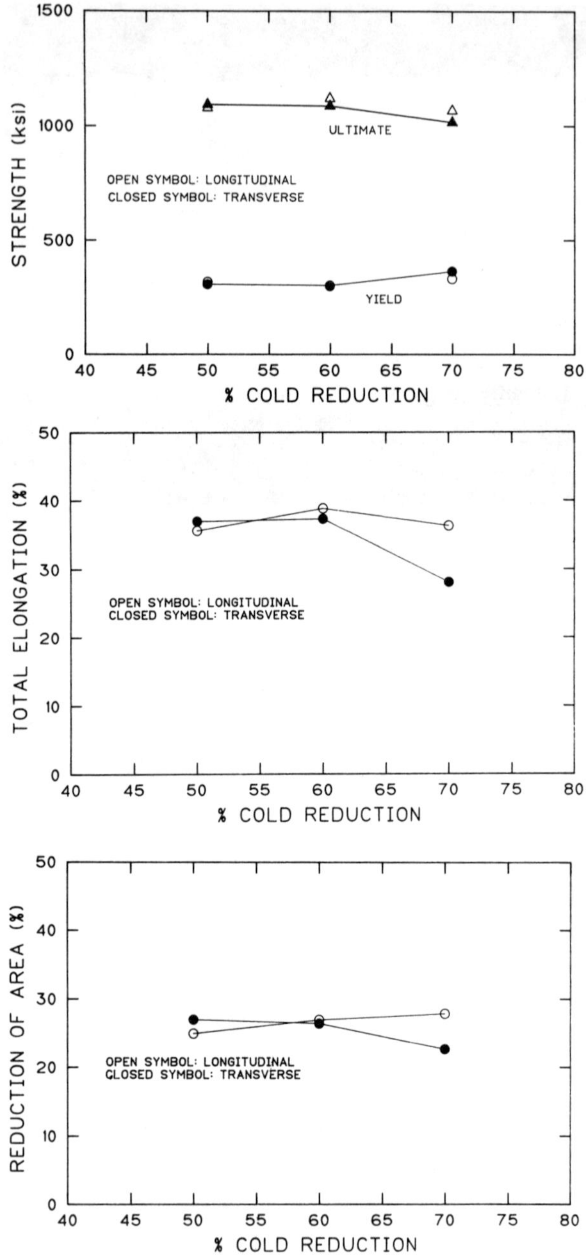

Figure 7 - Tensile properties at room temperature of direct cast sheet of IC-50 as a function of percent cold rolling prior to annealing at 1100°C. Specimens were tested in longitudinal and transverse orientations.

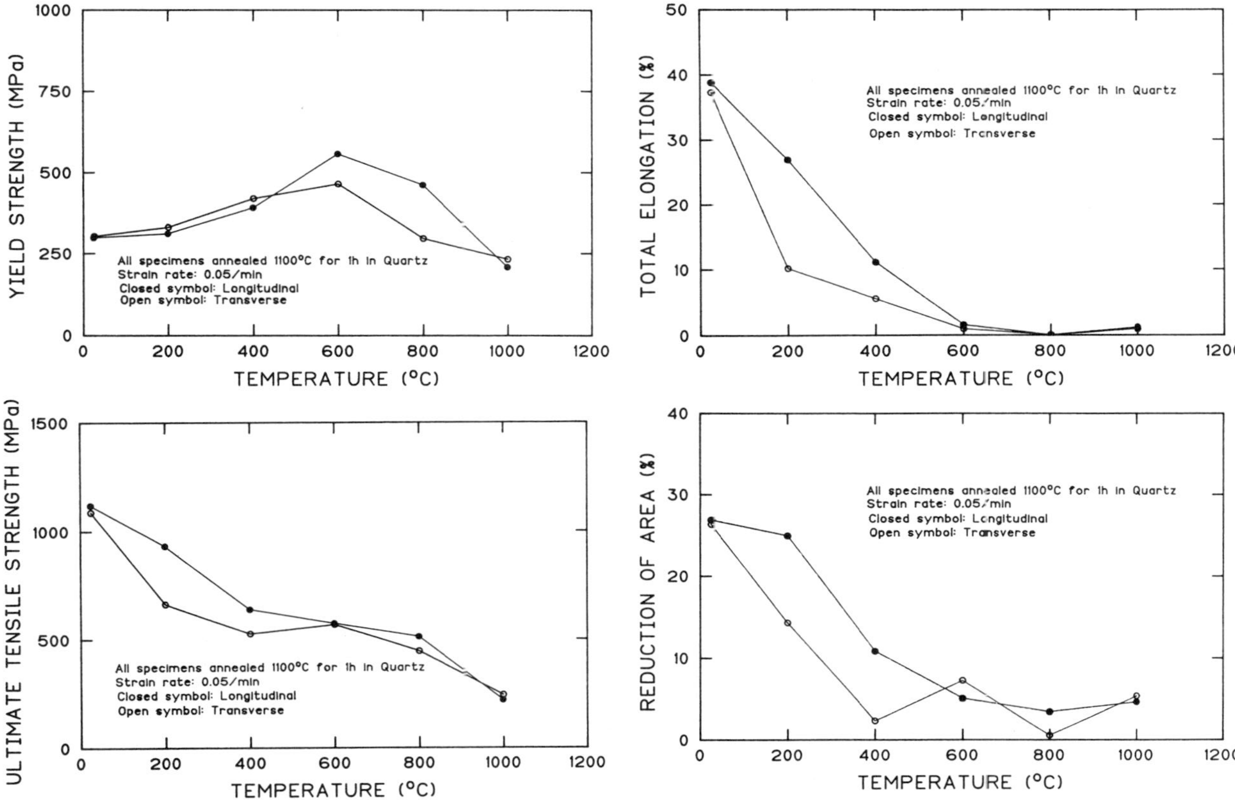

Figure 8 - Tensile properties as a function of test temperature for direct cast sheet of IC-50 with 60% cold rolling. Specimens were tested in longitudinal and transverse orientations.

diam were cast using a 250-kg heat. Figure 9 shows the coil of direct cast 12.5-mm-diam bar of IC-50. The bar produced by this technique shows two oscillation marks unique to the horizontal casting process. These marks identified as primary and secondary witness marks in Fig. 10 result from the two direction modes of solidification. In nickel aluminide alloys, these marks are shallow in terms of radial extension into the cast product and present no difficulty during subsequent processing of the alloy. Chemical analyses of the direct cast bar of IC-50 are compared with the target chemistry in Table IV. The agreement between the two analyses is excellent and the oxygen content of this air-melted heat is 0.0014 wt %. The as-cast microstructure of the 12.5-mm-diam bar of IC-50 is shown in Fig. 11. The structure consists of columnar grains with some fine shrinkage porosity near the center of the bar. The secondary arm spacing of the bar was measured in both the as-cast condition and after an annealing treatment. The results are shown in Table V. This table shows that annealing treatment reduces the interface area, increases the secondary arm spacing, and thereby reduces the internal energy and makes the structure more workable.

The bar was cold swaged both as-cast and after an annealing treatment. The work hardening behavior of the bar is represented by hardness curves in Fig. 12. The nickel aluminide alloy, IC-50, work hardens rapidly with 20% reduction and further reduction occurs at no additional increase in hardness. The thermomechanical processing steps of swaging 60%, annealing, swaging 60%, followed by annealing at 1100°C for 1 h produces equiaxed fine-grain structure, Fig. 13. The grain size values for the bar processed in as-cast and cast plus annealed condition at the edge and center were 20, 20 and 16, 18 mm, respectively. The bar of this microstructure could be further swaged and drawn into wire of various sizes.

Tensile properties of bar in the as-cast condition.and after thermomechanical processing are compared in Fig. 14. Refinement in structure by thermomechanical processing develops the expected yield, ultimate tensile strength, and ductility values. The as-cast material has lower strength values than wrought material up to 800°C and the highest at 1000°C. This is an expected result based on a diffusion control mechanism becoming important at higher test temperatures. Ductility values are lower for the cast material for the entire test temperature range. The wrought material values improve significantly at RT and 1000°C. The lower intermediate temperature values are explained (7) in terms of dynamic oxygen embrittlement in nickel aluminide free of chromium, such as IC-50. Chromium is added to alloys IC-218 and IC-221 to reduce the dynamic oxygen embrittlement effect. However, these alloys have not yet been cast into bar by this method.

Based on ductility data of cast material at 1000°C for IC-50, it is obvious that conventional hot-working techniques cannot be readily applied to this alloy. Thus, we believe that near-net-shape techniques such as direct sheet or bar casting are presently the best methods for fabrication of nickel aluminide alloy, IC-50. The chromium-containing and more recently iron-containing aluminides can be processed by conventional techniques such as casting followed by hot forging, hot extruding, and hot rolling. The powder metallurgy-based near-net-shape fabrication techniques for these alloys are described in other papers (11,12).

Summary and Conclusions

Near-net-shape continuous casting of sheet and bar of ductile ordered nickel aluminide intermetallic alloys is described. The IC-50 alloy was cast into 1.0- to 1.3-mm-thick sheet and 8- and 12.5-mm-diam bars. Sheet was cast from air-induction-melted 900-kg heats. Sheet and bar microstructures were evaluated in the as-cast condition and after thermomechanical processing. Tensile properties of both products were evaluated as a function of test temperature. The following observations and conclusions are made from this work:

1. The direct casting processes for sheet and bar developed for well-established alloy systems can be extended to newly developed ductile ordered intermetallics of nickel and aluminum. The casting process has to be optimized to improve the surface quality of both sides of the sheet.

2. Both sheet and bar can be thermomechanically processed at room temperature with intermediate anneals at 1100°C to produce equiaxed fine-grain microstructures.

Figure 9 - Photograph showing the coil of direct cast 12.5-mm bar of IC-50. The direct casting was carried out at Steel Casting Engineering, Orange, California.

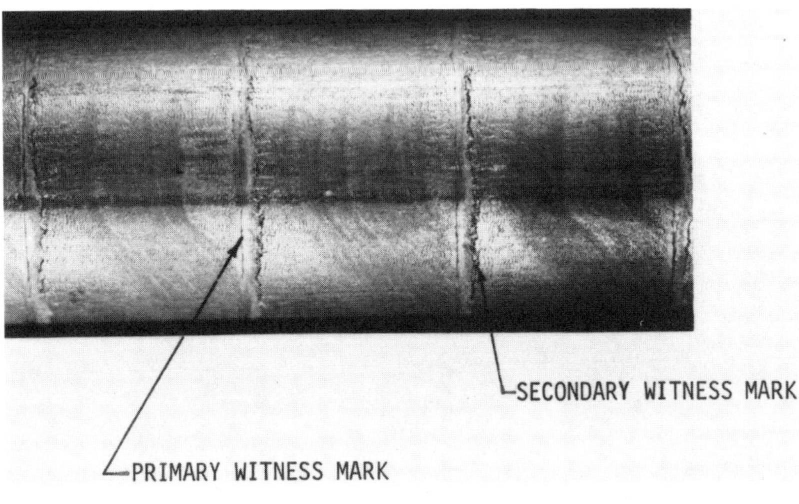

Figure 10 - Photograph showing primary and secondary witness marks on direct cast bar of IC-50. Magnification: 3.25×.

Table IV. Chemical Analysis of the Direct Cast Bar of Nickel Aluminide Met the Target Composition and Provided Low Residual Elements

Element	Weight Percent	
	Target	Check
Al	11.3	11.17
Zr	0.6	0.51
B	0.02	0.019
C		0.01
S		0.0008
Si		0.05
N		0.0008
O		0.0014

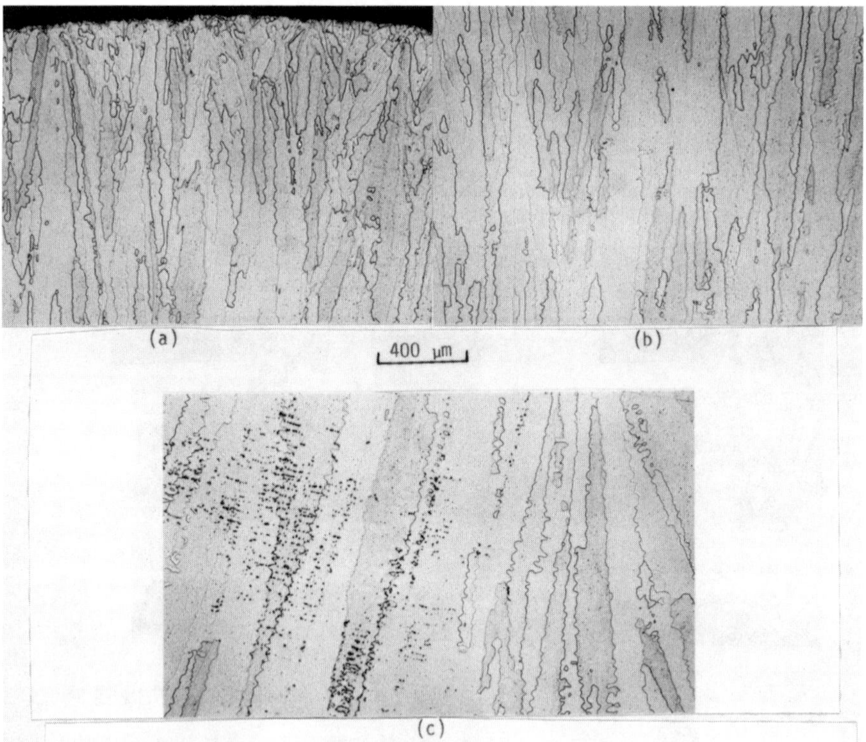

Figure 11 - Optical microstructure of transverse section of as-cast plus annealed bar of IC-50. (a) Edge, (b) one-half radius, and (c) center. Note the variation of solidification structure from edge to center. Annealing at 1100°C for 1 h appears to have produced somewhat cleaner microstructure. Some indications of solidification shrinkage are evident in the center. Magnification: 50×.

Table V. Effect of Annealing on Secondary Dendritic Arm Spacing of Direct Cast 12.5-mm-diam Bar of IC-50

Location	Spacing (μm)	
	As Cast	Cast and Annealed[a]
Transverse		
Edge	25	42
1/2 Radius	24	39
Center	38	47
Longitudinal		
Edge	30	
1/2 Radius	27	
Center	31	

[a]Annealed at 1100°C for 1 h.

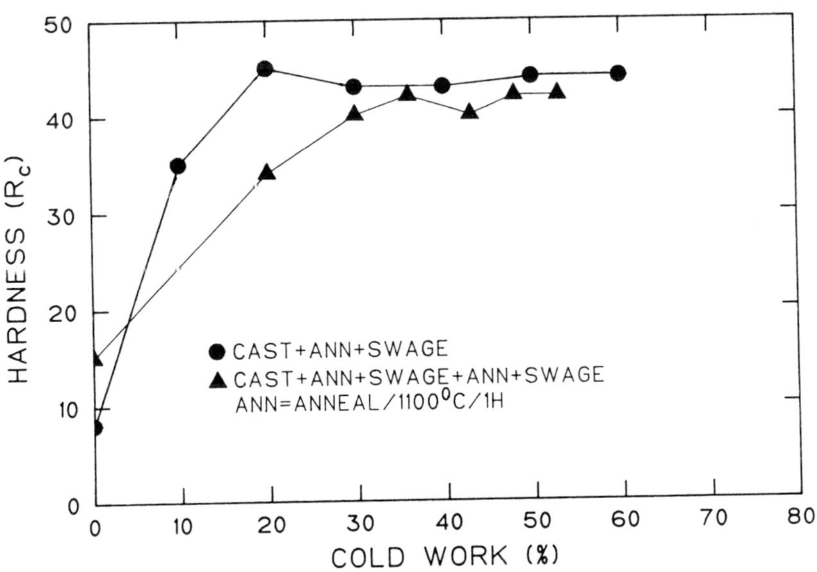

Figure 12 - Plot showing hardness change during swaging of direct cast bar of IC-50.

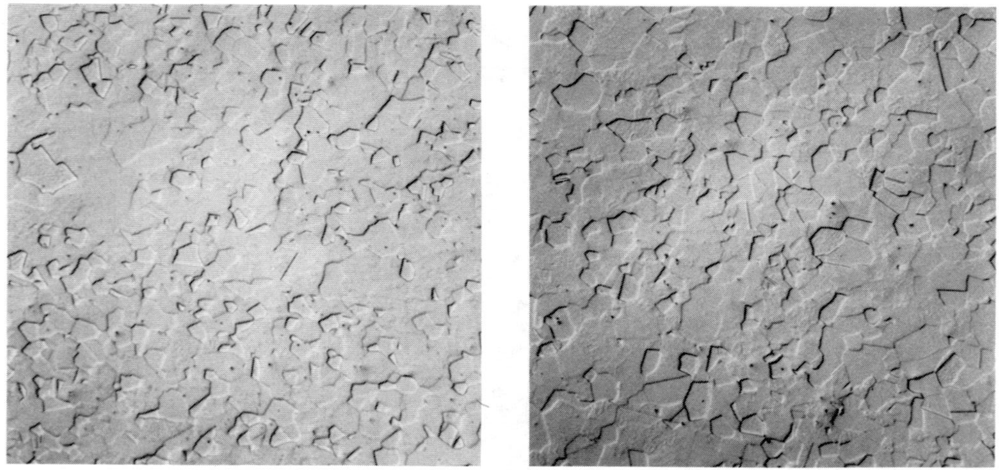

(a) (b)

Figure 13 - Optical micrographs of the direct cast bar of IC-50 after processing. (a) and (b) are edge and center of bar which was annealed, in cast condition, swaged 60% annealed at 1100°C for 1 h, swaged again 60%, and reannealed at 1100°C for 1 h.

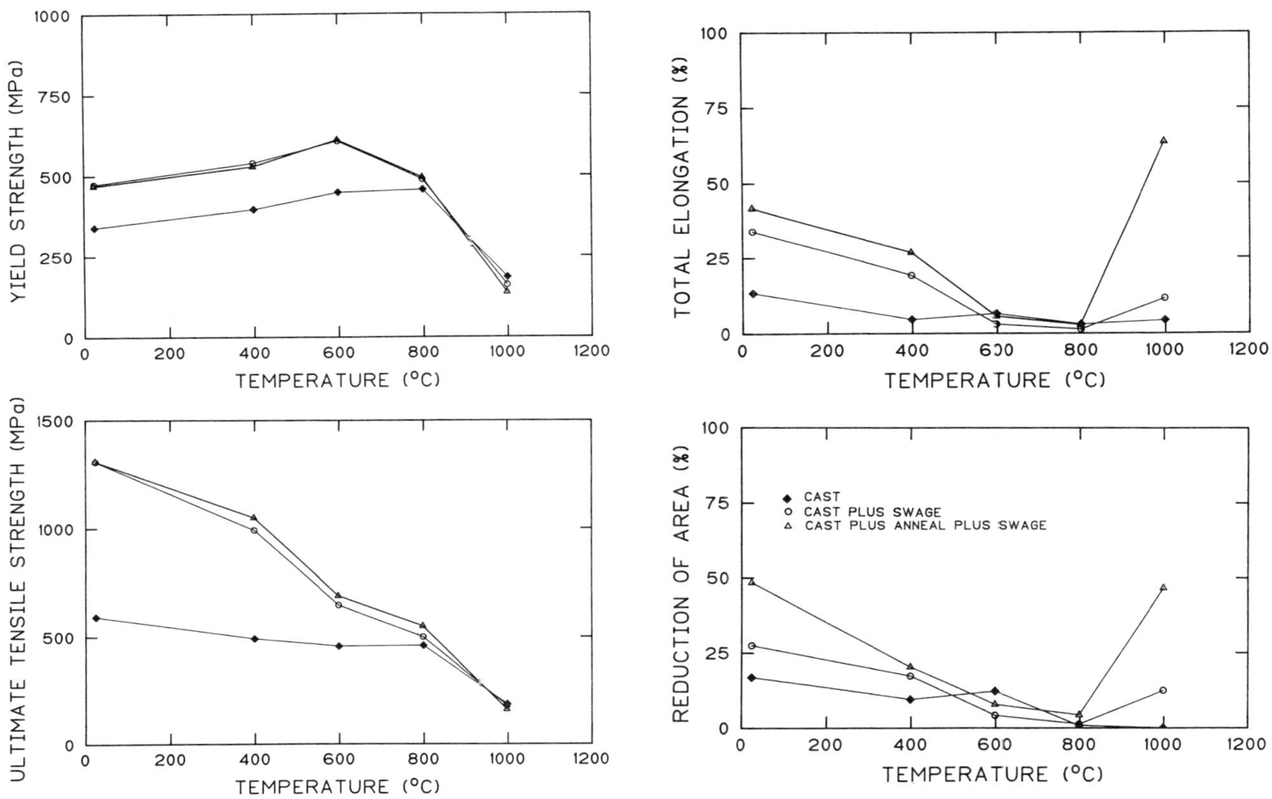

Figure 14 - Tensile properties for direct cast bar of IC-50 as a function of test temperature in as-cast condition and after thermomechanical processing.

3. Thermomechanical processing produces expected mechanical properties in the nickel aluminide alloy (IC-50). Tensile strength and ductility properties are both enhanced significantly in thermomechanically processed material as compared to as-cast material.

4. Based on the ductility data of cast bar at 1000°C for IC-50, it is concluded that conventional hot-working techniques cannot be readily applied to this alloy. Thus, we believe that near-net-shape techniques such as direct sheet or bar casting are presently the best methods for fabrication of nickel aluminide alloy IC-50. The chromium-containing aluminides and more recently iron-containing aluminides can be processed by conventional techniques such as casting followed by hot working.

Acknowledgment

The author wishes to extend his special thanks to Steel Casting Engineering for its participation; the Office of Technology Applications of the Oak Ridge National Laboratory for supporting part of the up-scaling effort for nickel aluminides using near-net-shape processing techniques; R. A. Baldwin for mechanical properties testing and plotting of the data; E. K. Ohriner and C. G. McKamey for paper review; and M. L. Atchley for preparing the manuscript.

References

1. C. T. Liu and J. O. Stiegler, "Ductile Ordered Intermetallic Alloys," Science, **226**, November 1984, p. 636.

2. S. Takeuchi and E. Kuramoto, "Temperature and Orientation Dependence of the Yield Stress in Ni_3Ga Single Crystals," Acta Metall., **21**, 1973, p. 415.

3. C. T. Liu and V. K. Sikka, "Nickel Aluminides for Structural Use," J. Met., **38**, 1986, p. 19.

4. K. Aoki and O. Izumi, "Improvement in Room Temperature Ductility of the $L1_2$-Type Intermetallic Compound Ni_3Al by Boron Addition," Nippon Kinzoku Gakkaishi, **43**, 1979, p. 1190.

5. C. T. Liu, C. L. White, and J. A. Horton, "Effect of Boron on Grain Boundaries in Ni_3Al," Acta Metall., **33**(2), 1985, pp. 213-229.

6. V. K. Sikka and E. A. Loria, "Deformation Behavior of Powder Metallurgy Ni_3Al-Hf-B," Proceedings of CIM's 25th Annual Conference of Metallurgists, 1986, pp. 293-303.

7. V. K. Sikka, C. T. Liu, and E. A. Loria, "Processing and Properties of Powder Metallurgy Ni_3Al-Cr-Zr-B for Use in Oxidizing Environments," to be published in Enhanced Properties of Structural Metals Via Rapid Solidification, ed. F. H. Froes, American Society for Metals.

8. V. K. Sikka, "Processing Technology for Nickel Aluminides," Proceedings of Mater. Res. Soc., 1987, Materials Research Society, Vol. 81, pp. 487-493.

9. V. K. Sikka, "Advances in Processing Techniques for Nickel Aluminides," to be published in First ASM Europe Technical Conference co-sponsored by SFM on "Advanced Materials and Processing Techniques for Structural Applications," Paris, September 7-9, 1987.

10. V. K. Sikka, "Engineering Processing and Properties of Nickel Aluminides," to be published in proceedings of Second International SAMPE Metals and Metals Processing Conference, Dayton, Ohio, 1988.

11. V. K. Sikka, "Superplasticity and Its Application to Near-Net Shaping of Intermetallic Alloys," Interdisciplinary Issues in Materials Processing and Manufacturing, Vol. 2 (1987), pp. 451-467.

12. V. K. Sikka, "Powder Processing of Ni_3Al," to be published in Modern Developments in Powder Metallurgy, Vols. 18, 19, 20, and 21, The American Powder Metallurgy Institute, Orlando, Florida.

A NOVEL CONCEPT FOR METAL DELIVERY TO THIN STRIP CASTERS

J. Herbertson, B.H.P., Central Research Laboratories, Australia

R.I.L. Guthrie, McGill University, Canada

Abstract

The delivery of liquid steel to the cooling/freezing zones of thin strip casting machines is a key issue that must be resolved if such machines are to compete with today's high tonnage slab casters, or the new thin slab casters, for the flat rolled sheet market. For application to tonnage production (e.g. carbon steels), strip casters capable of producing steel strip between 2 and perhaps 20mm thickness, at rates of 50 to 100 tph are needed. A novel metal delivery system in which the liquid is fed normal to the mould face under conditions of controlled flow, pressure, temperature, and chemical environment, is described. Controlled metal distribution, giving meniscus stability and uniform shell growth, is achieved using reticulated ceramics such as filters, with appropriate permeability profiles. The proposed delivery system can be applied to single and double wheel or belt systems, or any combinations thereof, and can be regarded as a combination tundish/mould arrangement.

Casting of Near Net Shape Products
Edited by Y. Sahai, J.E. Battles,
R.S. Carbonara and C.E. Mobley
The Metallurgical Society, 1988

INTRODUCTION

The commercial and technical implications of thin strip casting machines capable of competing with today's slab casters have long been appreciated. They include the elimination of the conventional hot mill together with its associated slab reheating furnaces and seven stand hot reduction rolling mills.

The recent status of thin slab and strip casting developments in the steel industry is summarised in Table 1. This table, while not an exhaustive compendium, was prepared as part of a presentation to the Canadian Steel Industry Workshop on the subject of Strip Casting of Steel, held in Hamilton, Canada, under the auspices of the Canadian Steel Industry Research Association, May 7-8th, 1987. Its purpose was to illustrate the main types of machines, and the scale of research being carried out. Changes since that time include cancellation of the final phase of the US/Bethlehem effort (A4), following successful commercial demonstration of the thin slab SMS/Concast machine. Figure 1 presents the same information but in graphical form, in terms of a plot of machine casting velocity (metres/min) versus product thickness (mm). As a frame of reference, Figure 1 also includes the lines which correspond to production rates of 50 and 100 tph/metre width.

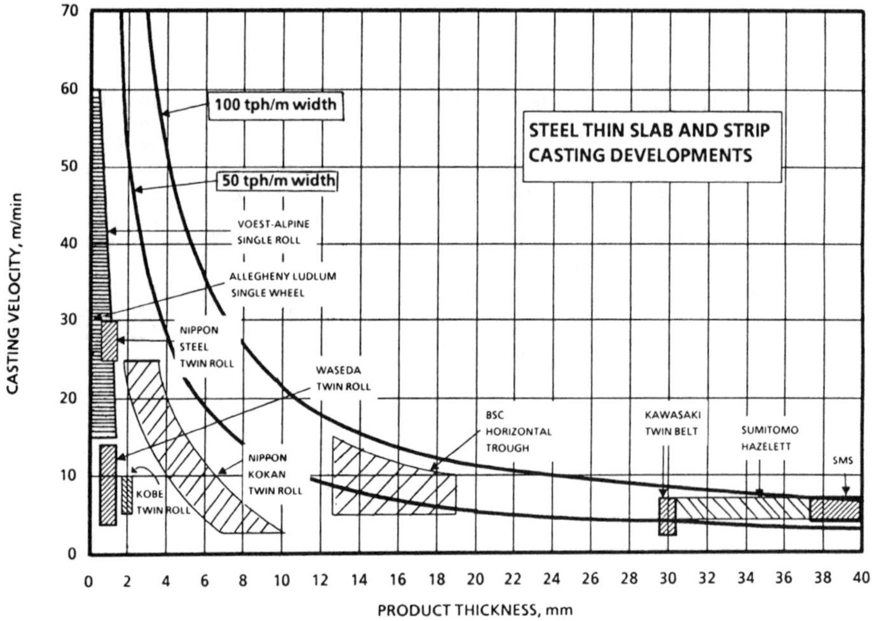

FIGURE 1: STEEL THIN SLAB AND STRIP CASTING DEVELOPMENTS

To the far right of this diagram lie the thin slab casting machines of S.M.S.-Concast, Sumitomo-Hazelett and Kawasaki, extending down to product thicknesses of 30mm with production rates of 100 tph or more.

In the central region is British Steel Corporation's contribution to strip casting in the 1970's; a horizontal trough machine producing experimental heats of various steel grades in the thicknesses of 12-19mm, at tonnage rates approaching 100 tph/m width.

To the far left of the diagram sit single and twin roll machines, mainly producing strip up to thicknesses of 2mm, but only at low tonnage rates to date. They include the Allegheny - Ludlum single wheel process for the commercial

TABLE 1 SOME RECENT STEEL THIN SLAB AND STRIP CASTING DEVELOPMENTS

Company	Scale of Operation	Dimensions Achieved Thickness X Width (mm)	Casting Speed (m/min)	Approx. Production Rate for 1 metre width (tph/m)	Ref.
(A) HAZELETT TYPE THIN SLAB CASTERS					
1) Sumitomo	30t melts ($30M)	(30-50) x 600	6	125	1
2) Krupp	10t melts	(50-70) x 180	6-8	235	2
3) Nucor	32t melts	aim: (25 x 38) x 1300			3
4) US/Bethlehem	30t melts ($30M)	aim: (12-35) x 1800			13
B) NON-HAZELETT THIN SLAB CASTERS					
1) Kawasaki	200kg melts Twin belt	30 x 100	2-6	75	4
2) Danieli		75			13
3) British Steel	(a) 4t melts (b) ½t melts	(50-75) x 500 (12-20) x (200-400)	upto 14 upto 15	441 90	5
4) SMS-Schloemann	Commercial (eg. to Nucor)	(40-50) - (1200 - 1600)	5-6	125	14
(C) TWIN-ROLL 'BESSEMER' CASTERS					
1) Nippon Kokan	250kg melts 400-600mm rolls	(1-20) x (100-400)	1 25	10	6
2) Nippon Steel	100kg melts 300mm rolls	1 x 200	30	13	7
3) Waseda University	200mm rolls	(1-1.5) x 100	4-14	9	8
4) Kobe	150mm rolls	2 x 70	5-10	9	9
(D) SINGLE ROLL MELT DRAG/PLANAR FLOW CASTERS					
7) Alleghany Ludlum	1.5t melts	(0.25 - 1.4) x 330)	15-15-	17	10
8) Voest Alpine	5t melts 630mm roll	(0.5-1.0) x 400	30.60	14	16
9) Armco/Westinghouse	200kg melts ($2.6M)	aim: (8-3.2) x 75			15

production of thin stainless steel sheet, and the experimental twin roll casters of NKK, Kobe, Nippon Steel and Waseda University.

Any thin strip casting machine capable of say 100 tph/m width requires high strip exit velocities to satisfy the material flow requirements. As seen from Figure 1, a caster producing 4mm thick strip at this rate requires a strip exit velocity of some 50 meters/minute.

Small tonnage applications for special steels such as stainless may be commercially viable. However, higher tonnage capabilities, say 50 to 100 tph, are necessary for wider industrial application of thin strip casters to carbon steels.

The purpose of this paper is to explore some of the limitations of current casters towards meeting the market place need for high speed casters, capable of producing steel strip at commercial tonnage rates of 100 tph/m width, in the 2-20mm product thickness range. A possible design solution is then presented and preliminary assembly work described.

CURRENT TECHNOLOGY

Conventional Slab Casters

Current slab casters operating with oscillating fixed moulds can approach velocities of 2m/min from the mould. The recent SMS-Concast breakthrough to 6m/min with their convex mould thin slab caster arrangement depended on metal delivery innovations, including SEN design, mould geometry, and mould lubrication. However, further improvements in fixed, oscillating mould technology can not hope to reach the very high casting speeds and thin sections necessary to essentially remove hot rolling.

Twin Belt Systems

The moving mould machines overcome these friction force, mould lubrication difficulties, the mould work faces moving at the same velocity as the adjacent freezing shell of steel. The Sumitomo-Hazelett machine and Kawasaki Twin Belt machines plotted in Figure 1 are capable of producing thin slab down to about 30mm thickness. However, further reductions in thickness are still hampered by the problem of delivering liquid steel to a wide, thin, gap between the moving belts, under conditions of uniform flow, controlled chemistry, superheat and high cleanness.

In the late 60's, metal cleanness represented a major stumbling block to the development of thin slab casting. Thanks to an integrated approach to clean steel practice, from furnace to mould, the pre-requisite levels of cleanness can now be achieved. Current best world practices are reflected in total oxygen levels as low as 10 ppm as seen from Figure 2(18).

Since strip casters imply higher surface area to volume ratios in the liquid being delivered to the work zone, the current Hazelett-Sumitomo delivery system requires inert gas shrouding as illustrated in Figure 3. To achieve the necessary control in metal delivery, a double tundish arrangement with an elaborate feed-back flow-control system is used. An alternative approach is the use of an enclosed pouring nozzle (e.g. Krupp), through which liquid steel is delivered to the roll gap. An advantage of this system is that the metal can be pressurised against the two work zones (upper and lower belts) to give potentially better surface quality. Apart from the problem of freezing between the nozzle tip and the belts, the potential for *hydrodynamic erosion* caused by the higher tangential velocities called for, places a much higher demand on the chemical and mechanical stability of the nozzle's refractories.

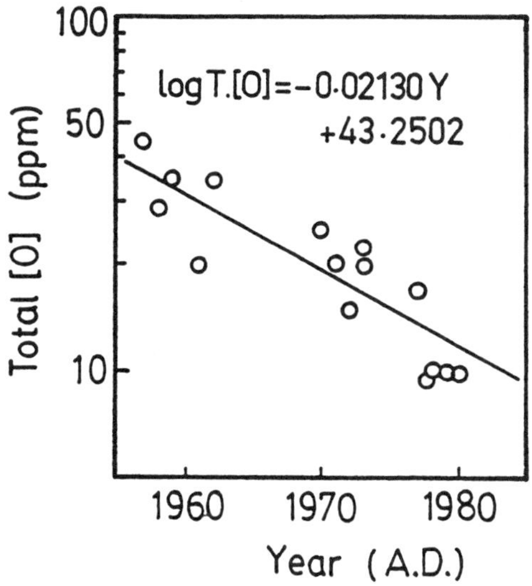

Figure 2 Improvements in steel cleanness (18)

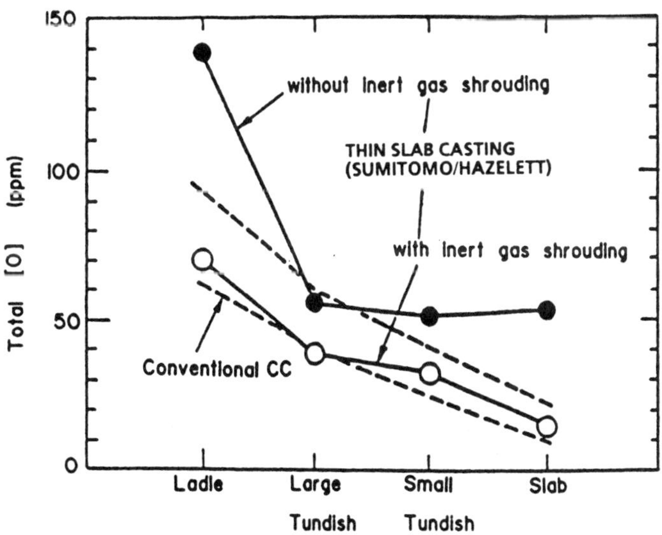

Figure 3 Effect of inert gas shrouding of caster inlet. Data from Hazelett thin slab caster operated by Sumitomo (1): Ladle 50t, large tundish 8t, small tundish 3t, casting rate 75 tph.

Further, metal delivery involving smaller cross sectional nozzles (e.g. slots) is more prone to blockage by any solid inclusions within a steel melt. Evidently this places added demands on upstream cleanness, and inclusion modification (i.e. transforming solid inclusions to liquid inclusions), and filtering may be necessary.

As shown in Figure (4), filters can contribute significantly to steel cleanness levels. Figure 5 illustrates the problems associated with buoyant inclusions entering the Hazelett-Sumitomo twin belt caster. One should note the significant rise in the distribution of inclusions near the top surface of the slab. Evidently, in extreme cases, inclusions that have already floated out to the top surface of the metal pool, will adversely affect the quality of the upper surface. A similar phenomenon can be observed in slab casters, where inclusions tend to accumulate at the 3/4 point of the slab.

Twin Roll Systems

Turning our attention to the metal delivery systems associated with twin roll casters, one may have downwards, upwards (as practised in the aluminum industry), or sideways, delivery of metal to the rolls.

However, it is first useful to comment on the relatively low productivity of the twin roll casters, vis a vis, the belt (block) and conventional casters.

TABLE 2 SOLIDIFICATION CONSTANTS - PILOT PLANT DATA

Company	Casting System	K_S (mm/min$^{1/2}$)	Reference
Sumitomo	Hazelett 30-50mm slab	20	1
Kawasaki	Twin Belt 30mm slab	10-20	4
Nippon Steel	Twin Roll 1mm strip	17	7

Note(1): $D = K_S \sqrt{t}$

where D = shell thickness, mm

t = time, min

Ks = solidification constant, mm/min$^{1/2}$

Note(2): K_S = 20mm/min$^{1/2}$ is also typical of conventional continuous casting

Table 2 presents experimental solidification rate constants reported from the various pilot plant operations. Experiments reveal that the increasing thickness of steel shell on a cold substrate essentially approximates a relationship of the form:

$$D_S = K_S \sqrt{t}$$

where D is the instantaneous shell thickness (mm), t represents time (minute), and K_S is the solidification rate constant, (mm/min$^{1/2}$). Figure 6 illustrates the roll diameters (meters) needed to produce strip of various thicknesses and outputs. These computations are based on a typical solidification rate constant, K_S of 20mm/min$^{1/2}$, a contact arc of 60°, and solidification of steel at the roll nip. We see that 12mm strip at 100 tph would require cooled rolls of 3 meters diameter

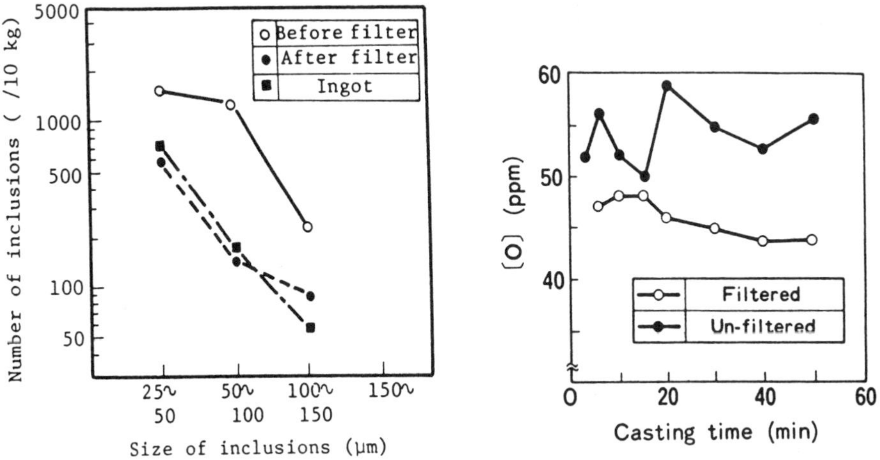

(a) Alumina/Zirconia foam filter; Al-Si killed steel (11) (b) Alumina/Zirconia foam filters: Al killed steel (12)

Figure 4 Inclusion filtration in steel

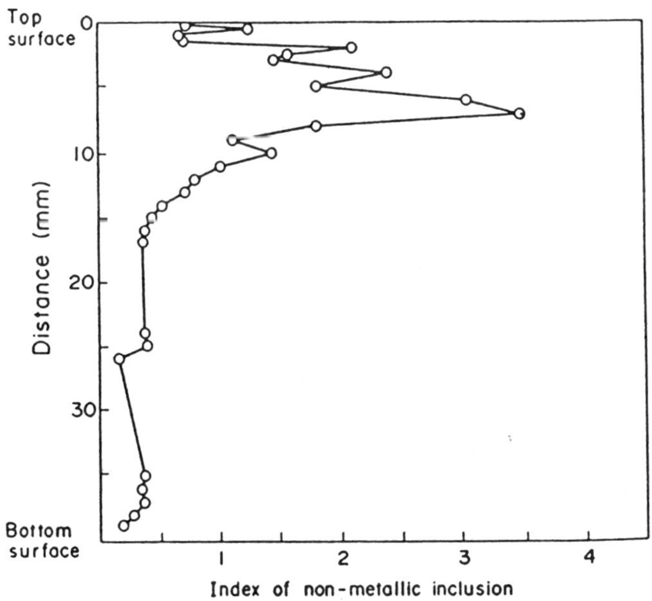

Figure 5 Distribution of inclusions in thin slab cast steel with a Hazelett machine at Sumitomo, Kashima Works (1).

(and 1 meter wide) to provide the necessary contact time with the chilled metal surface. Such rolls are significantly larger than those that have been used in pilot plants. Major roll diameter scale-up poses problems of side dam containment and roll life. The combination twin roll/belt caster systems, allowing solidification after the roll nip, can overcome some of these disadvantages. Single roll casters are suited to strip thicknesses below 1 mm (16).

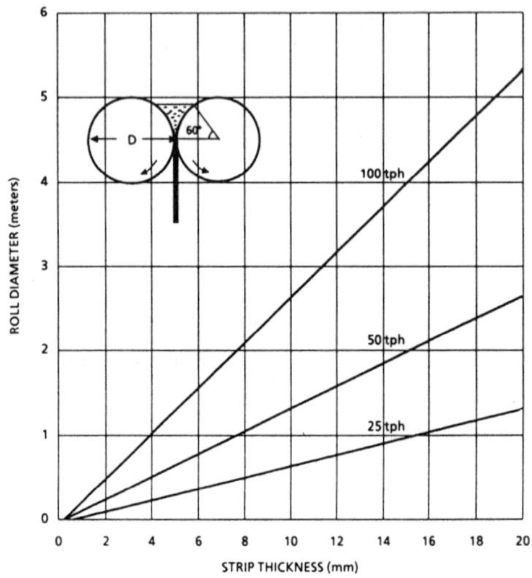

Figure 6 Relationship between roll diameter and exit strip thickness for twin roll casters.

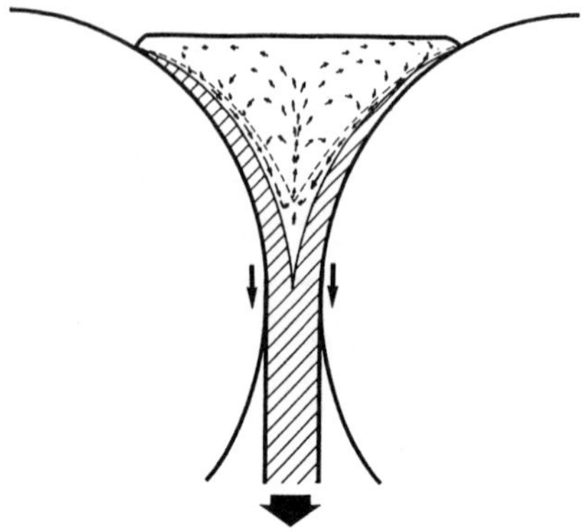

Figure 7 Flow patterns in twin roll casting. Boundary layer entrainment will give a tendency for recirculating flows.

In terms of metal delivery problems in twin roll strip casting, a significant recirculating flow of liquid metal in the pool established between the work rolls can occur as a result of the drag action of the rolls. This is shown in Figure 7. From the values of the kinematic viscosity of steel and the solidification rate constants for strip casting (Table 2), one can calculate that the magnitude of this recirculating load may approach half the nett tonnage throughput of the caster. Any inclusions entering the system which escape the freezing shells, will tend to concentrate at either the central impingement region - resulting in center-line defects, or be swept back up to the top of the pool and out to the sides - resulting ultimately in surface defects. The combined effects of irregular recirculating flows and localised entry flows of superheated liquid can result in non-uniform shell formation from the meniscus to the roll nip.

Cold Shuts

Common to all forms of metal delivery is the question of 'cold shuts', and how these may be avoided. Evidently, lapping and non-wetting of the metal, or its oxide skin, and the cooling work pieces can prove critical. The question of cold shuts relates to meniscus control. Physical movement, partial solidification or reoxidation of the meniscus, especially at very high casting velocities, will lead to poor strip surface quality. The tendency for 'cold shuts' decrease with:

1) increasing superheat, 2) lower metal/mould heat transfer coefficients and 3) more regular metal delivery. One solution, adopted successfully by British Steel Corp. in the 70's, was to have a separate thermally insulating layer interposed between the travelling bottom mould and the liquid steel, so as to delay the initial chill and thereby avoid cold shuts. This is a technical solution which becomes progressively less economic as the strip thickness decreases and the surface area to volume ratio increases. Further, the B.S.C. pilot plant used a relatively crude liquid steel delivery system based on an overflowing weir. With much improved metal delivery, it should be possible to cast directly onto a chill mould substrate, particularly is view of the excellent bottom surface quality achieved with the metallic glass, rapid solidification processes. These would be to the extreme left on Figure 1, at about 100 μm strip thickness.

PROPOSED METAL DELIVERY SYSTEM

Metal feedage problems to the solidification zone are recognised as an important focus for process development attention in many of today's thin section casting programmes (21, 22). Figures 8 and 9 illustrate the essential features of the metal delivery systems now proposed for belt and twin roll casters respectively. To identify the basic elements of the metal delivery systems in Figures 8 and 9, the following code is used: 1. Liquid metal supply, 2. Reticulated ceramic nozzle, 3. Meniscus, 4. Constrained liquid pool, 5. Solidifying shell, 6. Substrate, 7. Strip product.

Thus liquid metal is delivered *towards* the freezing fronts rather than *tangentially*, through an extended reticulated ceramic nozzle system positioned close to a moving cold substrate. A liquid metal pool is constrained under positive pressure in the narrow gap between the delivery nozzle and the substrate, a shell of solid metal forming against the cold substrate's surface. The extended reticulated nozzle also incorporates end, and side walls, which are close enough to the moving substrate to stop the escape of liquid metal. Alternatively, side dams can be incorporated into the mould. With a clearly defined liquid/substrate front, a constrained and pressurised liquid pool, close superheat control, and flow control using variable permeability reticulated ceramic nozzles, the meniscus can be fixed in position without premature freezing or reoxidation, and the soldifying shell can form against the cold substrate under conditions which provide good surface quality. The proposed casting process is essentially friction free since the substrate moves at the casting velocity, and the stationary reticulated ceramic nozzle and containment walls are 'self-lubricated' by the internal flow of liquid metal. When metal quality is in question, the steel can be filtered prior to distribution to the work face of the mould, as illustrated in Figure 8 B.

The advantage of a horizontal block, or belt, caster, over twin roll arrangements, is that they are essentially equivalent to infinite radius roll casters in which extended contact lengths are readily incorporated. The thickness of solidified shell that emerges from the enclosed metal delivery zone depends on the contact time (nozzle length/substrate velocity). Some thickness of liquid is carried along with the solid shell and this will contribute to the final strip thickness.

Figures 9A, B and C suggest machine configurations whereby this basic metal delivery system might be incorporated into new, or existing, casters. Figure 9 A shows a twin roll caster. By pressurising, distributing, and controlling metal delivery towards the freezing substrates, some of the undesirable features of current twin roll casters (delivery through submerged entry nozzles and attendant cross flows, surface lap marks, inclusion accumulations, difficulties with end wall (or end-roll) melt containment, etc.) can be improved.

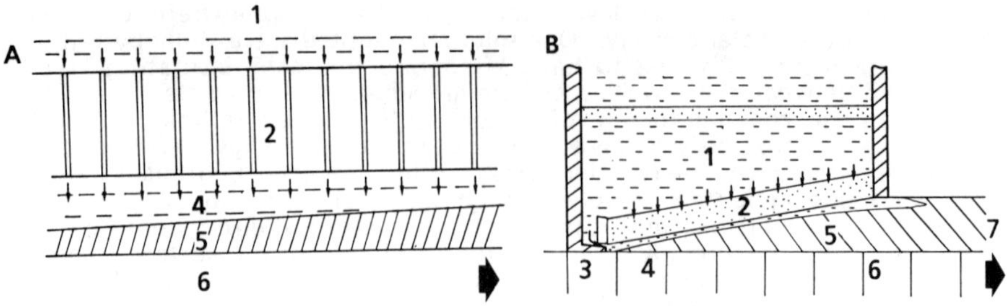

Figure 8 Proposed metal delivery system for high speed, high tonnage, thin strip casting of steel. (A and B).

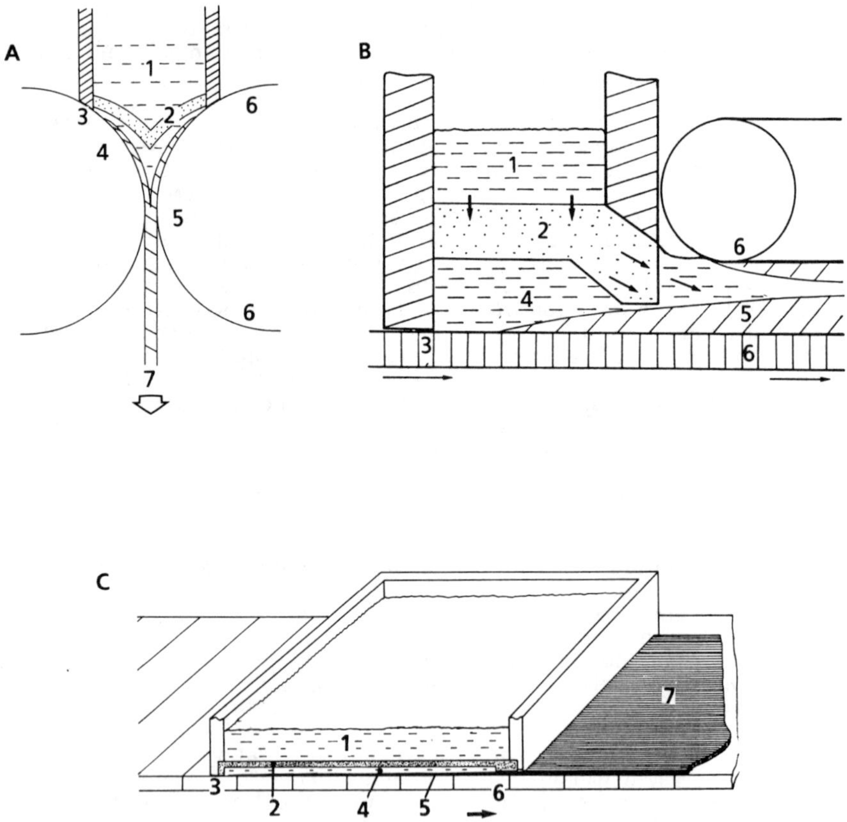

Figure 9 Possible delivery systems for incorporation into new, or existing, casters.

Figure 9B illustrates metal delivery wherein a significant fraction of the metal freezes after substrate delivery. Such a configuration could represent an adaptation of the metal delivery system to thin strip production via Hazelett-type machines, as an alternative to the preceding arrangement in Figure 9A.

Finally, Figure 9C illustrates the essentials of metal delivery for the production of thin strip from a block caster, wherein only a small fraction of liquid remains as the metal exits the extended nozzle system. (i.e. equiv. to Fig. 8B).

Figure 10 has been prepared to illustrate, in a quantitative manner, relationships between the thickness of metal exiting the delivery zone, the final strip thickness, production rate and nozzle contact length. These computations are based on a freezing rate constant, k_s, of 20 mm/min$^{1/2}$. Referring to Figure 10, the abscissa represents the required final thickness of cast strip and is plotted against a series of ordinates for various production rates through the apparatus.

As the production rate increases,the resident distance during which molten steel is in contact with the chilled substrate in the work zone increases. Curves are plotted for various shell thicknesses and lines radiating from the origin showing fixed percentages of solidification.

To demonstrate this diagram's use, consider a strip which is to have a final thickness of 2 millimeters. The 100 percent solidification line is at a resident distance of about 1.05 meters for a production rate of 100 tons per hour per metre width of product strip.

As an example of less than full solidification under the metal delivery nozzle consider if a final strip thickness of 10 millimeters is required. For a 4 millimeter shell thickness the resident distance approaches 0.9 meters for a flow production rate of 100 tons per hour per metre width of product strip. Thus an extended metal delivery nozzle 1 metre in length could be used to produce 100 tph of either fully solidified 2mm strip, or say 40% solidified strip of 10mm thickness.

The fully enclosed metal delivery system means that the conventional metal transfer operation between tundish and mould, and attendant reoxidation problems, is circumvented. The use of reticulated ceramic for the metal distribution system at the point of solidification guarantees a very clean product, as these ceramics also act as inclusion filters.

By delivering the liquid metal to the substrate through a relatively large area of reticulated nozzle, laminar flow can be achieved. Due to recent developments with ceramics for steel filtration applications, reticulated ceramics are now available in sizes and shapes that make the proposed metal delivery/casting systems practicable. A wide range of permeabilities are available, giving unique opportunities for flow control.

While the above remarks imply one sided liquid metal delivery to a horizontal belt/block caster or to a single roll caster, the metal delivery concept is equally valid for single roll, twin roll or twin belt casting machines. In essence, with twin roll or twin belt casters, one should aim to distribute metal normal to each of the solidifying shells, prior to their coming together as one product.

In summary, a metal delivery/casting concept is proposed for providing laminar flow, friction free casting, in a fully enclosed delivery system that would guarantee a super clean product. Control of fluid and heat flow to the meniscus and solidifying shell is the basis of achieving high surface quality.

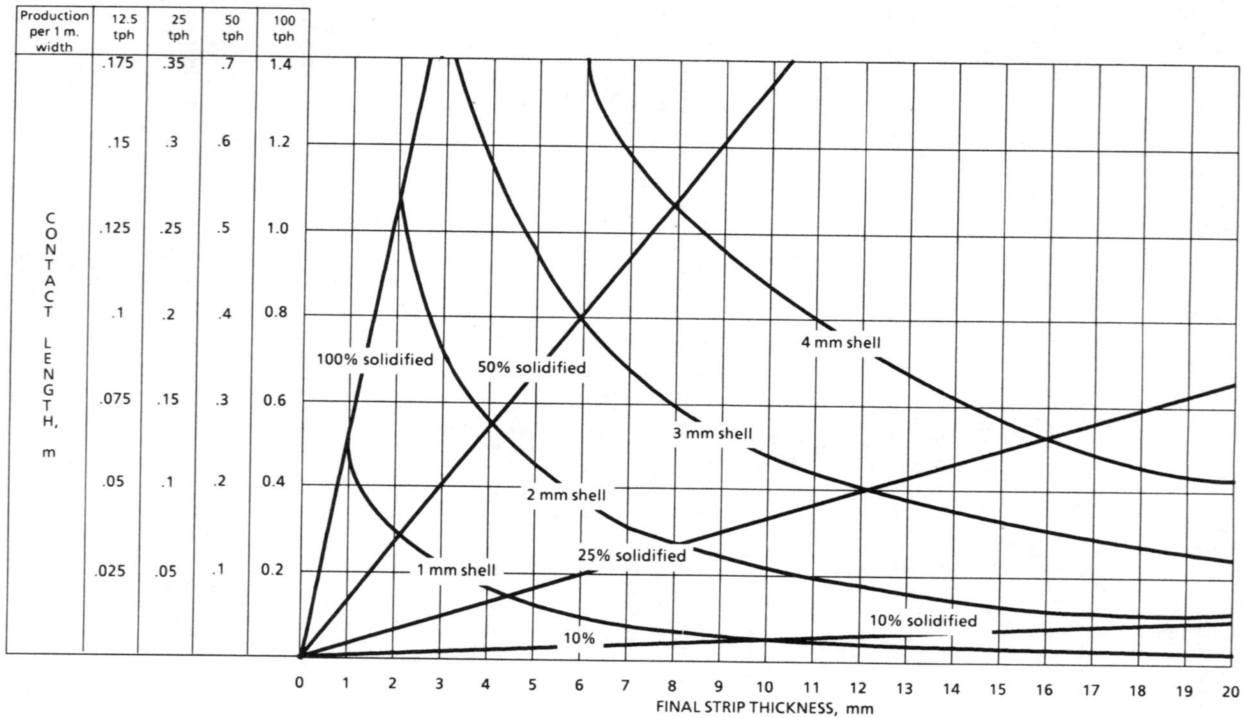

Figure 10 Relation between mould contact length, find strip thickness and percent of metal solidified at nozzle exit, for various production rates.

EQUIPMENT DESIGN

Figure 11 provides a photograph of the laboratory equipment at McGill University designed to test this method of metal delivery. The system involves a thin stainless steel belt, with direct cooling of its lower surface by water sprays. The metal delivery box (tundish) rides on the top surface of the belt, the bottom of which forms the top surface of the mould region, or work zone.

In order to assess the cooling water requirements for this caster, when contacted with molten aluminium, some experimental tests were carried out (19), in which a thermocouple welded to the top surface of a stationary stainless steel belt (cooled from below with water sprays), recorded the surface temperature transients of the belt on filling the pouring box with molten aluminum. Using the FAST-P program (20), transient variations in the thermal resistance between the belt and substrate were deduced. Figure 12 shows that good thermal contact was established during initial contact of the aluminum with the polished steel belt but that solidification and shrinkage of the solidified aluminum shortly thereafter, lead to a six fold increase in thermal resistance to $6x10^{-4}$ m^2K/W. These changes in thermal resistance are reflected in changes in belt surface temperature versus aluminum contact time shown in Figure 13.

The full mathematical description of the process requires incorporation of this data into a total fluid/heat transfer analysis, which is the subject of another publication(Ref. 17). Preliminary experimental results indicate that the surface quality of (aluminum) strip produced by this delivery system is very good. Larger scale work on steel melts is now planned at the 1 tonne level, at the Central Research Laboratories of BHP, Australia.

CONCLUSIONS

A novel metal delivery system has been proposed for high speed thin strip casting systems. By delivering metal towards the mould surfaces through reticulated nozzles, the goal of fully enclosed delivery of a super clean steel from a ladle to chilled work surfaces under controlled flow conditions may be achieved. Variants on such delivery systems raise interesting design possibilities for existing caster systems, and the possibility of producing product in the 2-20 mm thickness gauge, at commercially significant tonnage rates.

Figure 11 View of single belt, laboratory scale, equipment.

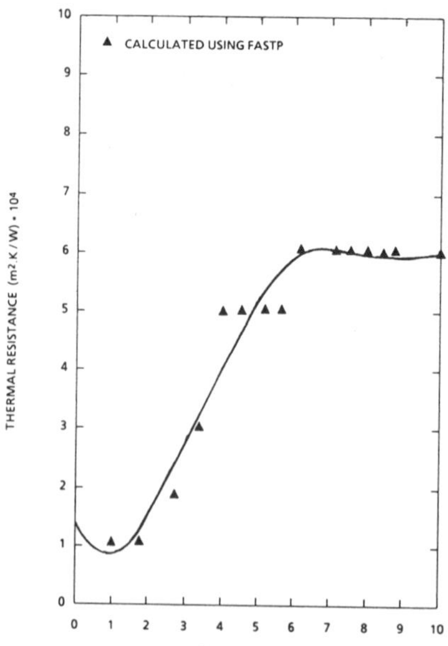

Figure 12 Variation in thermal resistance between S.S. belt and aluminum as a function of the fraction of metal strip solidified.

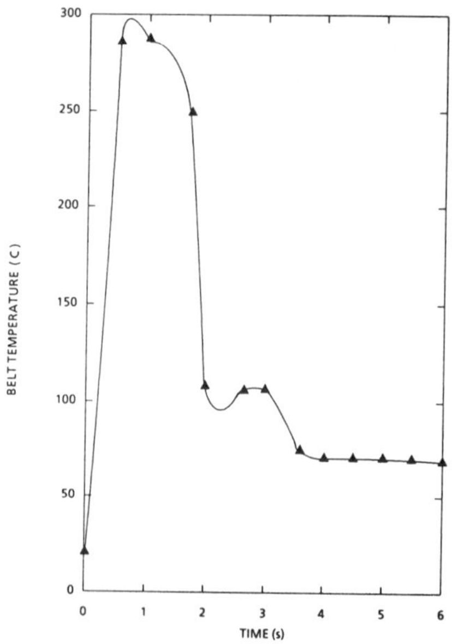

Figure 13 Temperature variation at the top surface of a static S.S. belt, water cooled from below following contact with liquid aluminum.

REFERENCES

1. K. Ushijima, T. Okazaki, M. Yoshihara, S. Shiode and M. Koide, 'Thin steel slab continuous casting process', 5th International Iron and Steel Congress, Washington, D.C., April 6-9, 1986, Steelmaking Proceedings, Vol. 69, ISS/AIME, pp. 827-832.
2. E. Buch and D. Figge, 'State of the art in casting steel billets on a twin belt caster', ibid, pp. 811-816.
3. M.D. Millett, 'Development of the Hazelett twin belt caster by Nucor Steel-General Comments', ibid, pp. 837-838.
4. S. Itoyama, H. Nakato, T. Nozaki, Y. Habu and T. Emi, 'Development and solidification characteristics of horizontal twin belt type thin slab caster', ibid, pp. 833-936.
5. H. Marr, G.D. Spenceley, V. Ludlow and J. Peace, 'Horizontal continuous casting of thin slabs at British Steel Corp. Teesside Labs', ibid, pp. 817-826.
6. K. Kawakami, J. Inagaki, S. Nishioka, H. Sato and M. Ohtsuki, 'Research on twin roll casting process', ibid, pp. 861-870.
7. A. Kasama, S. Mizoguchi, I. Miyazawa, M. Ito and T. Sugai, 'A consideration on the rapid solidification of steel-solidification of steel by twin roll caster', ibid, pp. 849-853.
8. T. Kusakawa, H. Yamamoto, Y. Yanagi, K. Endo and K. Oka, 'Characteristics of iron and steel thin sheet made by strip casting, ibid, pp. 855-859.
9. C. Yoshida, K. Taniguchi, T. Nagagawa, M. Sudo and T. Nozaki, 'Characteristics of rapidly solidified cast iron and high carbon steel', pp. 855-859.
10. J.F. Grubb, D.B. Love, A. Murthy and J.D. Nauman, 'Casting specialty steel strip', ibid, pp. 841-847.
11. H. Ichihasi, T. Okazaki, M. Yoshihara, S. Shiode and M. Koide, 'Application of the filter for clean steel', ibid, pp. 943-950.
12. T. Komai, H. Takeuchi, Y. Nuri and I. Sawada, 'Application trials of filtration technology to production of clean steel, ibid, pp. 951-955.
13. T. Komai, et al, 'Thin slab casting-has its time come?', 33 Metal Producing, Vol. 23, March 1985, pp. 43-49.
14. Concast Technology News, Vol. 25, 3/1986, pp. 108.
15. Newswatch: Westinghouse/Armco working to develop continuous caster for thin strip, Iron Age, April 5, 1985.
16. K. Schwaha et al., Strip casting of low carbon steel at Voest Alpine, Steel Times, July 1987, pp. 324-325.
17. I. Taggart, M. Hasan, J. Herbertson, R. Guthrie; 'A mathematical model of a novel high speed thin strip casting system', to be published.
18. M. Sasabe: Nishiyama Memorial Seminars, No. 90, 1983, Iron and Steel Institute of Japan, Tokyo, pp. 21-56.
19. Roberto Tavares, Nov. 1987, Int. McGill Report: Metallurgical Engineering 'Application of FASTP to the strip casting process'.
20. M. Brown and F. Mucciardi, 'Head and mass transfer simulations on a personal computer', Proc. Int. Symposium on Computer Software in Chemical and Extractive Metallurgy, C.I.M., Montreal, Aug. 1988, (in preparation).
21. 'Steelmakers pursue thin strip casting', Iron Age, August 1987, pp. 14-18.
22. M.R. Moore, 'Development of new continuous casting processes for tonnage steels', I&SM, December 1983, pp. 22-25.

THE ROLE OF SUBSTRATE SURFACES IN CONTINUOUS CASTING

R. E. Maringer

Transmet Corporation
Columbus, Ohio

Abstract

As continuous casting evolves, the various processes making up this technology are increasing in number and in capability. It is becoming clear that many of the newer processes such as chill block melt spinning, melt drag, planar flow casting, and melt extraction have a great deal in common with more established processes, such as the Hunter or Hazelett processes. As a prime example, each of these processes involves the rapid transfer of large amounts of heat (primarily the heat of fusion) from the casting into the substrate. Observations of the behavior of rapid solidification processes, such as chill block melt spinning, suggest that the mechanical, physical, and chemical nature of the substrate surface plays a vital role in determining how heat is transferred across the casting-substrate interface. This in turn controls the quality of both the process and the product. An understanding of these interfacial effects is vital to the continued development of this whole array of interrelated processes and products.

Casting of Near Net Shape Products
Edited by Y. Sahai, J.E. Battles,
R.S. Carbonara and C.E. Mobley
The Metallurgical Society, 1988

Introduction

In many casting processes, and in particular many continuous casting processes, molten metal is brought into contact with a metallic mold or substrate. The purpose of the substrate is to give shape to the casting and to provide a path for the removal of heat. For continuous casting these substrate casting methods fall into two groups. These are:

A. Static mold systems - Those in which the casting moves with respect to the substrate.

B. Moving mold systems - Those in which the casting moves along with the substrate.

Most of the more-or-less conventional slab, bar, or rod casters fall into Group (A). Almost all of these methods utilize a reciprocating mold. Casting speeds are of the order of centimeters per second.

In Group B there are also two categories. These are:

B1. Moving melt systems - Those in which the molten metal and the casting move together until solidification is complete.

B2. Decantation systems - Those in which that part of the casting which has solidified is moved away from the source of molten metal.

Category B1 is made up of methods such as Hunter, Hazelett, the Southwire process, and some less familiar variants we call open mold casting (1) and flow casting (2). These methods are capable of operating at speeds of centimeters to a meter or so per second. Category B2 is made up of methods such as chill block melt spinning, melt extraction, melt drag, and planar flow casting. These methods are capable of operating at speeds of 50 meters per second or more. It is no coincidence that Category B2 methods are those often exploited in the field of rapid solidification. By casting a thin product and quickly stripping away excess molten metal (decantation) it is possible to achieve extremely high rates of solidification. The free surface of such a casting represents the shape of the liquidus - solidus interface as it existed just prior to decantation. Such surfaces were studied extensively many years ago by Rutter and Chalmers (3) and Pond and Kessler (4) as a means of understanding the nature of solidification. The substrate side of the casting represents the surface on which nucleation and growth of the solid began.

For many of these continuous casting processes, the interface area is so small, or it has been so altered by oxidation or other reactions, that it is difficult to assess the events that took place at the interface. It is often equally difficult to assess the progress of the solidification front, since this is usually covered by subsequent solidification and solid state transformations. Both of those difficulties can be circumvented through the study of the methods used to cast thin strip (e.g. melt drag or planar flow casting). Strip can be cast wide enough so that broad areas of both substrate or free cast surfaces are available for direct observation. Casting artifacts which are difficult to recognize on a tiny melt spun ribbon or melt extracted fiber are often magnified in size and number on cast strip. As a consequence, much of the following discussion derives from strip casting studies. We contend that the conclusions drawn from those studies apply to the whole broadened class of substrate quenching methods.

In all of those processes, the initiation of solidification begins when molten metal is brought into contact with the surface of the substrate. The key to all that follows is what happens at the interface between the substrate and the casting. The assumption which underlies the discussion which follows is that, when atoms of a molten metal are brought into intimate contact with atoms of the substrate, a bond is formed. It is through this bond that heat is conducted. When solidification occurs and the temperature decreases, the casting will shrink. This will develop a stress sufficient to break the bond. Once the bond is broken, conductive heat transfer effectively ceases.

If this argument is accepted, then it is clear that the substrate surface must play a predominant role in determining the character of the bond, the amount and rate of heat transfer, and the nature of both the casting process and the cast product.

Discussion

Bonding and Release

When the molten metal contacts the substrate surface, it is assumed that a strong bond is formed. Obviously, if there are volatile or reactive materials on the surface of the substrate, these will interfere with the bonding process. Such things as "mold release agents" are specifically designed to prevent bonding. For the most part, many of the mechanisms which interfere with bonding have been discussed elsewhere (5) and will not be discussed here except as they may fall into context for the subject.

It must be recognized that, except under very unusual circumstances, neither the chemistry of the substrate surface nor that of the melt surface is representative of the chemistry beneath those surfaces. Both surfaces are covered with reaction products, usually oxides, which have mechanical, physical, and chemical properties of their own. The situation is sketched schematically in Fig. 1a. The problem is to bring atoms of the melt into intimate contact with atoms of the substrate. The existence of a surface layer of a different chemistry on the substrate is not normally a part of the problem. However, if the surface layer on the melt has a substantially higher viscosity than the melt itself, (e.g. consider MgO on Mg, ZnO on Zn, Al_2O_3 on Al, FeO on Fe) then it is virtually impossible to achieve intimate bonding between the melt and the substrate. The surface layer must be penetrated. This is shown schematically in Fig. 1b.

In this context, it is common to say that the molten metal "wets" or "doesn't wet" the substrate, but one can be easily misled by this terminology. Conventional wetting is an equilibrium between the forces of attraction among atoms of the melt with themselves and with atoms of the substrate, all of which are at the same temperature. In casting on a substrate, atoms of the melt at high temperatures are forced into contact with atoms of the solid substrate at low temperatures, with solidification occurring long before any force equilibrium can be achieved. The important thing is that intimate contact be achieved. If melt surface chemistry interferes with this by making it mechanically difficult to achieve intimate contact, the resulting failure to bond has little to do with the concept of wetting.

When a bond is formed, it is clearly a strong bond. A chill block melt spinning experiment can demonstrate that a relatively static melt can be accelerated in a millisecond or so to a solid ribbon going 50 m/s or more. This represents an acceleration of thousands of "gs". The centrifugal force generated during such an experiment is not sufficient to break this bond, nor is it necessary. The same experiment done at 1 m/s produces a ribbon which easily breaks its bond. The main force which contributes to release must be that due to thermal contraction. Once solidification has occurred, and the temperature of the casting at the interface begins to decrease, a shrinkage stress develops between the substrate and the casting. It can be shown (6) that shrinkage stresses of one or two MPa/C can easily develop on cooling. Thus a decrease in the temperature of only 10 or 20 C can generate very high stresses indeed.

Once the casting begins to cool, a differential stress begins to build between the casting and the substrate. This, of course, takes some finite time. During this time, stress relaxation processes act to reduce the developing stresses. Release then becomes a race between the rate of generation and the rate of relaxation of stresses. The strength of the bond also plays a role. The time between the initiation of solidification and the breaking of the bond is called the release time (t_r). Because the casting is bonded to and moving along with the substrate, the distance traveled during the time t_r is the spatial equivalent to t_r and is called the release distance (d_r). Any factor which increases the rate or amount of stress relaxation which can occur will increase both t_r and d_r.

In a pure metal, the normal cooling curve should resemble the schematic curve shown in Figure 2a. First the melt temperature decreases rapidly as the superheat is absorbed. Then the heat of crystallization is released causing a thermal arrest. Finally, on completion of solidification, the solid begins to cool and a release stress begins to develop. After a small ΔT, and consequently a small t_r and d_r, release occurs. At higher substrate speeds, the time scale is decreased, and the cooling curve is expected to resemble the schematic shown in Figure 2b. The melt is supercooled to below the melting temperature. At some temperature nucleation of the crystal phase begins, and release of the heat of crystallization causes the temperature to rise. Thus at least the initial solidification takes place as the temperature is increasing (recalescence). Once again, release occurs when the temperature of the casting decreases some small amount below the solidification temperature. It is difficult to think of either of these solidification scenarios as rapid quenching. Rapid solidification is a far more apt terminology (7).

For an alloy with an appreciable liquidus solidus range, the picture differs somewhat, for solidification takes place over a range of temperatures (Fig. 2 c and d). Solidification is not completed until the temperature drops below the solidus. Within the liquidus-solidus range a release stress cannot be developed because stresses are easily relieved in the remaining liquid. Once temperatures drop below the solidus, release stresses develop as before. Release temperatures can be considerably lower than those observed for pure metals of comparable melting point (liquidus) and t_r and d_r can be considerably increased. For an amorphous alloy, (Fig. 2e), there is no heat of crystallization involved, so the rate of change of temperature is monotonic, and release stresses can develop only after the viscosity of the glass has developed enough to support a stress. This <u>is</u> a valid use of the term "rapid quenching".

Once sufficient stress has developed, the bond between the substrate and the casting must be broken. Since surface layers can form on substrates under all but the most ideal circumstances, it must be assumed that the bond formed is between the casting and the substrate surface layer. This is shown schematically in Figure 3. The bond must break at its weakest point, which could be any of the following:

(a) Within the casting.
(b) At the casting - substrate surface layer interface.
(c) Within the substrate surface layer.
(d) At the substrate surface layer - substrate interface.
(e) In the substrate.

If the bond breaks within the casting, material is transferred to the substrate, changing the substrate chemistry and geometry. This kind of behavior is sometimes observed, and must be countered by some sort of vigorous substrate surface cleaning or maintenance program if continuous casting is to proceed.

Ideally, the bond should break at (b) the casting - substrate surface layer interface. In this case, no damage is done to either the substrate or the casting, and surface maintenance problems are at a minimum.

Alternatives (c) (d) and (e) all involve loss of substrate material and eventual degradation of the substrate surface.

The substrate must conduct heat away from the casting fast enough so the substrate temperature itself does not rise too far. As the temperature of the substrate approaches that of the casting, the rate of heat transfer must approach zero. Stress relaxation will dominate, and release of the casting from the substrate will not occur. A simple experiment can demonstrate the importance of this aspect of substrate character.

If a droplet of molten metal (tin, for example) is allowed to fall on a copper plate, the resulting splat will adhere briefly to the plate, then break free. If a thin foil of copper (say 0.1 mm thick) is draped over the copper plate and the experiment is repeated, the tin will adhere strongly to the foil and will not break free, even when both foil and splat return to room temperature. The foil cannot conduct heat away because of its lack of adequate mass. It can only become very hot, and this defeats the release mechanism.

The significance of the experiment lies in what it implies in substrate surface preparation. If the substrate surface contains irregularities such as asperities or mechanical laps, and these are to be expected on many machined or mechanically polished surfaces, these localized areas will not be capable of conducting heat away from the surface at a sufficiently high rate to permit release. For ideal heat transfer, <u>all</u> of the surface to which the melt can bond must be capable of removing heat very rapidly. All of the laps, asperities chips, etc. must therefore be removed to achieve "surface integrity". This does not mean that the surface must be flat. Machining a surface (with machining marks in the direction of casting) will produce a contour very well suited to breaking through the surface layer of the melt, giving room for trapped gases to escape, giving room for surface layer residue to move to, and nucleating solidification uniformly across the surface. Unfortunately a machined surface usually has a low integrity. The ideal surface seems to be one in which high surface integrity is achieved while retaining the macroscopic undulation caused by machining.

The Solidification Model

For many of the solidification studies, the results can be described by the relation:

$$D = K\ t^{1/2}$$

Where D is the thickness of the cast product, t is the time the cast product is in intimate contact with <u>both</u> the substrate and the melt, and K is a coefficient which involves mainly the thermal properties of both the melt and the substrate.

For a rotating substrate, such as in melt drag, the time t that the casting is in intimate contact with both the melt and the substrate is given by X/V where X is the length of contact of melt and substrate and V is the velocity of the substrate. The data in Fig. <u>4</u> are for tin cast on a water-cooled copper substrate. The significance of the X/V relationship is that, within limits, it is possible to cast any given thickness under a variety of experimental conditions. The limits involve very small values of X or very high values of V when casting thin material, and maintenance of intimate contact at large values of X and small values of V, when casting thick material.

At moderate values of casting speed (i.e. 1-5 m/s), it is possible to cast strip by an open-tundish melt drag system as shown in Figure 5. If a pour box with a fixed orifice is used, molten metal can be supplied at a fixed rate. If the volume of the tundish is small, an equilibrium will be achieved very quickly, with the rate of strip being cast (i.e. g/s) being equal to the rate of molten metal being poured (i.e. g/s). If the strip being cast is of fixed width, and the substrate speed is held constant, then the average strip thickness must remain constant. Under those circumstances, the depth of the melt in the tundish is the equivalent to X, the distance over which the casting is in intimate contact with both the substrate and the melt. Since the solid strip is being drawn out of the pool of molten metal (the decantation process) at relatively slow speed, only solid metal is being removed. Neglecting any small temperature gradient which might exist between the surface and the substrate side of the strip, the temperature of strip must be very close to the melting temperature of the metal. Thus all the heat removal necessary for the solidification of the strip must occur in an area defined by X and the strip width and in a time defined by X/V. The amount of material being cast per unit time is the same as the amount of material being solidified per unit time. From this, the average rate of solidification is known either as $g/cm^2/s$ or $cm^3/cm^2/s$.

If the pour rate is kept the same, and the substrate velocity is doubled, the strip thickness will be halved. According to equation 1, in order to maintain the balance, X will decrease to half its previous value. In other words, the same weight of strip is being solidified per unit time but in half the area. Thus the rate of heat transfer (and the average rate of solidification) is inversely proportional to the thickness of the strip, while the thickness of the strip is inversely proportional to the velocity V of the substrate. For a fixed feed rate system such as CBMS or MD, this means that, other things being equal, the solidification rate is directly proportional to the velocity V of the substrate. In a system where the rate of heat removal is approximately proportional to the rate of change of temperature the rate of change of temperature is directly proportional to substrate velocity. This is exactly what was observed in the CBMS of 316L Stainless Steel, by Bewlay and Cantor (8).

It was shown above that, if the substrate gets too hot, release is inhibited. There is another consequence of substrate temperature which is easily shown experimentally. If strip is cast as shown in Figure 5, and the substrate is not cooled, the substrate temperature will rise continuously. To obtain the data shown in Figure 6, X was measured during the course of the experiment in which tin strip was cast at a constant rate and constant thickness. If X represents the distance over which all the superheat and heat of crystallization is transferred into the substrate, then the rate of heat transfer and the rate of solidification decreases steadily as the temperature of the substrate increases. The data suggest that solidification rates can change by a factor of 5 or more during the course of a short experiment, depending upon the surface temperature of substrate at the time of initial contact with the molten metal. The surface temperature of a rotating substrate cycles widely, and will depend upon a number of factors related to how much heat is supplied and how fast it can be dissipated by the substrate. Even in a cooled system, a thermal gradient will exist between the substrate surface and the cooling medium. Thus, other things being equal, the surface of a stainless steel substrate may operate at a higher temperature than the surface of a copper substrate. Many of the differences in product and process character which are ascribed to the substrate chemistry may in fact be solely due to the substrate temperature with its associated effects on nucleation, solidification rate, and release.

Summary

The model describing solidification on a substrate assumes that a strong bond is formed when molten metal is brought into contact with the surface of an appropriate substrate. Once solidification begins, subsequent cooling can generate thermal stresses strong enough to break the bond. Consideration of this model and of a variety of experimental results permits the identification of 4 major characteristics of the substrate which should be considered. These are given below:

Surface Geometry. The main concern is to have a surface integrity which permits heat to be transferred through the interface in a rapid and uniform fashion. Surface contours should be such as to encourage rupture of any surface layer on the melt, to allow the escape of trapped gases, to uniformly nucleate solidification, to render surface layer fragments ineffective, and to minimize turbulence.

Surface Chemistry. The surface must be chemically uniform to promote bond and release uniformity. It must be free of volatile and reactive components.

Surface Physics. The substrate must have the mechanical and physical properties necessary to withstand the thermal cycling which accompanies casting. It must have the dimensions and thermal properties which permit heat to be transferred away from the surface so that the surface temperature never approaches the solidus temperature of either the casting or the substrate itself.

Surface Maintenance. The substrate surface, especially where material transfer takes place, must be maintained to achieve uniformity during continuous casting.

References

1. U.S. Patent No. 405, 914, F.W. Schultz, "Wiresolder Machine," June 25, 1889.

2. U.S. Patent No. 4,646,812, Robert E. Maringer, "Flow Casting," issued March 3, 1987.

3. J.W. Rutter and B. Chalmers, "A Primitive Substructure Found During Solidification of Metals," Can J. of Phys 31 15-39 (1953).

4. R.B. Pond and S.W. Kessler, "Model for Dendrite Growth Form in Metals and Alloys," J. Metals 3, Trans (1951) 1156-62.

5. S.C. Huang and H.C. Fiedler, "Effects of Wheel Surface Conditions on the Casting of Amorphous Metal Ribbons," Metal. Trans. A (1981) 1107-1112.

6. R.E. Maringer, "Solidification on a Substrate," Mat. Sci. and Eng. 89 (1988).

7. R.E. Maringer and A.N. Patel, "Rapid Solidification Versus Rapid Quenching," Progress in Powder Metallurgy" 43 (1987) 669-681.

8. B.P. Bewlay and B. Cantor, "Photocalorimetric Cooling Rate Measurements on 3166 Stainless Steel Rapidly Solidified by Melt Spinning," Int. J. Rapid Solidification, 2 (1986), 107-123.

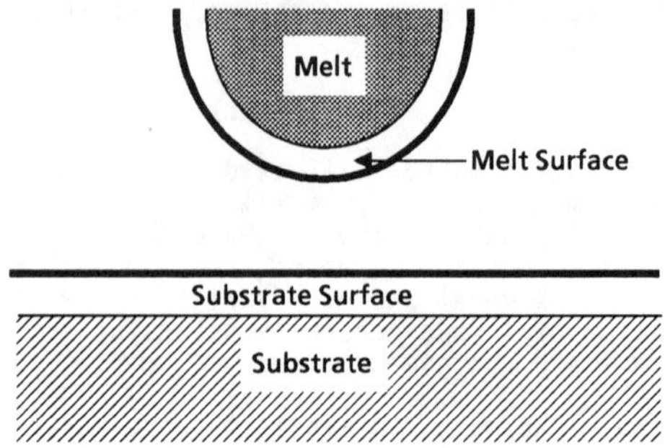

a) Schematic representation of melt and surfaces

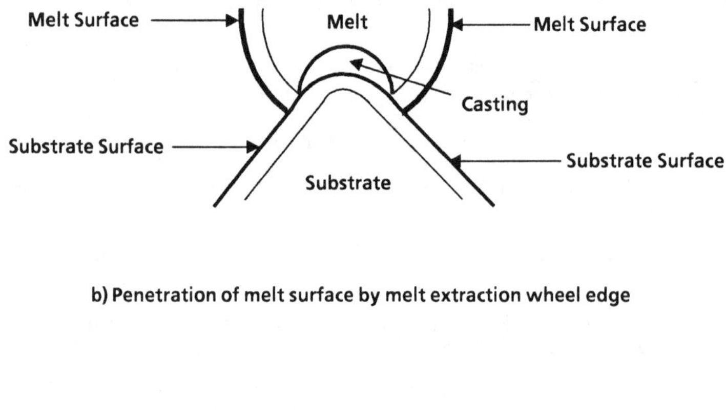

b) Penetration of melt surface by melt extraction wheel edge

FIGURE 1. SURFACE RELATIONSHIPS IN CASTING ON A SUBSTRATE

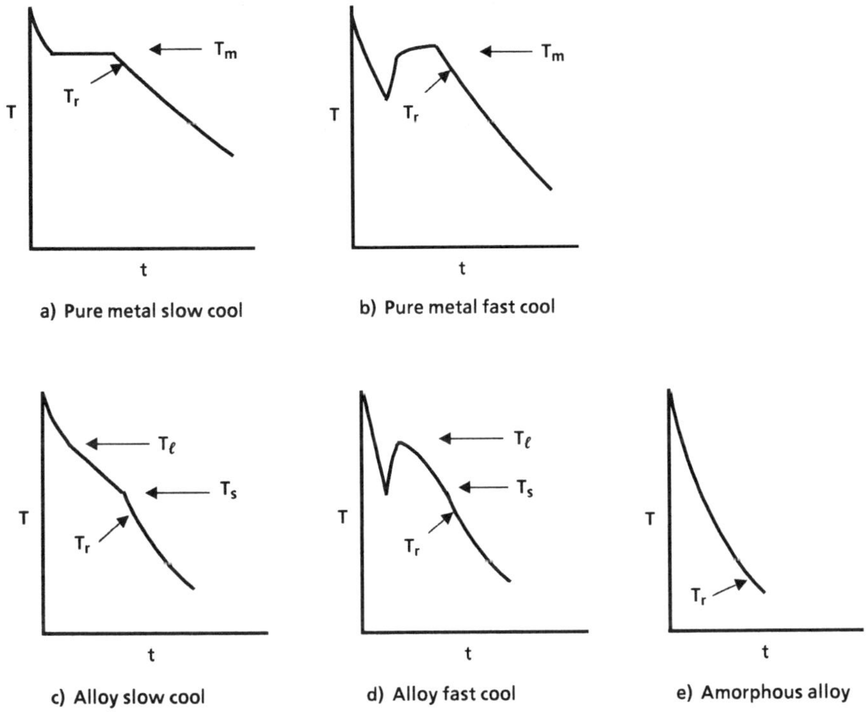

FIGURE 2. SCHEMATIC COOLING CURVES FOR METALS AND ALLOYS

T_m = melting temperature
T_ℓ = liquidus temperature
T_s = solidus temperature
T_r = release temperature

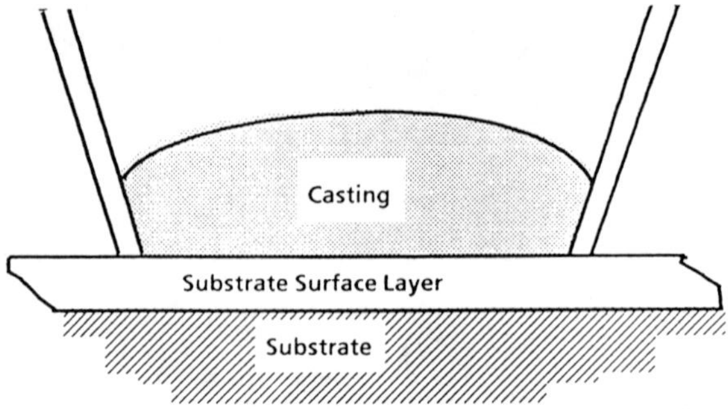

FIGURE 3. SCHEMATIC DIAGRAM OF CASTING BONDED TO SUBSTRATE SURFACE LAYER

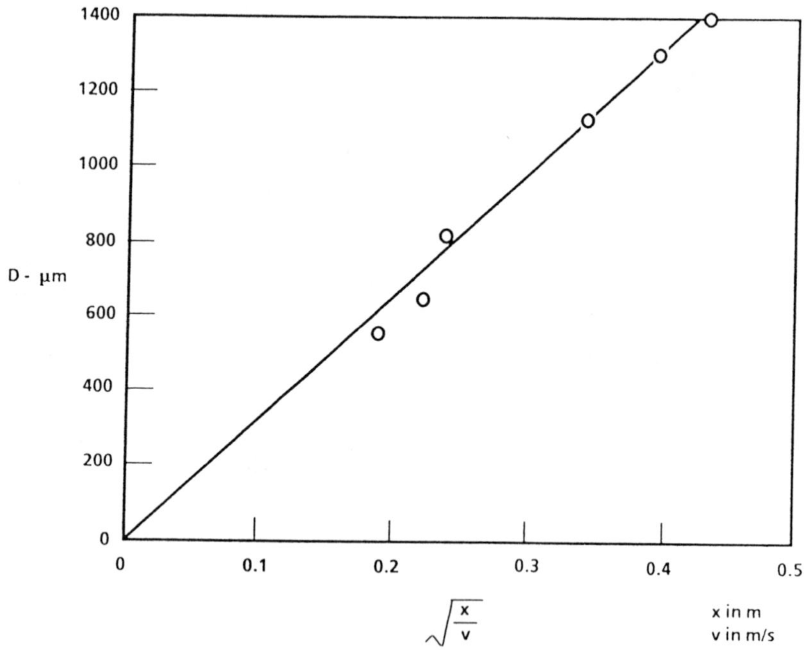

FIGURE 4. DEPENDENCE OF THICKNESS (D) ON CONTACT LENGTH (X) AND SUBSTRATE VELOCITY (V) FOR TIN ON WATER-COOLED COPPER

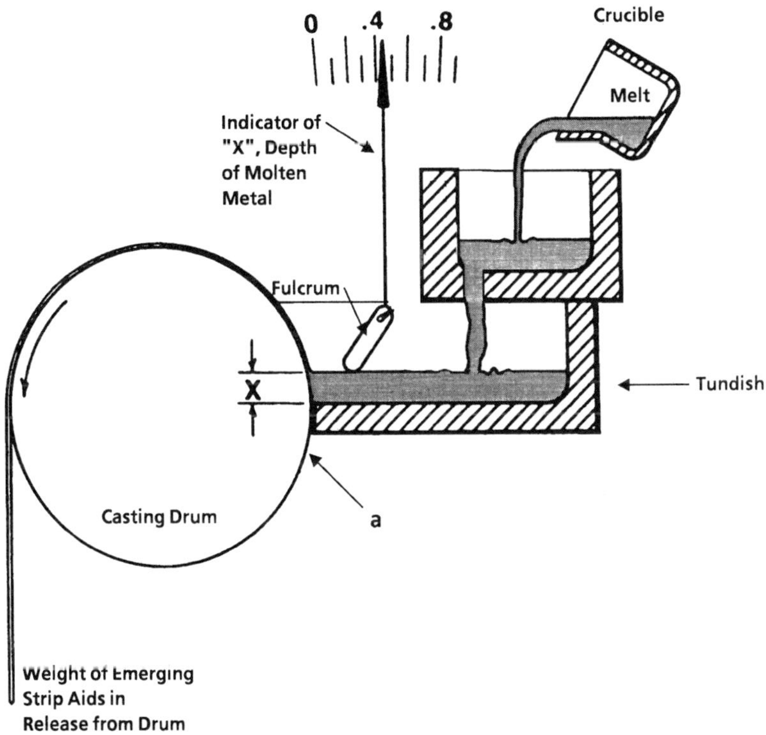

FIGURE 5. DETAILS OF CASTING SYSTEM

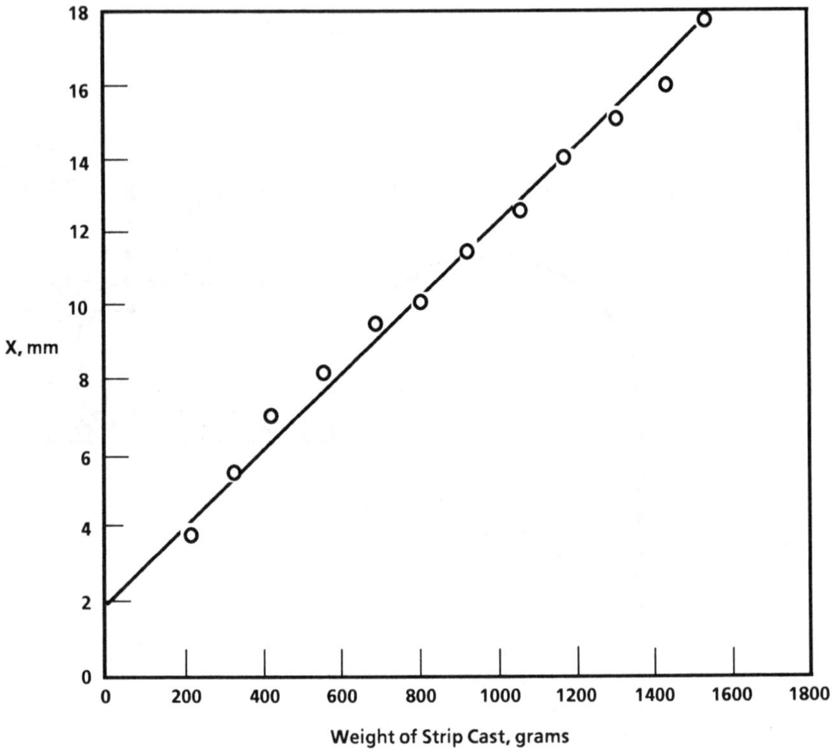

FIGURE 6. INCREASE IN HEAD HEIGHT (X) IN TUNDISH DURING CASTING OF TIN STRIP AT A RATE OF 60 G/S ON AN UNCOOLED SUBSTRATE

STABILIZATION OF A MOLTEN-METAL/COOLANT-GAS INTERFACE BY ELECTROMAGNETIC FIELDS

D. M. Rote, J. R. Hull, and T. Wiencek

Argonne National Laboratory
9700 S. Cass Avenue
Argonne, IL 60439

Abstract

One of the critical issues affecting the cooling of a liquid-metal free surface with a high-velocity gas is the threshold gas velocity for wave formation. In electromagnetic levitation casting, the presence of high-frequency magnetic fields affects the onset and character of waves formed on the interface. A series of experiments was conducted in which nitrogen gas flowed at uniform velocity in a rectangular channel over the surface of a shallow horizontal dish containing liquid tin. The threshold gas velocity was determined when 10-kHz magnetic fields were applied to the surface. The magnetic fields were oriented both parallel and perpendicular to the gas flow. The threshold velocity increased as magnetic field strength increased. Substantial suppression of wave formation was observed with the magnetic field perpendicular to the gas flow, which contradicts previous theoretical predictions.

Casting of Near Net Shape Products
Edited by Y. Sahai, J.E. Battles,
R.S. Carbonara and C.E. Mobley
The Metallurgical Society, 1988

Introduction

It is generally believed that a large reduction in the cost of producing metal sheet could be achieved if the sheets could be continuously cast in near net shape. The use of electromagnetic (EM) levitation in induction melting (1) and continuous casting of aluminum (2) has led to the suggestion that EM levitation of molten steel could achieve continuous casting of near net shape while maintaining or possibly improving the metallurgical properties of the steel (3). One option for EM casting is the suspension of a vertical, molten-metal sheet in several alternating magnetic fields of different frequency (4). A second option involves EM levitation in a horizontal configuration (5,6).

Efficient cooling of the sheet is a problem common to both configurations. When the metal sheet leaves the EM caster, the surface of the sheet, together with a sufficiently large fraction of the interior, must be solidified to the point that mechanical support of the sheet is possible. The difficulty with "moldless" casting using EM fields is that the large heat transfer rate to the chilled, solid mold wall is no longer available. Further, any EM heat generated in addition to the latent and sensible heat of the metal must also be removed.

To minimize the size of the magnet and maximize material throughput, the molten sheet must be cooled quickly. To augment radiation cooling, a fast-flowing stream of inert cooling gas is introduced on both sides of the molten sheet parallel to the sheet surface. High gas velocities are preferred because they generally result in higher rates of heat transfer. In practice, the gas velocity is limited because the hydrodynamic configuration introduces the possibility of flow-induced wave formation via mechanisms such as Kelvin-Helmholtz-type (KH-type) instability (7). At sufficiently high gas velocities, the instability can tear the molten sheet apart.

The presence of an alternating magnetic field at the interface between the conducting liquid and a conducting gas generally suppresses wave formation. This paper describes analyses and experiments that investigate the behavior of this stabilization process.

Theory

A major mechanism for wave formation at a gas/liquid interface is KH instability. The instability arises because the pressure perturbations exerted by the gas can do work on the interface. A number of authors have examined the KH problem with constant magnetic fields (e.g., Refs. 7 and 8).

Garnier and Moreau (9) solved for the KH instability conditions in the presence of alternating magnetic fields. With a horizontal interface, and subscript 1 representing the lower fluid and subscript 2 representing the upper fluid, their analysis leads to the stability condition

$$\frac{\rho_1\rho_2}{\rho_1+\rho_2}k^2(U_1 - U_2)^2\cos^2\theta_1 \le gk(\rho_1 - \rho_2) + Tk^3 + F(\theta,k) \ , \qquad (1)$$

where ρ is the density, U is the velocity, g is the acceleration of gravity, T is the fluid surface tension, k is the magnitude of the wave vector of the disturbance, θ_1 is the angle between k and U_1 (gas flow velocity), and F represents the stabilizing effect of the magnetic field, which is given by

$$F(k,\theta) = (B^2/2\mu)k^2\cos^2\theta \; F(x) \;, \qquad (2)$$

where B is the magnetic field strength at the interface, μ is the magnetic permeability of the liquid, θ is the angle between the wave vector and the magnetic field, and F(x) is a factor that corrects for diffusion of the magnetic field into the liquid. F(x) is near unity for high electrical conductivities and decreases with conductivity. The stabilizing effect is largest for disturbances parallel to the magnetic field and is zero for disturbances perpendicular to the magnetic field.

Equation 1 indicates that waves are most likely to form with wave vectors in the same direction as the gas flow. According to Eq. 2, a magnetic field perpendicular to the gas flow should not affect wave formation.

The analysis of Garnier and Moreau (9) does not totally correspond with the EM caster designs (3-6) in which flux compression is used to establish local stability. The uniform and constant magnetic field strengths assumed by Garnier and Moreau (9) to be present at the surface of the molten metal provide a stabilizing influence when magnetic tension arises from the distortion of the field lines as the surface of the molten metal deforms. In the EM caster cases, stabilization also occurs because the field strength changes as the surface deforms. A restoring force is exerted locally on the surface by the magnetic fields, in proportion to the magnitude of the deformation at that location.

Vertical Interface

Hull (10) showed that for thicknesses typical of EM casting, the two molten metal surfaces can be treated independently. We thus consider a liquid slab separated from a gas slab by a boundary at $x = 0$, denoting the molten slab by subscript 2 and the coolant gas by subscript 1. We approximate the molten metal and coolant gas as incompressible, inviscid fluids, with the mean stream flow in the z direction with velocity W. We are concerned only with the purely molten state and assume that the densities of the liquid and gas are homogeneous in the y and z directions.

We consider the high-conductivity limit, so that the magnetic field generates a pressure $P_m = B^2/(2\mu)$ in the x direction that acts at the surface of the molten metal. In the stationary state, this pressure exactly cancels the metallostatic pressure P_o along the interface between the liquid metal and the gas. In addition, conservation of magnetic flux between the molten slab and the magnet requires the pressure to change as

$$P_m = P_o - \gamma(x_s - x_{so}) \;, \qquad (3)$$

where x_s is the position of the free surface, x_{so} is the surface position at equilibrium, and γ is the pressure gradient due to the magnetic restoring force acting at the surface. If the free surface of the molten slab moves into the gas half-space (i.e., toward the magnet in a real system), the magnetic flux in the gas half-space (i.e., between the magnet and the slab) compresses, and the magnetic pressure increases. A corresponding pressure decrease occurs for movement of the surface away from the magnet.

The calculation follows along the lines of that in Chandrasekhar (7). The basic hydrodynamic equations governing the perturbation are

$$\rho\frac{\partial u}{\partial t} + \rho W\frac{\partial u}{\partial z} = -\frac{\partial}{\partial x}(\delta P) + T\left[\left(\frac{\partial^2}{\partial y^2} + \frac{\partial^2}{\partial z^2}\right)(\delta x_s)\right]\delta(x-x_s) - \gamma\delta x_s\delta(x-x_s) , \quad (4)$$

$$\rho\frac{\partial v}{\partial t} + \rho W\frac{\partial v}{\partial z} = -\frac{\partial}{\partial y}(\delta P) , \quad (5)$$

$$\rho\frac{\partial w}{\partial t} + \rho W\frac{\partial w}{\partial z} + \rho u\frac{\partial W}{\partial x} = -\frac{\partial}{\partial z}(\delta P) - g\delta\rho , \quad (6)$$

$$\frac{\partial u}{\partial x} + \frac{\partial v}{\partial y} + \frac{\partial w}{\partial z} = 0 , \quad (7)$$

$$\frac{\partial}{\partial t}(\delta x_s) + W_s\frac{\partial}{\partial z}(\delta x_s) = u(x_s) , \quad (8)$$

$$\frac{\partial}{\partial t}(\delta\rho) + W\frac{\partial}{\partial z}(\delta\rho) = -u\frac{\partial\rho}{\partial x} , \quad (9)$$

and represent, respectively, conservation of momentum in each coordinate direction, incompressibility, the continuity of the free surface from one medium to the next, and the equation of state after homogeneity in the y and z directions has been considered. The variables u, v, and w are the perturbed velocities in the x, y, and z directions, respectively; T is the fluid surface tension; ρ is density; g is acceleration of gravity; P is pressure; $\delta(x-x_s)$ is the Dirac delta function; and δP and δx_s represent perturbations from the equilibrium state of pressure and interface position, respectively. For a normal mode analysis, the perturbed variables are assumed to have the form

$$u, v, w, \delta\rho, \delta P, \delta x_s \sim \exp[nt + ik_y y + ik_z z] , \quad (10)$$

where k is a wave vector for the perturbation, and n is the time constant for growth or decay of the perturbation. After we substitute into Eqs. 4-9 and take derivatives, we solve the resulting set of equations for u. At a free surface

$$\Delta_s[\rho(n+ik_zW)Du - ik_z\rho u(DW)] = [-ik_zg\Delta_s(D\rho) + k^2(\gamma+k^2T)]\left(\frac{u}{n+ik_zW}\right)_s , \quad (11)$$

where

$$k^2 = k_y{}^2 + k_z{}^2 ,$$

$$D = \partial/\partial x ,$$

and Δ_s represents the jump that a quantity experiences at an interface. Away from a surface

$$(n+ik_zW)(D^2-k^2)u - ik_zu(D^2W) + \frac{(D\rho)}{\rho}[(n+ik_zW)(Du) - ik_zu(DW)]$$

$$- \frac{ik_zg}{\rho}\left(\frac{ik_zu(DW)}{(n+ik_zW)^2} - \frac{uD^2\rho + (Du)(D\rho)}{n+ik_zW}\right) = 0 . \tag{12}$$

At this stage of the analysis, both W and ρ may have an x dependence.

Let the stream velocities W_1 and W_2 be constant, and let densities ρ_1 and ρ_2 be constant, that is,

$$DW_1 = DW_2 = 0 \text{ and}$$

$$D\rho_1 = D\rho_2 = 0 .$$

Away from the boundaries, Eq. 12 reduces to

$$(D^2 - k^2)u_1 = 0, \qquad (x < 0) , \tag{13}$$

$$(D^2 - k^2)u_2 = 0, \qquad (x > 0) , \tag{14}$$

with solutions

$$u_1 = A(n + ik_zW_1)\exp(k_xx) , \qquad (x < 0) , \tag{15}$$

$$u_2 = A(n + ik_zW_2)\exp(-k_xx) , \qquad (x < 0) . \tag{16}$$

For the solutions to satisfy Eqs. 13 and 14, $k_x = k$. At $x = 0$, Eq. 11 reduces to

$$\Delta_s[\rho(n+ik_zW)Du] = k^2(\gamma+k^2T)\left(\frac{u}{n+ik_zW}\right)_s , \tag{17}$$

$$\rho_2(n+ik_zW_2)^2 + \rho_1(n+ik_zW_1)^2 + (\gamma+k^2T)k^2/k_x = 0 . \tag{18}$$

The characteristic equation is

$$n^2(\rho_1+\rho_2) + n[2ik_z(\rho_1W_1+\rho_2W_2)] - k_z{}^2(\rho_1W_1{}^2+\rho_2W_2{}^2) + k(\gamma+k^2T) = 0 , \tag{19}$$

which is a quadratic equation for the time constant n. For the solution to be stable, the discriminant of Eq. 19 must be negative, resulting in the stability condition

$$(W_1 - W_2)^2 < \frac{(\rho_1 + \rho_2)k(\gamma + k^2T)}{\rho_1 \rho_2 k_z{}^2} . \tag{20}$$

Minimizing the right-hand side of Eq. 20 with respect to k_z results in

$$k = k_z .$$

Further minimization with respect to k results in

$$k_c = (\gamma/T)^{1/2} , \tag{21}$$

Substituting this result into Eq. 20 yields

$$(W_1 - W_2)^2 < 2(\rho_1 + \rho_2)(\gamma T)^{1/2}/(\rho_1\rho_2) . \tag{22}$$

Unless this condition is met, the vertical interface will be unstable in the KH sense.

Horizontal Interface

For a horizontal interface at z = 0 and a velocity U for the fluids, a similar set of equations can be solved for the threshold velocity V_c. Equations 4-6, 8, and 9 become, respectively,

$$\rho\frac{\partial w}{\partial t} + \rho U\frac{\partial w}{\partial x} = - \frac{\partial}{\partial z}(\delta P) + T[(\frac{\partial^2}{\partial x^2} + \frac{\partial^2}{\partial y^2})(\delta z_s)]\delta(z-z_s) - \gamma\delta z_s\delta(z-z_s) , \tag{23}$$

$$\rho\frac{\partial v}{\partial t} + \rho U\frac{\partial v}{\partial x} = - \frac{\partial}{\partial y}(\delta P) , \tag{24}$$

$$\rho\frac{\partial u}{\partial t} + \rho U\frac{\partial u}{\partial x} + \rho w\frac{\partial U}{\partial z} = - \frac{\partial}{\partial x}(\delta P) - g\delta\rho , \tag{25}$$

$$\frac{\partial}{\partial t}(\delta z_s) + U_s\frac{\partial}{\partial x}(\delta z_s) = w(z_s) , \tag{26}$$

$$\frac{\partial}{\partial t}(\delta\rho) + U\frac{\partial}{\partial x}(\delta\rho) = - w\frac{\partial\rho}{\partial z.} , \tag{27}$$

The stability condition analogous to Eq. 22 is

$$(U_1 - U_2)^2 < \frac{\rho_1 + \rho_2}{\rho_1 \rho_2} \left[\frac{g(\rho_2 - \rho_1) + \gamma}{k} + kT\right] . \tag{28}$$

When the gas flows over the liquid, the gravitational term is stabilizing. When the gas flows under the liquid, this term is destabilizing. This aspect is related to Rayleigh-Taylor instability (6). Minimization with respect to k results in

$$(U_1 - U_2)^2 < (2/\rho_g)\{[g(\rho_2 - \rho_1) + \gamma]T\}^{0.5} . \tag{29}$$

Experiment

Apparatus and Method

The apparatus consisted of a system to control and monitor the gas temperature and flow parameters; a high-frequency, C-shaped magnet with a large air gap; and a rectangular glass test section designed to simulate the gas-cooling channel of an EM caster. Figure 1 is a photograph of the entire system.

Nitrogen gas was supplied from a compressed-gas cylinder fitted with a special high-volume regulator. The gas lines were temperature controlled from the gas cylinder to the round-to-rectangular transition segment of the test section. An entrance diffuser and flow-straightening vanes produced uniform flow. Gas velocity was measured with a movable hot-film anemometer both upstream and downstream of the test chamber. Measurements indicated that flow was uniform across the width and height of the test section entrance and exit.

The magnet yoke and poles were fabricated from tape-wound cores. The coil was wound on the yoke and designed to match the output of a 10-kHz, 30-kW motor/generator set. A copper shield plate at the top of the pole edge helped to confine the magnetic field to the region between the poles above the liquid metal. Further copper shielding on the outer surfaces of the cores and coil prevented leakage flux around the magnet.

The glass test section consisted of a rectangular gas channel (61 cm long × 8.9 cm wide × 1.0 cm deep) and a rectangular glass dish (25 cm long) containing a pool of liquid metal about 0.9 cm deep that was heated from below. A C-shaped, copper, eddy-current return path was located below the liquid metal and was separated from it by a thin glass insulating plate. The return path was connected electrically to the liquid metal by tungsten electrodes welded to the short arms of the C. (Copper electrodes dissolved too quickly in the tin, and stainless steel electrodes did not sufficiently wet the tin.)

The magnet was placed over the glass dish so that a magnetic force could be exerted downward on the liquid metal surface. Two orientations were possible: a magnetic field parallel to gas velocity ($B \parallel V$) and a magnetic field perpendicular to gas velocity ($B \perp V$). With $B \perp V$, the jaws of

Figure 1 - Experimental apparatus, showing power strip line connecting motor/generator (left, back) to the magnet, which is supported above glass test section. Gas flows from cylinder (right) through a heated cable to the test section and exits through a round pipe (between the water-cooling lines for the magnet on the left side of the bench).

the magnet could be placed around the sides of the glass test section (Fig. 2). With $B \parallel V$, the jaws had to be above the top glass plate of the gas channel. The magnetic field strength could be adjusted to values almost as high as that needed for levitation in a horizontal caster. However, the magnet design meant that the magnetic restoring force that countered surface disturbances in the liquid metal was much lower in magnitude, and both the field and restoring forces were much less spatially uniform. In both orientations, significant vertical components of B were present near the magnet pole faces.

The amplitude of the waves was measured by a capacitance detector, shown schematically in Fig. 3, consisting of a probe wire mounted horizontally above the liquid metal surface and perpendicular to the gas flow, in the middle of the magnetic field region. The wire and metal surface formed the two "plates" of a capacitor, with capacitance increasing with the distance between the wire and the surface. The capacitor was contained in an electronic circuit that consisted of a dc power supply, amplifier, and return electrode to the liquid metal, made by connection to the stainless steel sheath of a thermocouple. The amplifier signal passed to a Fourier-transform wave analyzer, which determined the power spectrum of the surface disturbances.

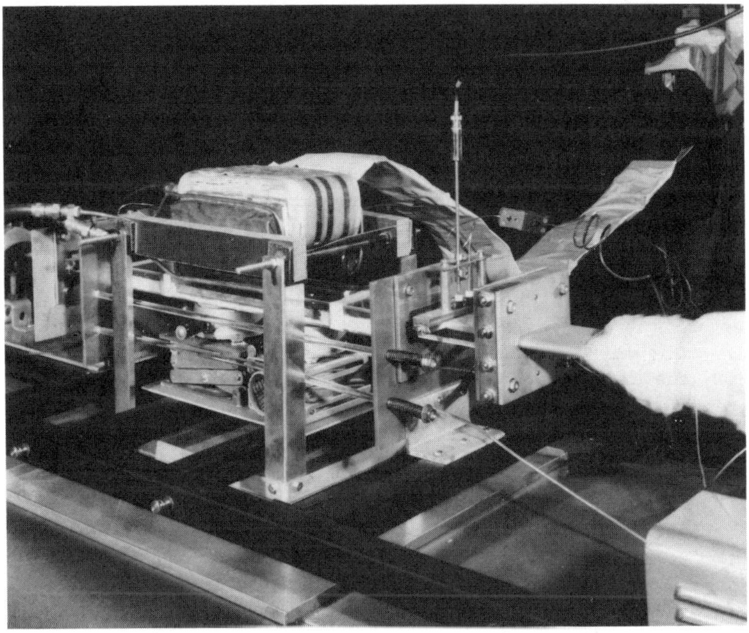

Figure 2 - Magnet supported above the glass test section in the B⊥V configuration. A movable, hot-film anemometer is shown (to the right) upstream of the test section.

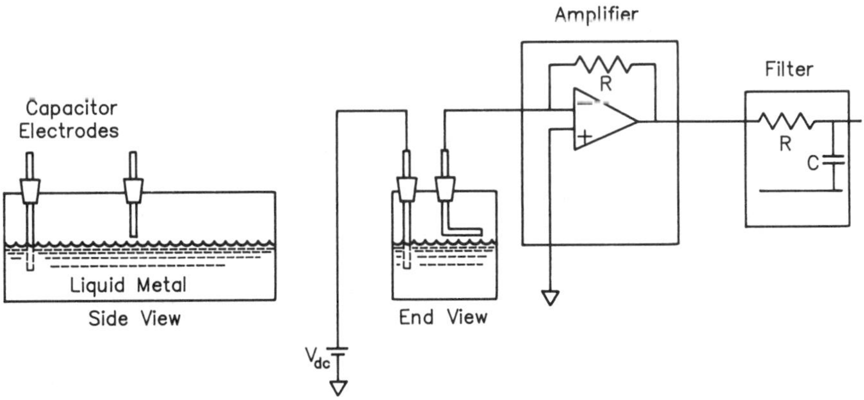

Figure 3 - Diagram of the capacitance wave detector.

An experiment typically began by heating the tin to a temperature 20-40°C above its melting temperature of 232°C. The magnetic field was set to a given value, the wave detector noise level was measured, and the nitrogen gas velocity was increased until the signal-to-noise ratio (SNR) of the wave amplitude to background noise level reached a value of 10. This SNR of 10 was taken as the criterion for wave action and corresponded to formation of very small but visible waves in the liquid tin. The gas velocity required to generate waves of this SNR is defined as the threshold velocity V_c. Visual confirmation of wave formation at this SNR held for all degrees of EM stirring that were observed. For each measurement, the gas velocity was held constant for an average of 10 s to determine the power spectrum of the waves. The signal and noise were averaged over a 0.5-Hz band centered about the dominant frequency for the wave at the threshold velocity. Figure 4 is a photograph of surface waves in the liquid tin created by the flowing nitrogen gas. The presence of such a wave under the wave detector would result in a SNR much larger than 10.

Three parameters were varied during the course of the experiments: (1) the orientation of B relative to the gas velocity V, (2) the vertical spacing (z) between the surface of the tin and the bottom edges of the magnet pole pieces, and (3) the depth of the gas channel (values of 10.2 mm and 7.0 mm were used). The quality of the tin surface varied throughout the experiments. Although each experimental day usually started with a clean tin surface, the surface became increasingly oxidized with time, even with a small, continuous flow of dry nitrogen gas over the surface.

Figure 4 - Waves on downstream liquid tin surface.

For B⊥V, most of the tin surface (8.9 cm × 25 cm) was exposed to the magnetic field (11 cm × 20 cm), with the field lines parallel to the 11-cm dimension. However, because the field was nonuniform, only about 10 cm along the magnet length (⊥B) was exposed to the maximum field strength. Consequently, when the nitrogen gas velocity was increased beyond the value of V_c for zero field strength, waves would form upstream and downstream of the maximum field region. The downstream wave amplitude was the limiting quantity in the experiment because, even with complete wave suppression in the maximum field region, the downstream wave amplitude in the nearly field-free region would become so large that the gas would blow liquid tin out of the dish into the downstream gas channel. This condition was exacerbated when B∥V because, in that case, only about one-third of the tin surface was exposed to the magnetic field.

The magnetic pressure exerted by the field on the tin surface depressed that portion of the surface exposed to the field and elevated that portion outside the field region. This effect was more pronounced for the B∥V case, owing to the smaller tin area to which the field was applied and to the greater field nonuniformity, compared to the B⊥V case.

The magnetic field was measured as a function of position above the tin surface by a small, movable search coil, with the search output voltage monitored by an rms voltmeter. These search-coil measurements were used to estimate the local magnetic pressure exerted on the tin surface in the vicinity of the wave detector probe wire.

The threshold velocity measurements were analyzed in terms of field strength, magnetic pressure at a point, magnetic pressure averaged over the vicinity of the probe wire, and gradient of the average magnetic pressure. The mean pressure in the vicinity of the wave detector is given by

$$\overline{P} = \frac{\overline{B^2}}{2\mu_o} \quad , \tag{30}$$

where μ_o is the permeability of free space ($4\pi \times 10^{-7}$ N/A^2). The bar indicates a mean square value averaged along the probe wire. The gradient of the mean pressure is given by

$$\gamma = \frac{d\overline{P}}{dz} \quad , \tag{31}$$

and is determined by measuring the magnetic field for different magnet heights above the tin at constant power supply outputs.

Results

The test section exhibited mechanical resonances at about 5 Hz and 7 Hz. Those resonances showed up as large signals on the wave detector whenever the test bench was jostled. In time, these oscillations decayed to almost the background level, as shown in Fig. 5(a).

The nonuniformity of the magnetic field resulted in EM stirring, which was visible to the eye for large magnetic field strengths as random surface disturbances that usually emanated from the sides of the liquid sheet. The surface disturbances do not show up well in still photographs or individual video frames because of their small amplitude. Such stirring was not present in levitation experiments in which the field was uniform (6). For both field orientations, the stirring affected the signal from the wave detector by increasing the background in the frequency range 0-15 Hz, with a maximum at about 5 Hz, as shown in Figs. 5(a)-(c). The corresponding spectra with a wave present (not necessarily at V_c) are shown in Figs. 5(d)-(f), respectively.

A strong magnetic field was capable of completely suppressing any KH-type disturbance at low gas velocities, as evident by comparing Figs. 5(a), (c), (d), and (g). In Fig. 5(g), at a gas velocity capable of producing a detectable wave [see Figs. 5(a) and (d)], the spectrum is indistinguishable from that with no velocity, as shown in Fig. 5(c). As the gas velocity increased, with or without the magnetic field present, the power spectra increased over the frequency range 0-20 Hz, with the dominant wave frequency at about 10 Hz. Without a magnetic field, the first visual surface disturbances were the 10-Hz waves, which were accompanied by a signal on the wave detector, as shown in Fig. 5(d). With a magnetic field, as the gas velocity increased, the surface disturbances associated with EM stirring began to change shape, extending farther from the edges, and began to increase the power spectra from the wave detector in the range of 0-20 Hz. Occasionally, an individual wavelet was able to propagate unattenuated across the magnetic field region. As the the gas velocity increased further, the 10-Hz waves became visible in the

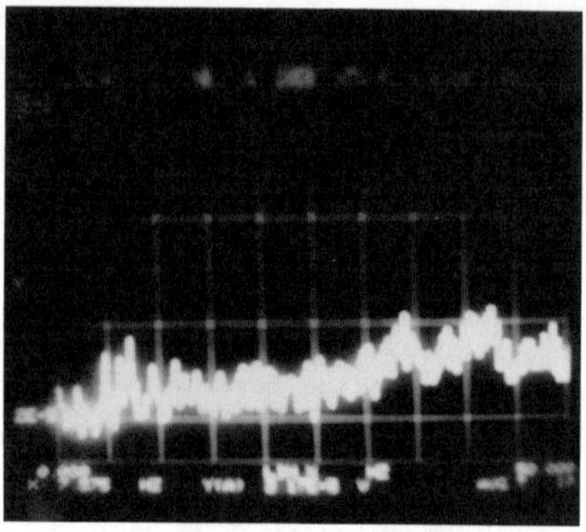

5(a) $B = 0$, $v = 0$

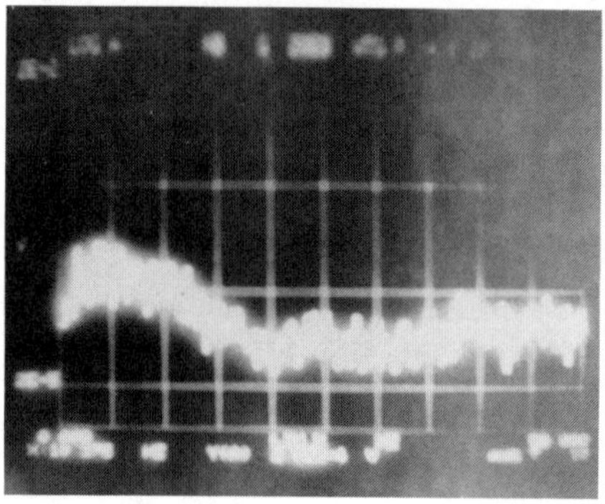

5(b) $B(\perp V) = 77$ G, $\gamma = 42$ Pa/m, $v = 0$

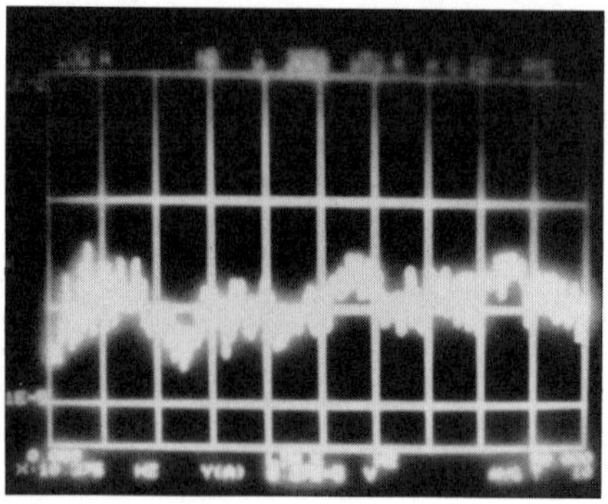

5(c) $B(\parallel V) = 75$ G, $\gamma = 21$ Pa/m, $v = 0$

magnetic field region, producing a dominant signal on the wave detector, as shown in Figs. 5(e) and (f).

The data for V_c versus γ for $B \perp V$ are shown in Fig. 6. The value of V_c, in the absence of any magnetic field, ranged from about 8.5 to 12 m/s, depending on the condition of the tin surface. Smooth clean surfaces had higher values of V_c. In general, V_c increased with γ, and the thickness of the gas channel had no detectable effect within the scatter of the data.

The data for V_c versus γ for $B \parallel V$ are shown in Fig. 7. These data were more difficult to interpret, largely because of the surface deformation effects referred to earlier. The results show the existence of two opposing effects, one tending to suppress wave formation (the γ-dependent effect that was clearly dominant in the case of $B \perp V$), and one tending to enhance wave formation (the surface-deformation effect not seen in the case of $B \perp V$). In the case where the tin surface was quite clean and smooth, the γ-dependent effect dominated at low values of $\bar{P}$ and γ, and the surface-deformation effect became increasingly important at higher values of P. In cases where the tin surface had become quite rough and scaly, on the other hand, the surface deformation effect appeared to neutralize the γ-dependent effect and even to dominate it at low values of γ and $\bar{P}$. That is, both the surface deformation effect and surface roughness tend to enhance wave formation, whereas the presence of the magnetic field and surface smoothness tended to enhance wave suppression.

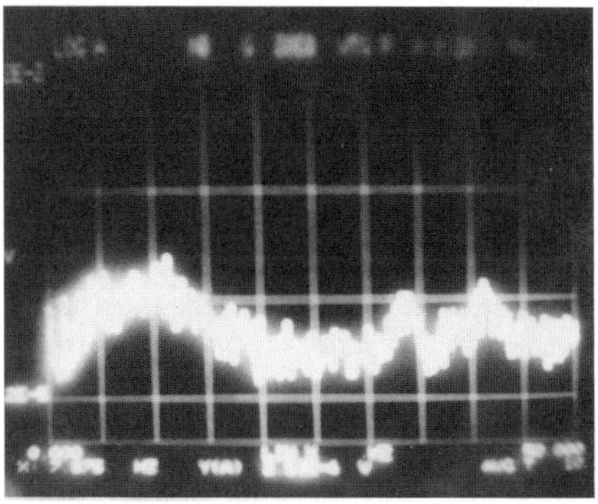

5(d) B = 0, v = 8.4 m/s

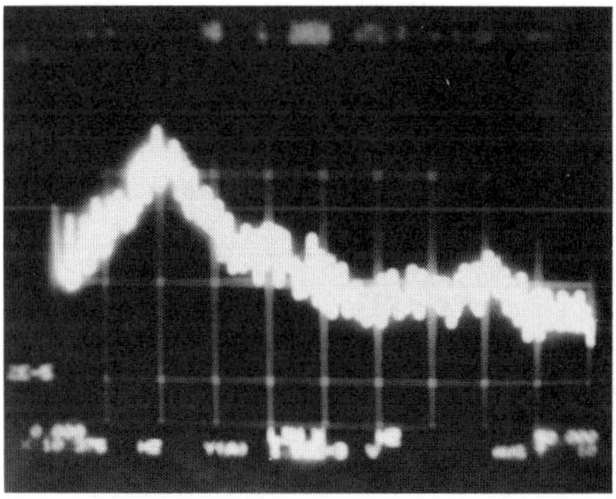

5(e) $B(\perp V)$ = 77 G, γ = 42 Pa/m, v = 14.0 m/s

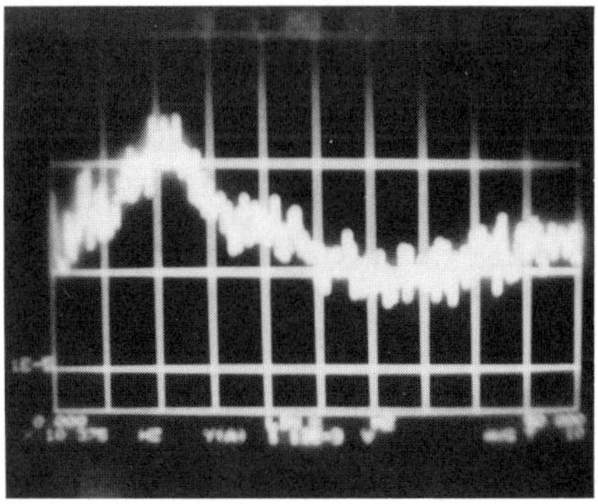

5(f) $B(\parallel V)$ = 75 G, γ = 21 Pa/m, v = 17.5 m/s

Discussion

A number of factors must be considered in relating the theory discussed previously to the experimental results. The finite extent of the tin pool, combined with the mechanical resonance of the system, would tend to establish the dominant wavelength. One might therefore expect Eq. 28, rather than Eq. 29, to best describe the behavior of the system. Further, at 10 kHz in tin, the skin depth is significant, and the magnetic pressure is not strictly a surface force, as assumed by the theory. Finally, the theory does not account for EM stirring, which was observed experimentally.

According to Eq. 29 and the conditions of the experiment, with no magnetic field present, the threshold gas velocity V_c would be 18.6 m/s, a result approached in the experiments (Fig. 7) only when very clean tin surfaces were present. Equation 28 predicts a straight line when V_c^2 is plotted against γ. The data tend to agree with this prediction (Fig. 6), although the agreement is not definitive.

To place the results of these experiments in perspective, the maximum field strength was 108 G, and the maximum restoring pressure was 61 Pa/m. About 330 G is needed to balance the force of gravity on a 6.4-mm sheet of steel. In addition, a value of $\gamma = 6.9 \times 10^4$ Pa/m is required to satisfy the Rayleigh-Taylor criterion for stable levitation of the liquid sheet [i.e., essentially a doubling of the magnetic pressure within a distance equal to one sheet thickness (6)]. Hence, the values of B and γ used in these experiments are well below those required for use in a horizontal levitation-type caster. The experimental results do give a preliminary indication that gas

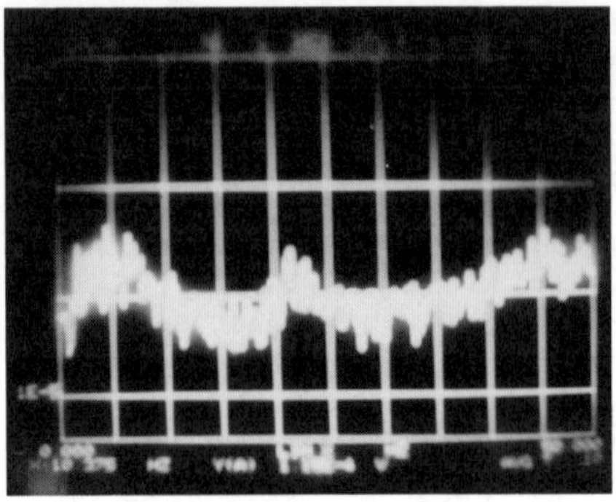

5(g) $B(\parallel V) = 75$ G, $\gamma = 21$ Pa/m, $v = 7.8$ m/s

Figure 5 - Power spectra of wave detector output (shown with a logarithmic scale) versus frequency from 0 to 50 Hz (shown on a linear scale). The vertical scales are ticked, starting at 2×10^{-5} V for (a), (b), (d), and (e), and at 1×10^{-5} V for (c), (f), and (g).

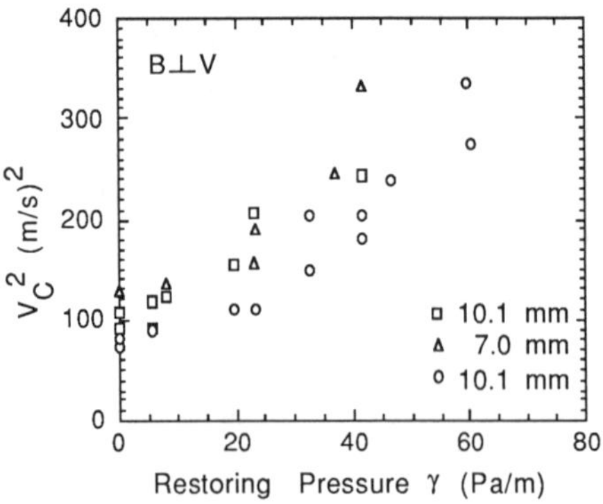

Figure 6 - Square of the critical velocity V_c versus restoring pressure γ for a magnetic field perpendicular to the gas flow.

velocities substantially higher than 20 m/s could be used in a horizontal EM caster. The results also suggest that small magnetic fields above the tin surface could be used to suppress formation of waves by the coolant gas during levitation.

Comparison of the square of the slopes of the upper two sets of data in Fig. 7 with the slopes of the data in Fig. 6 indicates that the magnetic field may suppress waves slightly more in the parallel configuration than in the perpendicular configuration.

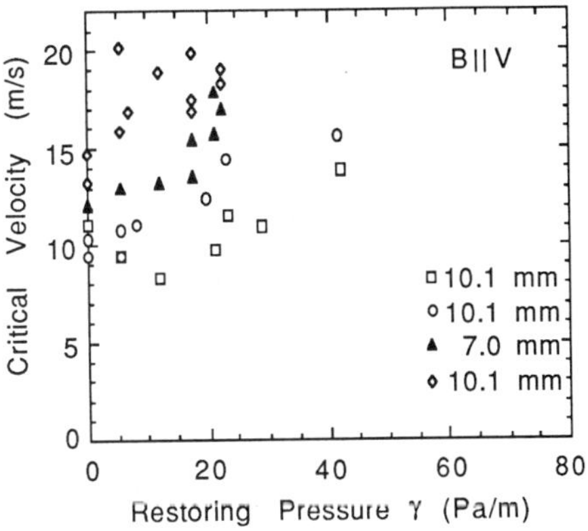

Figure 7 - Critical velocity V_c versus restoring pressure γ for a magnetic field parallel to the gas velocity

Conclusions

The threshold gas velocity V_c was calculated by linear stability analysis for the case in which a molten sheet is constrained by magnetic fields that provide a surface-restoring force on the sheet and a cooling gas is directed parallel to the sheet surface. The theory presented indicates that V_c should increase as the restoring magnetic pressure gradient increases.

The experimental results show that wave formation at the coolant-gas/liquid-metal interface is strongly suppressed by relatively modest values of magnetic pressure and the magnetic pressure gradient produced by 10-kHz excitation. Although V_c increases with magnetic pressure gradient, as predicted by theory, the values of P and γ in the experiments are well below those expected to pertain in stable levitation of a liquid metal sheet.

The suppression was observed for both B⊥V and B‖V. In the latter orientation, the results, although somewhat confounded by experimental artifacts, suggest that the suppression effect is larger when B‖V than when B⊥V. The degree of surface smoothness also appears to significantly affect V_c. Unoxidized liquid tin surfaces show substantially higher V_c values. Within the limited parameter range of the experiments, dependence on the depth of the nitrogen gas channel is not evident.

Acknowledgments

This work was supported by the U.S. Department of Energy, Office of Industrial Programs, under Contract W-31-109-Eng-38. The authors are indebted to N. Gopalsami for providing the wave-amplitude measurement system.

References

1. W. A. Peifer, "Levitation Melting -- A Survey of the State-of-the-Art," J. Metals, 17 (1965) 487-493.

2. Z. N. Getselev, "Casting in an Electromagnetic Field," J. Metals, 23(10)(1971), 38-39.

3. J. E. Battles et al., Argonne National Laboratory, unpublished information (1988).

4. L. R. Turner, R. J. Lari, and W. F. Praeg, "Approaches to Vertical Electromagnetic Moldless Casting of Steel," (Paper presented at this conference, 1988).

5. J. R. Hull et al., "Horizontal Electromagnetic Casting of Thin Metal Sheets," U.S. Patent No. 4,678,024 (July 7, 1987).

6. J. R. Hull et al., "Design and MHD Stability for Horizontal Electromagnetic Continuous Casting," (Paper presented at this conference, 1988).

7. S. Chandrasekhar, Hydrodynamic and Hydromagnetic Stability (London: Oxford University Press, 1961), Ch. XI.

8. R. A. Gerwin, "Hydromagnetic Surface Waves in a Conducting Liquid Surrounded by a Flowing Gas," Phys. Fluids, 11 (1968) 1699-1708.

9. M. Garnier and R. Moreau, "Effect of Finite Conductivity on the Inviscid Stability of an Interface Submitted to a High-Frequency Magnetic Field," J. Fluid Mech., 127 (1983) 365-377.

10. J. R. Hull, "Kelvin-Helmholtz Instability and Cooling Limitations of a Vertical Molten Sheet Constrained by Alternating Magnetic Fields," submitted to ASME J. Heat Transfer (1987).

PLENARY LECTURE

Session Chairman: **R.S. Carbonara**
SEA Inc.
Worthington, Ohio

AN OVERVIEW OF COIL CASTING: PRODUCTION OF FLAT COILABLE PRODUCT DIRECTLY FROM MOLTEN METAL

Richard K. Pitler

Consultant
5718 King of Arms Drive
Gibsonia, PA 15044

Abstract

The history of attempts to cast strip directly from molten metal is reviewed with emphasis on more recent efforts to develop processes for ferrous alloys. The forces driving these developments are discussed. Various processes such as single roll, double roll, and belt-type casters are compared. Metal delivery systems, casting techniques, and products cast are described to the extent that data are available. The potential advantages and disadvantages of these processes are discussed in relation to casting rates, sensitivity to process parameters, defect generation, product size range, etc. Some predictions are made concerning future developments and their potential impact on ferrous metal production.

Casting of Near Net Shape Products
Edited by Y. Sahai, J.E. Battles,
R.S. Carbonara and C.E. Mobley
The Metallurgical Society, 1988

Introduction

The origins of processes designed to cast sheets and strip directly from molten metal are about one hundred and fifty years old. Several excellent papers (1,2,3,4,5,6) have reviewed this history from the early work of Sellers (7) and Laing (8) on lead, to the work of Bessemer (9,10) and Norton (11) on steel using a double roll technique, to the roll and belt system of Daniels (12) and to the work of Strange and Pim (13) on a single roll apparatus. Lang (14), Staples (15) and Strange (16,17) also obtained patents on single roll casters. In 1891, Norton (11) recognized that, for his double roll caster, uniform metal feed and adequate deoxidation were requisites for the production of good quality sheets. Dancy (18) has pointed out that lack of adequate control systems and refractory materials were overwhelming obstacles to the further development of these novel concepts into practical commercial processes. Both Dancy (18) and Cramb (2) have noted that after this early development, interest in the casting of flat product remained relatively dormant until the late 1920's and 1930's when the early development work by Clarence Hazelett (4,5) sparked renewed interest in casting coils of strip.

Hazelett's work evolved from single and double roll systems to belt casters over a period of thirty-plus years. Today, commercial production of copper anodes by the use of Hazelett casters and of aluminum, zinc and lead strip by Hazelett casters with in line rolling is now relatively common. Even so, commercial installations for non-ferrous alloy strip did not appear until the late 1940's and, most currently, operating machines were installed between 1960 and 1980 (5). However, no commercial installations for producing coils of ferrous alloys resulted from this work.

Current Developments

In the late 1970's and 1980's interest in ferrous alloy coil casting was revived. It continues today at an increasing rate, as indicated by the frequency of conferences devoted to the subject, the amount of patent activity and the number of organizations actively working on development projects in the field. A brief review of only the U.S. Patents issued in the year 1987 showed at least seventeen related to methods for producing coilable strip, assigned to thirteen different companies or individuals in six different countries. Table I is a compilation from recent literature (2,6,19,20,21,22,23,24,36) of organizations working in the field and the processes being investigated by them. At least six different processes are under investigation by thirty-five separate organizations in nine different countries. Undoubtedly, there is more activity in the world still unpublished or as yet unrecognized. Today, interest in the direct casting of coils of ferrous alloys is probably higher than at any time in the past.

Figures 1 through 5 show schematic views of five of the coil casting processes currently under investigation.

Incentives Driving These Developments

A number of reasons could be cited for this renewed interest in casting coilable ferrous alloy strip. These reasons can be divided into two categories: economic incentives and technical incentives. These are listed in Table II. The economic advantages of lowered capital costs, operating costs and energy costs are immediately obvious. If conventional casting, reheating and hot rolling operations could be replaced by a single reliable casting operation to produce coils ready for further processing, then capi-

Table I. Ferrous Alloy Coil Casting Activity

Process	Investigator	Country
Single Roll	Allegheny Ludlum	U.S.A.
	Armco-Westinghouse	U.S.A.
	Battelle	U.S.A.
	United Technologies	U.S.A.
	Vacuumschmelze	Germany
	Voest Alpine	Austria
Double Roll	Armco-Inland-Weirton-Bethlehem	U.S.A.
	Concast	Switzerland
	CSM	Italy
	Hitachi Zosen	Japan
	IRSID	France
	Kawasaki Steel	Japan
	Kobe Steel	Japan
	Krupp	Germany
	Nippon Kokan	Japan
	Nippon Metal	Japan
	Nippon Steel	Japan
	Nippon Yakin	Japan
	Nisshin-Hitachi	Japan
	Thyssen Grillo Funke	Germany
	Voest Alpine	Austria
	Waseda University	Japan
Belt and Roll	Battelle	U.S.A.
	Mannesmann-Demag	Germany
Inside the Ring	LTV	U.S.A.
Powder Spraying	Aurora Metals	U.K.
	Drexel University	U.S.A.
	Mannesmann-Demag	Germany
	M.I.T.	U.S.A.
	Osprey Metals	U.K.
	Sandvik	Sweden
	Spray Forming Development	U.K.
	Sumitomo	Japan
	University of Swansea	U.K.
Electromagnetic Levitation	Argonne National Laboratory	U.S.A.

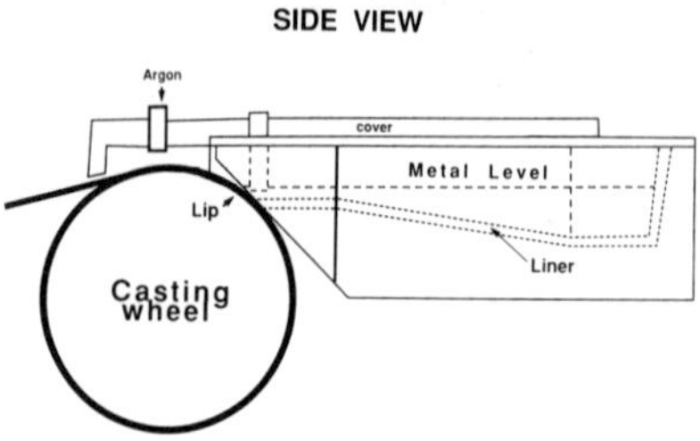

Figure 1 - Schematic Diagram of Single Roll Caster (25).

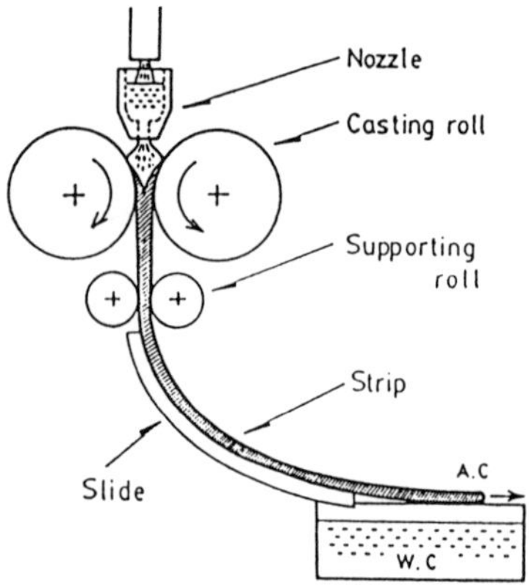

Figure 2 - Schematic Diagram of Double-Roll Caster (27).

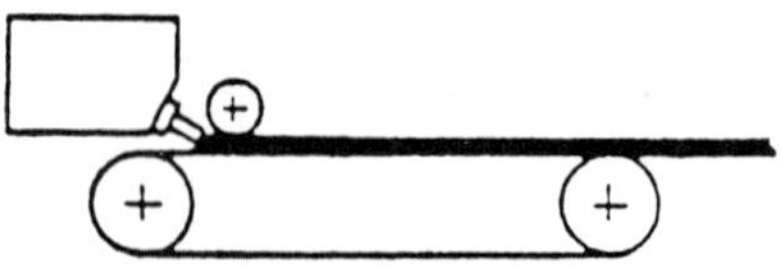

Figure 3 - Schematic Diagram of Belt-Roll Caster (6).

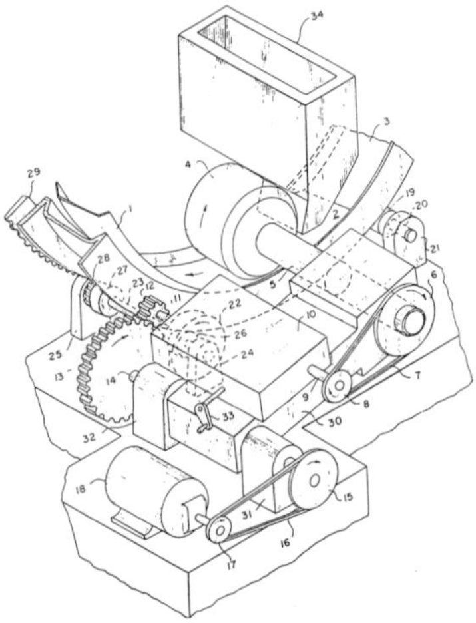

Figure 4 - Schematic Diagram of Inside-the-Ring Caster (35).

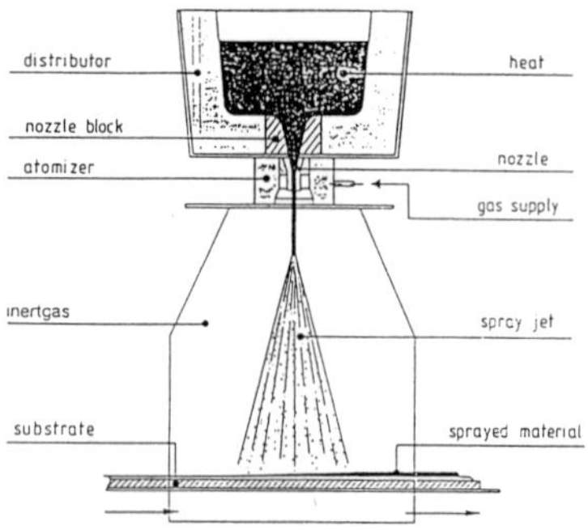

Figure 5 - Schematic Diagram of Spray Caster (39).

Table II. Incentives Driving Developments on Coil Casting Ferrous Alloys

A. Economic

- o Reduced Capital Investment
- o Lowered Operating Cost
- o Lowered Energy Requirements
- o Improved Yield
- o Smaller Increments of Efficient Capacity Possible
- o Heavier Coils Possible
- o Reduced Manufacturing Cycle Time

B. Technical

- o Improved Molten Metal Cleanliness
- o Improved Molten Metal Temperature Control
- o Improved Refractory Systems
- o Improved Process Control Systems
- o Availability of Mathematical Modeling Techniques
- o Recognition of Property Improvements Attainable by Rapid Solidification Techniques

tal required for buildings and equipment, manpower for operations and energy requirements would be substantially lower.

Perhaps less obvious is the potential for adding smaller increments of flat product capacity. Efficient increments of 300,000-500,000 tons/year might be possible rather than large capacity additions of the order of 3,000,000 tons/year required to justify investment in a new hot strip mill. This provides the benefits of flexibility to any expansion plan. It also significantly reduces the investment required of anyone not already producing flat products to become a flat product producer. Additional flexibility is also possible in the production of varying coil widths and weights to meet customer requirements. The heavier coils would be a possibility since, with the elimination of the need for slab reheating for hot rolling, reheat furnace limitations on slab length would no longer exist. In addition, gage variations resulting from end to end temperature variations in the hot strip mill, which can occur when rolling long slabs, would be eliminated.

All of these economic incentives have existed for a long time. Why, then, is it seemingly more attractive today to allocate resources to these developments than was the case in past years? The answer lies in the technical advances that have occurred during the intervening years. The technology developed and the experience gained over the past twenty-five years in achieving and maintaining molten metal cleanliness through raw material selection, refractory practice, melting techniques and atmosphere control systems have removed many of the obstacles faced by earlier workers in this field. Certainly we now have techniques, which were unavailable before, for measuring and maintaining uniform metal temperature. There exist today powerful mathematical modeling techniques (19,25,26,27,28,29,30,31) which can be used to save substantial time and money in planning experiments and in designing equipment. The development of computer-aided process control systems now makes more achievable the precise control of process parameters required to insure the necessary reliability and reproducibility of these casting systems. Furthermore, the recognition and demonstration of the property enhancements possible through rapid solidification techniques make these processes, with cooling rates between 10^2 and 10^6 °K/sec (25,27,32) of even more technical interest; for they offer the possibility of significantly improved products.

Advantages and Disadvantages of Casting Systems

As shown in Table I, there are many coil casting processes under investigation. The single roll (19,25,33,34), double roll (3,6,27), belt and roll combination (6,23), inside-the-ring (28,35) and spray casting (32,37,38,39) techniques are described elsewhere in more detail than is possible in this paper. Although twin belt casters are in use for non-ferrous strip and development work is being carried out on thin slab (10mm-50mm) casting of ferrous alloys using this method (5,6), no projects were identified using twin belt systems for casting coilable thicknesses of ferrous alloys. This is probably because of the problems with feeding high melting point liquid metal uniformly into this type of system.

Table III summarizes some of the attractive features and some of the drawbacks of five casting systems.

The single roll system has a "free" surface which does not contact a chilled substrate. This surface is subject to defects from slag carry-over and non-uniform solidification if precise process control is not exercised. Obviously, a single chill surface will not permit as wide a range of thicknesses to be produced as two chill surfaces. This limits not only the castable thickness range, but also the production rate of a single roll system compared to processes having two chill surfaces. Careful handling of the cast strip, having one very hot side, is critical to prevent upper surface cracking. Providing an effective atmosphere control system is also more difficult in the relatively open single roll configuration. However, the single roll process has the advantage of a simple open tundish metal feeding system and fewer alignment and synchronization problems compared to processes having two chill surfaces. This makes dealing with the control aspects of the process easier. Also, it is relatively easy to change the strip width by changing the tundish width. It is possible to eliminate center porosity and center segregation by proper process controls. It has been demonstrated that good quality cast strip of carbon steel (19,25), silicon steel, stainless steel, a nickel-iron alloy, and a nickel-base superalloy (25,41) can be cast by this technique. More recently, strip of the intermetallic, Ni_3Al, has been cast and successfully cold rolled from the as cast condition (40).

Double roll systems eliminate the problems associated with a "free" surface, provide easy handling and control of the hot strip and allow for the possibility of limited hot deformation within the casting machine. They can achieve a wider range of thicknesses and higher production rates than single roll systems. Atmosphere control is simpler in this more closed system. Edge containment, however, is a more difficult problem for double roll processes. Good product requires excellent synchronization of roll speeds, metal flow and heat removal since complete solidification should occur at or near the nip of the rolls to avoid center porosity and center segregation or surface cracking (42). Therefore, the control requirements are more stringent than for a single roll system. Carbon steel, cast iron and stainless steel (42,43) have been reportedly cast by this method. It is believed that silicon steels and nickel-iron alloys have also been cast.

The inside-the-ring process, like the double roll process, provides two chill surfaces, a wide thickness range, higher production rates and the possibility of limited hot reduction within the casting machine. It

Table III. Comparison of Coil Casting Processes

Process	Advantages	Disadvantages
Single Roll	o Simple tundish system o Easy to vary width o Simpler synchronization o Less porosity and segregation tendency	o "Free" surface problems o Limited thickness range o Lower production rate o More difficult atmosphere control o More sensitive to hot strip handling
Double Roll	o Two chill surfaces o Easier hot strip handling control o Limited hot deformation possible o Heavier thicknesses possible o Easier atmosphere control	o Difficult edge containment o Width not easily varied o Precise synchronization required o Uniform metal feeding more difficult
Belt-Roll	o Two chill surfaces o Good hot strip support system o Limited hot deformation possible	o Precise synchronization required o Atmosphere control more difficult o Width not easily varied o Uniform metal feeding more difficult
Inside-the-Ring	o Two chill surfaces o Heavier thicknesses possible o Simple tundish system o Limited hot deformation possible	o Atmosphere control more difficult o Precise synchronization required o Width not easily varied o Difficult edge containment o Complicated strip removal and associated yield losses
Spray Casting	o Higher cooling rates o Minimum segregation o Easy atmosphere control o Insensitive to alloy composition o Clads and composites possible	o Residual porosity o Rough surfaces o Difficult to obtain uniform cross width thickness o Yield losses from non-uniform thickness and overspray

has a simpler metal feeding arrangement, but providing atmosphere control is more difficult as is usually the case in more open systems. It requires the same precise synchronization of chill surface speed and casting rate to achieve complete solidification at the location necessary to avoid centerline segregation and porosity. The method of edge containment results in a "turned up" edge which can lead to edge cracking and yield losses as the strip is translated out of the plane of the caster and flattened prior to coiling. Strip widths are not easily changed.

Although the belt-roller combination is noted in the literature (6) and in patents (23), no information was found regarding experimental data on the casting of ferrous alloys.

Spray casting results in higher cooling rates (10^4-10^6 °K/sec) (32) than the other systems (10^2-10^4 °K/sec) (25,27,42) and therefore can produce a finer, more segregation-free microstructure. This increases the potential for property improvements and makes the process relatively insensitive to variations in alloy solidification characteristics. Atmosphere control is simple although alloys sensitive to pickup of nitrogen, the usual atomizing gas, might require the use of a more expensive gas such as argon. Multi-layered clads and composites can be readily produced by this method. There is some residual porosity which, thus far, has been eliminated by a hot reduction step. As-sprayed surfaces are rough and require smoothing. Thickness uniformity across the width is difficult to achieve because of the nature of the spray pattern; the strip near the edges being thinner. This, and overspraying, cause yields from molten metal to strip product to be lower than for some of the other processes. Low carbon steel, alloy steel and silicon steel have reportedly been made by this method (3,38). Other more complex alloys undoubtedly either have or will soon be experimentally produced this way.

Future Trends

It is difficult to predict how successfully each of these developments will achieve commercial reality. They all have significant potential for reducing the cost of ferrous alloy flat product production and for providing improved property levels in a variety of alloy systems. It is most likely that no one casting system will be the best for all alloys at all production rates. The spray casting process seems most suited to alloys which are difficult to coil cast by other methods, which have a high potential for property improvement by rapid solidification techniques and which are produced in modest quantities. It is an ideal process for making clad and composite products.

The two-sided systems appear best suited for high tonnage production rates of alloys requiring thicker cast coils, either because more subsequent reduction is required to obtain the necessary final properties or because the market demands that range of thicknesses. Because the single roll system appears simpler, it can be expected to be utilized for those coil thicknesses within its capability. Tonnages produced by this process will probably be intermediate between those for the spray caster and the two-sided casters.

All of these processes will probably eventually be coupled with a simple hot rolling step. The double roll, belt-roll combination and the inside-the-ring systems have the potential for accomplishing this within the casting machine. The spray casting and single roll systems would require an external one or two stand hot rolling mill. All of these processes are carried out at relatively slow speeds compared to hot strip mill rolling speeds. These slower speeds, and the existence of techniques for restoring some heat uniformly (e.g., transverse flux induction heating), make an in-line hot rolling step very feasible. Such a step could certainly minimize internal porosity and segregation as well as improve surface quality and thickness uniformity.

I believe that, with the technologies available today, the technical problems associated with each of these processes can be solved. Each can be developed to produce acceptable commercial product. The pertinent question is; how economical will these solutions be? For it will be economics that will determine which process will be used and, to what extent and on which alloys it will be used.

Potential Impact

The commercialization of coil casting for ferrous alloys will impact the industry in several ways. New and improved materials will become available as a new tool for controlling metallurgical structures in flat products is used. Since the cost to a new producer entering the flat product business will be lowered, I can envision more producers making more products tailored to specific market segments; a result of being able to install economically smaller increments of efficient flat product capacity. This should lead to an industry less concentrated as to product line and geography, with production sites closer to markets. Scrap and pre-reduced pellets feeding electric furnaces for melting, followed by refining vessels, will probably provide the molten metal to these casters. But do not look for this to happen quickly. The adoption of Hazelett's machines by the non-ferrous industries took place over a 20-30 year period. Optimistically, these new coil casting systems for ferrous alloys will be commercialized in less time. However, although there may be some uncertainty concerning the timing of the success of these developments, I have no doubt about their ultimate success. The initial advancements are encouraging, the required technologies appear to be available and the potential benefits driving these development programs forward are simply too great to go unrealized.

References

1. H. Warlimont and K. Emmerich, "Rapidly Cast Cristalline Thin Sheet Materials," Advanced High-Temperature Alloys, S. M. Allen, R. Pelloux and R. Widmer eds. (American Society for Metals, Metals Park, Ohio 1986), 73-84.

2. A. W. Cramb, "New Developments in the Continuous Casting Steel," to be published Trans. of the Iron and Steel Society, (August 1988).

3. M. C. Flemings and N. J. Grant, "Technology Development of Thin Strip Metal Casting - Feasibility Study," (U.S. Dept. of Interior, Washington, D.C., January 1984), DOE/ID/12444-T1.

4. R. D. Hammer, "The Hazelett and Rotary Strip Casting Machines for Continuous Casting of Aluminum," Journal of the Institute on Metals, 87 (1958-59), 219-226.

5. J. B. Allen and W. Szczypiorski, "Thin Slab Casting of Ferrous and Non-Ferrous Metals on the Hazelett Twin-Belt Caster," International Symposium on Near-Net-Shape Casting, (89th Annual Meeting of Canadian Institute of Mining and Metallurgy, Toronto, May 6, 1987), Paper 218.

6. W. Reichelt, W. Kapellner and R. Steffen, "Near-Net-Shape Casting of Flat Products," Symposium on Casting and Solidification, (Winterschied uber Hennef/Seig, October 4-9, 1987).

7. G. E. Sellers, U.S. Patent 1908, December 17, 1840.

8. J. Laing, U.S. Patent 3023, March 30, 1843.

9. H. Bessemer, U.S. Patent 49,053, July 25, 1865.

10. H. Bessemer, "On the Manufacture of Continuous Sheets of Malleable Iron and Steel Direct from Fluid Metal," Journal of the Iron and Steel Institute, (London, October 7, 1891), 23-41.

11. E. Norton, "Notes on Experiments in Rolling Fluid Steel," Iron Age, (April 11, 1940), 50-51.

12. F. H. Daniels, U.S. Patent 359,348, March 15, 1887.

13. E. H. Strange and C. A. Pim, U.S. Patent 905,758, December 1, 1908.

14. E. M. Lang, U.S. Patent 112,054, February 21, 1871.

15. W. G. Staples, U.S. Patent 989,075, April 11, 1911.

16. E. H. Strange, British Patent 24,320, October 23, 1910.

17. E. H. Strange, U.S. Patent 993,904, May 30, 1911.

18. T. E. Dancy, "Strip Casting - Where Do We Go From Here?," Iron and Steelmaker, 1987, no. 12:25-28.

19. K. Schwaha et al, "Strip Casting of Low Carbon Steel at Voest Alpine," Steel Times, 25(7)(July 1987), 324-325.

20. P. H. Dauby ed., Near Net Shape Casting, (A Publication of the Iron and Steel Society, Inc., Warrendale, PA 1987.)

21. Nakanori et al, U.S. Patent 4,702,300, October 27, 1987.

22. F. Willim, U.S. Patent 4,648,439, March 10, 1987.

23. R. E. Maringer, U.S. Patent 4,646,812, March 3, 1987.

24. G. J. McManus, "Steelmakers Pursue Thin Strip Casting," Iron Age, 3(8) August 1987), 14-18.

25. R. K. Pitler, "Direct Casting of Coilable Ferrous Alloy Strip," Advanced High-Temperature Alloys, S. M. Allen, R. Pelloux and R. Widmer eds., (American Society for Metals, Metals Park, Ohio 1986), 1-10.

26. R. J. O'Malley and M. E. Karabin, "A Theoretical and Experimental Study of the Roll Casting Process," International Symposium on Near-Net-Shape Casting of Strip, (89th Annual Meeting of Canadian Institute of Mining and Metallurgy, Toronto, May 6, 1987), Paper 189.

27. A. Kasama et al, "A Consideration on the Rapid Solidification of Steels - Solidification of Steel by Twin-Roll Caster," Near Net Shape Casting, (A Publication of the Iron and Steel Society, Inc., Warrendale, PA 1987.)

28. V. Stanek and J. Szekely, "A Mathematical Model of a Drum and Ring Horizontal Strip Casting Process," Metallurgical Transactions B, 14B, (September 1983), 487-493.

29. J. D. Nauman, M. M. Malecki and D. J. McMahon, "Modeling of ALC's Strip Casting Process," Engineering Foundation 4th Conference on Modeling of Casting and Welding Processes, (Palm Coast, Florida, April 17-22, 1988).

30. Y. G. Kim, B. Farouk and D. Apelian, "Modeling of a Twin-Belt Strip Casting Process," Ibid.

31. P. C. Mathur, D. Wei and D. Apelian, "Modeling of the Solidification Process During Spray Casting," Ibid.

32. A. R. E. Singer, "The Challenge of Spray Forming," Powder Metallurgy, 25(4)(1982), 195-200.

33. R. S. Carbonara and R. A. Gleixner, "Limitations of the Single Roller Direct Strip Casting Process," International Symposium on Near-Net-Shape Casting of Strip, (89th Annual Meeting of Canadian Institute of Mining and Metallurgy, Toronto, May 6, 1987), Paper 221.

34. J. D. Nauman and D. B. Love, "Controlling the Shape of Cast Stainless Steel Strip," Ibid., Paper 222.

35. C. C. Gerding, U.S. Patent 3,811,491, May 21, 1974.

36. J. R. Hull et al, U.S. Patent 4,678,024, July 7, 1987.

37. A. R. E. Singer, "The Principles of Spray Rolling of Metals," Metals and Materials, 4(1970), 246-257.

38. A. R. E. Singer and W. N. Jenkins, "An Assessment of Spray Forming as a Potential Continuous Route to Steel Strip," P. B. Dauby ed., Near Net Shape Casting, (A Publication of the Iron and Steel Society, Inc., Warrendale, PA 1987), 87-92.

39. W. Reichelt et al, "A New Method to Produce Near Net Shape Initial Material for the Production of Strip and Tubes," Paper Presented at Metallurgists Conference, Dusseldorf, October 10, 1986.

40. Private Communication, Allegheny Ludlum Corporation.

41. J. F. Grubb et al, "Casting Specialty Steel Strip," P. B. Dauby ed., Near Net Shape Casting, (A Publication of the Iron and Steel Society, Inc., Warrendale, PA, 1987), 51-57.

42. K. Kawakami et al, "Research on Twin-Roll Casting Process," Ibid., 71-80.

43. T. Kusakawa et al, "Characteristics of Iron and Steel Thin Sheet Made by Strip Casting," Ibid., 81-85.

Acknowledgments

Several people who assisted in the preparation of this paper deserve recognition. Mrs. Mary Guenther typed the manuscript, Mr. Dale Deemer helped in preparing the figures and Mrs. Eileen Gallagher and Mr. P. J. Viccaro assisted in securing many of the references. Their efforts are sincerely appreciated.

FUNDAMENTAL ASPECTS IV: ELECTROMAGNETIC CONFINEMENT

Session Chairman: **J.R. Hull**
Argonne National Laboratories
Argonne, Illinois

MATHEMATICAL MODELING OF ELECTROMAGNETIC CONFINEMENT

J.D. Lavers   M. Ramadan Ahmed

Department of Electrical Engineering
University of Toronto
Toronto M5S 1A4 CANADA

ABSTRACT

The problem of predicting the shape and stability of the molten metal free surface in electromagnetic confinement applications is considered. Basic principles relating to electromagnetic confinement are reviewed and practical limitations are established. Modeling an electromagnetic problem in which the conducting body has a deformable free surface is discussed. Particular attention is given to establishing the e.m. modeling capability that is presently available for 2-D and 3-D problems. The application of a Boundary Element Method to the e.m. confinement problem is described and an iteration to obtain the free surface shape is developed. Successful applications of this model to predict free surface contours are presented.

INTRODUCTION

Electromagnetic confinement of molten metal is increasingly being used in a broad range of metallurgical applications. These range from relatively mature applications such as the electromagnetic mould for casting aluminum and bronze [1,2] to emerging or proposed applications such as: (i) levitation and/or confinement in the entry region between the tundish and a horizontal casting mould, (ii) edge confinement [3] and control [4] within a strip or plate caster, and (iii) control of shape (eg. thickness) to near final form specifications [5]. In virtually all such applications, an alternating electromagnetic field is used to *induce* current to flow within the molten metal so as to create the desired confinement forces. The confinement forces work against a static head that, in many applications, is primarily gravitational in origin. A major problem in most of these applications is to design a coil configuration such that the developed electromagnetic pressure is compatible with the desired free surface contour. From an electromagnetic viewpoint, analysis is difficult since the *unknown* free surface contour critically influences the magnetic field distribution, and thus the electromagnetic pressure distribution.

This paper seeks to fulfill several objectives. First, basic principles relating to electromagnetic confinement are reviewed and limitations (equipment, achievable static head, etc.) are established. Modeling an electromagnetic problem in which the conducting body has a deformable free surface is then discussed. Particular attention is given to establishing the e.m. modeling capability that is presently available for 2-D and 3-D problems. It is sug-

Casting of Near Net Shape Products
Edited by Y. Sahai, J.E. Battles,
R.S. Carbonara and C.E. Mobley
The Metallurgical Society, 1988

gested that when modelling deformable bodies, the Boundary Element Method (BEM) is particularly attractive, especially when the Impedance Boundary Condition (IBC) can be used to eliminate one or more conducting regions. The application of the BEM to the e.m. confinement problem is described and an iteration to obtain the free surface shape is developed. Two particular applications are considered: (i) the crucible induction furnace, for which ample data are available for the shape of the meniscus under various operating conditions, and (ii) the CREM process [6,7] for controlling the meniscus in aluminum d.c. casting.

ELECTROMAGNETIC FIELDS & FORCES

Many of the new electroheat applications involve the use of electromagnetic pressures to shape or contain molten metal. These applications exploit the same phenomenon that creates the meniscus in the crucible induction furnace. The design problem reduces to that of placing a confinement coil or coils such that a desired free surface shape can be obtained or such that metal can be confined against a specified static head. The shape of the free surface is governed primarily by the balance of electromagnetic pressures against pressures due to gravity; transfer of momentum due to molten metal flow may also be of critical importance. Therefore, at each point on the free surface, the force balance may be summarized as follows:

$$-\bar{\nabla}p = -\gamma\bar{g} + \frac{1}{4}\mu\bar{\nabla}\left(\bar{H}\cdot\bar{H}^{*}\right) + \frac{1}{2}\gamma\bar{\nabla}\left(\bar{v}\cdot\bar{v}\right) \tag{1}$$

where γ is the mass density, g is the gravitational constant, $\bar{H}$ is the magnetic field strength and $\bar{v}$ is the fluid velocity. The superscript * denotes a complex conjugate quantity. In many problems, the radius of curvature over most, if not all, of the free surface is large and surface tension effects become of secondary importance. **Electromagnetic Fields.**

The magnetic field strength $\bar{H}$ and the electric field strength $\bar{E}$ can be obtained by solving Maxwell's equations which, when the sources are time harmonic of frequency ω, can be written as:

$$\bar{\nabla}\times\bar{H} = Y\bar{E} \qquad Y = \sigma - j\varepsilon\omega \tag{2a,b}$$

$$\bar{\nabla}\times\bar{E} = Z\bar{H} \qquad Z = +j\mu_0\mu_r\omega \tag{3a,b}$$

where the relative permeability $\mu_r = 1$ in nonmagnetic media and the electrical conductivity $\sigma = 0$ in nonconducting regions. Knowing the electric field strength $\bar{E}$, the conduction current density $\bar{J}$ is simply:

$$\bar{J} = \sigma\bar{E} \tag{4}$$

When the electromagnetic problem can be described in terms of a 2D model, it may be convenient to describe $\bar{H}$ in terms of a magnetic vector potential $\bar{A}$:

$$\bar{H} = \frac{1}{\mu_0\mu_r}\bar{\nabla}\times\bar{A} \tag{5}$$

with $\bar{A}$ satisfying the following constraint, or gauge condition:

$$\bar{\nabla}\cdot\bar{A} = 0 \tag{6}$$

For time harmonic induction problems:

$$\bar{J} = -j\omega\sigma\bar{A} \tag{7}$$

In the case of 3D problems, however, it is not at all clear that the magnetic vector potential provides a simplification [8]. The difficulty lies in the choice of an appropriate gauge condition.

In 2-dimensions, time harmonic electromagnetic field problems have been solved by Finite Difference [9,10], Finite Element [11,12], Volume Integral [13,14] and Boundary

Integral [15-19] methods. Today, the method of choice in 2D is probably the Finite Element method. Several excellent packages, running on machines ranging from mainframes to AT-class PCs, are available for this class of problem. On the other hand, the development of reliable computational methods for 3D eddy current problems remains an area of very active research. While Finite Element methods have been proposed [20-22] for 3D eddy current problems, there remain difficulties with the specification of the problem geometry and with the treatment of the very important free space region. These problems, together with advantages to be gained from parallel computation [23], has resulted in considerable interest being shown in Boundary Element methods for 3D problems [24-28].

As the shape of the free surface is not known *a priori* in most electromagnetic confinement problems, the e.m. field problem must be iteratively solved subject to the constraint provided by (1). Electromagnetic problems of this class have recently begun to receive attention in the literature [10,29-32]. For the most part, (volume) Integral Equation [32] and (point) Finite Difference [10] methods have been used to compute the required electromagnetic fields in confinement applications. However, these two methods are awkward to use for this class of problem for several reasons. First, the *entire* solution grid for the conducting region must be modified at each step in the iteration for the free surface shape. Second, both Volume Integral and Finite Difference methods model do not easily model *smooth* surfaces. Finally, point methods (Finite Difference or Finite Element) *must* include the exterior free space region in the solution of any induction e.m. problem. This considerably increases the magnitude of the numerical problem. For these reasons, formulations that include *only* the interface boundaries offer significant advantages when modelling electromagnetic confinement problems. Such formulations are discussed in a subsequent section of this paper.

Electromagnetic Forces & Pressure.

At any point within a metallic conductor where the time harmonic current density is $\overline{J}$ and the magnetic field strength is $\overline{H}$, the time average electromagnetic volume force density is:

$$\overline{f} = \frac{\mu}{2}\mathrm{Re}\left(\overline{J}\times\overline{H}^{*}\right) \tag{8}$$

These forces are confined to a surface layer having a thickness of approximately 2δ, where δ is the electromagnetic penetration depth:

$$\delta = \sqrt{\frac{2}{\mu\omega\sigma}} \tag{9}$$

In confinement applications, δ is generally chosen to be small relative to the dimensions of the molten metal region in order to ensure that the confinement field is relatively *stiff.* However, a small δ also complicates the Volume Integral and Finite Difference/Element solution of the electromagnetic field problem since a fine solution grid, concentrated near the surface of the conducting region, must be used. Conversely, a small electromagnetic penetration depth considerably simplifies the Boundary Integral solution of the problem.

It is a relatively simple matter to expand (8) in order to obtain:

$$\overline{f} = -\frac{\mu}{4}\overline{\nabla}\left(\overline{H}\cdot\overline{H}^{*}\right) + \frac{\mu}{2}\mathrm{Re}\left[\left(\overline{H}\cdot\overline{\nabla}\right)\overline{H}^{*}\right] \tag{10}$$

The first term in (10) represents a conservative force that is derived from an electromagnetic pressure p_m:

$$p_m = \frac{\mu H^2}{4} \tag{11}$$

This force can support a static head (eg. as in a confinement application), but cannot do work (eg. as in stirring the molten metal). The nonconservative contribution to the force field is given by the second term in (10). It is this force component that results in fluid motion in stirring applications. It is important to note, however, that in a cylindrical geometry, the second term in (10) also contributes to the *total* electromagnetic pressure when δ is large relative to the conductor dimensions; i.e. when there is full penetration of the magnetic flux.

Regardless of the solution method used to determine the electromagnetic field distribution in a confinement application, the free surface of the molten metal must be such that the constraint (1) is satisfied. In those applications where surface tension and momentum transfer from the moving metal can be neglected, the constraint simply represents a balance between gravitation and electromagnetic pressures. An iteration can therefore be implemented in which the shape of the free surface is adjusted such that the required balance is achieved. In many important applications, this balance can be achieved in relatively few iterations since the free surface assumes a simple shape. In the case of the crucible induction furnace, for example, the meniscus can be accurately described by a parabolic surface of revolution.

In applications where fluid velocity (electromagnetically driven or otherwise) contributes significantly to the pressure balance, the free surface problem becomes very computationally intensive. This is especially the case when the flow is turbulent. In such cases, four additional partial differential equations (two for the flow fields and two for turbulence) must be solved. The question of turbulent, electromagnetically driven flow has been widely discussed in the literature [33-34]

HARDWARE AND SOFTWARE LIMITATIONS RELATED TO ELECTROMAGNETIC CONFINEMENT

A. Hardware Limitations.

1. Maximum Electromagnetic Pressure. Fundamental limits exist regarding the maximum *practical* force or pressure that can be achieved via an induction process. These limits are governed by the maximum allowable source current density that can be achieved without overheating the conductors of the excitation coil. Specifically:

(a) **Current Density.** *Without any cooling*, the allowable current density in the excitation coil would be limited to approximately 5-8 A/mm^2. *With water cooling*, a normal current density would be in the order of 15-20 A/mm^2. *Under optimal conditions*, this might be pushed to 35-40 A/mm^2. The maximum current density in the coil is set by the available cooling; the cooling must be sufficient to limit the temperature rise of the coil conductors to not more than 30-40 deg. C above ambient.

(b) **Magnetic Field Strength**. The magnetic field strength is directly proportional to coil ampere turns and is inversely proportional to coil length. When relatively few turns are involved, the field strength falls off with distance from the coil. It follows directly from the limit on allowable current density that *a typical magnetic field strength* would be in the order of 1000 A/cm. The best that might be achieved with normal coil designs in air might be a field strength in the order of 3000 A/cm. To put this in perspective, 3000 At/cm is approximately the field strength on the surface of a 1 cm radius conductor carrying a current of 19 kA!. At a distance of 1 cm from the conductor surface, the field strength has been reduced to 1500 A/cm.

(c) **Ferrostatic Head for Confinement**. The electromagnetic pressure is proportional to the square of the magnetic field strength. The maximum field strength of 3000 A/cm would yield an electromagnetic pressure that is equivalent to a ferrostatic head of approximately 76 cm. This assumes that a uniform field strength of 3000 A/cm could, in fact, be obtained *at the surface* of the molten metal. Under optimal conditions with present technology, it is doubtful whether this static head could be achieved. It is estimated that *practical values* for the ferrostatic head that could be supported electromagnetically would be in the order of 10-25 cm.

2. Field Shaping. Very real limits exist regarding the extent to which electromagnetic fields can be shaped to match desired design contours for the confinement force field. The fields have a very strong spatial dependence on both the shape and the placement of the source conductors. In many applications, space constraints will limit both the size and the number of source conductors. The net result is that the confinement fields will tend to have weak regions or holes unless the coil design is carefully considered. Molten metal will quickly find and exploit these weak regions in the confinement field.

3. Water Cooling. To achieve the upper limit of 3000 A/cm, the primary conductors must be water cooled. Furthermore, the conductors must be in very close proximity to the molten metal. Typically, the separation between the water-cooled conductors and the molten metal would be in the order of a centimeter or less. This close proximity has an obvious safety implication and it is often not possible to include a refractory layer between the molten metal and the water-cooled conductor. Additionally, there is a significant heat transfer from the molten metal that: (i) has an impact on the surface cooling of the metal, and (ii) adds to the cooling requirements for the inductor.

4. Secondary Effects. The electromagnetic confinement forces can be accompanied by strong local heating due to stray fields. This heating occurs within the molten metal in the confinement region as well as in adjacent structural members. Additionally, very strong electromagnetic stirring forces can also occur. Despite the fact that the stirring forces are very local in nature, being confined to *corner* surface regions by skin effects, such forces can produce flow fields that extend well beyond the confinement region. This stirring may have solidification implications. In some geometries, the e.m. stirring forces may in fact be beneficial (i.e. the e.m. stirring forces could be designed to either augment or reduce the net stirring, as desired).

B. Simulation Software & Analysis.

This section will only focus on limitations that apply to simulation software that is presently available for electromagnetic field problems.. The identified limitations include the following:

(a) The electromagnetic field problems that arise in many confinement applications are 3-dimensional (e.g. the e.m. mold for slab casting). While the development of 3-D eddy current computational methods is presently a very active research area, this activity has yet to result in the common availability of simulation software packages.

(b) Solution of 3-D eddy current problems is expensive and time consuming. Rather than being cast in a single variable (eg. a vector potential in the case of Finite Elements, a current density in the case of Volume Integrals), a 3-D simulation will require 3 unknowns in the eddy current region and a minimum of 1 unknown in free space. Further, because of skin effect, a fine grid with a large number of nodes will generally be required. Such problems easily extend a VAX 780 to the limit (several days of run time) and are normally run on a supercomputer (Cray, Cyber).

(c) For confinement, the real problem is to design a coil arrangement so as to achieve a desired free surface profile. The electromagnetic free surface problem is only now being seriously considered. One principal limitation to date has been the lack of an efficient means of solving shallow penetration depth eddy current problems, even in two dimensions. Reported solutions to the problem of predicting the meniscus in a crucible induction furnace using Finite Differences [10] and Volume Integrals [32], for example, have been very expensive.

BOUNDARY ELEMENT FORMULATION OF EDDY CURRENT PROBLEMS

In recent years, considerable interest has been shown in the Boundary Integral/Element Method of formulating eddy current problems [15-19,24-28,35-38]. By formulating such problems in terms of the surfaces bounding the individual regions, the problem dimensionality is reduced by one. To illustrate, Figure 1 shows a typical electromagnetic mould configuration, with a confinement coil (**3**), a field shaping shield (**2**) and a column of metal (**1**). For the purpose of this illustration, the liquid and solidified regions in the metal column are not shown. With conventional formulations (i.e. Volume Integral Equation, Finite Difference, Finite Element) of this 2D electromagnetic field problem, a full 2D solution grid (including the free space region in the case of the FD and FE formulations) would have to be used. In the case of a Boundary Integral/Element formulation, only the 1D surface regions (**1**) and (**2**) need be considered. Further, the induction coil is treated in terms of an *incident* magnetic field of known strength $\overline{H}^i$. This is the field that the coil would produce *without any other conducting regions being present.* The ability to separate the *incident* and *reaction* fields is a significant advantage when designing confinement coils.

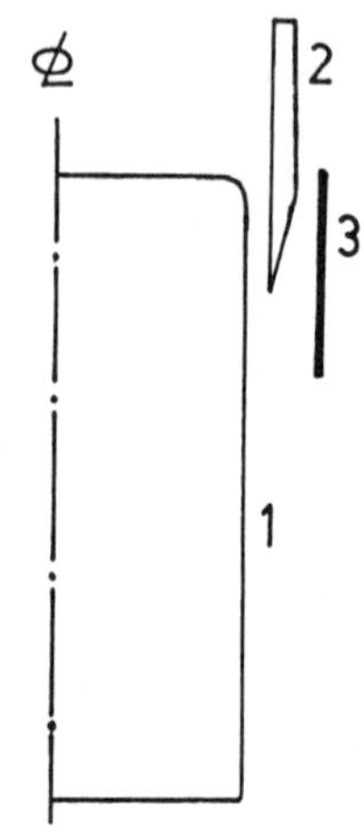

Figure 1. Typical geometry for an electromagnetic mould.

Full details on the formulation and implementation of Boundary Integral and Boundary Element solutions for a broad range of eddy current problems can be found elsewhere [16-18,24-28,35-38]. The purpose of the present section is simply to suggest the procedure that is followed when formulating such problems.

To simplify matters, consider a problem having a single conducting region denoted by Ω and a free space region denoted by Ω–. Within Ω and Ω–, the governing Maxwell equations (2) and (3) can be combined to obtain vector Helmholtz equations for $\overline{H}$ and $\overline{E}$. These equations are of the form:

$$\overline{\nabla} \times \overline{\nabla} \times \overline{H} - \beta^2 \overline{H} = 0 \tag{12}$$

$$\overline{\nabla} \times \overline{\nabla} \times \overline{E} - \beta^2 \overline{E} = 0 \tag{13}$$

$$\beta^2 = YZ \tag{14}$$

where Y and Z are defined by (2b) and (3b), respectively. Note that the magnitude of β^2 is different in each of the regions Ω and Ω–, thus defining separate Helmholtz equations for each region.

An *integral equation* representation of either $\overline{H}$ or $\overline{E}$, at arbitrary points of observation $\overline{\xi}$ *within* Ω and Ω–, can be obtained by applying the vector Green's theorem:

$$\int_V \left[\overline{U} \cdot \overline{\nabla}_\eta \times \overline{\nabla}_\eta \times \overline{G} \; - \; \overline{G} \cdot \overline{\nabla}_\eta \times \overline{\nabla}_\eta \times \overline{U} \right] dv_\eta$$
$$= \int_S \left[\overline{G} \times \overline{\nabla}_\eta \times \overline{U} \; - \; \overline{U} \times \overline{\nabla}_\eta \times \overline{G} \right] \cdot \hat{n}_\eta \, dS_\eta \tag{15}$$

where η is the point of integration, the subscript η denotes an operation taken with respect to the point of integration and $\overline{U}$ represents the vector field of interest. It is convenient to define the function $\overline{G}$ as follows:

$$\overline{G}(\overline{\xi},\overline{\eta}) = g(\overline{\xi},\overline{\eta})\,\hat{c} \tag{16a}$$

$$g(\overline{\xi},\overline{\eta}) = \frac{1}{4\pi R}\exp\left[-\beta R\right] \tag{16b}$$

$$R = |\,\overline{\xi}-\overline{\eta}\,| \tag{16c}$$

where $\hat{c}$ is an arbitrary unit vector, $g(\overline{\xi},\overline{\eta})$ is the fundamental solution [39] of the vector Helmholtz equation and β has a value defined by the material constants for regions Ω or $\Omega-$, as appropriate. The detailed application of (15) to derived integral expressions for $\overline{H}(\overline{\xi})$ or $\overline{E}(\overline{\xi})$ in axisymmetric geometries, for example, has been described by M.R. Ahmed [36]. Note that at this stage, $\overline{H}$ and $\overline{E}$ are defined for a points of observation $\overline{\xi}$ that are *within* Ω and $\Omega-$, but not *on* the boundary between these two regions.

By letting the points of observation $\overline{\xi}$ within Ω and $\Omega-$ move to the interface surface between the two regions *and* by taking proper account of the limiting behaviour of *single* and *double* layer potentials [15,16,35,36,39], Boundary Integral Equations for the tangential components of $\overline{H}$ and $\overline{E}$ on the interface surface can be derived. Allow, for example, the point of observation $\overline{\xi}$ within the free space region $\Omega-$ to move to the interface boundary. In the case of a 2D axisymmetric geometry (eg. crucible induction furnace, e.m. mould for cylindrical billets), the resulting Boundary Integral Equations are of the form [36,40]:

$$1/2\,E_\phi(\overline{\xi}) = E^i{}_\phi(\overline{\xi}) + L^{(1)}_{\Omega-}\,E_\phi(\overline{\eta}) + \frac{1}{Y_{\Omega-}}L^{(2)}_{\Omega-}\,H_t(\overline{\eta}) \tag{17}$$

$$1/2\,H_t(\overline{\xi}) = H^i{}_t(\overline{\xi}) + \frac{1}{Z_{\Omega-}}L^{(2)}_{\Omega-}\,E_\phi(\overline{\eta}) + L^{(1)}_{\Omega-}\,H_t(\overline{\eta}) \tag{18}$$

A similar pair of equations can be derived by starting with a point of observation $\overline{\xi}$ within the conducting region Ω. In that above equations, the superscript i denotes an incident field, η is a position vectors for the point of integration, $Y_{\Omega-} = -j\omega\varepsilon_0$ and $Z_{\Omega-} = j\omega\mu_0$. Similarly, the subscript ϕ represents the azimuthal direction on the conductor surface while the subscript t represents a surface tangential direction that is orthogonal to ϕ.

In the above equations, the integral operators $L^{(1)}$ and $L^{(2)}$ have the following form [36,40]:

$$L^{(1)}\,\overline{U}(\overline{\eta}) = \oint_s \left[\,\left\{\,\hat{n}_\eta \times \overline{U}(\overline{\eta})\,\right\} \cdot \left\{\,\overline{\nabla}_\eta \times \overline{G}(\overline{\eta},\overline{\xi})\,\right\}\,\right] dS_\eta \tag{19}$$

$$L^{(2)}\,\overline{U}(\overline{\eta}) = \oint_s \left[\,\left\{\,\hat{n}_\eta \times \overline{U}(\overline{\eta})\,\right\} \cdot \left\{\,\overline{\nabla}_\eta \times \overline{\nabla}_\eta \times \overline{G}(\overline{\eta},\overline{\xi})\,\right\}\,\right] dS_\eta \tag{20}$$

where $\overline{G}$ is given by (16), with β being defined by the free space material properties and the unit vector $\hat{c}$ being in either the ϕ or the t direction, as appropriate.

In the case of most electromagnetic confinement problems, the electromagnetic penetration depth δ is small relative to the dimensions of the molten metal region. This constraint is necessary in order to obtain a relatively *stiff* confinement field. Fortunately, when δ is small, modelling this electromagnetic field problem is simplified considerably when a Boundary Integral formulation is used. In particular, at shallow penetration depth values, the tangential components of the electric and magnetic field can be related to one another, on the surface of the conductor, by using the Leontovich Impedance Boundary Condition (IBC) [41]:

$$\hat{n} \times \overline{E} = Z_s\,[\hat{n} \times \hat{n} \times \overline{H}] \tag{21}$$

where Z_s is the surface impedance defined as follows:

$$Z_s = \tfrac{1}{2}\,(1-j)\,\omega\mu\delta \tag{22}$$

Using (21) and (22) to eliminate H_t from (17), one obtains:

$$1/2\, E_\phi(\bar{\xi}) = E^i_\phi(\bar{\xi}) + \left[L^{(1)}_{\Omega-} - \frac{1}{Y_{\Omega-} Z_s} L^{(2)}_{\Omega-} \right] E_\phi(\bar{\eta}) \tag{22}$$

Similar formulations can be derived for problems involving multiple conductors [36,38,42], for cases where δ is not small relative to conductor dimensions [16-18,35-38] and for geometries that have translational symmetry [15,35,37-38,42] or are 3D [24-2].

A Boundary Integral formulation, and particularly the IBC formulation given by (22), has several significant advantages when applied to electromagnetic confinement problems:

(a) Being a boundary formulation, a solution grid is only required on the conductor surface(s). Therefore, it is very simple to modify the solution grid as the iteration for the free surface shape proceeds. This is not the case when methods such as Finite Differences, Finite Elements or Volume Integrals are used. This advantage it true for any boundary formulation, IBC or otherwise.

(b) The surface electric and magnetic fields are obtained with equivalent accuracy. With other methods (FD, FE, VI), one field is obtained as the derivative of the other, thus compromising the accuracy of the derived forces and pressures.

(c) When the fields and/or forces are required within the interior of the molten metal, these must be separately computed at each desired point when a Boundary Integral/Element solution is used. However, this additional computational cost is offset by the fact that the resulting values are of higher accuracy than would be obtained by using FD, FE or VI solutions on a *comparable* grid.

(d) When the IBC can be used, the computational cost is reduced significantly since:

  (i) The number of integral equations is reduced by one half as is the number of field unknowns.

  (ii) The IBC integral equation given by (22) only involves the parameters of the free space region and thus is not computationally intensive.

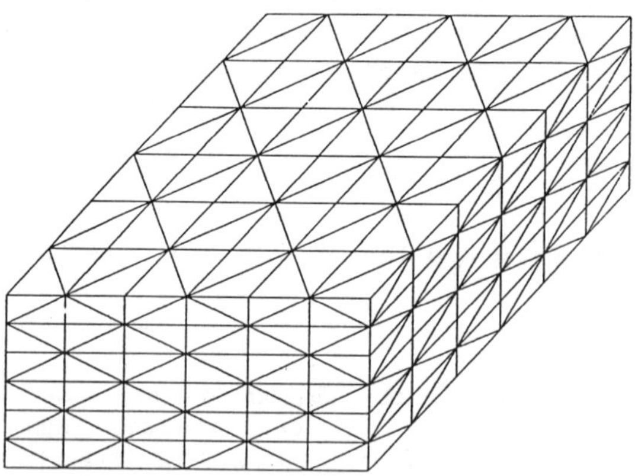

Figure 2. Typical Boundary Element solution grid for a 3D model of a conducting slab.

Boundary Integral Equations such as (22) can be cast in discrete form by:

(a) Subdividing the conductor surface in some appropriate fashion. Figure 2, for example, shows a typical distribution of triangular patches that was used with a 3D model of a conducting slab [27]. For 2D problems, a conducting surface is defined by it's contour and that contour is subdivided into line segments.

(b) As in Finite Elements, appropriate basis functions are used to approximate the field behaviour within each line segment (in 2D) or surface patch (in 3D). Typically, constant, linear and quadratic interpolating polynomials have been used as basis functions.

(c) With a Boundary Integral Equation formulations, it may be necessary to pay particular attention to the singular integrals that arise.

The resulting discrete form of the Boundary Integral Equations is termed a Boundary Element formulation of the specified electromagnetic field problem.

APPLICATION OF BOUNDARY ELEMENT METHODS TO E.M. CONFINEMENT PROBLEMS

This section describes several applications of the Boundary Element method to predict electromagnetic fields, forces and free surface boundaries.

The power induced in a conducting sphere of radius R by a uniform, time harmonic magnetic field is shown in Figure 3 as a function of the R/δ ratio for several Boundary Element formulations. As the closed form solution for this problem is known [36], it provides a convenient means of testing the various formulations. Particular points to note in this figure include the following:

(a) Good accuracy can be obtained by using relatively few elements.

(b) Accuracy is a function of the number and type of elements used.

(c) The accuracy can be quite sensitive to the formulation used. When using the *combined* formulation, the solution errors increase significantly with increasing values of R/δ, On the other hand, the accuracy of the *separate* and the IBC formulations remains relatively constant at large values of R/δ.

Further details on this application of the method can be found elsewhere [36,40].

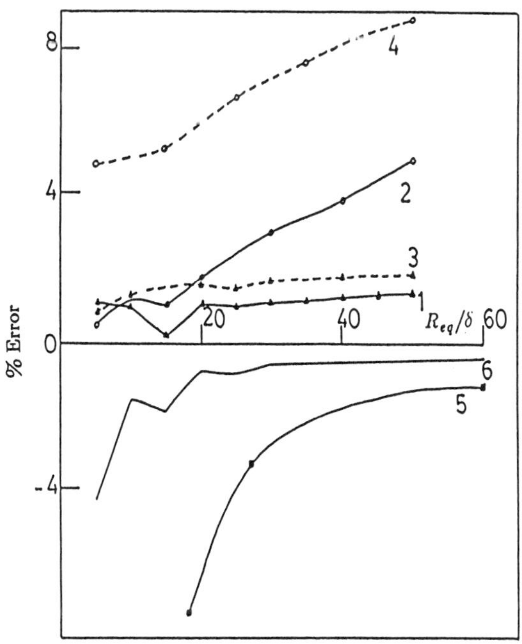

Figure 3. Error in power induced in a conducting sphere when using various Boundary Element models [40].

1. Separate BE formulation, quadratic elements, 65 nodes.

2. Combined BE formulation, quadratic elements, 65 nodes.

3. Separate BE formulation, linear elements, 150 nodes.

4. Combined BE formulation, linear elements, 150 nodes.

5. Separate BE formulation with singularity problems.

6. Impedance Boundary Condition, 65 nodes.

The pressure distribution along the axial surface of the molten metal column in the electromagnetic mould geometry of Figure 1 is shown in Figure 4. The pressure distribution predicted by the Boundary Element method is compared with the distribution obtained previously by a Volume Integral (i.e. Coupled Circuit) approach [43]. In performing this analysis, it was necessary to assume a particular free surface contour for the molten metal. It will be noted that the pressure distributions predicted by the two methods are in close agreement when the axial distance from the free surface is greater than 4 cm. Close to the free surface, however, there is a considerable deviation between the two predicted distributions. After considerable testing, it has been concluded that this deviation is due to the limitations of the Volume Integral approach in modelling the curved boundary of the free surface. In particular, it was necessary to use a step-wise approximation. It is believed that this example illustrates an important advantage of the Boundary Element approach in cases where the electromagnetic pressure distribution is a sensitive function of conductor shape.

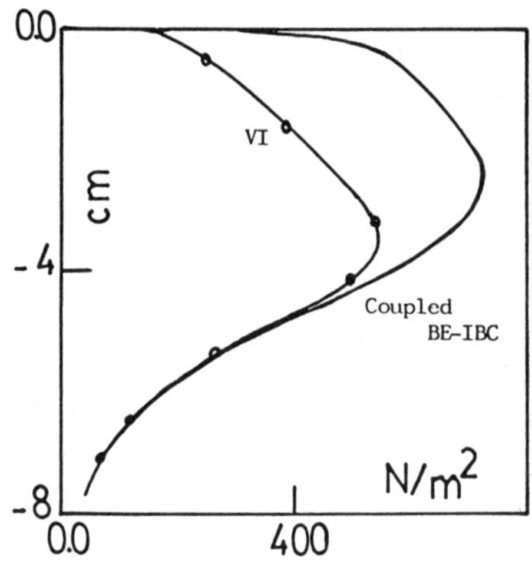

Figure 4. Electromagnetic pressure distribution, at 3 KHz, acting on the vertical side of an aluminum cylinder in an e.m. casting process. Comparison between Boundary Element solution and Volume Integral solution [43].

Figures 5 and 6 show the measured and predicted shapes of the meniscus in a laboratory scale crucible induction furnace as a function of melt level and frequency. For the test furnace, the crucible I.D. was 12.4 cm, the coil I.D. was 19.5 cm, the coil length was 18 cm and the melt was aluminium. The inside bottom surface of the crucible was aligned with the bottom of the coil. Melt levels of 9 cm (Figure 5) and 13.5 cm (Figure 6), relative to the bottom of the crucible, were considered. The supply frequencies were 1330 Hz and 10 KHz.

In this example, the shape of the meniscus was computed iteratively. Initially, the melt was assumed to have a sharp cornered, right cylindrical shape. The Boundary Element model was then used to determine the corresponding electromagnetic pressure distribution. By balancing the electromagnetic and static pressures, under the constraint that equal volumes of metal are displaced above and below the zero power melt level, an initial estimate of the meniscus shape can be obtained. This estimate is then refined during subsequent iterations. For most problems, in the order of 3 to 5 iterations have been required to obtain the final shape of the equilibrium free surface. The convergence criteria is that the change in the displacement volume of the melt, for two successive iterations, is not more than 2%.

The induction furnace example is important for several reasons. First, it clearly illustrates that the shape of an electromagnetically confined free surface can be efficiently and accurately predicted using the models that have been described in this paper. Second, it has been shown previously [44] that the shape of the free surface can have a profound influence on the electromagnetic *stirring* forces that occur in any confinement application. Unless the shape of the molten metal region is accurately known, the stirring forces and consequent flow predictions can be seriously in error.

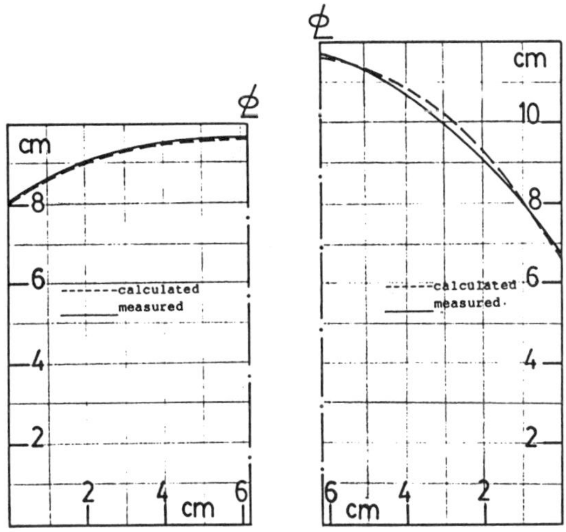

Figure 5. Meniscus contours, measured and predicted, for the free surface in a crucible induction furnace with an aluminum melt having a level of 9 cm. Supply frequencies are 1330 Hz (RHS) and 10 KHz (LHS).

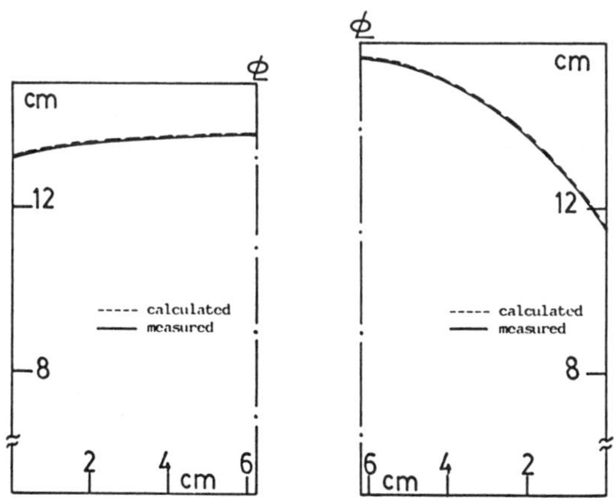

Figure 6. Meniscus contours, measured and predicted, for the free surface in a crucible induction furnace with an aluminum melt having a level of 13.5 cm. Supply frequencies are 1330 Hz (RHS) and 10 KHz (LHS).

As a final example, the shape of the free surface in a CREM aluminum casting mould [6,7] is shown in Figure 7. In this application, a line frequency induction coil is used to control the free surface meniscus, and thus the *wetting* point of the metal on the *nonconducting* mould. In this particular case, the cast billet is 320 mm in diameter while the induction coil has an I.D. of 183 mm and a cross section of 43 mm × 43 mm. Vives [7] presents results for the measured meniscus shape as a function of coil power. It would appear from the Vives results that in all cases, the top of the meniscus and the top of the coil were aligned. It should be noted that in using coil power, as opposed to coil current, as the electrical parameter that is related to meniscus height, it becomes particularly difficult to make good, quantitative comparisons between the measured data and the values calculated using any electromagnetic model. In this particular example, certain assumptions had to be made regarding the coil ampere turns and the coil losses.

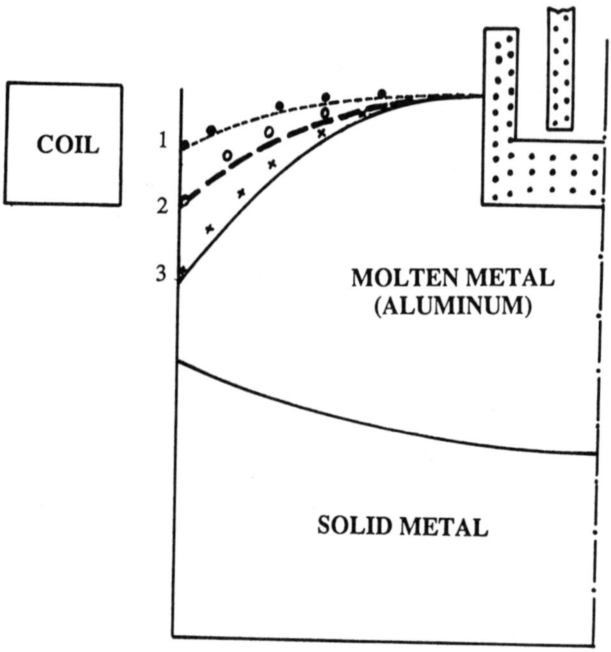

Figure 7. Measured [7] and predicted meniscus profiles for the CREM aluminum casting process.

1	2.2 kw	___ calculated	xxx measured
2	1.4 kw	-- -- calculated	ooo measured
3	0.79 kw	----- calculated	...... measured

As is illustrated in Figure 7, the Boundary Element electromagnetic model provides excellent agreement with the measured meniscus shapes for this application. Once again, it was found that in the order of 3 to 5 iterations were required in order to obtain the final meniscus shape. In predicting the meniscus height, particular attention had to be given to the effect of the submerged nozzle. The meniscus height and the induced power are summarized in TABLES I and II for the three input power levels used by Vives. It should be noted that in the case of 790 Watts input, Vives calculated (on the basis of measured current densities) that 200 Watts would be induced in the molten metal. The predicted value of 212 Watts is in excellent agreement with this measured result.

TABLE I

MENISCUS HEIGHT		
	measured	calculated
2.2 kw	z = 6.4 cm	z = 6.6 cm
1.4 kw	z = 4.3 cm	z = 4.3 cm
0.79 kw	z = 2.0 cm	z = 2.0 cm

TABLE II

INDUCED POWER		
	measured	calculated
2.2 kw	-	405 watts
1.4 kw	-	318 watts
0.79 kw	200 Watts	212 watts

Much of the analysis presented by Vives for the CREM casting process was based on measured distributions of magnetic flux density and electric current density in the molten metal. As noted by Vives, "The direct experimentation on aluminum alloys is difficult, due to the limited working time of the sensors". The results presented in this section indicate, however, that the shape of the conducting region can now be reliably predicted with excellent accuracy. Knowing the problem geometry, the magnetic fields and forces can be computed. Consequently, the present reliance on measured field distributions can be significantly reduced.

The previous examples have all featured 2-dimensional geometries. Recently, however, Boundary Element models have been used to solve 3-dimensional electromagnetic field test problems [24-28,45-46]. One particular case was that of a conducting slab in a uniform, time harmonic magnetic field (Figure 8). This is quite similar to a test geometry for which measured data is available [47]; in the latter case, however, magnetic pole pieces were used to establish the uniformity of the source field. With respect to electromagnetic confinement, this particular model can be used to examine the field and force distributions in the corner regions of electromagnetic and CREM moulds. The predicted distribution of H_y along the contour $y = -3$ cm, $z = +7$ cm is compared with the measured data in Figure 9 [27,46]. The two distributions have been normalized in order to eliminate a 20% deviation in the surface values (at $x = +/-7$ cm due to the magnetic pole pieces used in the experimental set-up.

The availability of reliable (albeit not inexpensive) 3D electromagnetic models is of significant importance in the area of electromagnetic confinement as many of the practical problems are, in fact, 3-dimensional. It must be emphasized, however, that the area of 3D field modelling is one of very active research and it should not be assumed that arbitrary 3D geometries can be easily modelled.

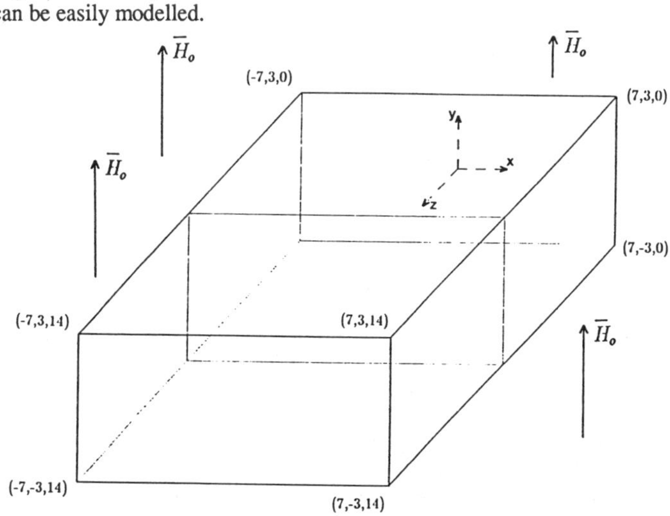

Figure 8. Test example geometry: conducting slab in a uniform, time harmonic magnetic field [27,47].

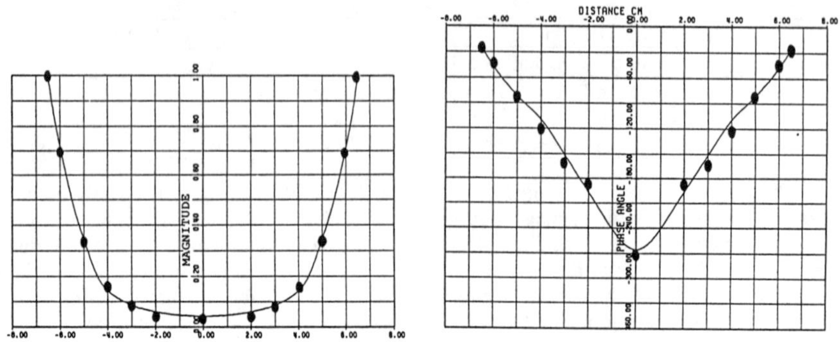

Figure 9. Distribution of predicted (solid) and measured values of H_y as a function of x along the contour $y = -3$ cm , $z = +7$ cm. Magnitude (LHS), Phase (RHS).

CONCLUSIONS

Issues relating to the mathematical modelling of electromagnetic confinement have been considered in this paper. It has been argued that from an electromagnetic point of view, the most significant problem in modelling such applications is to determine the shape of the free surface. At the most fundamental of levels, the electromagnetic fields must support molten metal against a static head in a stable fashion. However, unless the shape of the free surface is known, the electromagnetic fields and forces cannot be reliable computed. It has been further argued that a natural modelling approach for this type of problem is to use Boundary Integral/Element methods for the electromagnetic fields. This approach to modelling electromagnetic field problems has been outlined and has been applied to a range of problems, each of which relates to aspects of the electromagnetic confinement application. It has been shown that the method allows the free surface meniscus in both the crucible induction furnace and the CREM casting process to be predicted with excellent accuracy. The practical limitations on the magnitude of the fields and forces that can be obtained in electromagnetic confinement have also been reviewed. It has been shown, for example, that the current carrying capacity of conventional conductors limits the static head that can be supported by an electromagnetic field to something in the order of 10-25 cm of steel. An additional major problem is that of obtaining a uniform force field, particularly near the end turn region of the confinement coil.

REFERENCES

[1] Z.N Getselev, "Method of Forming Ingot in a Process of Continuous and Semi-Continuous Casting of Metals", U.S. Patent 4,014,379, March 29, 1977.

[2] G.K. Gaule, J.C. Yarwood, D.E. Tyler, "Multi-turn Coils of Controlled Pitch for Electromagnetic Casting", U.S. Patent 4,570,699, February 18, 1986.

[3] E.A. Olsson, "Method and Apparatus for Excluding Molten Metal from Excaping from or Penetrating into Openings or Cavities", U.S. Patent 4,020,890, May 3, 1977.

[4] B.G. Lewis, "Electromagnetic Shaping of Thin Semiconductor Ribbon Strip Cast onto a Chilled Block", U.S. Patent 4,562,878, January 7, 1986.

[5] H.R. Lowry *et al*, "Continuous Metal Casting Apparatus", U.S. Patent 4,662,431, May 5, 1987.

[6] Ch. Vives, B. Forest, J.P. Riquet, U.K. Patent Application GB 2 164 280, March 19, 1986.

[7] Ch. Vives, B Forest, "CREM: A New Casting Process. I: Fundamental Aspect", *Proc. AIME Light Metals Conf.*, pp. 769-778, Denver, CO, Feb. 24-26, 1987.

[8] S.J. Polak, A.J.H. Wachters, J.S. van Welij, "A New 3-D Eddy Current Model", *IEEE Trans. Magnetics*, **MAG-19**, pp. 2447-2449, 1983.

[9] K. Reichert, "A Numerical Method to Calculate Induction Heating Installations", *Elektrowaerme Int.*, **26**, pp. 113-123, 1968.

[10] W. Erdmann and A. Muhlbauer, "Praxisnahe Berechnung Des Elektrischen Und Magnetohydrohynamischen Verhaltens Von Induktionstiegelofen", *Elektrowarme International*, **43**, No. B6, pp. B270-B277, 1985.

[11] M.V.K. Chari, P.P. Silvester, eds., *Finite Elements in Electric and Magnetic Field Problems*, New York: Wiley, 1980.

[12] A. Konrad, "Eddy Currents and Modeling", *IEEE Trans. Magnetics*, **MAG-21**, pp. 1805-1910, 1985.

[13] P.P. Silvester, "The Accurate Calculation of Skin Effect in Conductors of Complicated Shape", *IEEE Trans. Power App. & Syst.*, **PAS-87**, pp. 735-742, 1968.

[14] R.F. Dudley, P.E. Burke, "The Prediction of Current Distribution in Induction Heating Installations", *IEEE Trans. Ind. Appl.*, **IA-8**, pp. 565-571, 1972.

[15] C. Mueller, *Foundations of the Mathematical Theory of Electromagnetic Waves*, Berlin: Springer-Verlag, 1969.

[16] T.H. Fawzi, *Two-Dimensional and Quasi-Two-Dimensional Induction Problems*, Ph.D. Diss., University of Toronto, 1973.

[17] T.H. Fawzi, K.F. Ali, P.E.Burke, "Boundary Integral Equations Analysis of Induction Devices with Rotational Symmetry", *IEEE Trans. Magnetics*, **MAG-19**, pp. 36-44, 1983.

[18] K. Ishibashi, "Eddy Current Analysis of Induction Heating by the Boundary Element Method", *Elect. Engng. in Japan*, **104B**, pp. 109-117, 1984.

[19] M.H. Lean, "Application of Boundary Integral Equation Methods to Electromagnetics", *IEEE Trans. Magnetics*, **MAG-21**, pp. 1823-1828, 1985.

[20] C.R.I. Emson, J. Simkin, "An Optimal Method for 3-D Eddy Currents", *IEEE Trans. Magnetics*, **MAG-19**, pp. 2450-2452, 1983.

[21] J. Simkin, "Eddy Current Modelling in Three Dimensions", *IEEE Trans. Magnetics*, **MAG-22**, pp. 609-613, 1986.

[22] A. Bossavit, J.C. Verite, "A Mixed FEM-BIEM Method to Solve 3-D Eddy Current Problems", *IEEE Trans. Magnetics*, **MAG-18**, pp. 431-435, 1982.

[23] C.W. Trowbridge, "Electromagnetic Computing: The Way Ahead?", *IEEE Trans. Magnetics*, **MAG-24**, pp. 13-18, 1988.

[24] O.V. Tozoni, I.D. Mayergoyz, *Analysis of 3-Dimensional Electromagnetic Fields*, Kiev USSR: Technika, 1974 (in Russian).

[25] I.D. Mayergoyz, "Boundary Integral Equations of Minimum Order for the Calculation of Three-Dimensional Eddy Current Problems", *IEEE Trans. Magnetics*, **MAG-18**, pp. 536-539, 1982.

[26] W.R. Hodgkins, J.F. Waddington, "The Solution of Three-Dimensional Induction Heating Problems Using an Integral Equation Method", *IEEE Trans. Magnetics*, **MAG-18**, pp. 476-480, 1982.

[27] S. Kalaichelvan, *A Boundary Integral Equation Method for 3-Dimensional Eddy Current Problems*, Ph.D. Diss, University of Toronto, 1987.

[28] H. Tsuboi, T. Misaki, "Three Dimensional Analysis of Eddy Current Distributions by the Boundary Element Method", *IEEE Trans. Magnetics*, **MAG-23**, pp. 3044-3046, 1987.

[29] S. Sagardia, *Electromagnetic Levitation Melting of Large Conductive Loads*, Ph. D. Diss, University of Toronto, 1974.

[30] M. Garnier, R. Moreau, "Stability of Molten Metal Free Surface in the Presence of an Alternating Magnetic Field", *Proc. IUTAM Symp. on Metallurgical Appl. of MHD*, pp. 211-216, Cambridge, UK, September 6-10, 1982.

[31] J.A. Shercliff, "Magnetic Shaping of Molten Metal Columns", *Proc. Royal Soc. London A*, **375**, pp. 455-473, 1981.

[32] A. Gagnoud, I. Leclercq, *IEEE Trans. Magnetics*, **MAG-24**, pp. 256-258, 1988.

[33] E.D. Tarapore, J.W. Evans, "Fluid Velocities in Induction Melting Furnaces. Part I: Theory and Laboratory Experiments", *Met. Trans.*, **7B**, pp. 343-351, 1976.

[34] N. El-Kaddah, J. Szekely, "The Turbulent Recirculating Flow Field in Induction Melting Furnaces: A Comparison of Theoretical Predictions with Measurements", *J. Fluid Mech.*, **133**, pp. 37-46, 1983.

[35] M. Taher Ahmed, *Application of the Boundary Element Method and the Impedance Boundary Condition to T-W and 2-D Eddy Current Problems*, Ph.D. Diss., University of Toronto, 1986.

[36] M. Ramadan Ahmed, *A Boundary Element Method for Eddy Current Problems Having Rotational Symmetry*, Ph.D. Diss., University of Toronto, 1987

[37] Shao Keran, Zhou Keding, "Boundary Element Solutions to Transient Eddy Current Problems", *Engng. Anal.*, **1**, pp. 182-187, 1984.

[38] M.H. Lean, "Dual Simple-Layer Source Formulation for Two-Dimensional Eddy Current and Skin Effect Problems", *J. Appl. Phys.*, **57(1)**, pp. 3844-3846, 1985.

[39] I. Stakgold, *Boundary Value Problems of Mathematical Physics: Vol. II*, New York: MacMillan, 1968.

[40] M. Ramadan Ahmed, J.D. Lavers, "A Comparison of Boundary Integral Formulations for 2D Axisymmetric Eddy Current Problems", *IEEE Trans. Magnetics*, **MAG-24**, pp. 138-141, 1988.

[41] T.H. Fawzi, M. Taher Ahmed, P.E. Burke, "On the Use of the Impedance Boundary Conditions in Eddy Current Problems", *IEEE Trans. Magnetics*, **MAG-21**, pp. 1835-1840, 1985.

[42] Zhou Pei-Bai, J.D. Lavers, "High Frequency Losses in Multi-Turn Coils Using a Boundary Element Model", *IEEE Trans. Magnetics*, **MAG-22**, pp. 1060-1062, 1986.

[43] J.D. Lavers, "Force Distributions Produced by an Electromagnetic Mould", *IEEE Trans. Ind. Appl.*, **IA-22**, pp. 586-593, 1986.

[44] J.D. Lavers, P.P. Biringer, "The Influence of System Geometry on the Electromagnetic Stirring Forces in Crucible Induction Melting Furnaces", *Proc. IUTAM Symp. on Metallurgical Appl. of MHD*, pp. 62-78, Cambridge, UK, September 6-10, 1982.

[45] S. Kalaichelvan, J.D. Lavers, "Singularity Evaluation in Boundary Integral Equations of Minimum Order for 3-D Eddy Currents", *IEEE Trans. Magnetics*, **MAG-23**, pp. 3053-3055, 1987.

[46] S. Kalaichelvan, J.D. Lavers, "On the Implementation of a Boundary Element Method for 3D Multiply Connected Eddy Current Problems" *IEEE Trans. Magnetics*, **MAG-24**, pp. 569-572, 1988.

[47] M.J. Balchin, J.A.M. Davidson, "Numerical Method for Calculating Magnetic Flux and Eddy Currents in Three Dimensions", *Proc. IEE*, **127**, Pt. A, pp. 46-53, 1980.

THREE-DIMENSIONAL CALCULATIONS OF FIELDS IN ELECTROMAGNETIC CASTERS

B. Q. Li and J. W. Evans

Department of Materials Science and Mineral Engineering
University of California
Berkeley, California 94720

Abstract

A method has been developed for the calculation of three dimensional magnetic fields in electromagnetic casters. This method is a surface integral technique that has similarities to the well known boundary element method employed for the solution of partial differential equations. It relies upon the introduction of imaginary magnetic surface charges; these imaginary charges simplify the otherwise complicated calculation of the 3-D electromagnetic fields. Computed results for a physical model of a CREM caster are presented and are in accord with expectations. This surface integral approach appears to offer a powerful means of computing the complicated 3-D fields encountered in real electromagnetic casters and similar devices.

Casting of Near Net Shape Products
Edited by Y. Sahai, J.E. Battles,
R.S. Carbonara and C.E. Mobley
The Metallurgical Society, 1988

Introduction

Electromagnetic casting is now widely practiced throughout the world, especially for aluminum. Various shapes of ingots such as cylinders, pipes, billets and slabs can be cast using this technique. A great deal of research has been carried out on the casting of cylindrical shapes (1-5). However, relatively little attention has been paid to the casting of slabs and billets. Recently Vives (6) presented experimental results from a study of the three dimensional fields and melt flow in a physical model of a caster producing metal of rectangular section. Until now there appear to have been no mathematical models for such three dimensional casters.

A satisfactory model for this type of caster should be capable of addressing two practical aspects of casting: the electromagnetic support of the molten metal meniscus and the electromagnetically driven flow. These important aspects of the caster are dependent on the accurate prediction of the electromagnetic fields involved. The fields in a caster producing a slab or billet of rectangular section are more complicated than those in a cylindrical caster in that the former are three dimensional in both the vectorial and geometric sense. That is, all three components of the magnetic and electric field vectors are generally non zero and they are dependent on all three spacial co-ordinates.

A feature of electromagnetic casters is that the skin depth is usually small compared to the dimensions of the caster. Therefore the currents induced within the metal can be considered as surface currents, $\vec{K}$, defined as

$$\vec{K} = \int_0^\infty \vec{J} d\xi \tag{1}$$

where $\vec{J}$ is the curent density.

This concept of surface currents (7) leads to considerable simplification of the field calculations as will be seen below. The alternative is the calculation of the three dimensional current densities throughout the bulk of the metal such as has been carried out by previous investigators (8,9). While these alternative procedures are theoretically correct they are computationally cumbersome compared to the method described in the present paper.

We now describe the surface integral formulation based on the surface current approximation (7), following which some sample results are presented for the geometry described by Vives (6).

Mathematical Formulation

Consider the system shown schematically in Fig. 1. It consists of the volume Ω, with surface $\partial\Omega$, the volume being occupied by a good electrical conductor (metal in the present instance). Ω is surrounded by a non-conductor Ω_0 (air in the present case) containing three dimensional, time dependent, electric and magnetic fields, $\vec{E}$ and $\vec{H}$ respectively. Within Ω_0 it will be convenient for us to regard each of these fields as the sum of an "exciting field" ($\vec{E}_0$ or $\vec{H}_0$) produced by some external agency, such as an inductor located in Ω_0 and a "reflected field" ($\vec{E}_r$ or $\vec{H}_r$), viz:

$$\vec{E} = \vec{E}_0 + \vec{E}_r \tag{2}$$

and

$$\vec{H} = \vec{H}_0 + \vec{H}_r \tag{3}$$

The exciting and reflected fields are shown schematically in Fig. 1. To elaborate on these definitions, $\vec{E}_0$ and $\vec{H}_0$ are the fields that would exist if Ω were non-conducting, while $\vec{E}_r$ and $\vec{H}_r$ are the differences between $\vec{E}_0$ (or $\vec{H}_0$) and the fields, $\vec{E}$ (or $\vec{H}$) when Ω is conducting.

The magnetohydrodynamic form of Maxwell's equations relevant to the present case are:

$$\nabla \times \vec{H} = \vec{J} \tag{4}$$

$$\nabla \times \vec{E} = -\frac{\partial \vec{B}}{\partial t} \tag{5}$$

$$\nabla \cdot \vec{B} = 0 \tag{6}$$

$$\nabla \cdot \vec{J} = 0 \tag{7}$$

where B is the magnetic induction $= \mu \vec{H}$ (8)

μ is the magnetic permeability

$\vec{J}$ is the current density

and t is time.

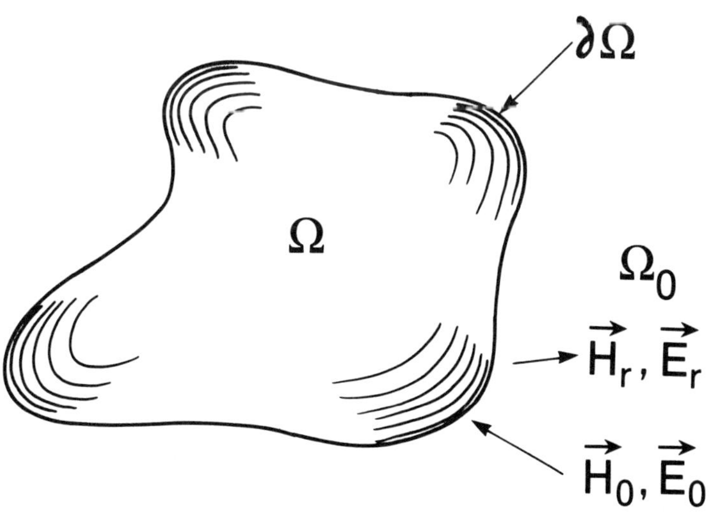

Figure 1 - Three-dimensional electromagnetic field in the presence of a conductor.

The current density is related to the electric field by

$$\vec{J} = \sigma \vec{E} \tag{9}$$

where σ is the electrical conductivity and the usual assumption has been made that the $\vec{V} \times \vec{B}$ correction to Ohm's law is negligible.

Let us consider the region Ω_0 and require that the exciting and reflected fields separately satisfy Maxwell's equations, hence

$$\nabla \times \vec{E}_r = -jw\mu\vec{H}_r \tag{10}$$

and

$$\nabla \times \vec{H}_r = 0 \tag{11}$$

(because there are no currents in Ω_0), similarly for $\vec{E}_0$ and $\vec{H}_0$. From equation (11) we see that the reflected magnetic field is irrotational and can therefore be expressed as a gradient of a scalar potential ϕ.

$$\vec{H}_r = -\nabla\phi \tag{12}$$

Any scalar potential can be thought of as produced by a number (infinite if necessary) of "charges" distributed in the region of interest. In the present case, because the reflected field originates from surface currents in the metal then these charges (which are best regarded as imaginary magnetic monopoles) are distributed on the surface of the metal, i.e. on $\partial\Omega$. We can then use classical results (e.g. from electrostatics) to calculate ϕ at any point q in Ω:

$$\phi(q) = \frac{1}{4\pi} \int_{\partial\Omega} \frac{\sigma_m(p)}{r_{pq}} d\Gamma_p \tag{13}$$

where $\sigma_m(p)$ is the imaginary magnetic charge density (charge per unit area) at point p in the surface and $\vec{r}_{pq}$ is the vector (of length r_{pq}) connecting points p and q.

In this surface current approximation we regard the metal as an excellent conductor which implies that there is negligible penetration of the magnetic field into the metal. Consequently the normal component of the magnetic induction at the surface is zero, i.e.,

$$\vec{n} \cdot (\vec{B}_0 + \vec{B}_r) = 0 \tag{14}$$

where $\vec{n}$ is the surface normal.

Taking the gradient of equation (13) at point q the magnetic induction normal to the surface can now be calculated using equations (8) and (12)

$$\vec{n}_q \cdot \vec{B}_r = \mu \vec{n}_q \cdot H_r = -\frac{\mu}{4\pi} \vec{n} \cdot \nabla \int_{\partial\Omega} \frac{\sigma_m(p)}{r_{pq}} d\Gamma_p$$

$$= -\frac{\mu}{2} \sigma_m(q) - \frac{\mu}{4\pi} \int_{\partial\Omega} \sigma_m(p) \frac{\vec{n}_q \cdot \vec{r}_{pq}}{r^3_{pq}} d\Gamma_p = -\vec{n}_q \cdot \vec{B}_0 \qquad (15)$$

where equation (14) has been exploited and the surface integral recast in a more convenient form (7,10). Finally a rearrangement and a second use of equation (8) yield the key result

$$2\pi \sigma_m(q) + \int_{\partial\Omega} \sigma_m(p) \frac{\vec{n}_q \cdot \vec{r}_{pq}}{r^3_{pq}} d\Gamma_p = 4 \vec{n}_q \cdot \vec{H}_0(q) \qquad (16)$$

If $\partial\Omega$ is broken up into small surface elements ("boundary elements") then the integral in equation (16) becomes a sum over the surface. Each term in this sum is an unknown magnetic surface charge density multiplied by a known geometric factor. The term on the right is known from classical electromagnetic theory given the inductor current and geometry. Applying equation (16) to each boundary element we therefore obtain a set of simultaneous linear equations that can be solved by standard numerical techniques for the unknown surface charges. The procedure is the well known indirect boundary element technique for solution of partial differential equations. Once the surface charges are known then equations (12) and (13) can be employed to calculate the magnetic field $\vec{H}_r$ and the induced surface current can be calculated from the boundary condition on the tangential component of the magnetic field

$$\vec{n} \times (\vec{H}_0 + \vec{H}_r) = \vec{K} \qquad (17)$$

Results and Discusion

Fig. 2 shows the surface mesh used to calculate the sample results to follow. The caster shown produces metal of 0.67 m. by 0.30 m. The symmetry of the design is considered in the calculations so only one quadrant of the caster need be considered. This caster is the one described by Vives (6) and the shape of the meniscus is assumed based on the geometry of the caster. There are a total of 1296 elements and 1364 nodes in the mesh shown in Fig. 2. Alternative meshes, some possibly with greater computational efficiency, are possible but this is not of concern here. Computations were completed in approximately 10 minutes on an IBM 3090 with vector processing capabilities. Note that the origin of the co-ordinate system is hidden from the observer and the lines marked "axis" on the figure are actually displaced from the true axes.

Fig. 3 shows the imaginary magnetic charge density, σ_m on the top surface of the metal. The viewpoint is different from that of Fig. 2; the observer is now behind the plane of the paper of Fig. 2 and in Fig.3 the lines marked "axis" are the true axes. As expected, the magnetic charge density is highest near the periphery of the metal where it is closest to the inductor.

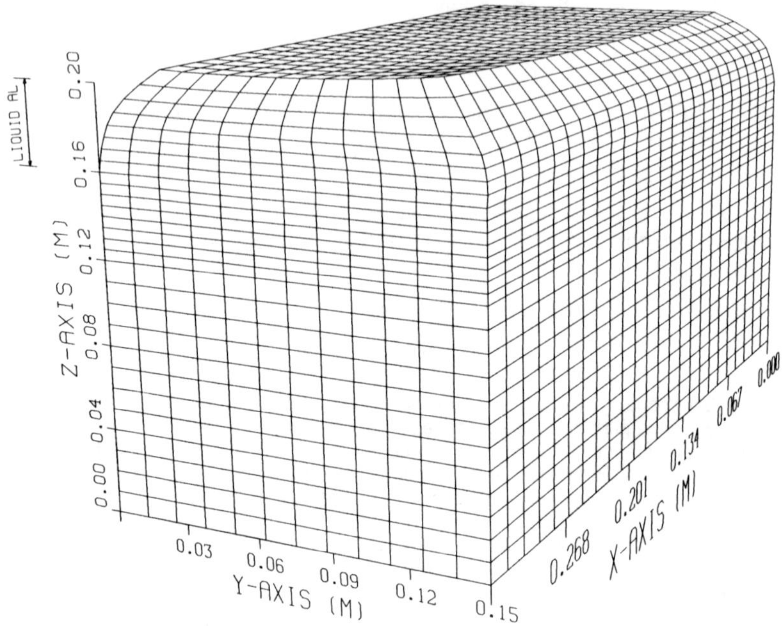

Figure 2 - Three-dimensional surface mesh of a quadrant of an electromagnetic caster used in numerical calculations. The ingot has a cross section of 0.67 m x 0.30 m and the liquid meniscus shape is assumed based on the geometric configuration used by Vives (6). Not shown in the figure is the inductor of a cross section of 0.04 m x 0.04 m. The inductor is located at 0.16 m in the Z-direction. The gap between the inductor and ingot is: 0.045 m in the X-direction and 0.045 m in the Y-direction.

The surface currents are depicted in Fig. 4, with the usual convention that the length of the arrow represents the magnitude of the current at the tail of the arrow. That magnitude is a current per unit length, i.e., is independent of the size of the surface element. The currents shown are instantaneous ones for the time when the current flow in the inductor is in the direction shown. The currents would reverse as the inductor current reverses. The three dimensional nature of these currents should be noted; even though the inductor current has zero components in the vertical direction, the currents induced in the metal have components in all directions. This is in contrast to the cylindrical caster where the current flows in only one direction (the tangential direction). In the near-corner regions, where the two orthogonal lengths of the inductor are joined, the magnetic field has components in all three dimensions, consequently its curl, i.e., the current from equation (4), can be expected to have non-zero components in all directions. Away from the corner the induced currents are nearly antiparallel to the inductor current, i.e., the current flows in only one direction, at least in the metal close to the inductor. Note, however, (Fig. 4b) that the current flowing over most of the top surface is opposite in direction to that on the sides.

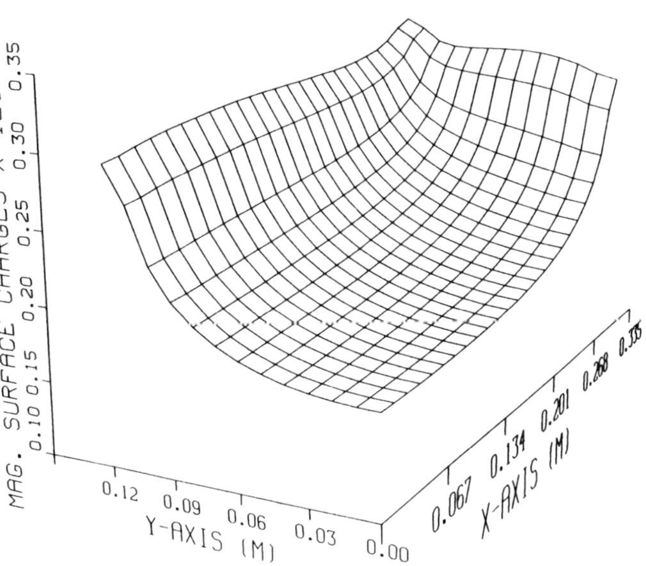

Figure 3 - Imaginary surface magnetic charge distributions on the top surface of the caster. Input current is 5000 A.

Surface currents for a differently shaped meniscus are depicted in Fig. 5. Note that, as in Fig. 4, the currents are three dimensional and very high at the corner of the meniscus. Currents diminish lower in the figure because the inductor is located well above the mid height of the cube of Fig. 5. The large currents along the upper edges of the meniscus should be contrasted with the much lower currents in the curved region of the meniscus in Fig. 4. In fact, as will be seen by comparing Figs. 5(a) and 5(b) the calculated surface current is discontinuous at the edge, an indication that a sharp edge is physically unrealistic. As in Fig. 4 the currents are parallel (or antiparallel) to the inductor currents, except near the join of the inductors. Hence two dimensional models such as that of Sakane et al. (5) might prove satisfactory away from the corners.

The instantaneous magnetic fields ($\vec{H}_0 + \vec{H}_r$) at the metal surface are shown in Fig. 6 for the geometry of Figs. 2-4. It will be recalled that these fields are tangential to the surface under the surface current assumption. As expected the magnetic field is strongest in the near corner region and is otherwise almost perpendicular to the inductor current. Inspection of Figs. 4 and 6 permits an assessment of the electromagnetic forces acting on the metal surface. These forces are orthogonal to both the current vector and the magnetic field vector (i.e. are perpendicular to the melt surface) and therefore difficult to depict in the manner of previous figures. Nevertheless it can be seen that these forces are high in the near corner regions and otherwise orthogonal to the inductor. Recognizing that this force must balance the hydrostatic pressure of the molten metal (with small corrections for the effects of surface tension and melt flow) it is clear that the meniscus shape used in the present calculations is not a stable one. The large inward forces in the near corner regions would deform the meniscus inwards producing a more rounded corner than assumed in the present calculations.

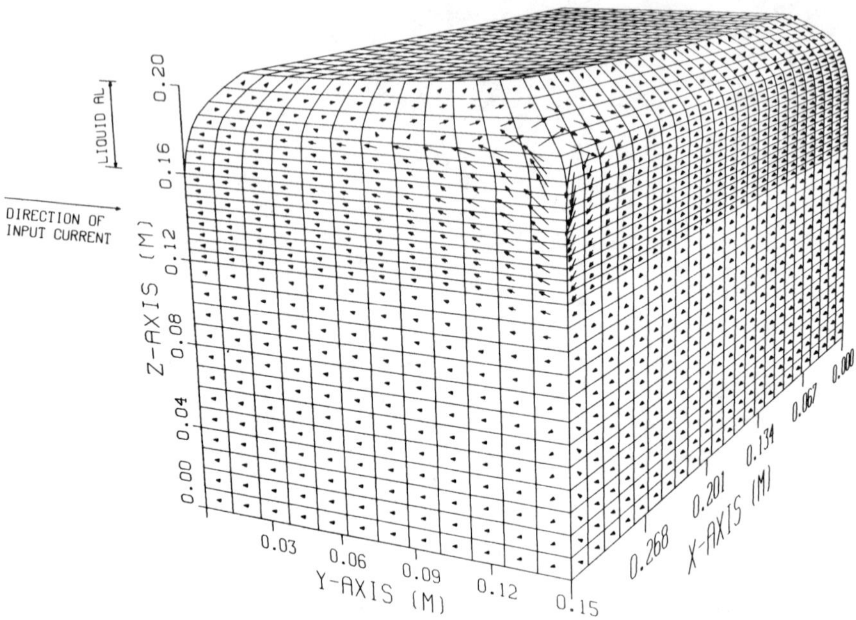

(a)

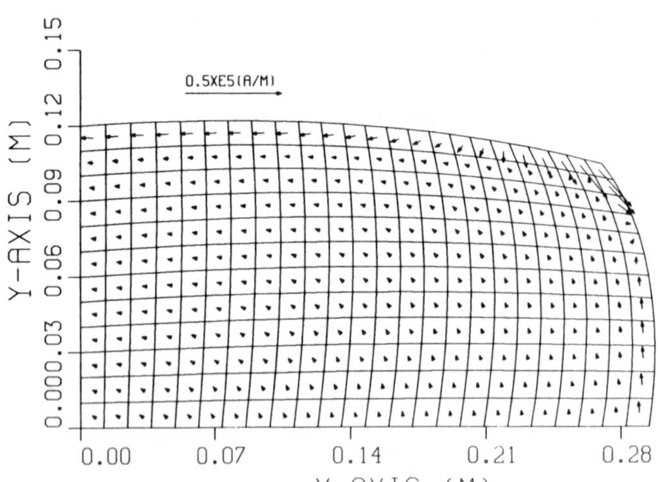

(b)

Figure 4 - Induced currents on the surface of the caster: (a) induced currents on the side faces and (b) induced currents on the top surface. Input current is 5000 A.

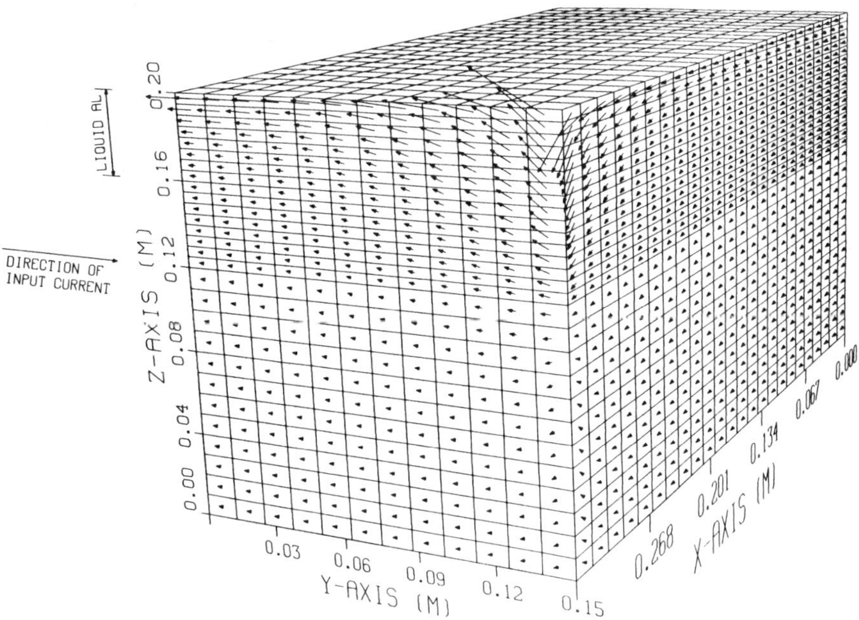

(a)

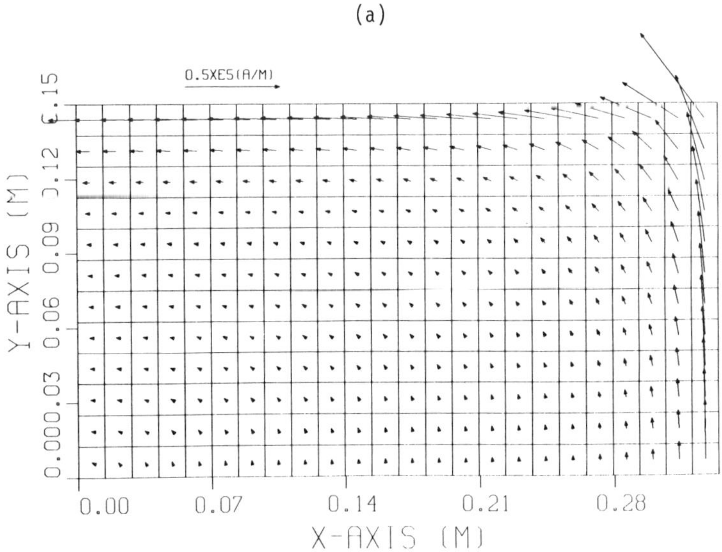

(b)

Figure 5 - Induced currents on the surface of a caster with uniform cross section of 0.670 m x 0.30 m : (a) induced currents on the side faces and (b) induced currents on the top surface. Input current is 5000 A.

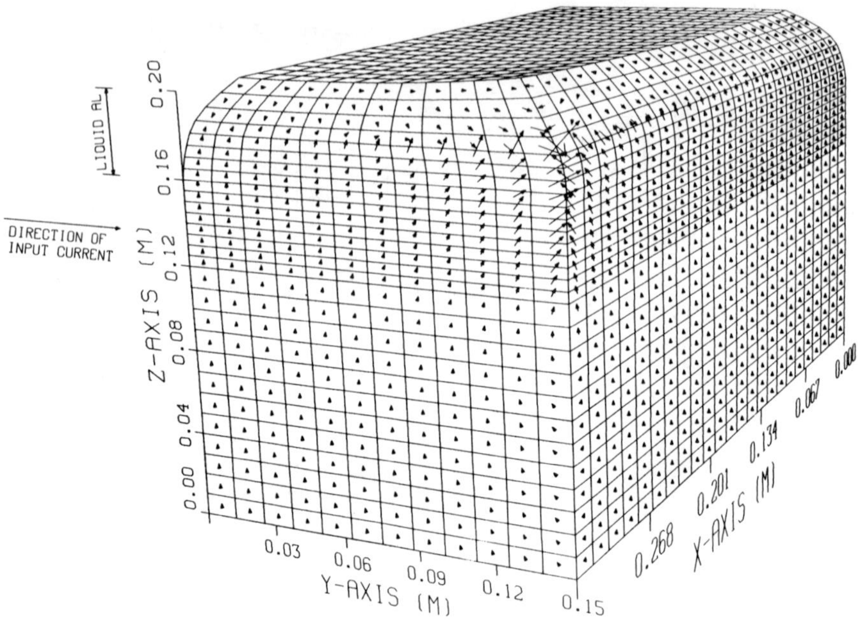

(a)

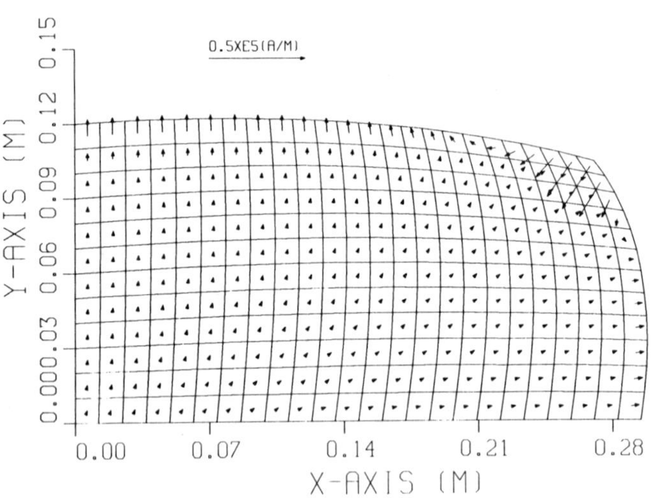

(b)

Figure 6 - Induced magnetic intensity on the surface of the caster: (a) induced magnetic intensity on the side faces and (b) induced magnetic intensity on the top surface. Input current is 5000 A.

Concluding Remarks

In this paper the surface current concept has been employed to develop a formalism for the solution of the three dimensional electromagnetic field problem in electromagnetic casters. The resulting algorithm entails the evaluation of a surface integral and is reminiscent of the boundary element method. Examples of fields calculated for a geometry representative of a real caster are presented. The computed fields are consistent with expectations and indicate pronounced three dimensional effects in the regions of the caster near a corner. Away from the corners a two dimensional assumption appears reasonable. Additional calculations to compute a stable meniscus shape and the electromagnetically driven flow are in progress.

Acknowledgement

Support of this research by the National Science Foundation under grant No. MSM - 8520783 is gratefully acknowledged.

References

1. Z. N. Getselev and G. I. Martynov, "Circulation of the Metal in the Liquid Phase of an Ingot Formed by an Electromagnetic Field", Magnitnaya Gidrodinamika, 1979, no. 4, 127-131.

2. C. Vives and R. Ricous, "Experimental Study of Continuous Electromagnetic Casting of Aluminum Alloys", Metall. Trans., 16B (1975) 377-405.

3. J. L. Meyer, N. El-Kaddah, J. Szekeley, C. Vives and R. Ricou, "A Comprehensive Study of the Induced Current, the Electromagnetic Field, and the Velocity Field in a Complex Electromagnetically Driven Flow System", Metall Trans., 18B (1987) 529-538.

4. J. L. Meyer, J. Szekeley, N. El-Kaddah, C. Vives and R. Ricou, "Electromagnetic and Fluid Flow Phenomena in a Mercury Model System of the Electromagnetic Casting of Aluminum", Metall. Trans., 18B (1987) 539-548.

5. J. Sakane, B. Q. Li and J. W. Evans, "Mathematical Modeling of Meniscus Profile and Melt Flow in Electromagnetic Casters", Metall. Trans., 20B (1988) to appear.

6. C. Vives, "Three-Dimensional Aspects of the Electromagnetic and Fluid Flow Phenomena in the CREM Process", Light Metal, (1988).

7. John D. Jackson, Classical Electrodynamics, (New York, NY, John Wiley & Sons, Inc., 1975), 334-383.

8. K. R. Davey and J. W. Barnes, "Three-Dimensional Eddy Current Problems Using the T-Ω Method and Fredholm Integral Equations", IEEE Trans. Mag., 19(2) (1983), 120-125.

9. C. W. Trowbridge, "Some Three Dimensional Eddy Current Algorithms Under Investigation in the UK", Computational Electromagnetics, ed. Z. J. Cendes (New York, NY, Elsevier Science Publishers B.V., 1986), 173-180.

10. C. A. Brebbia, J. C. F. Telles and L. C. Wrobel, "Boundary Element Techniques", (Berlin, Springer-Verlag, 1984) 47-107.

MATHEMATICAL MODEL FOR THE SHAPING OF

MOLTEN METAL BY AN ELECTROMAGNETIC FIELD

Nagy El-Kaddah and F. A. Acosta-Gonzalez

Department of Metallurgical and Materials Engineering,
The University of Alabama, Tuscaloosa, Al 35487.

Abstract

A mathematical formulation is presented for calculating the electromagnetic field and the shape of the free surface of molten metal in an alternating magnetic field. A fast and efficient method is used to calculate the electromagnetic field quantities, and hence the magnetic pressure distribution at the free surface. In essence it involves the solution of the integral form of the Maxwell equations for the electromagnetic field quantities at the surface of the molten metal, assuming an exponential decay of the electromagnetic field in the direction normal to the free surface. The validity of this method was critically examined by comparing the computed results for problems with simple geometry, a sphere and a cylinder, in a uniform magnetic field with the exact analytical solutions. It was found that this method gives accurate results provided that the magnetic interaction parameter, $R\omega$, is greater than 10. The free boundary problem is represented by writing a variational statement of electromagnetic, gravitational, and surface tension energies and the resultant equations were solved simultaneously with the electromagnetic field equations. The model is demonstrated for two cases, levitation melting and the confinement and shaping of a falling liquid column confined by external parallel conductors.

Casting of Near Net Shape Products
Edited by Y. Sahai, J.E. Battles,
R.S. Carbonara and C.E. Mobley
The Metallurgical Society, 1988

Introduction

Since the development of electromagnetic casting of aluminum during the early 1970's /1/ in which aluminum slabs are cast without contact with a mold, there has been a growing interest in the development of similar technology for near net shape casting of metals. Recently, General Electric has developed an electromagnetic levitation continuous casting (GELEC) process for production of copper rods /2/. At present, work is also in progress at Argonne National Laboratory on development of an electromagnetic strip casting process for steel in which an electromagnetic field is used for shaping and levitation of a molten metal into a strip /3/.

The common feature of these processes is that molten metal is placed in an alternating magnetic field, generated by the passage of an electric current through an induction coil of suitable geometry. In such a system, the interaction of the current induced in the melt with the associated magnetic field will give rise to an electromagnetic force field, also called Lorentz force, which may be resolved into stirring forces and an irrotational component (magnetic pressure) whose effect may be visualized as a pressure increase at the boundaries of the metal. The confinement and shaping of the molten metal is thus provided by the magnetic pressure, as it has to balance both the hydrostatic and surface tension pressures at the free surface.

In general, the problem of electromagnetic shaping of molten metal involves: the determination of the electromagnetic force field and the unknown shape of the free surface. These two aspects of the problem are strongly coupled as the induced electromagnetic field, and hence the electromagnetic force distribution on the free surface depends on geometry and location of the melt within the induction coil.

Only recently has the electromagnetic shaping of molten metals been theoretically studied, notably by Shercliff /4,5/, Sneyd and Moffatt /6/, Mestel /7,8/ and Gagnoud and Garnier /9/. Two basic systems were investigated: electromagnetic levitation melting /6-9/, Fig. 1(a), and the shaping of falling molten columns /4,5/ confined by a quadrupole inductor, Fig. 1(b). In these studies the electromagnetic pressure distribution on the free surface is either specified /4,5/ or evaluated using the analytical method developed by Sneyd /10/ for slightly distorted cylindrical and spherical conductors in the limit of high frequency. Exception is the work of Gagnoud and Garnier for levitated molten droplets.

In this paper a mathematical model is developed which can be used for confinement and shaping problems such as electromagnetic casting, as well as electromagnetic levitation problems. The model is demonstrated for two classes of problems: levitation melting of molten droplets and electromagnetic shaping of a falling liquid column confined by external parallel conductors.

Model Formulation

Calculation of the Electromagnetic Force Field

Figure 2 shows a sketch of a molten metal with an irregular boundary placed in a two-dimensional magnetic field which may be generated by passing alternating current through either a coaxial circular coil, as in levitation melting, Fig. 1(a), or line conductors, as in column shaping, Fig. 1(b).

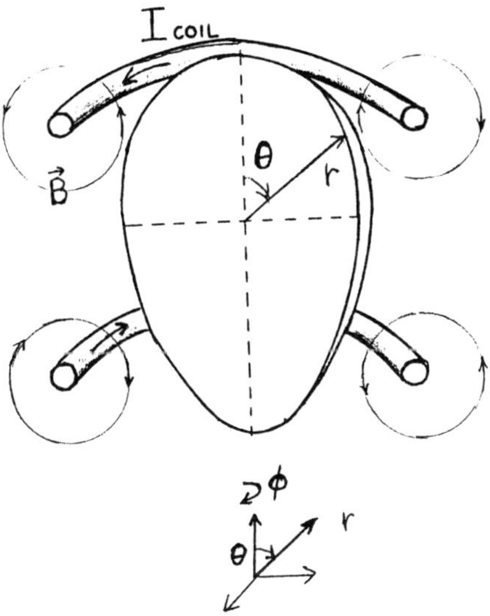

Figure 1(a). Sketch of a typical levitation melting coil.

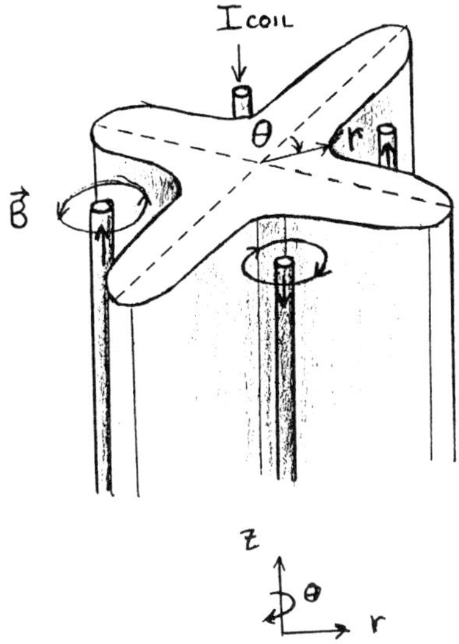

Figure 1(b). Sketch of a molten metal column shaping device.

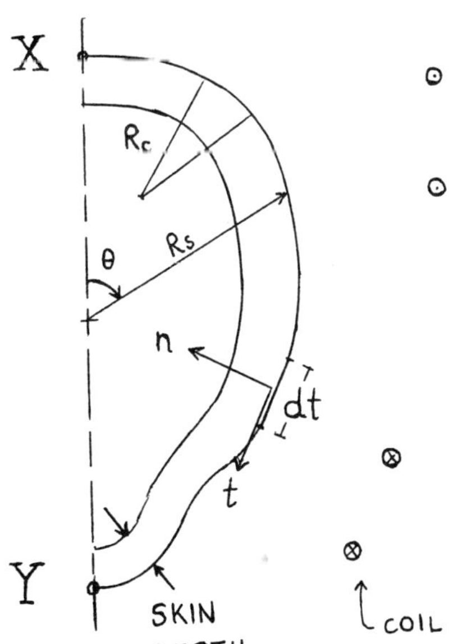

Figure 2. Sketch of a deformed metal in an electromagnetic field showing coordinate systems.

The magnetic field expressed in terms of the vector potential, A ($B = \nabla \times A$), at the boundaries, which is in the direction of the applied current, may be obtained by Ampere's law as:

$$A = \frac{\mu_0}{4\pi} \left[\int_{melt} \frac{J_{melt}}{r} d\upsilon + \int_{coil} \frac{J_{inductor}}{r} d\upsilon \right] \tag{1}$$

where J is the current density in a volume element $d\vartheta$ at a distance r from the point at which the vector potential is being evaluated.

The first term on the right hand side of this equation represents the induced vector potential as a result of the induced current in the melt, while the second term represents applied vector potential as a result of the inductor. From Faraday's law the relationship between the induced currents in the melt and the vector potential is as follows:

$$J = - j \omega \sigma A \tag{2}$$

where ω is the angular frequency and σ is the electric conductivity of the melt.

If it is assumed that the frequency of the applied field and/or the conductivity of the melt are large enough so that "the skin depth", $\delta = (2 / \omega \mu_o \sigma)^{0.5}$, is much smaller than the radius of curvature of the free surface, see Fig. 1, the electromagnetic field quantities in the melt may be described by plane-wave solution of the Maxwell equations, and the induced currents in the normal direction to the free surface may be expressed as:

$$J(n) = \mathrm{Re} \{J_s e^{(j \omega t - (1+j) n/\delta)}\} \tag{3}$$

where J_s is the complex current density at the free surface and n is the distance into the melt.

From equations (1) to (3), the induced currents at the surface of the melt may be expressed as:

$$J_s = -j \omega \sigma \mu_0 \left[\int_{boundary\ line} C_s J_s f_s dt + \sum_{k=1}^{coil} f_{s,k} I_k \right] \tag{4}$$

where C_s and f_s are given by:

$$C_s = - \left(\frac{\delta}{1+j}\right) \left[e^{-(1+j)R/\delta} - 1\right] \tag{5}$$

and

$$f_s = \oint \frac{dl_s}{4r} \tag{6}$$

where I is coil current, R is the tangential radius of curvature, and dl and dt are line elements along the direction of the current and the melt boundary respectively.

The problem is now reduced from a volume integral of current distribution in the melt, to a line integral involving only the surface currents along the melt boundary. The integral, Eq. (4), has to be solved numerically. By dividing the boundary curve XY, see Fig. 2, into line segments, Δt, the integral equation (4) will take the following form:

$$J_s = - j\omega\sigma\mu_0 \left[\sum_{i=1} C_{i,s} J_{s,i} f_{s,i} \Delta t_i + \sum_{k=1}^{coil} f_{s,k} I_k \right] \tag{7}$$

Solution of Eq. (7) gives the induced current at the surface of the melt. Once the current distribution is obtained, the vector potential is readily calculated, Eq.(2), and hence the magnetic pressure distribution at the free surface, P_m:

$$P_m = 2 \text{ Re } \{A_s \bar{A}_s / \delta^2\} \tag{8}$$

where $\bar{A}_s$ is complex conjugate of A_s.

Finally, for levitation problems the total lifting force in axial direction which determines the location of the melt within the inductor is given by:

$$F_l = \frac{1}{2} \text{ Re } \left[\sum_i B_{a,s} \int_0^R J_{s,i} \ (I_1 n_i^2 + I_2 n_i + I_3) \ dn_i \right] \tag{9}$$

where $B_{a,s}$ is the applied magnetic field at the surface and I_1, I_2 and I_3 are geometrical factors associated with the transformation of the volume integral of the lifting force exerted on the melt ($J \times B_a$) from general coordinate system, r and Θ, to a line integral in the surface coordinate system, n and t, see Fig 2.

<u>Calculation of the Melt Shape</u>

The static equilibrium of the electromagnetic, gravitational, and surface tension forces on the free surface may be expressed by:

$$(P_m + \gamma K + \rho g z) = \text{constant} \tag{10}$$

where γ is the surface tension, K is surface curvature, and z is distance in the direction of the gravitational field.

Moffat and Sneyd /6/ have shown that the variation of the gravitational, surface tension and the magnetic potential energies of the system for small displacement of the melt is given by:

$$\delta U = \delta \left[\int (P_m + \gamma K + \rho g z) \, dv \right] = 0 \qquad (11)$$

If the melt boundary line is divided into N line segments, see Fig. 2, Eq. (12) may be written in terms of displacement distance normal to free surface, dn, as:

$$\delta U = \sum_{i=1}^{N} \left(\frac{\partial U}{\partial n} \right)_i . \delta n = 0 \qquad (12)$$

subjected to the constraint that the volume of the melt remain constant:

$$\delta V = \sum_{i=1}^{N} \left(\frac{\partial V}{\partial n} \right)_i . \delta n = 0 \qquad (13)$$

Using Lagrangian multipliers, Eqs. (12) and (13) can be written as:

$$\frac{\partial}{\partial n_i} (U + \lambda V) = 0 \qquad i = 1, 2, \ldots, N \qquad (14)$$

where

$$\lambda = - \frac{\int (P_m + \gamma K + \rho g z) \, dS}{\int dS} \qquad (15)$$

is Lagrange multiplier, and S is surface area of the free surface.

Physically, equation (14) describes the force acting on the melt as a function of position of the free surface. To determine the surface coordinates of the equilibrium shape of the melt, Eq. (14) was approximated to a set of non-linear algebraic equations, which was solved simultaneously with magnetic field equations, using Newton's method.

Computed Results

The formulation presented above enables us to predict the electromagnetic field and the unknown shape of the free surface of molten metal for any given inductor design, applied current and frequency, and system dimensions, subject to the assumptions employed in the model. The most critical assumption is the exponential decay of the electromagnetic field quantities, Eq. (3), used in the calculation of the electromagnetic field, which is generally valid at high values of the magnetic interaction parameter.

$$R\omega = \mu_o \sigma \omega R^2$$

To investigate the validity of this assumption, the results of the model were compared with the results from the analytical solutions over a wide range of $R\omega$. The systems examined are a sphere and a circular cylinder in a uniform magnetic field. The results of the model were also compared with the results from the full numerical solution of the Maxwell equations.

In the following, we shall critically examine the validity of this assumption by comparing the model predictions with the results from the analytical and full numerical solutions, for a sphere and a circular cylinder in a uniform magnetic field /11/, over a wide range of $R\omega$. The numerical method used for solving the Maxwell equations is fully documented in references /12-14/. Computed results are also presented for two shaping problems; namely, shaping of falling molten column in a quadrupole magnetic field and levitation melting to demonstrate the flexibility of the model.

A Sphere in a Uniform Axial Magnetic Field

Figures 3(a) and 3(b) show a comparison between the three methods over a broad range of $R\omega$. The dimensionless amplitude of surface vector potential, $A_s^* = A_s/(B_o\ \delta)$, at 90 degrees from the axis of symmetry is given in Fig. 3(a), while the corresponding phase shift is shown in Fig. 3(b). It is seen that the model predictions are in perfect agreement with the numerical and the analytical results regarding both the amplitude and the phase shift, provided that $R\omega$ is larger than 15. It is also seen that the deviation from the exact analytical solution increases with decreasing $R\omega$ and becomes more pronounced when the skin depth is comparable with the radius of the sphere, $R\omega$ around 2, as would be expected. However, the maximum error is less than 10%.

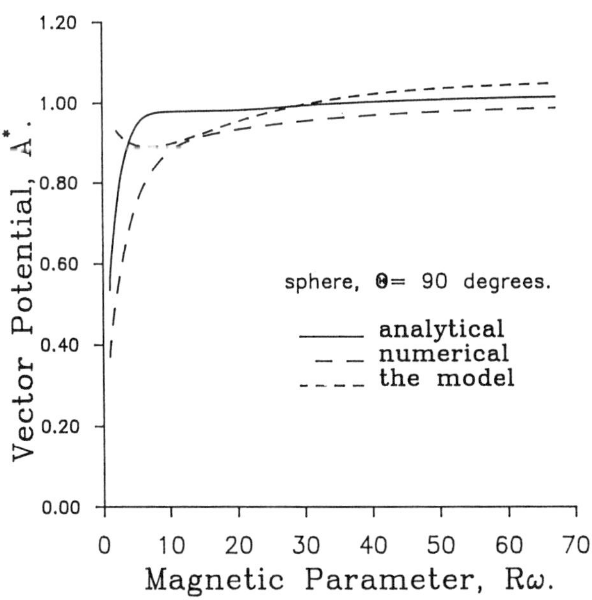

Figure 3(a). Computed dimensionless amplitude of vector potential at the surface of a sphere, $\Theta = 90$.

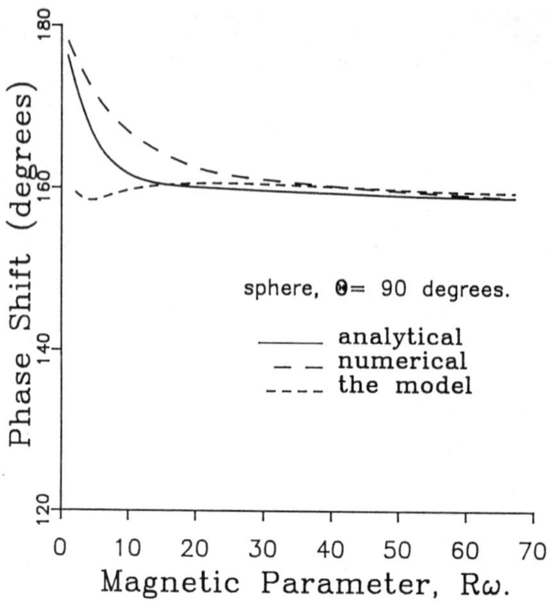

Figure 3(b). Computed phase shift of vector potential at the surface of a sphere, Θ = 90.

Figure (4) shows the variation of the vector potential along the surface of the sphere at $R\omega$ equal 20. The agreement between the approximate, numerical, and analytical solutions is excellent.

The results of this study may be summarized by stating that the approximate method predicts reasonably well the magnetic field distribution at the boundary of spherical shapes, provided that the skin depth is equal to or smaller than the radius of the sphere.

An Infinitely Long Circular Cylinder in a Uniform Transverse Magnetic Field

Figures 5(a) and 5(b) show the effect of $R\omega$ on the dimensionless surface vector potential at 90 degrees to the applied field. Inspection of these two figures shows that the approximate solution approaches both the analytical and numerical solutions at the limit of high frequency. The maximum error is about 5% at $R\omega$ greater than 20, and progressively increases with decreasing $R\omega$ to about 15% at $R\omega$ equal to 4.

Figure 6 shows the azimuthal variation of the vector potential at the surface of the circular cylinder for $R\omega$ equal to 20. The agreement between the analytical and approximate solutions is again reasonably good.

The results of this work suggest that the applicability of the model for predicting the magnetic pressure distribution for cylindrical geometries is limited to problems with $R\omega$ greater than 10, i.e. the skin depth is less than one-half the radius of curvature.

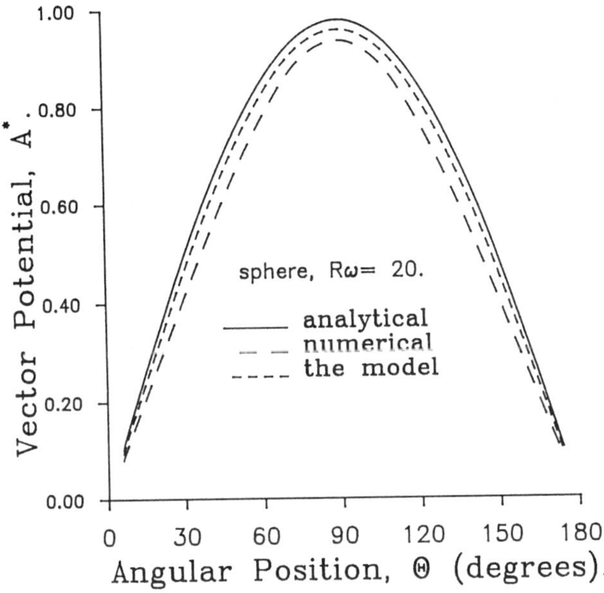

Figure 4. Computed angular distribution of vector potential at the surface of a sphere, Rω = 20.

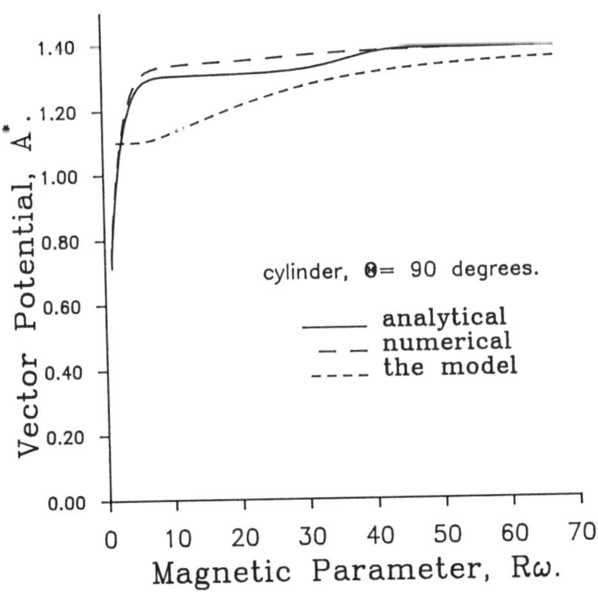

Figure 5(a). Computed dimensionless amplitude of vector potential at the surface of a cylinder, Θ = 90.

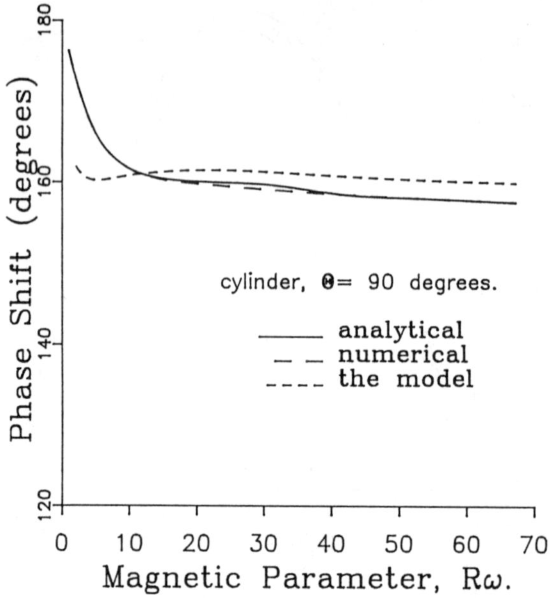

Figure 5(b). Computed phase shift of vector potential at the surface of a cylinder, Θ = 90.

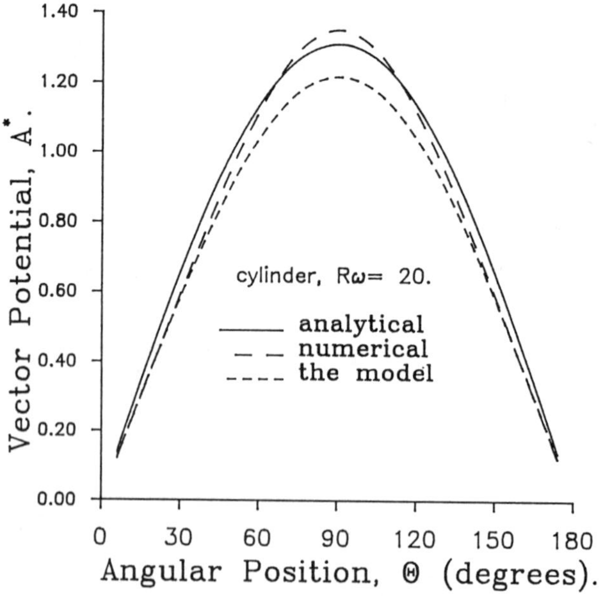

Figure 6. Computed angular distribution of vector potential at the surface of a cylinder, Rω = 20.

Levitation of Molten Metal Droplets

The first application of the shaping model is concerned with the determination of equilibrium position and shape of molten metal droplets in a magnetic field generated by a coaxial circular coil, Fig. 1(a). A selection of the computed results will be presented for a coil configuration and input parameters which correspond to the experimental conditions used by one of the authors to study carburization of levitated molten iron droplets /12,15,16/.

Figure 7 shows the computed shape and the equilibrium location of one gram of molten iron for a coil current and frequency of 250 Amps and 450 kHz, respectively ($R\omega$ = 31). The position of the levitated solid sphere is also shown in the same figure. It is seen that the melt is positioned close to the bottom coil and the equilibrium location of the droplet is slightly lower than that for solid sphere. It is also seen that the melt is slightly deformed and the droplet has a pearl-like shape similar to what is usually observed in actual levitation experiments. This is because the magnetic pressure on the lower half of the droplet has to be larger than that on the upper half in order to support the melt against gravity, Fig. 8.

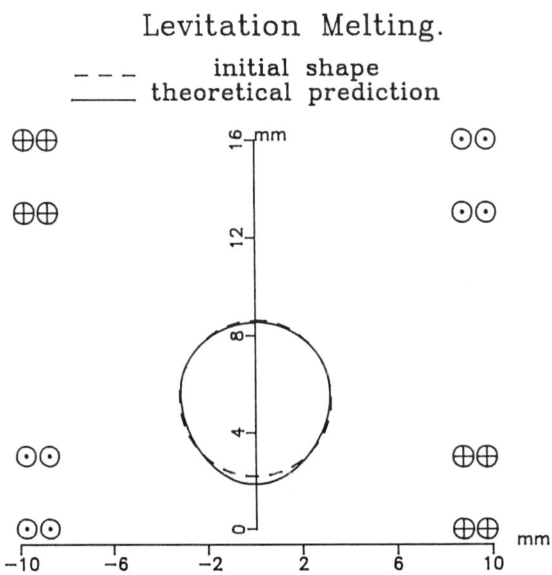

Figure 7. Computed equilibrium shape of a levitated 1 g. molten iron droplet contrasted against the equilibrium position of a solid sphere.

Figure 9 shows a comparison between the theoretically predicted shape and the measured profile of a rapidly quenched droplet /16/. It is seen that the model predicts accurately not only the profile of the levitated droplet but also the curvature of the surface at the top and the bottom of the melt. The perfect agreement between the theory and experiment clearly demonstrates the accuracy of the model.

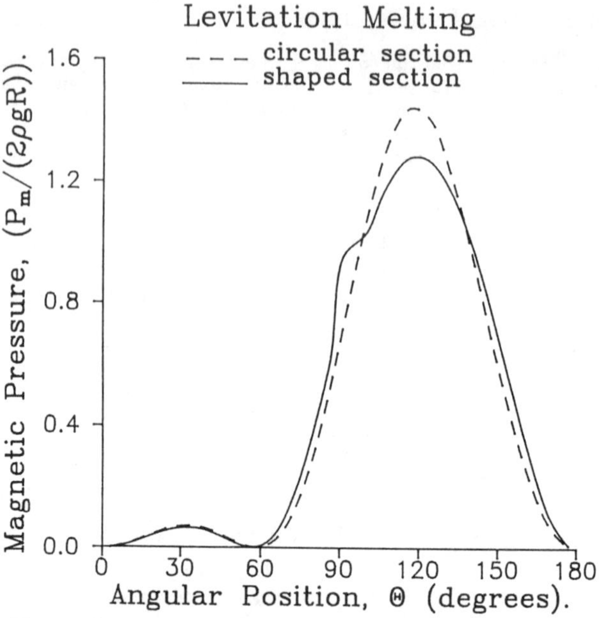

Figure 8. Computed dimensionless magnetic pressure distribution along the surface.

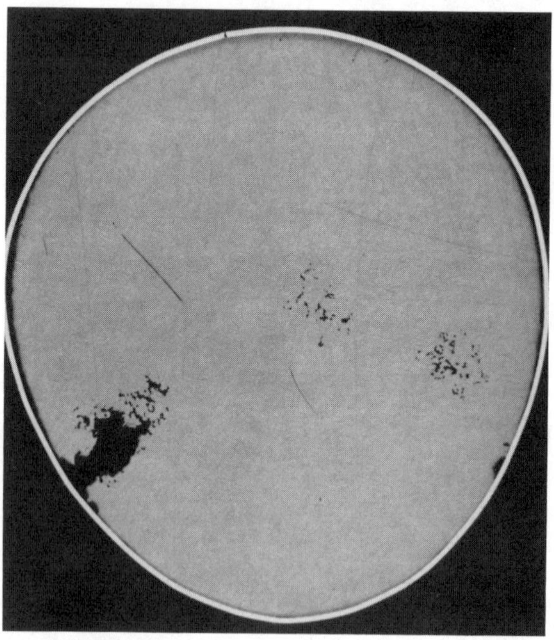

Figure 9. Comparison between theoretically predicted and experimentally measured shape of a levitated iron droplet; measurements Ref. 15, ▬▬ theoretical prediction.

Shaping of Falling Molten Column

Calculations were carried out for the experimental conditions reported by Etay and Garnier /17/ in studying the shaping of mercury column using a quadrupole inductor. The experimental details are given in their paper, and only essential aspects of the experimental setup and the operating parameters are given here. In essence, it consists of a closed loop mercury circuit in which a round mercury jet, 10 mm in diameter, is allowed to fall through an inductor. The latter is comprised of four parallel line conductors similar to that shown in Fig. 1(b). Experiments were performed using a 175 kHz high-frequency power supply. Under these experimental conditions, $R\omega$ for this system is about 40.

Figure 10 shows a comparison between the experimentally measured and theoretically predicted shape of the mercury column in a magnetic field intensity of 980 Gauss. It is seen that the model predictions and the experimental measurements are again in good agreement.

The strong coupling of the electromagnetic and the free boundary problems is illustrated in Fig. 11, which shows the magnetic pressure distribution on the surface normalized with respect of surface tension pressure for the shaped column and a circular solid cylinder having the same cross-sectional area. It is seen that both the absolute value and the distribution of the magnetic pressure are strongly affected by the shape of the melt boundary.

Discussion and Conclusions

A mathematical model for the shaping of molten metals by an electromagnetic field has been developed. The model is based upon an approximate solution of the electromagnetic field equations and permits the determination of the unknown shape of the melt free surface in terms of the inductor configuration and the applied current and frequency. The model is a general one and can be applied to problems involving either shaping and confinement of molten metals or levitation melting.

The main advantages of using the approximate method over the numerical techniques /12-14/ for calculating the electromagnetic field are the flexibility of handling irregular geometries and for solving the electromagnetic field quantities only at the boundary of the melt with a significant savings in computer time. However, this method has a limitation in that it cannot be used when "the skin depth" is larger than the characteristic dimension of the melt. The range of applicability was critically examined, and it was found that the magnetic interaction parameter, $R\omega$, has to be greater than ten if an accurate solution of the electromagnetic field is to be obtained. Here, it should be noted that in most electromagnetic shaping and confinement applications, this condition is usually satisfied.

The flexibility and the accuracy of the model was demonstrated for two classes of problems; namely, electromagnetic levitation of molten metal droplets and the shaping a falling liquid column confined by external four parallel conductors. The theoretical predictions concerning the shape of the melt were found to be in perfect agreements with experimental measurements.

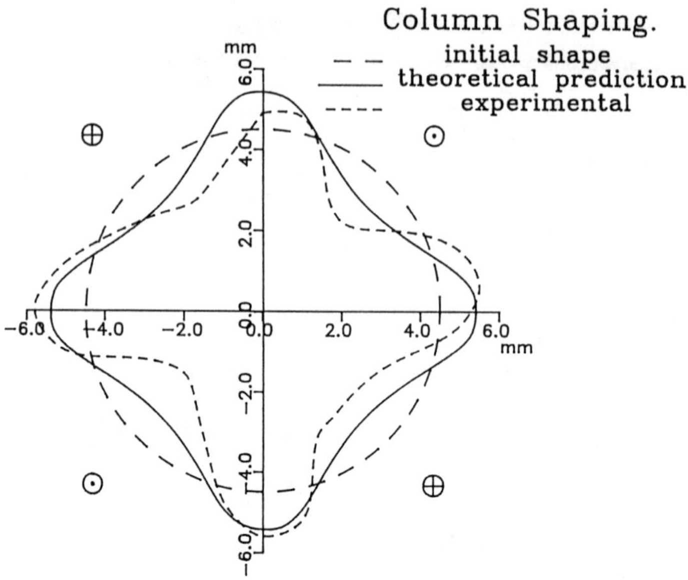

Figure 10. Comparison between theoretically predicted and experimentally measured shape of a mercury liquid column, Ref. 17.

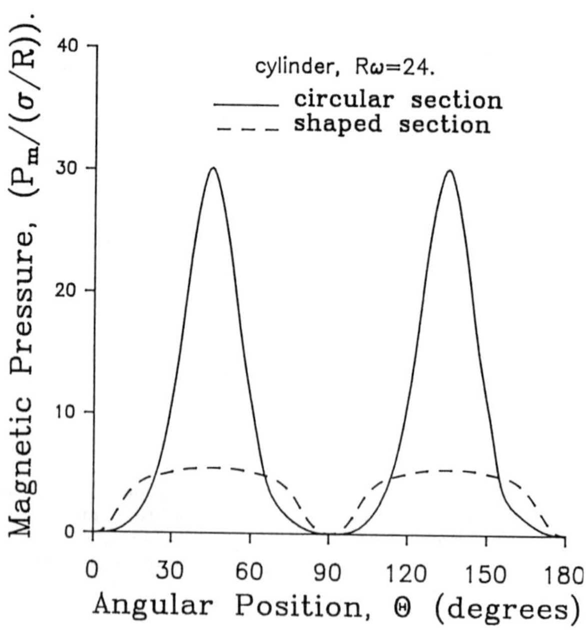

Figure 11. Computed dimensionless magnetic pressure distribution along the surface.

Acknowledgements

The authors are grateful to NSF-ALABAMA EPSCOR Science and Engineering Program and Inductotherm Corporation for their support of this research.

References

1. H.A. Meier, *Light Metals*, 1977, p. 64.

2. H. R. Lowry, *Iron and Steelmaker*, Apr. 1988, p. 16.

3. A. W. Cramb, *Iron and Steelmaker*, May. 1988, p. 65.

4. J. A. Shercliff, *Proc. R. Soc. Lond.*, 1981, vol. A 375, p. 455.

5. J. A. Shercliff, *Proc. of 3rd. Int. Seminar* on "Liquid Metals Flows and Magnetohydrodynamics", Bat-Sheva, Israel, 1981, p. 414.

6. A. D. Sneyd and H. K. Moffat, *J. of Fluid Mech.*, 1982, vol.117, p. 45.

7. A. J. Mestel, *J. of Fluid Mech.*, 1982, vol. 117, p. 27.

8. A. J. Mestel, *Proc. of Symp.* on "Metallurgical Applications of MHD", The Metals Society, 1982, p. 197.

9. A. Gagnoud and M. Garnier, *IEEE Trans. on Magnetics*, 1985, vol. 21, p. 1886.

10. A. D. Sneyd, *J. of Fluid Mech.*, 1979, vol. 92, p. 35.

11. W. R. Smythe, "Static and Dynamic Electricity", McGraw-Hill Publ. Company, 1950.

12. N. El-Kaddah and J. Szekely, *Met. Trans. B*, Sep. 1983, vol. 14B, p. 401.

13. J. L. Meyer, J. Szekely, N. El-Kaddah, C. Vives and R. Ricou, *Met. Trans. B*, Sep. 1987, vol. 18B, p. 539.

14. J. L. Meyer, N. El-Kaddah and J. Szekely, *IEEE Trans. on agnetics*, 1987, vol. 23, p. 1806.,

15. N. El-Kaddah, Ph.D. Thesis, Imperial College, London, 1976.

16. N. El-Kaddah and D. G. R. Robertson, *Met. Trans. B*, 1978, vol. 9B, p. 191.

17. J. Etay and M. Garnier, *Proc. of 3rd. Int. Seminar* on "Liquid Metals Flows and Magnetohydrodynamics", Bat-Sheva, Israel, 1981, p. 422.

CASTING TECHNIQUES IV:
THIN SLAB CASTING

Session Chairman: **W. Dubroff**
EG&G Idaho Inc.
Idaho Falls, Idaho

HORIZONTAL AND VERTICAL TWIN-BELT CASTING

OF 1-INCH-THIN SECTIONS OF SHEET STEEL*

M. R. Moore,+ S. S. Daniel,+ and M. R. Ozgu++

+ USX Corporation, USS Division, Technical Center
4000 Tech Center Drive
Monroeville, Pennsylvania 15146

++ Bethlehem Steel Corporation
Homer Research Laboratories
Bethlehem, Pennsylvania 18016

Abstract

Innovative horizontal and vertical twin-belt casting systems were developed and tested in casting trials to provide alternative approaches to solving the crux technical problem of feeding liquid steel into the thin mold gap so as to achieve the excellent as-cast surface quality necessary to allow subsequent direct rolling of top-quality sheet. The horizontal approach involved an innovative feeding system to feed the 1-inch-thick caster entry, but the twin-belt Hazelett machine itself was largely conventional. In contrast, the opened-up entry of the vertical twin-belt machine allowed the use of conventional large tube and flux pouring systems, but the machine itself was complex and deformed the casting down to the 1-inch thickness. Extensive modeling work was done to support both approaches.

* This project was partially funded by the U. S. Department of Energy, Office of Industrial Programs, under Cooperative Agreement DE-FC07-84ID12545.

Casting of Near Net Shape Products
Edited by Y. Sahai, J.E. Battles,
R.S. Carbonara and C.E. Mobley
The Metallurgical Society, 1988

Background

United States Steel Corporation (now called USX Corporation) and Bethlehem Steel Corporation entered into Cooperative Agreement No. DE-FC07-84ID12545 with the Department of Energy (DOE), Office of Industrial Programs, Idaho Operations Office, effective September 28, 1984. Major specialist subcontractors were Harbison-Walker Refractories Group, Dresser Industries, Inc. (refractory systems), Hazelett Strip-Casting Corporation (twin-belt casting machines) and Westinghouse Electric Company (MHD systems). The program objective was to develop a new casting technology to produce the total range of both hot- and cold-rolled sheet steels with a major energy savings (USA 1990 estimated savings of up to 0.191 QUAD) and with attractive capital and operating cost savings (up to $35/ton and $20/ton, respectively) over existing steel slab casting technology. A key decision was the approach to be used to achieve these program objectives and the choice lay between strip casting (at around 0.1 inch thick) and thin-section casting (TSC) (at around 1.0 inch thick). The TSC approach was chosen over the strip casting approach because of two critical requirements in (1) the scope of the technology (application to the total range of hot- and cold-rolled sheet steels was desired) and (2) the project time scale (successful development of the technology at a pilot-plant scale was desired within five years).

The chosen TSC approach was to cast thin steel sections of 1 to 1-1/2 inches in thickness at casting rates of 20 to 60 tons per hour per foot of cast width in compact twin moving-mold casters which are commercial for nonferrous casting. These thin sections would then be hot-rolled to about 0.1 inch in thickness. The resulting steel hot band could be commercially applied and also subsequently cold-rolled and heat-treated to meet performance specifications. This technology, although revolutionary, could still be retrofitted into both integrated steel plants where hot rolling will be done in the finishing trains of existing hot-strip mills and into minimills where new types of hot mills could be used for the hot rolling. The original project time scale was five years with project completion planned for September 1989.

Introduction

The research program was based on the recognition that the critical technical problem to be solved was the development of a system to feed liquid steel into the thin mold gap of the moving-mold casting machine so as to achieve excellent surface quality in the as-cast thin section which would allow subsequent direct rolling of excellent-quality sheet products.

To maximize the chances of success, multiple, independent approaches to the critical feeding problem were developed. These approaches were labeled the horizontal and vertical approaches, but the differences between these were more fundamental than just the inclination of the TSC casting machine. The horizontal approach depended on refractory nozzle and fluid-mechanics/heat-transfer developments to allow successful feeding of the 1-inch thin-section entry of the near-horizontal TSC machine. In this case, no deformation of the casting strand was required and a relatively conventional twin-belt casting machine could be used. However, conventional caster pouring systems could not be used because of the small 1-inch entry size, and innovative pouring systems had to be developed. The vertical approach depended on strand mechanics and developments to allow successful deformation of the initial solid strand shell from an opened up thickness down to the required 1-inch thickness in the V-bell mouth upper end of the vertical casting machine. In this case, conventional pouring tubes could be used and no special feeding nozzle was required but an innovative twin-belt casting machine had to be developed to achieve the required deformation of the solidifying strand. As a contingency, MHD enhancements of the horizontal feeding system were also attempted.

The original research program was split into two major phases. In the Phase I, several different liquid-steel feeding approaches were to be developed and tested in small moving-mold casters (the horizontal dynamic simulator and the prototype vertical caster, which both cast 17-inch-wide by 1-inch-thick sections) so as to identify the most promising approach. In the Phase II, the preferred approach was to be tested in a larger scale casting pilot plant which would allow commercial rolling of the as-cast sections to sheet product and subsequent commercial trial of this sheet product. In February 1987, as a result of the announced success of the competitive SMS Schloemann/Concast TSC development (1)*, USS, Bethlehem, and the DOE jointly decided to limit Phase I to "orderly phase-down" casting campaigns (to be completed in September 1987) and to cancel Phase II. The objective of this paper is to give an overview of all the work done under this project. Future publications (2,3) will provide more details of the results obtained in modeling work and in casting trials on each of the two casting machines.

The Two Crux Technologies

Two totally different approaches to solving the crux liquid-steel feeding problem were followed. In the horizontal approach, the casting thickness was held constant at 1 inch thick and the problem of feeding the slot-like, high-aspect ratio entry to the casting machine had to be solved (this feeding geometry obviated the use of conventional pouring-tube/mold-flux systems). In this case, the feeding system was complex and innovative, but the casting machine design was largely a conventional nonferrous twin-belt Hazelett machine (with some modifications to accommodate steel casting). In the vertical approach, the casting thickness was opened up to around five inches thick at the machine entry and was subsequently deformed down inside the casting machine. In this case, the feeding system was conventional (large pouring tube and mold flux) and the problem of feeding at the 1-inch thickness was avoided. However, the twin-belt casting machine design was complex and innovative and unlike any existing nonferrous Hazelett machine.

The horizontal feeding concept involved feeding the 1-inch-thick entry region of the twin-belt horizontal casting machine via a full-width, thin-walled nozzle with an outside thickness dimension of just under 1 inch. This nozzle was inserted into the front end of the horizontal machine and fitted precisely between the casting belts and the edge dam trains (Figure 1). The objective was to run the feeding train, the nozzle, and the casting machine totally full of liquid steel so that no distinct gravity-driven pouring stream existed and the whole system operated like a flow-through tube. It was recognized that this "closed-pool" feeding approach to horizontal TSC casting was much more complex and higher risk than the more conventional "open-pool" feeding approaches being pursued in competitive near-horizontal TSC projects such as Sumitomo (4) and Nucor (5). (In these "open-pool" feeding approaches a liquid-steel pool level exists within the near-horizontal twin-belt machine and a distinct pouring stream feeds into this pool.) However, it was judged that this "closed-pool" approach was the only pouring system

---

*See References.

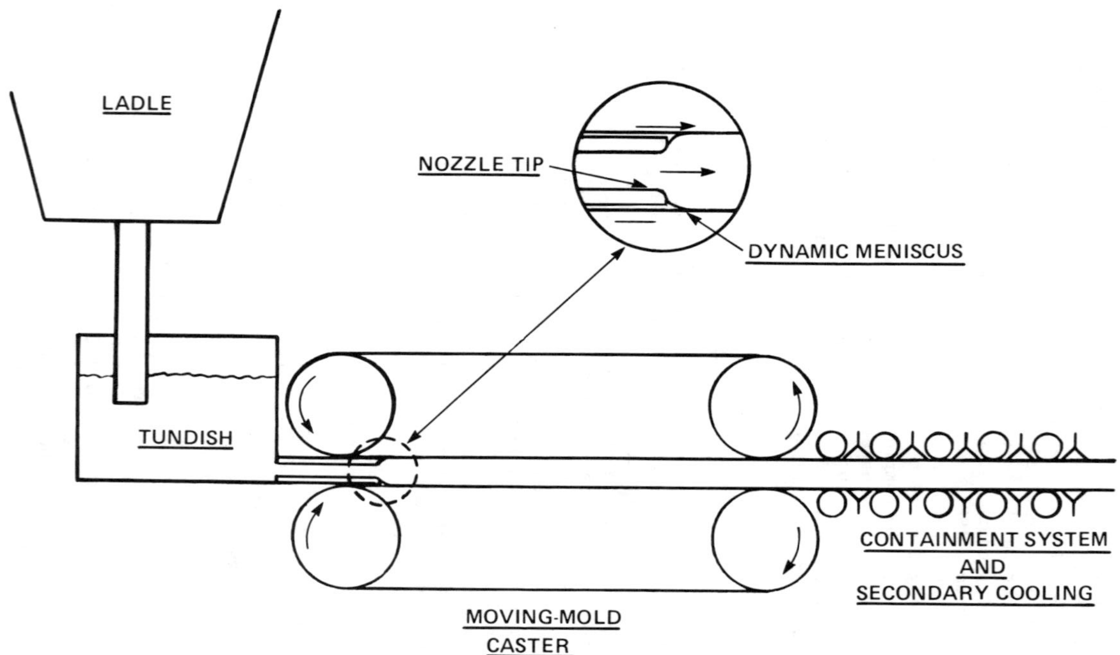

Figure 1 - Schematic Diagram Illustrating the Horizontal Feeding and Casting Approach.

with the potential to eliminate the cast product surface-quality problems caused by pouring turbulence and gas entrainment which are especially severe when the casting thickness is around 1 inch. It was judged that this was the only way to achieve the as-cast surface quality necessary to allow direct hot rolling of sheet for top-quality applications from as-cast TSC sections around 1 inch thick. This closed-pool approach is conceptually similar to the "pressure casting" method used by Krupp (6) in the twin-belt casting of steel billets.

This closed-pool approach required two major developments—(1) a wide, thin-walled nozzle which would perform with minimum thermal distortion and wear, and (2) methods of stabilizing the very small liquid-steel menisci which would be present across the typically 0.020-inch gaps between the nozzle tip and the moving belts when the system ran full of liquid steel. The nozzle presented challenges in both material development and fabrication methods. Stabilizing the very small menisci presented challenges in both mechanical methods of setting and controlling the very small nozzle-to-belt gaps and in very precise control of tundish liquid-steel level (Figure 2) (because of the limited overpressure stability ranges of the top-belt and bottom-belt menisci). These areas were the focus of modeling and hardware development effort in the two years prior to actual horizontal casting trials.

In developing this "closed-pool" horizontal feeding approach two simplifications were used in the small H-DS casting machine trials in that (1) the machine was set up at a near-horizontal inclination (six degrees down) so that casting could start up in the simpler "open-pool" mode and later change to desired "closed-pool" mode when steady-state conditions had been reached, and (2) casting widths were limited to 17 inches wide so that a single-piece nozzle could be used (plans for casting wider commercial sections were based on the "nesting" of several narrower nozzle modules, as wide nozzles could not be fabricated in one piece). A key issue in assessing the viability of the crux "closed-pool" pouring system was its performance during extended casting times. For this reason, the H-DS casting line was set up at USS—Universal where the 15-ton heat size and the downstream TSC product-handling equipment allowed casting of up to 500-foot lengths over casting times of about 15 minutes.

The <u>vertical feeding concept</u> involved opening up the entry of the twin-belt casting machine to a thickness of around five inches so that conventional-type pouring systems could be used (large pouring tubes and mold flux). The increased-thickness dimension of the casting was then deformed down to the desired 1-inch thickness in a V-bell mouth region on the front end of the twin-belt machine. Subsequent to this entry region, the belts were parallel at a fixed 1-inch thickness mold spacing to allow the completion of solidification within the machine (Figure 3). In the deformation process, the product broadface deformation is a straightforward bending into the 1-inch-thick parallel mold. However, the product narrow-face deformation is complex, and this end wall must be crushed down to the 1-inch-thickness dimension. Competitive vertical twin-belt TSC projects (such as Kawasaki) (7,8) utilize a refractory mold end wall in the V-bell mouth to try to suppress the formation of any end-wall solid shell thus avoiding any end-wall crushing. Because of serious questions about the operability, cost, and service life of such refractory end walls, USS/Bethlehem selected a different end-wall approach which, although high risk, could potentially overcome such problems. It is important to note that these assessments were made before the details of the SMS/Concast vertical TSC approach were known. SMS/Concast did not use twin belts but used a conventional oscillating copper mold with a broadface bulge at the meniscus level to allow for the entry of a large pouring

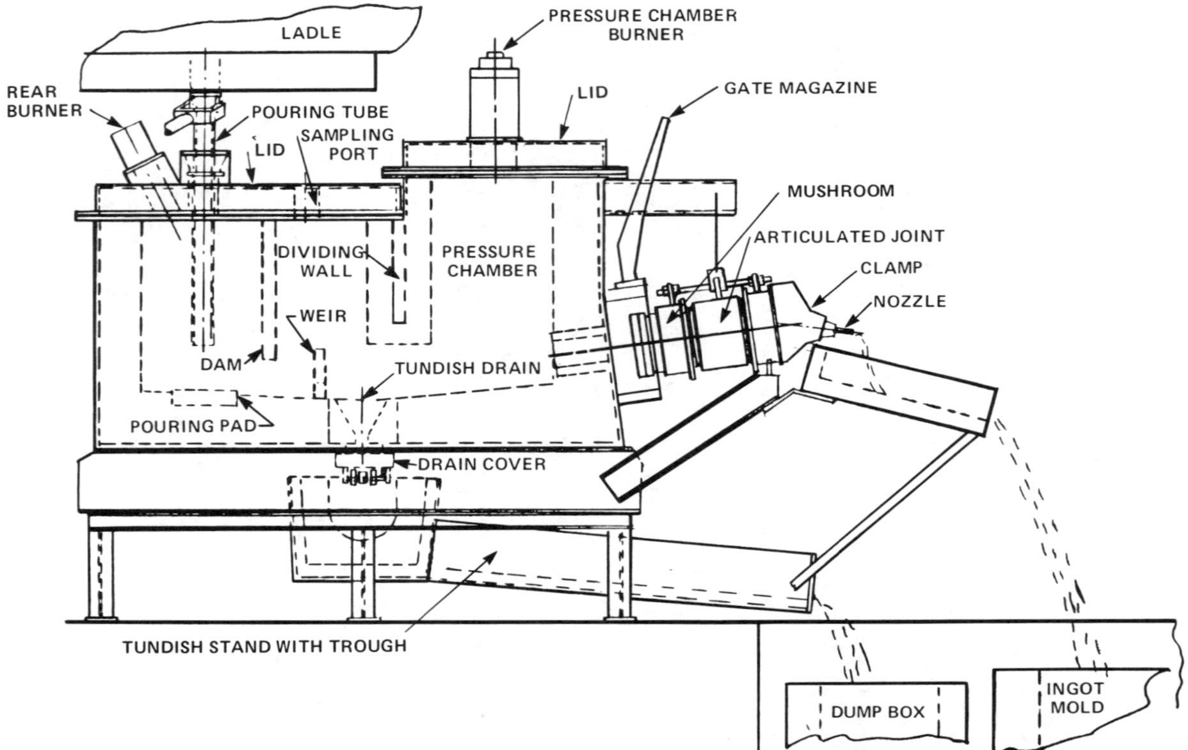

Figure 2 - Horizontal Closed-Pool Feeding Hardware Setup for a Pouring Trial Prior to Casting Trials.

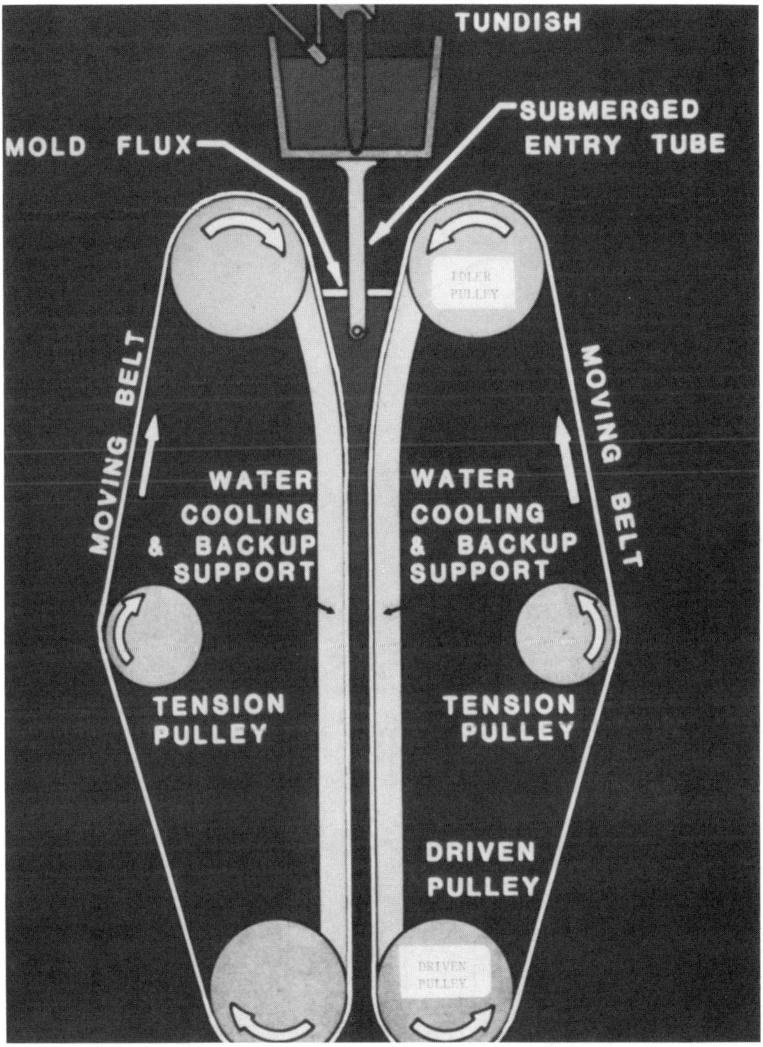

Figure 3 - Schematic of Prototype Vertical Twin-Belt Caster Mold and Feeding System.

tube. The mold broadface bulge converged down the mold length resulting in a thin-parallel-side mold cavity at the bottom of the mold (in the case of moving-belt molds such a broadface bulge is not feasible as it would require major belt deformations as the belt bulge is flattened out and then reformed on each cycle of the moving belt). Thus, the SMS/Concast vertical TSC approach was fundamentally different to the USS/Bethlehem and Kawasaki vertical twin-belt approaches in that broadface rather than end-wall deformation was utilized, and conventional-type slab casting hardware was utilized rather than totally new twin-belt hardware. (The success of the SMS/Concast TSC project eventually led to the curtailment of the subject USS/Bethlehem TSC work.)

In the USS/Bethlehem approach, end-wall shells were deliberately formed on water-cooled copper end-wall molds which were specially shaped to form end-wall shells which would easily deform during passage through the V-bell mouth (Figure 4). This deformable end-wall approach required

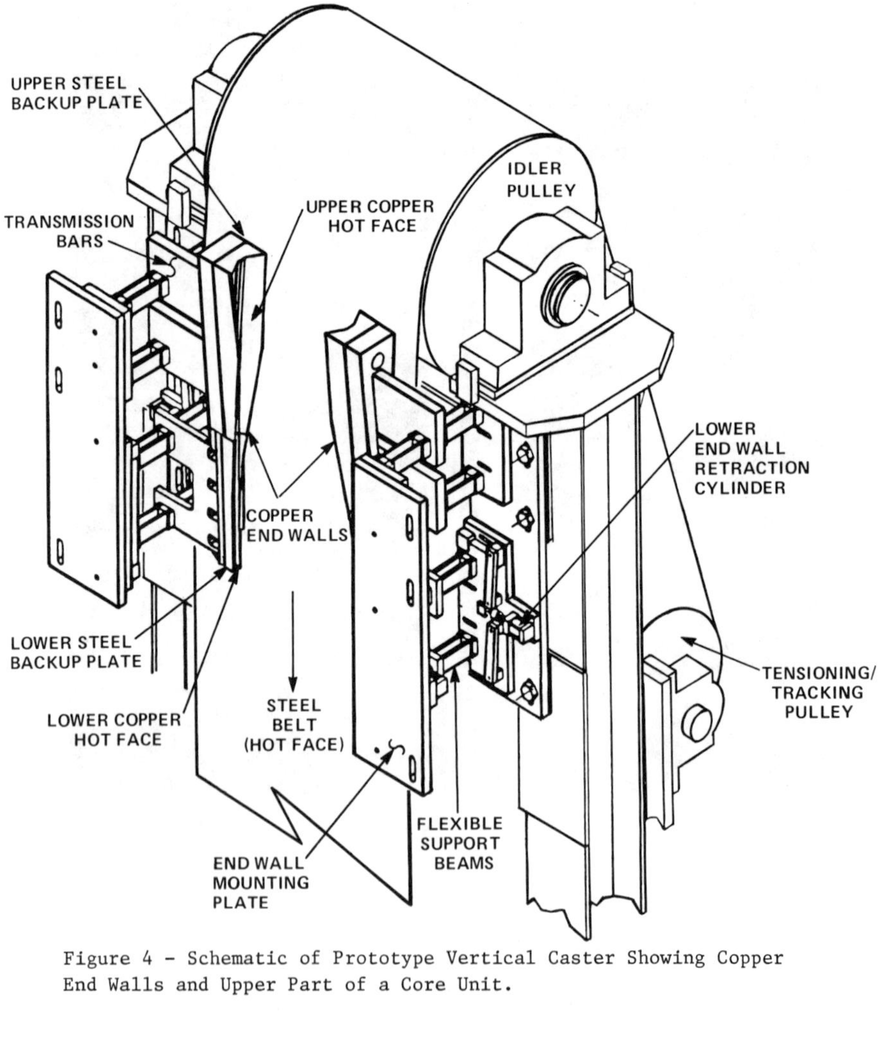

Figure 4 - Schematic of Prototype Vertical Caster Showing Copper End Walls and Upper Part of a Core Unit.

(1) the development of a design basis for the complex end-wall mold shapes from models of the high-temperature deformation of solidifying steel shells, and (2) the development of twin-belt machine concepts for belt support and cooling to handle both the casting deformation loads in the V-bell mouth and the ferrostatic loads throughout the vertical machine. These areas were the focus of modeling and hardware development activities in the two years prior to actual vertical casting trials. In contrast to the horizontal case, the most critical test of this deformable end-wall technology is during start-up and early casting times when the solidified mold shells are relatively cold and difficult to deform. Thus a simplified prototype vertical casting line was set up at Bethlehem with a small heat size (1200 pounds) and short (few minutes) casting times.

Project Activities

The project began in September 1984 and actual casting trials commenced in January 1987. During the 2-1/4 years prior to the casting trials, three major activities occurred in parallel—(1) the development of innovative feeding systems for both horizontal and vertical TSC casting machines, (2) the design, construction, and installation of two small TSC casting machines to provide the critical "test-of-concept" for these innovative feeding systems (the H-DS machine at USS—Universal and the prototype vertical caster machine at Bethlehem), and (3) support activities involving both the development of MHD systems as a contingency to enhance the horizontal feeding system and the static casting and processing of 1-inch-thick plates to confirm that satisfactory sheet properties could be obtained via conventional hot and cold rolling. These three activity areas are overviewed below.

In the development of both horizontal and vertical feeding systems, the activities included (a) the development of multiple concepts and the selection of the preferred approach; (b) extensive two-dimensional, steady-state mathematical modeling that resulted in comprehensive models for both coupled heat-transfer/fluid-flow effects in the horizontal closed-pool nozzle entry region (9) and coupled solidification/deformation effects in the V-bell mouth (10) (typical model results are shown in Figures 5 and 6); (c) water modeling of the vertical tundish and pouring-tube system and a complex, moving-belt model of the horizontal closed-pool meniscus region (some results are shown in Figures 7 and 8); (d) refractory developments by Harbison-Walker that were focused on the wide, thin-walled horizontal nozzle and included both materials selection and the development of new injection-molding nozzle-fabrication techniques; (e) small liquid-steel test rigs—these included dipping and spinning tests of candidate refractory materials, a rig to measure dynamic contact angles for various liquid-steel/refractory systems, a rig to test the articulated-joint components of the horizontal feeding train (Figure 9) and a static ingot-crushing rig to simulate the V-bell mouth deformation (Figure 10); and (f) finally, full-scale liquid-steel pouring trials were run using low-carbon aluminum-killed steel and the actual horizontal and vertical feeding hardware but without casting machines being involved. These full-scale pouring trials allowed the development of all steelmaking, preheating, and automatic-tundish level-control practices prior to actual casting trials being attempted.

The installation of the two twin-belt casting machines involved very different problems (both were designed to cast 17- by 1-inch TSC sections).

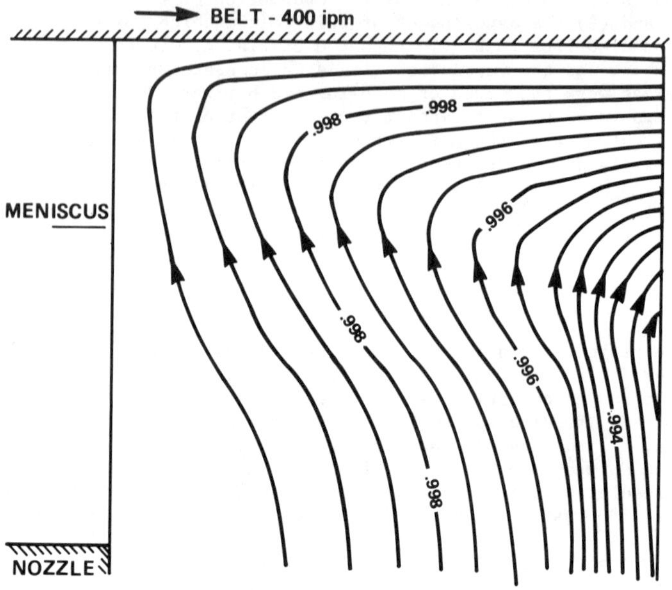

Figure 5 - Mathematical Model Results Predicting Liquid-Steel Flows Near Nozzle-to-Belt Meniscus in Horizontal Closed-Pool TSC Casting.

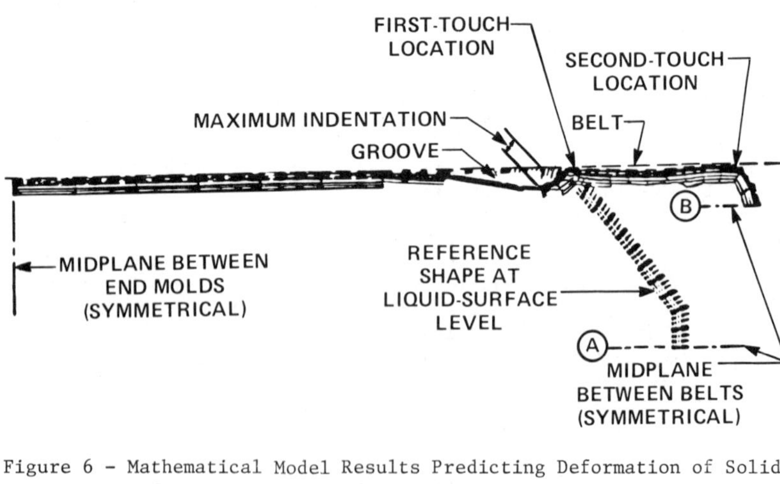

Figure 6 - Mathematical Model Results Predicting Deformation of Solid Shell and Surface Grooves Formed During Shell Travel Meniscus Level ("A") to Bottom of V-Bell Mouth ("B") in Vertical TSC Caster.

POURING CONDITIONS VERSUS SURFACE WAVES

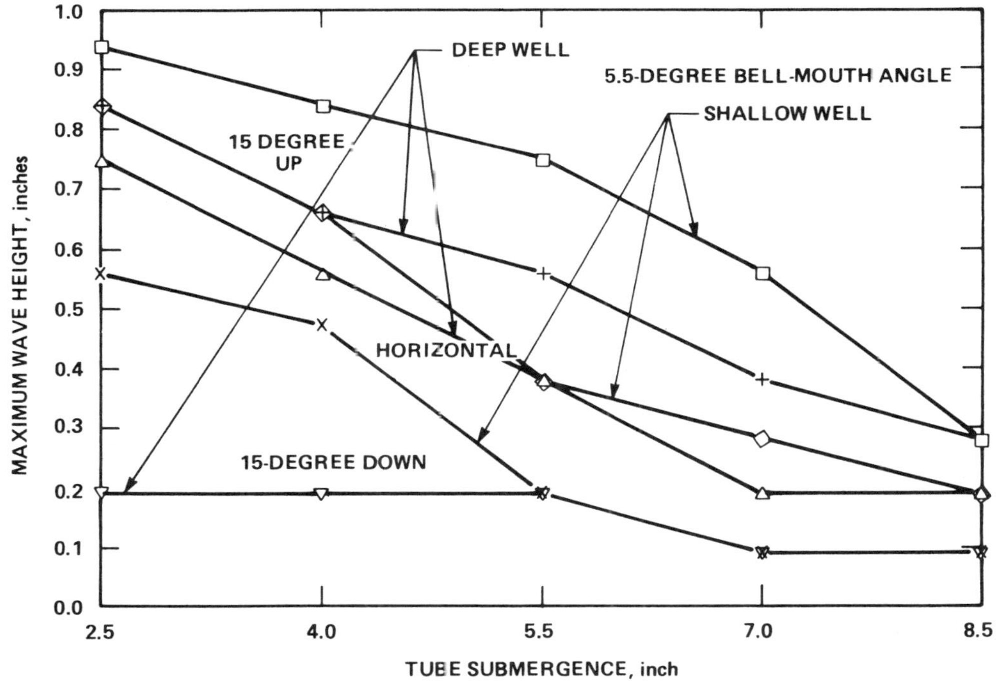

Figure 7 - Water Modeling Studies of the Pouring Tube Used to Feed the Vertical TSC Caster.

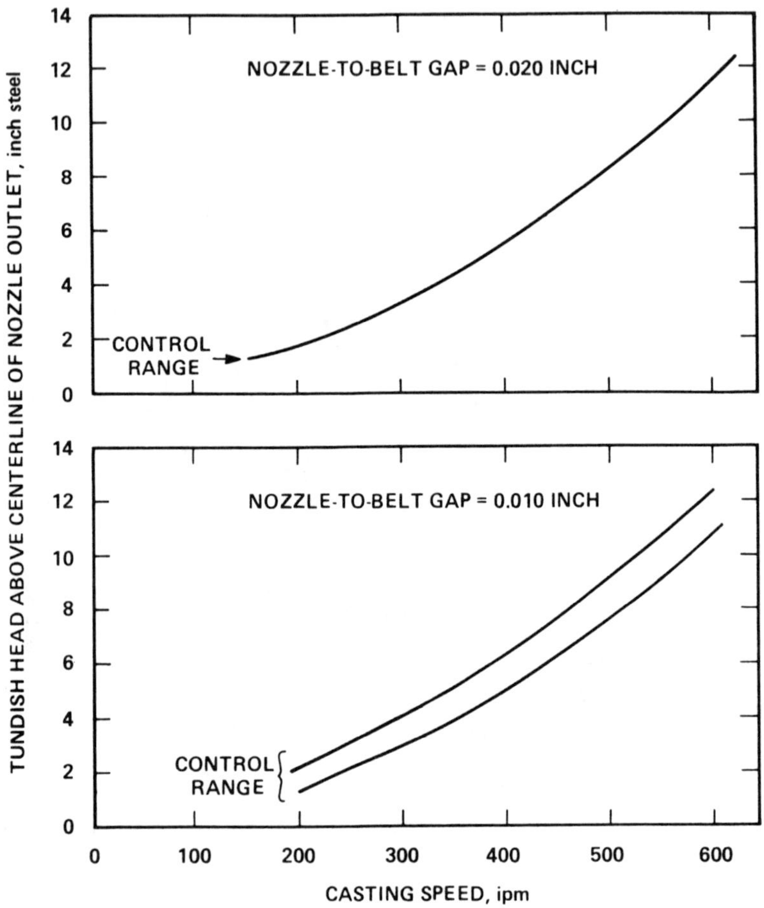

Figure 8 - Tundish Level Control Ranges Over Which Both Upper and Lower Menisci Between Nozzle Tip and Moving Belt Are Stable in Horizontal Closed-Pool TSC Casting of 1-Inch-Thin Sections (Results From Moving-Belt Water Model).

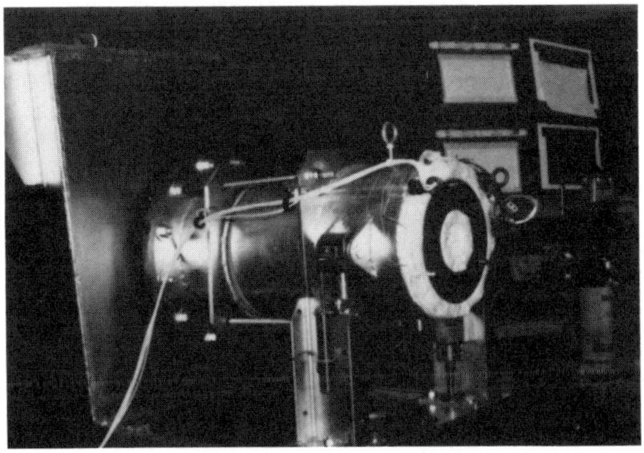

Figure 9 - Rig to Test Movement of Articulated Joint During Liquid-Steel Pouring.

Figure 10 - Static Ingot Crushing Rig to Simulate V-Bell Mouth Deformation in the Vertical TSC Caster.

The horizontal machine at USS—Universal was constructed by Hazelett (Model 20.75) and was largely conventional and based on existing non-ferrous Hazelett machines (11) with some modifications for steel casting (Figure 11). The major task here was engineering and installing a total casting line around this machine with the complex product-handling and controls necessary to allow the required long casting times (up to 30 minutes). The horizontal machine (machine length, 7.5 feet) was designed to cast at speeds up to 600 ipm by running liquid core out of the belts through a close-coupled "foot-roll" section of innovative design to complete solidification in a close-spaced support roll rack (total metallurgical length, 22 feet) (Figure 12). The horizontal machine could operate at various angles of inclination to the horizontal, but six degrees was always used.

The vertical machine (Figure 13) posed just the opposite problems. As short casting times (a few minutes) were satisfactory, the casting line was simple and product-handling rudimentary. However, the vertical machine was totally original and new designs for the V-bell mouth and belt cooling and support had to be developed because of the increased ferro-static and deformation loads on the vertical belts (Figure 14). The vertical machine was designed and constructed by Bethlehem with Hazelett fabricating the belt "core" hardware for belt support and cooling. The vertical machine was designed to cast at lower speeds (up to 240 ipm) such that solidification was completed within the twin-belt mold length of 130 inches.

Support activities included both MHD enhancement of the horizontal feeding system and static casting and processing of 1-inch-thick plates. Westinghouse was the MHD subcontractor and developed an MHD system to increase the stability range of the liquid-steel menisci between the nozzle tip and the moving belts. The chosen MHD approach used MHD body forces induced in the liquid steel close to the meniscus surface from electric current which traveled transversely across the casting belt from side contact brushes mounted outbound of the edge dam trains in the vicinity of the nozzle tip. The MHD project included (a) the development of multiple MHD concepts and the selection of the preferred approach; (b) extensive mathematical modeling coupling MHD forces with fluid-flow effects; and (c) the use of various test rigs including one which poured low-temperature liquid metals (Gallium/Indium) and demonstrated MHD enhancement of liquid-metal meniscus stability. The static-casting support work involved the casting of 1-inch-thick plates with metallurgical structure similar to those expected in TSC product. Some static-cast plates were hot- and cold-rolled, and others were subjected to bending and unbending tests to simulate the hot processing which would occur in an actual TSC casting line which would use coiling and uncoiling to handle TSC products (12). The resulting hot-rolled sheet product was extensively tested and some was cold-rolled to produce cold-rolled sheet samples.

By January 1987, all the above activities had been completed and the two small casting machines were ready to provide the critical "test-of-concept" casting trials for both horizontal and vertical feeding approaches. As a result of the success of the competitive SMS/Concast TSC technology, casting plans were abbreviated and phase-down casting campaigns of nine-months duration were run on each machine. Low-carbon aluminum-killed steel grades were cast throughout these campaigns. On the prototype vertical caster 14 casting trials were attempted and two different designs of copper end walls were tested. On the H-DS machine, 30 casting trials were attempted and closed-pool pouring was achieved.

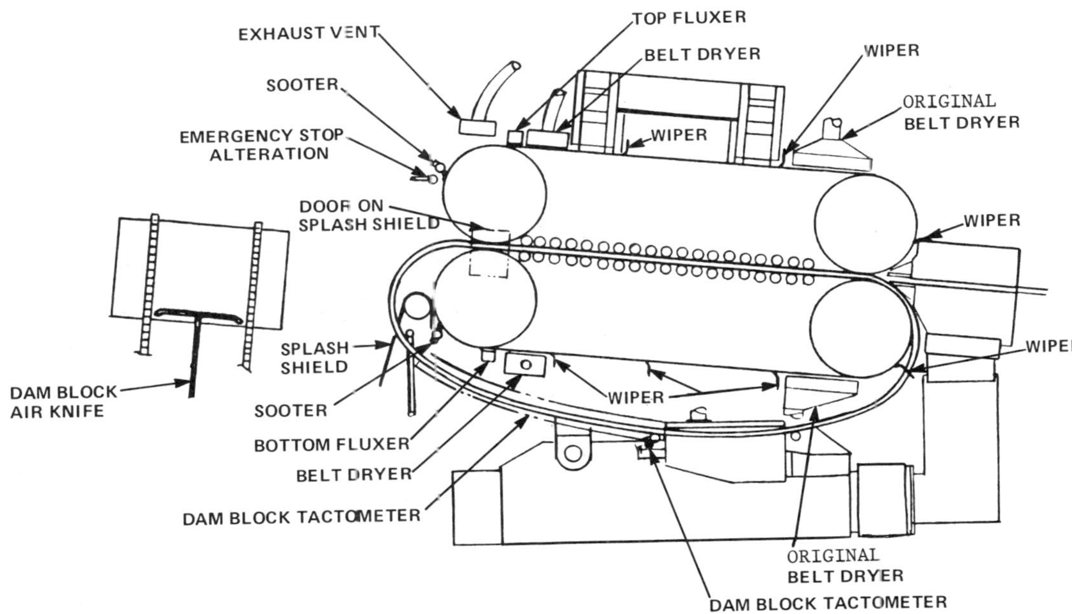

Figure 11 - Additions and Modifications to the Hazelett Casting Machine.

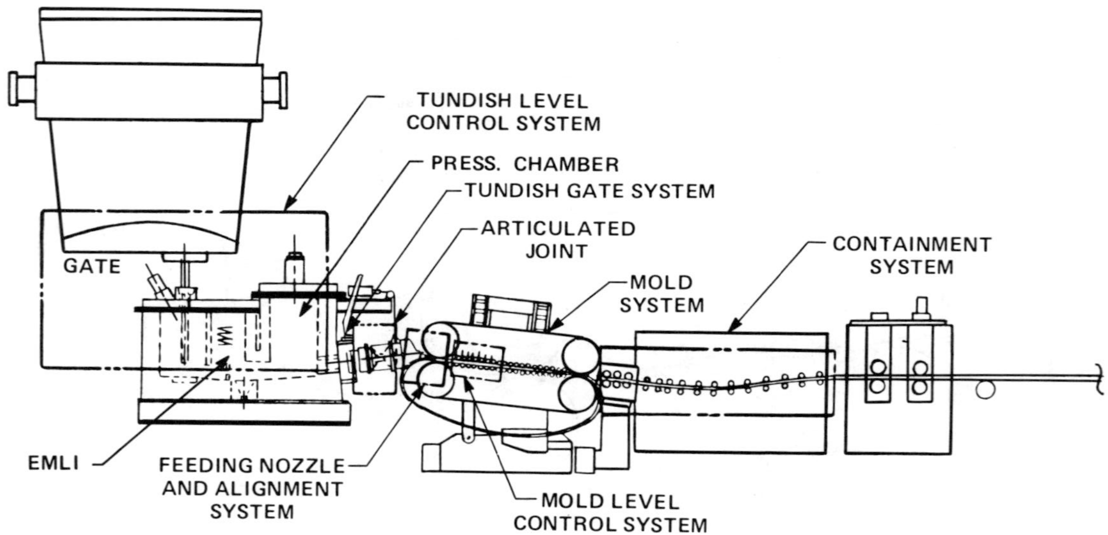

Figure 12 - A Schematic of the H-TSC Process Showing Its Main Subsystems.

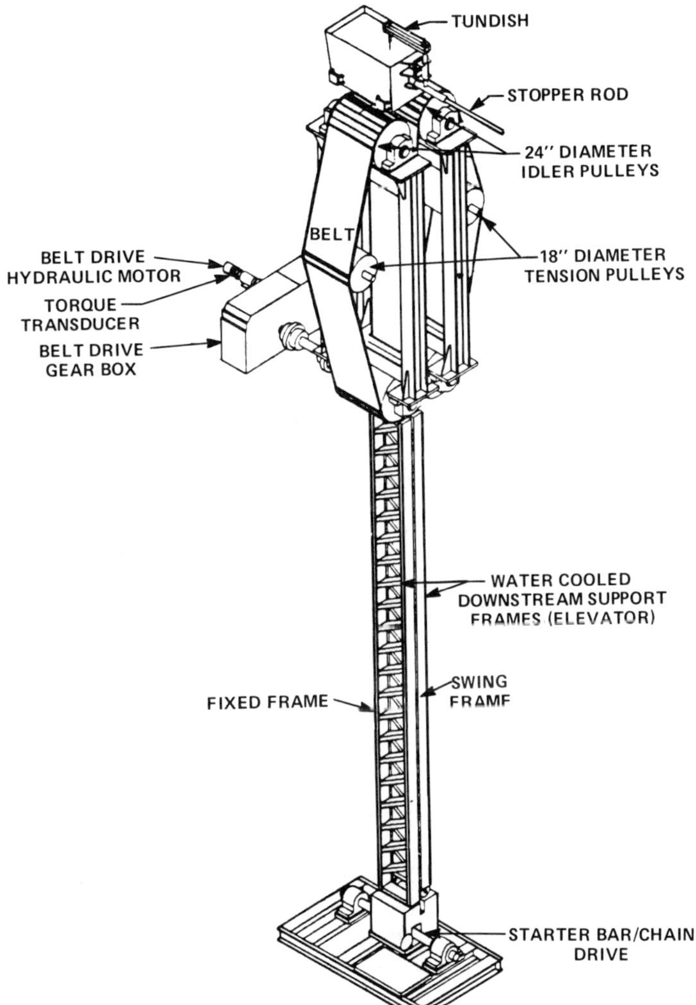

Figure 13 - Schematic of Prototype Vertical Caster Showing Belt Drive, Tundish, Mold and Downstream Support Structure.

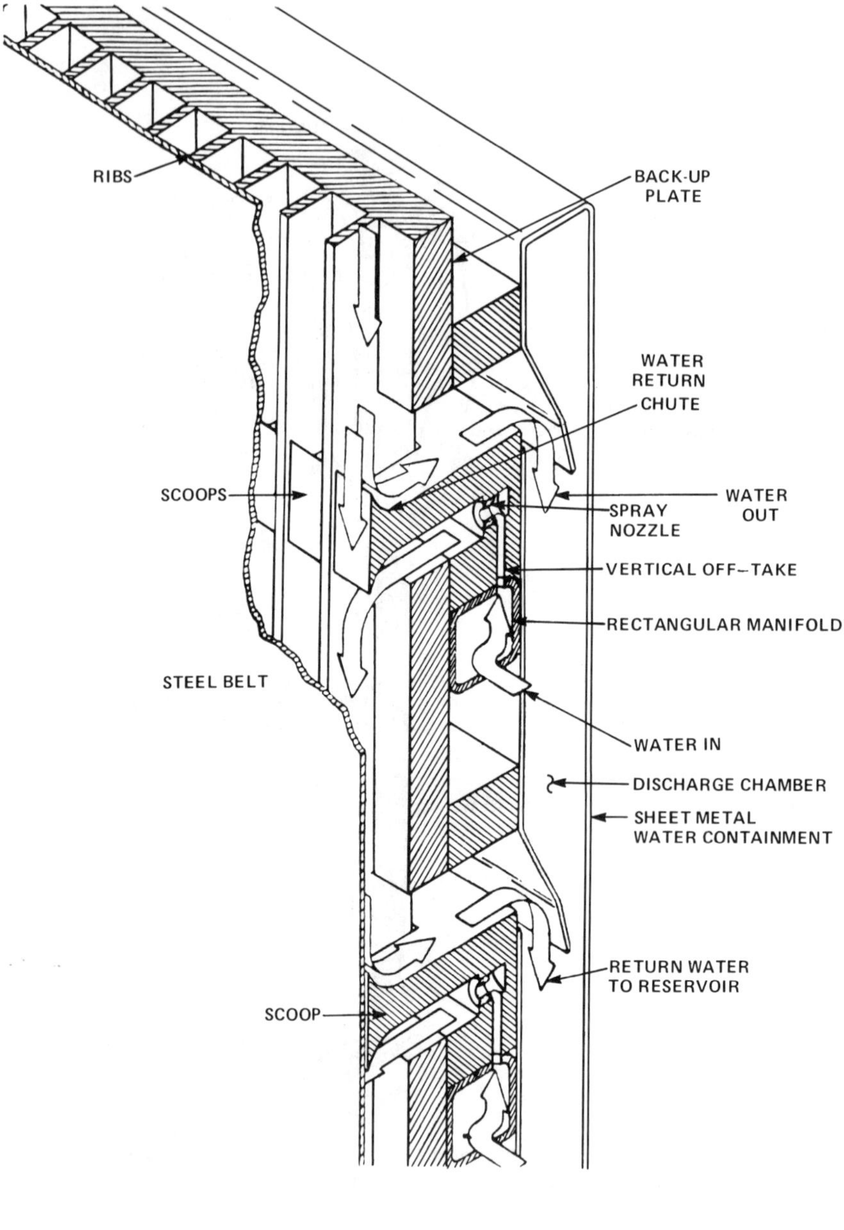

Figure 14 - Schematic of Belt Cooling Water Application and Removal.

Silica nozzles were used throughout all 30 H-DS casting trials as the liquid-steel erosion of the silica was not significant during the limited H-DS casting time (other nozzle materials were available, but the shortened phase-down casting campaign did not allow time for their use). The nozzle design was modified during the casting campaign to minimize nozzle distortion at the nozzle tip. In all casts, extensive instrumentation and data-logging systems were used to record all important casting variables and cast TSC product quality was evaluated as the critical dependent variable. Samples of TSC cast product were hot-rolled off-line to allow testing of the resulting hot-rolled sheet and some material was subsequently cold-rolled.

Project Results

The horizontal TSC approach was successful in the sense that closed-pool casting was achieved during several casts on the H-DS machine. However, at the end of the H-DS casting campaign in September 1987, this closed-pool feeding technology was only partially developed and many unsolved problems remained. The first 17 H-DS casting trials were used to debug the casting line and develop operating practices in the simpler open-pool casting mode. Closed-pool casting was attempted in the subsequent 13 casts (after starting each cast in the open-pool mode) and 7 casts were terminated prior to the transition to closed-pool conditions. The remaining six casts successfully achieved closed-pool conditions. Under open-pool conditions, the longest cast was 341 feet and the highest average casting speed was 410 ipm; under closed-pool conditions, the comparable values were 151 feet and 445 ipm.

Encouraging results were that (1) during the best cast, when the transition was made from open-pool to closed-pool conditions, a dramatic improvement in surface quality occurred and the turbulence and pouring stream marks clearly visible on open-pool TSC product were essentially eliminated (Figure 15); (2) the control system and sensors* allowed stable closed-pool casting conditions to be maintained over extended casting periods without variations in tundish head or nozzle-to-belt gaps causing collapse of the liquid-steel menisci; (3) the control system* also achieved a straightforward and repeatable transition from open-pool to closed-pool casting conditions; (4) the total horizontal feeding system* (including tundish, preheaters, control systems, sensors, articulated joint and nozzle) was generally trouble-free and operated well without any steel skulling or alumina clogging problems within the feeding train; this feeding system was successful in casting low-inclusion-content TSC sections with total oxygen contents around 20 ppm (Figure 16); (5) the post-machine foot rolls and containment racks* allowed high-speed casting (casting speeds of 450 ipm were reliably achieved and on occasion speeds reached 500 ipm) with liquid core extending beyond the twin-belt machine itself; and (6) satisfactory sheet properties were obtained by conventional hot and cold rolling of as-cast product samples from these H-DS casts (this confirmed earlier rolling trials on static-cast 1-inch-thick plates) (Figure 17).

Discouraging results were that (1) all closed-pool casting attempts were terminated early due to problems with nozzle tip breakage; (2) this nozzle tip breakage resulted from the formation of steel skulls on the nozzle tip which periodically grew to contact the solidifying shell which was moving with the belts (Figure 18). The nozzle skull was then removed

* NOTE: These developments are the subject of U. S. patent applications.

Open-Pool Feeding

Closed-Pool Feeding

Figure 15 - Comparison of Effects of "Open-Pool" and "Closed-Pool" Feeding Systems on As-Cast Surface Quality (Bottom TSC Slab Surface Shown).

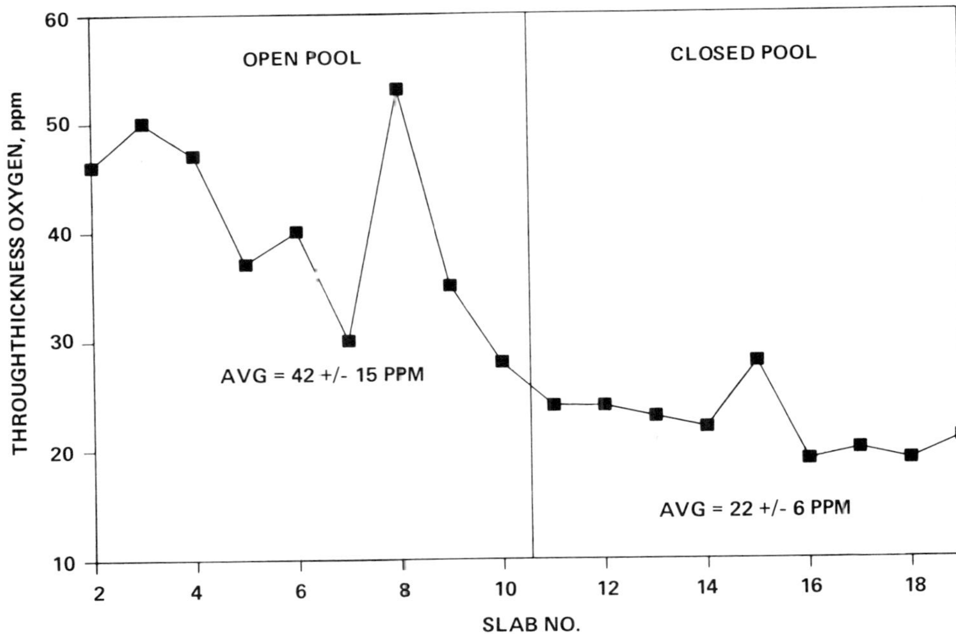

Figure 16 - Variaticn of Oxygen Content Along Cast Length (Cast No. H027)

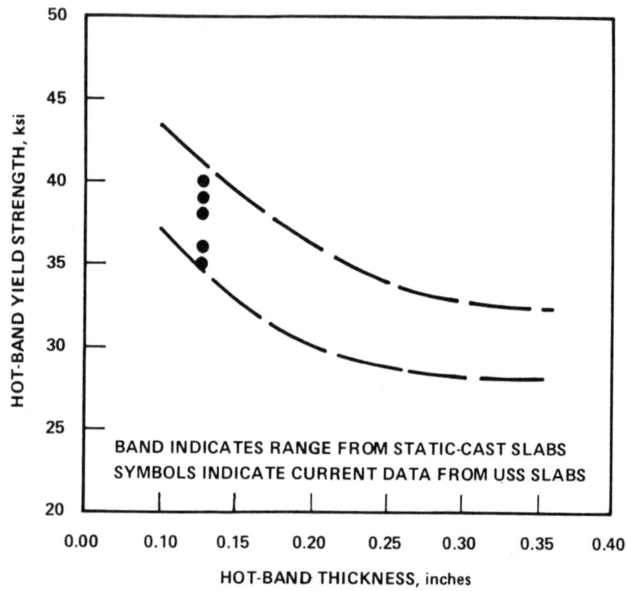

Figure 17 - Hot-Band Yield-Strength Data from Horizontal Simulator and Static-Cast Slabs.

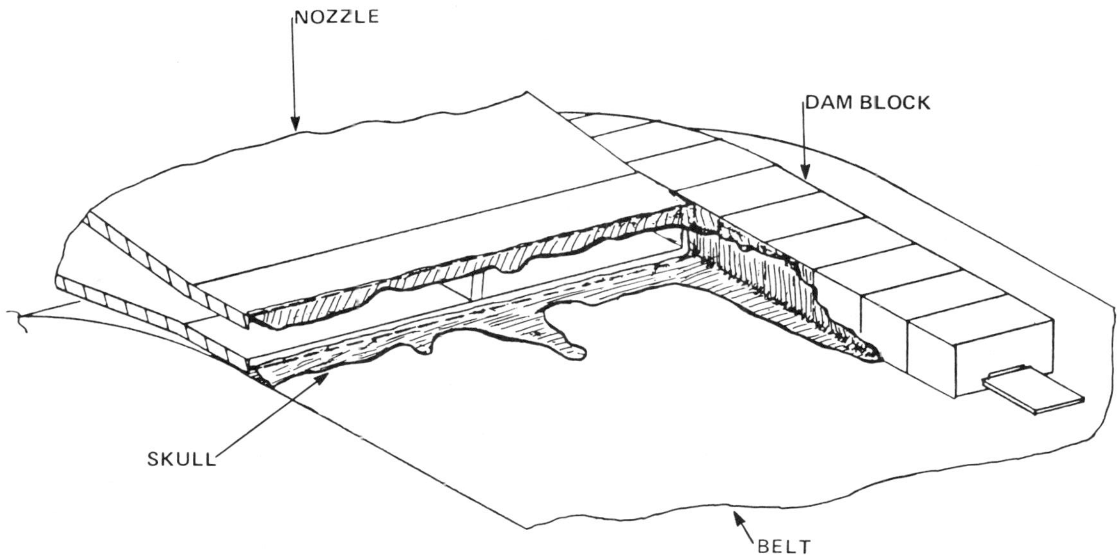

Figure 18 - Schematic of Nozzle Skull Formation During Closed-Pool Casting

from the nozzle tip and included into the cast product surface (Figure 19). Repetitive removal of these nozzle skulls eventually broke parts of the nozzle tip away causing uncontrolled flooding of the resulting large nozzle-to-belt gaps; (3) none of the countermeasures attempted in this limited series of casting trials (including increased casting speed, increased liquid-steel superheat and belt coatings) had any effect on this nozzle skulling problem; (4) as predicted by water-modeling results, very tight nozzle-to-belt gaps (under 0.015 inch) proved necessary to prevent liquid-steel flooding of the gap and this required very time-consuming precast setup and adjustment. For such a system to be commercially practiced, some type of robotic system with feedback logic and position sensing would be required to keep manual adjustment to a minimum; (5) the excellent surface quality achieved in the best closed-pool cast could not be repeated and generally poor, irregular surfaces were produced in subsequent closed-pool casts. It is suspected that nonpouring system changes made to allow higher casting speeds resulted in increased belt distortion and rough, severely cracked as-cast surfaces; and (6) belt distortion was a serious problem and it was not possible to develop belt coatings to overcome this during the phase-down casting campaign.

MHD enhancement of the horizontal feeding system was unsuccessful in that the effort was terminated when experimental results showed that it was not possible to achieve any significant improvement in the stability of the liquid-steel menisci in the H-DS machine via MHD effects. Although both mathematic modeling and results from a small liquid-metal (Gallium/Indium) test rig showed MHD stabilization was possible using the preferred approach of electric currents passed transversely across the belt width, the critical issue of upstream and downstream current spreading was overlooked. In the mathematical models, such current spreading was assumed to be minor. In the Gallium/Indium rig, only 6-inch-wide belts were used so significant current spreading did not occur. However, a full-width current spreading rig was subsequently built to check this critical point for the case of the 29-inch-wide belt (17-inch-wide casting) H-DS machine. Results from this rig showed that current spreading over the 17-inch width was significantly greater than had been assumed previously. This seriously decreased the electrical efficiency of the system and meant that no significant MHD stabilization of the liquid-steel menisci in the H-DS machine could be achieved.

The vertical TSC approach using deformable end walls was not successful as the prototype vertical caster machine always failed during the start-up phase of all casting trials and steady-state casting was never achieved. The critical requirement for achieving successful vertical casting was for the end-wall solidified shells formed high in the V-bell mouth at the meniscus level to deform down to the 1-inch thickness as they traveled down through the V-bell mouth. In practice, this did not occur and the end-wall shells jammed at the meniscus level and tore away from the broadface solidified shells which moved as planned down through the V-bell mouth region (Figure 20). When these torn shell regions reached the bottom of the stationary end-wall molds (a few inches into the parallel 1-inch-thick mold section), breakouts occurred on one or both end walls. It should be noted that these transient start-up effects could not be predicted by the steady-state solidification/deformation mathematic model. Because the vertical TSC caster never achieved steady-state casting, mathematical model predictions of surface grooving resulting from solidified shell deformation could not be validated. Attempts to improve end-wall shell deformation, including decreased end-wall heat transfer, increased casting speed, modified start-up procedures, and end-wall molds of different shape, did not prevent the end shells from jamming in the V-bell mouth region. It is possible that deformable end walls may be made

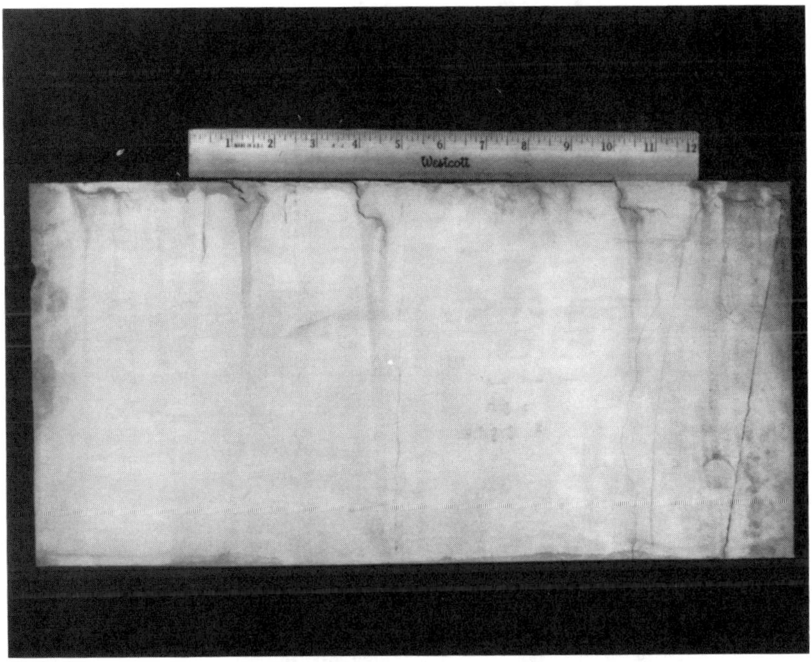

Figure 19 - Nozzle Skulls Included in TSC As-Cast Product (Top TSC Slab Surface Shown).

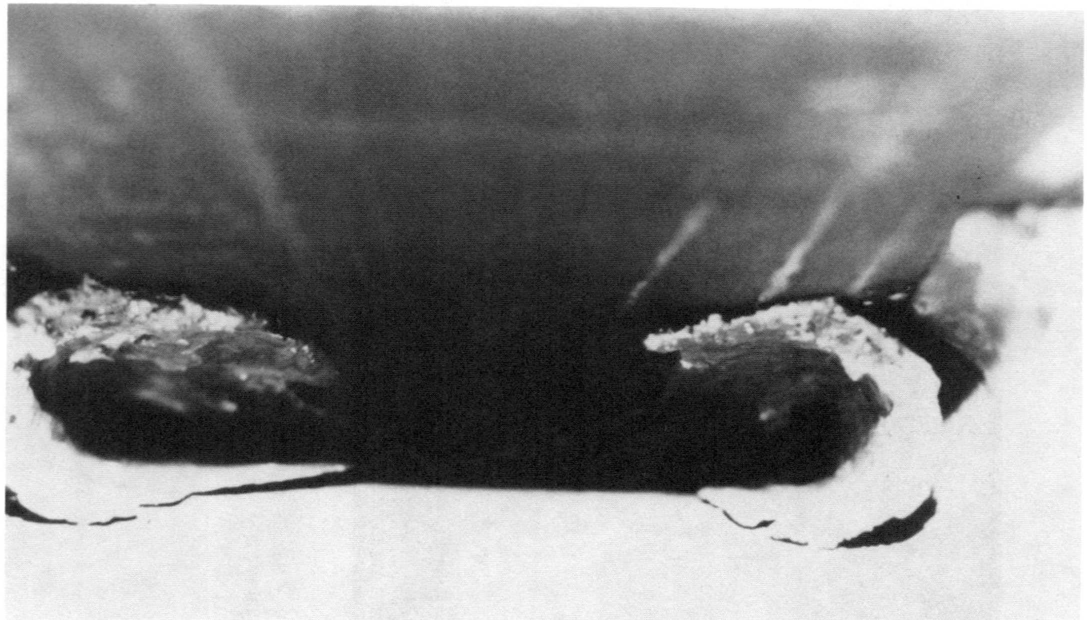

Figure 20 - Top View of V-Mouth Showing Shells After Fifth Trial With Prototype Vertical Caster.

to work via improved lubrication (including optimum vibration) and/or by decreasing the overall deformation by casting sections thicker than 1 inch, and refractory end walls have been successfully used by others (notably Kawasaki). However, none of these variables were tested during the vertical prototype casting trials.

Static casting and processing of 1-inch-thick plates showed that satisfactory sheet properties would be obtained by conventional hot and cold rolling of 1-inch-thick TSC cast product. In addition, hot bending and unbending trials on 1-inch static-cast plates showed that it would be feasible to hot coil and uncoil as-cast TSC products as a method of connecting new TSC caster installations with existing hot-strip-mill finishing trains (12). To ensure that no product defects (such as transverse surface cracks) will be induced from such bending and unbending, for low-carbon aluminum-killed sheet steels coiling/uncoiling temperatures should be kept above 1800°F and coiling/uncoiling strains should be kept below 6 percent (Figure 21) (for 0.1/0.12% carbon steels these coiling/uncoiling requirements were more restrictive and required strains less than 4%). These requirements should be relatively easy to meet in the design of coiling/uncoiling equipment for TSC casting lines but some cast product edge heating may be required. It is important to note that no TSC cast 1-inch-thick slabs were actually coiled and uncoiled during these experiments. However, because the metallurgical structure of these 1-inch static-cast slabs was similar to that of the 1-inch-thick H-DS cast TSC product, the results are expected to be representative.

Conclusions

Although many very interesting technical problems were identified as candidates for further study during the subject project, the potential benefits of any such developments were not judged sufficient to justify further work. There are two key points here. First, the fixed-mold-type TSC technologies [such as the SMS development and other similar TSC developments by Mannesmann-Demag (13) and Danieli (14)] have the major advantage of being based on an evolution of proven, slab casting hardware rather than requiring the development of totally new twin-belt technologies for steel casting. Second, the area of application where the higher casting speed of twin-belt TSC technology is a significant advantage is rather limited (see Table I).

The vertical twin-belt TSC approach showed no technical promise during the prototype vertical casting trials within the range of variables investigated. It is possible that deformable end walls may be made to work via improved lubrication (including optimum vibration) and/or decreasing the overall deformation by casting sections thicker than 1 inch, and refractory-type end walls have been successfully used by others (notably Kawasaki). However, it is questionable whether the vertical twin-belt approach can offer any significant casting-speed advantage over fixed-mold-type TSC technology because of the high ferrostatic pressure that the casting belts must sustain in the vertical configuration [though Kawasaki claims casting speeds up to 15 m/min (15)].

The horizontal twin-belt TSC approach using closed-pool feeding showed some technical promise during the H-DS casting trials and does offer a potential casting-speed advantage over fixed-mold-type TSC technology. However, this potential casting-speed advantage is of limited commercial significance and the crux technology of closed-pool pouring is complex, potentially high cost, and has significant unresolved problems. The unresolved problems include (1) the elimination of the steel skulls

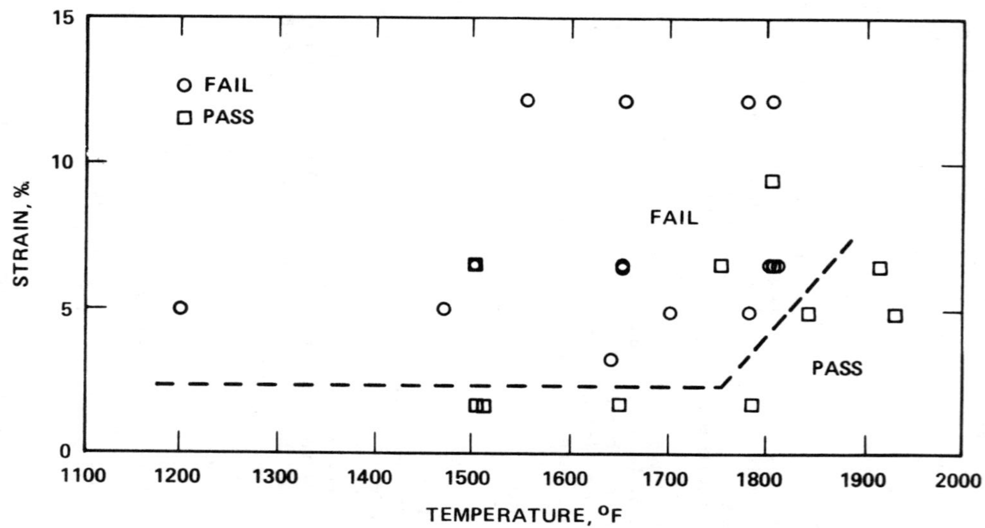

Figure 21 - Strain Versus Temperature From Bending Tests on 1-Inch Static-Cast Slabs

TABLE I. APPLICATION OF TSC CASTING TECHNOLOGIES

HOT-ROLLING PROCESS	POSSIBLE TSC CASTING TECHNOLOGIES: FIXED-MOLD TYPE	POSSIBLE TSC CASTING TECHNOLOGIES: MOVING-MOLD TWIN-BELT TYPE	COMMENTS
EXISTING HOT-STRIP-MILL FINISHING TRAIN	MULTIPLE STRANDS REQUIRED.	MULTIPLE STRANDS REQUIRED.	SATISFACTORY FIXED-MOLD-TYPE TSC CASTER PRODUCTIVITY CAN BE ACHIEVED VIA CASTING 2-INCH-THICK AND/OR ADDITIONAL CASTING STRANDS.
LOWER PRODUCTIVITY CONVENTIONAL HOT MILL	*SINGLE STRAND POSSIBLE. CANNOT LINK CASTING/ ROLLING.	SINGLE STRAND POSSIBLE. CAN LINK CASTING/ROLLING	MILL CANNOT DIRECT-ROLL IN CASE OF FIXED-MOLD-TYPE TSC. AS-CASTING SPEED TOO LOW.
LOWER PRODUCTIVITY PLANETARY HOT MILL	SINGLE STRAND POSSIBLE. CAN LINK CASTING/ROLLING.	SINGLE STRAND POSSIBLE. CAN LINK CASTING/ROLLING.	PLANETARY MILL CAN ACCOMMODATE LOW CASTING SPEED.

[box] = AREA OF ADVANTAGE FOR MOVING-MOLD TWIN-BELT-TYPE TSC (LINKED CASTING/ROLLING CAN OFFER HIGHER COIL WEIGHTS AND SLIGHTLY LOWER COSTS).

NOTE: *CONFIGURATION FOR NEW NUCOR TSC MINIMILL.

which continually form on the nozzle tip; (2) the elimination of the irregular rough, cracked, surfaces which always formed in later closed-pool casts (probably involves methods of minimizing belt distortion); (3) the development of practical methods of setting and maintaining the tight nozzle-to-belt gaps (under 0.015 inch) which were shown to be required (or the development of approaches which would allow larger gaps to be used); (4) the development of practical methods of fabricating and operating the wide nozzles required when casting wide commercial TSC products; and (5) the development of belt coatings which both optimize initial heat transfer to the belts and do not generate gas bubbles (either by reaction or release of adsorbed gases) which form surface depressions on the cast product.

It would now appear that fixed-mold-type TSC technology evolved from slab casting will be satisfactory for thin slab casting, and the next challenge in casting research is to develop the revolutionary technology which will be required for strip casting where the high casting speeds required to achieve acceptable productivity levels will mandate the use of moving-mold systems.

Acknowledgments

This project would not have occurred without strong support from research management at both USS and Bethlehem Steel. Support and counsel from A. M. Rathbone, Director - Research (USS) and R. W. Bouman, Manager - Process Research (Bethlehem Steel) is gratefully acknowledged. Valuable technical consulting was provided by J. H. Cox (formerly Bethlehem Steel) and critical technical contributions to the project were made by T. W. Lewis (Harbison-Walker Refractories) and J. B. Allyn (Hazelett Strip Casting Corporation). In addition, the contributions of the total project teams at all five companies are gratefully acknowledged as is the project funding from the U. S. Department of Energy, Office of Industrial Programs.

Technology Licenses

USX Corporation and Bethlehem Steel Corporation, together with their subcontractors under Cooperative Agreement No. DE-FC07-84ID12545 with the Department of Energy, have filed and are owners of patent applications on certain developments made during the course of work performed pursuant to said Cooperative Agreement. Under the Agreement, nonexclusive licenses are available for reasonable terms from the owners of any patents issuing on such applications.

References

1. H. F. Marten and H. Fastert, Electric-Furnace Proceedings, Dallas, Texas, 44 (December 9 through 12, 1986).

2. S. S. Daniel, et al, to be published.

3. M. R. Ozgu, et al, to be published.

4. K. Ushijima, et al, Proceedings of Seaisi Thailand Conference, Bangkok, Thailand (May 18 through 20, 1987).

5. Steel Times International (March 1987).

6. E. Buch and D. Figge, AIME/ISS Steelmaking Proceedings, Washington, 69 (1986).

7. European Patent Application (Publication No. 0-147-474).

8. Japanese Patent Applications (55-133803, 60-108101, 60-124446).

9. A. Vassilicos, to be published.

10. J. F. McDermott and S. S. Daniel, Proceedings of ABAQUS Users Conference, Newport, Rhode Island (June 1 through 3, 1988).

11. J. B. Allyn and W. Szczypiorsky, CIM 89th Annual Meeting, Toronto, Ontario, Canada (May 3 through 7, 1987).

12. M. R. Ozgu, R. R. Pradhan, and D. N. Rego, Proceedings of Fourth International Conference on Continuous Casting, Brussels, Belgium (May 17 through 19, 1988).

13. 33 Magazine (December 1987), 11.

14. Danieli News, No. 107 (1987).

15. Japan Metal Bulletin (March 10, 1988), 2-3.

THE DESIGN AND IMPLEMENTATION OF THE IMRI STRIP CASTER

J. Masounave, J. Blain, E. Essadiqi and S. Caron

Industrial Materials Research Institute
National Research Council Canada
75 De Mortagne
Boucherville, Quebec, Canada J4B 6Y4

Abstract

Increased competition in the steel producing market has led to a renewed interest in strip casting technologies. Current market trends would favor technologies capable of producing strips from 1 to 10 mm in thickness with a minimum of post processing.

To this end, IMRI has started a research effort aimed at producing a laboratory scale strip caster. This machine is being used to study the metallurgical aspect of twin-roll casting and to better understand the dynamics of twin-roll casters.

The "vertical twin roll" design was chosen for its simplicity, security, versatility, low cost and flexibility in thickness of the produced material. This paper gives a detailed discussion of the IMRI twin-roll strip caster. Particular attention is given to the thermal simulation of the process. The parameters examined include : the contact angle between the melt and the rolls, the heat transfer coefficient H, the roll speed, and the amount of superheat used when casting. The influence of these parameters on the strip exit temperature is discussed, as well as the problems associated with real time control of this process.

Typical microstuctures of strip-cast alloys are also presented and discussed with the aim of reducing post processing. Alloys were either strip cast in the IMRI twin-roll machine or on a moving-plate caster. Typical microstructures of 304 stainless steel, 1020 and 1010 plain C steels, copper alloys and aluminum alloys are presented. The influence of solidification times on microsegregation is discussed. The predictions from the computer model are in good agreement with experimental results. Heat tranfer at the melt - mold interface is very high during twin-roll casting when compared to planar single mold casting. This is caused by the presence of roll separating forces.

Casting of Near Net Shape Products
Edited by Y. Sahai, J.E. Battles,
R.S. Carbonara and C.E. Mobley
The Metallurgical Society, 1988

Introduction

After suffering a long depression period the steel industry is gradually regaining its dynamism and is once again looking at ways of improving its product and reducing its production costs. Patented in 1865 (1) but soon forgotten because of its technical complexity, strip casting of metals has recently resurfaced and is once again becoming a hot topic. In Canada, steel makers have recognised the immense advantages that could be realised by casting plates in thickness ranging from 1 to 10 mm, and are putting together a research program. The major goal of this program is to evaluate different methods of strip casting and to find the one best suited to casting steel sheet 1 to to 2 meters wide at a rate of 100 tons per hour.

As part of these efforts, IMRI has started a research program on strip casting (Table I). First a vertical twin-roll strip caster has been designed and built. Using this machine we are looking at the dynamics of strip casting and trying to identify the best way of controlling this somewhat unstable process. Among the questions we are trying to answer are:

What casting rates are possible?
What are the as-cast microstructures and mechanical properties?
Will hot or cold rolling be needed after strip casting?
What types of control sensors are needed?
What are the main material and engineering problems?
Is the process economically feasible?

Using aluminum and copper alloys, we have shown the feasibility of the process. Every alloy system selected is being or will be characterized in terms of microstructure and mechanical properties.

Table I. An Overview of the Evolution of the IMRI Strip Casting Project

1- Evaluation of different casting technologies in order to select the one best suited for IMRI's goals
2- National seminar sponsored by CSIRA to start national research program on strip casting
3- Design and construction of the IMRI twin-roll caster
4- Casting trials for aluminum and aluminum silicon alloys on solid copper wheels
5- Casting trials for aluminum silicon and copper based alloys on solid 4340 steel wheels
6- Design and construction of water-cooled casting wheels
7- Casting trials for copper alloys, stainless and carbon steels on the water-cooled rolls

The IMRI Twin-Roll Caster

Table II gives an overview of the current work being done world wide on strip casting. An examination of this table reveals the strengths and weaknesses of the different methods of producing metal strips:

a) Single-roll casting yields a strip with varying and non-symmetric microstructure through the thickness. Strip thickness is limited to no more than 3 or 4 mm and the casting rates are low due to limited heat transfer between metal and rolls. Planar flow casting is able to increase the casting rates but is still limited as to strip thickness and asymmetry of microstructure.

b) Twin-belt casters can produce, at very high rates, plates with minimum thickness of at least 10 mm. Because of belt warping due to thermal stress, it seems difficult to obtain thinner strips. Furthermore, the mechanical complexity of such machines cannot be overlooked.

c) Finally, twin-roll casters are able to cast strips in the right range of thickness (1 to 10 mm) with good surface finish. The challenge with this type of machine is to be able to produce at high enough rates and with minimum roll separation force to prevent roll damage.

TABLE II. Strip-Casting Technologies for Steels

Organization	Plate dimension Width•thickness (mm)	Casting processes	Casting speed (m/min)
Japan			
NSC	260 x 1	twin roll	
NKK	100 x (2 to 5)	twin roll	2 - 25
Sumitomo	600 x (20 to 80)	twin belt (Hazelett)	4 - 6
Kawasaki	200 x (0,1 to 0,4)	twin roll	180 - 720
	100 x 30	twin belt	0,5 - 12,5
	600 x (40 to 50)	twin belt	15 (anticipated)
Kobe	270 x (1 to 2)	twin roll	5 - 10
Nippon Metal	315 x (1 to 3)	twin roll	
United States			
Bethlehem/USS	1800 x (12 to 15)	twin belt (Hazelett)	9 - 12
Nucor	1300 x 40	twin belt (Hazelett)	
Allegheny-Ludlum	300 x (1 to 2)	single roll	
Armco/Westinghouse	75 x (0,7 to 3)	single roll	
National/Battelle	1800 x (0,25 to 4,5)	single roll	
LTV	250 x (2 to 5)	continuous casting	
SMS-CONCAST	1600 x 50	continuous casting	6
Europe			
BSC	400 x (1 to 25)	twin roll	
	300 x 30	casting train	10 - 13
CSM	300 x (1 to 25)	twin roll	
Krupp	200 x (20 to 70 mm)	twin belt (Hazelett)	10
IRSID	200 x (1 to 25)	twin roll	
Thyssen	1 mm	single roll	
Mannesmann Demag	600 x 20 mm	spray rolling process	
Schloemann Siemag (SMS-CONCAST)	300 x 30 mm	standard equipment with convex mold	

The aluminum industry has shown that twin-roll casting is a metastable process (2) and that controlling the machine is a difficult task indeed. They have usually solved the problems by casting with steel cylinders at low speed and under high loads. This solution while effective in the case of aluminum alloys has to be rejected in the case of copper and steel alloys. The temperature of the plate would rapidly destroy the roll surface because of cracking.

For all of the above reasons, we decided to build a twin-roll machine at IMRI. The rolls were placed vertically because of the advantages this geometry brings:

a) Reduced risk of breaktrough
b) No need to bend the strip after it exits the machine
c) Reduced metal head at injection nozzle
d) More compact machine design

Experimental procedures

Twin-roll caster

Figure 1 shows a schematic representation of the machine. The rolls are 10 cm wide and 60 cm in diameter. Two types of rolls can be used, either OFHC copper or 4340 steel. The metal is fed to the rolls through a ceramic nozzle. The height of the nozzle opening varies depending on the experiment from 3 to 15 mm. Constant metallostatic pressure is maintained in the nozzle through the use of an overflow chamber. The bulk of the melt is maintained in a 200 kg capacity tundish and fed to the machine through a ceramic valve. Experiments have shown that this design gives us a temperature control of the melt to within ±2°C.

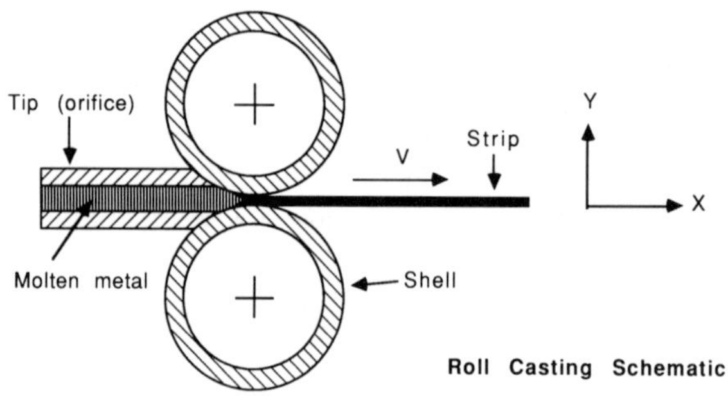

Figure 1 - Twin-roll strip caster (schematic)

The roll speed can be continuously varied from 0 to 16 RPM or from 0 to 80 RPM if a reduction gear box is removed. The maximum torque and load obtainable by this design are respectively 3400 N-m and 500 kN.

Therefore reduction between the rolls during casting is, by design, always limited. The tundish capacity and the fact that the rolls are not currently water cooled limits the possible strip length to about 30 m for 2 mm thick of brass. Roll temperature does not exceed 50°C throughout a test so that the strip solidification is not affected. Water-cooled rolls are being designed and will be used for steel strip casting.

The chemical composition of the alloys cast and the experimental conditions in the case of the steel rolls are given in Tables III and IV respectively. The physical properties of the above materials are summarized in Table V. The cast strips were 2, 1 and 0.5 mm in thickness and 10 cm in width for A380 alloy and 2 mm in thickness and 10 cm in width for brass 2600 alloy.

Moving-plate caster

In order to be able to generate some data for steel alloys before the twin-roll caster was completed, a horizontal moving-plate caster was also built. With this machine, carbon and stainless steel strips were produced by pouring the alloy directly onto the plate while the plate was moving with a velocity of 0.5 m/s. The chemical composition and the experimental conditions are given in Tables VI and VII.

Table III. Chemical Compositions (wt.%) of Aluminum and Copper Alloys

Material	Si	Cu	Zn	Fe	Mg
1060 Al Alloy	.07	--	0.01	0.18	--
380 Al Alloys	9.16	3.25	2.41	0.84	0.05
Brass	--	70	30	--	--

Table IV. Casting Conditions for Aluminum and Copper Alloys Cast on Steel Rolls

Speed (RPM)	Liquid Temperature T°C	Contact Angle, θ	Strip Thickness, mm	Heat Transfer Coefficient H $W/m^2 \cdot K$
14.5	610	7.5	0.5	10000
7.5	617	7.5	1.0	8500
5.0	617	11.0	2.0	8500
13.0	995	11.0	2.0	10000

Table V. Physical Properties of 380 and 1060 Al Alloys (3,4)

Material	A380	Brass
Density, ρ (kg/m^3)	2740	8530
Liquidus T_L (°C)	572*	955
Solidus T_S (°C)	550*	915
Latent heat (kJ/kg)	389	203
Specific heat, Cp (J/kg•K)	963	375
Thermal conductivity,K (W/m•K)	96.2	120

*these values have been determined experimentally.

Table VI. Chemical Composition (wt.%) of Steels Cast on Moving Plate

Steel	C	Mn	Si	Cr	Ni	P	S	Al	N_2
Carbon steel 1010	0,08	0,38	0,08	0,01	0,02	0,008	0,011	0,02	0,0016
Carbon steel 1020	0,18	0,7	0,01	0,03	0,01	0,005	0,01	0,02	0,0018
Stainless steel 304	0,05	1,81	0,53	18,3	8,39				

Table VII. Experimental Conditions of Plain C 1010, 1020 and Stainless Steel 304

Material	Melting Temperature (°C)	Pouring Temperature (°C)
Plain C 1010	1630	1550
Plain C 1020	1620	1540
Stainless Steel 304	1500	1480

Computer modelling of the process

Because the process is metastable and because one of our goals is to automate the strip production, a computer program was written to simulate the process. This two-dimensional simulation predicts heat flow and temperature within the cast strip and rolls. Work is also under way in collaboration with Noranda Research Center to incorporate a mechanical simulation but only the thermal analysis will be presented here.

The simulation is based on the explicit finite difference method. The formulation used in cylindrical coordinates is as follows:

For the element's tangential face:

Conduction $$\left(\frac{K\Delta\theta}{\Delta R}\right)\left(R_o + \frac{\Delta R}{2}\right)\left(T_1 - T_o\right) = Q \qquad (1)$$

Convection $$\left(H_1\Delta\theta\right)\left(R_o + \frac{\Delta R}{2}\right)\left(T_1 - T_o\right) = Q \qquad (2)$$

Radiation $$\varepsilon\sigma\Delta\theta\left(R_o + \frac{\Delta R}{2}\right)\left(T_1^4 - T_o^4\right) = Q \qquad (3)$$

For the element's radial face:

Conduction $$\left(\frac{K\Delta R}{R_o\Delta\theta}\right)\left(T_2 - T_o\right) = Q \qquad (4)$$

Unit volume temperature variation with time:

$$\Delta T = \frac{Q\Delta t}{\rho\ \Delta\theta\ R_o\Delta R\ Cp} \qquad (5)$$

where: H is the heat transfer coefficient, ε is the material's emittance, σ is the Stefan-Boltzmann constant, Q is the heat, Δt is the time increment and ΔT is the temperature increment.

The current program can be used to simulate either single or twin-roll casting. The phase change and the solid fraction are obtained by using the equivalent specific heat method (5) and by assuming a linear relationship between temperature and the amount of solid phase present. Although not rigorously correct, these hypotheses enable adequate simulation of the process even in the absence of complete thermodynamic data for the cast material.

The model permits the use of temperature-dependent material properties such as thermal conductivity, specific heat and density. However, experience has shown that using constant values equivalent to those near the melting point decreases the software runtime by an order of magnitude without significantly changing the simulation results. Thus these values were kept constant and are given in Table V. The heat transfer coefficient, H, can also be made variable along the contact angle between the strip and the rolls. We kept it constant for our simulation because the predictions were in good agreement with the results and because there is a lack of accurate data on the subject.

The boundary conditions used for the mesh are straightforward (Figure 2):

- The surface of the mesh which corresponds to the strip center has a heat flux of zero.

- The heat flux through the bottom of the mesh (the interface between the strip and the roll) is set by the heat transfer coefficient and the temperature difference between the roll surface and the strip surface.

- The input temperature is set equal to the temperature of the melt inside the injection nozzle.

- The radial temperature gradient is assumed to remain constant after the last node.

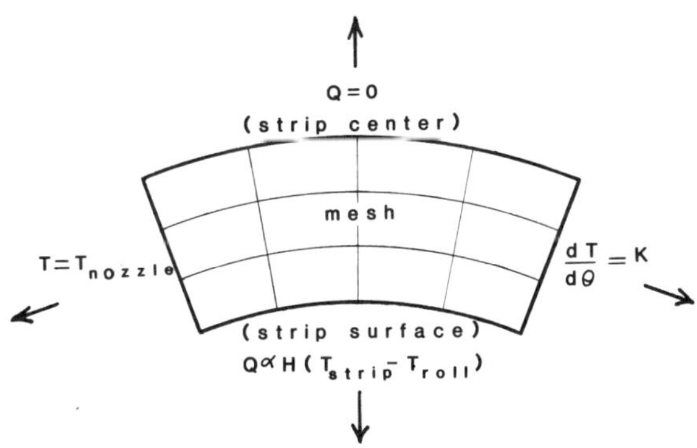

Figure 2 - Boundary conditions for the mesh

Because some of our strip casting is also done on moving plates, another program using the same techniques but in rectangular coordinates was developed. This program was used to determine the heat transfer coefficient corresponding to the cooling rates observed during the plate solidification.

Results and discussion

Simulation results

In order to verify the validity of the model, simulations of existing twin-roll machines were done. Data was obtained from aluminum producers (6) and showed that once the heat transfer coefficient (the only unknown value) was correctly set, the model predicted with accuracy both the solidification front position between the rolls and the correct strip exit temperature. These were verified for various strip thicknesses and roll speeds (6).

The first results from our machines to be simulated were for the strip casting of commercially pure aluminum. Figure 3 gives a comparison of the actual results with the ones predicted by the model. On this graph giving the roll speed during casting as a function of the melt injection temperature it can be seen that the two results agree very well if the heat transfer coefficient is set to 1500 $W/m^2 \cdot K$. This value is quite low and it is not much better than what can be obtained for a single-roll machine or a planar flow caster. This is caused partly by the fact that a protective graphite film was applied to the copper rolls prior to casting. This film probably reduced the heat transfer between the strip and the rolls. Through examination of slow motion videos of the experiments, we were also able to observe slip between the strip and the roll due to the lubricating properties of the graphite.

Another point to mention is that in the case of pure aluminum the transition between liquid coming out of the machine and solid strip coming out is very sharp because of the absence of a solidification range. Many of the tests ended when the strip became solid before the roll pinch. If this happens, the strip temperature drops sharply, the roll separation force increases and then the machine is blocked. Thus the value of H already cited applies only to cases where strip production was sustained. Therefore, in all these cases, the roll separating force was very low, which explains the poor heat transfer at the interface between strip and roll.

To better understand this phenomenon, the program was also used to generate Figure 4. This time the strip exit temperature is given as a function of roll speed for a given heat transfer coefficient, contact angle and injection temperature. It can be noted that for casting speeds below a given critical value, the exit temperature starts to drop at a very fast rate. In the case of our machine this explains why the rolls often jammed. This phenomenon has been observed in commercial machines (aluminum) (2) and is responsible for the formation of surface defects on the cast plates when the casting speed is too close to the critical value. The usual way to avoid these defects has been to operate the machine at speeds well below the critical value. The disadvantages of doing this are numerous. The two single most important ones being :

a) the production rate is lessened
b) the roll separating forces have to be very high.

With the current state of the art in materials technology it is inconceivable to consider the same approach for producing steel. So a much finer control of the process must be devised. Our aim has thus been to try, first manually and then through computer control, to operate the machine at speeds higher than to the critical speed. As the dynamics of our machine become clearer, we will be able to operate closer and closer to the critical speed and thus get strips of higher quality.

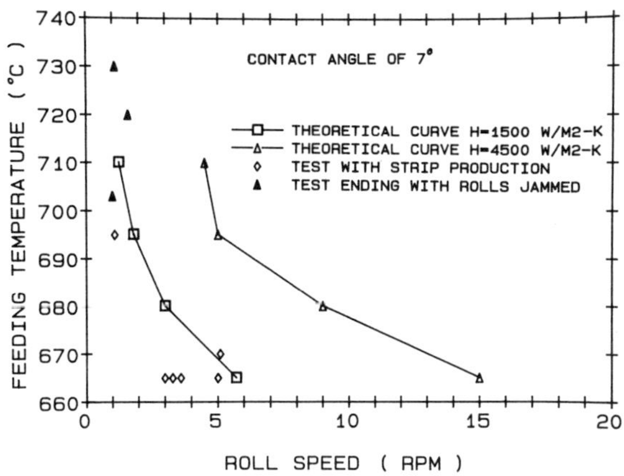

Figure 3 - Comparison between experiments and computer simulation for the casting of pure aluminum

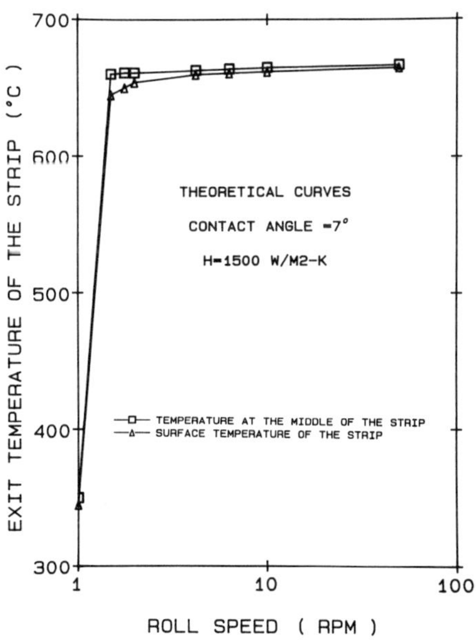

Figure 4 - Model prediction of the exit temperature of a strip of pure aluminum.

Automating the process

From the previous discussion it is clear that in order to be successful, twin-roll strip casting must be automated. Up to now, our machine has been controlled manually during the aluminum and copper casting. However, we have also been monitoring different process parameters. These parameters can be separated into two groups : the control variables which can be adjusted during production (speed, nozzle temperature) and the feedback variables that act as warning signals (force, torque, exit temperature, etc.).

Our main control variable up to now has been the roll speed and our main feedback variable the roll separation force. A typical illustration of one of our tests is given in Figure 5. At the onset of the test, the roll speed is set to the maximum. In this way, the first metal to come out of the rolls will be liquid. After 10 to 30 sec the nozzle injection temperature has stabilised and the roll speed is gradually decreased. In the present test, aluminum strip, rigid enough to be handled, started coming out of the machine at 12 RPM. The graph shows a clear relationship between the roll torque and the casting speed. The minimum speed used in this run was 5 RPM. This speed yielded torque measurements of 100 N•M and a roll loading not higher than 1 kN. During this three minute test approximatly 25 to 30 meters of strips were produced.

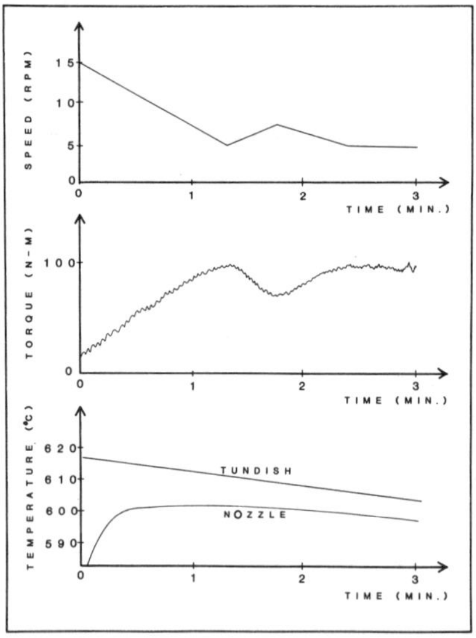

Figure 5 - Typical runout casting of 25 m of A380 2mm strip.

This test is typical of the results we have obtained to date. It clearly shows that the process can be controlled by using the roll speed to set the condition. Either the torque or the roll separating force can be used as a feedback signal to set the speed. We believe that for our purpose, the separating force is a more accurate signal. This force in effect signals the presence of solid (rigid) material between the rolls. On the other hand the torque is more a measure of where this solid material is in relation to the roll pinch.

Although annoying, the existence of the critical speed also gives us another very good feedback parameter. By continuously measuring the strip exit temperature, one can control the casting rate so as to maintain the strip quality at a constant level. We are currently installing a two color pyrometer and will shortly be able to try out this concept.

Microstructure results

Twin Roll.

A380 Aluminum Alloy. A typical structure of a transverse section of a 2 mm strip close to the surface of the wheel at a quarter thickness and at the centre is shown in Fig. 6a-c. It consists of dendrites of aluminum solid

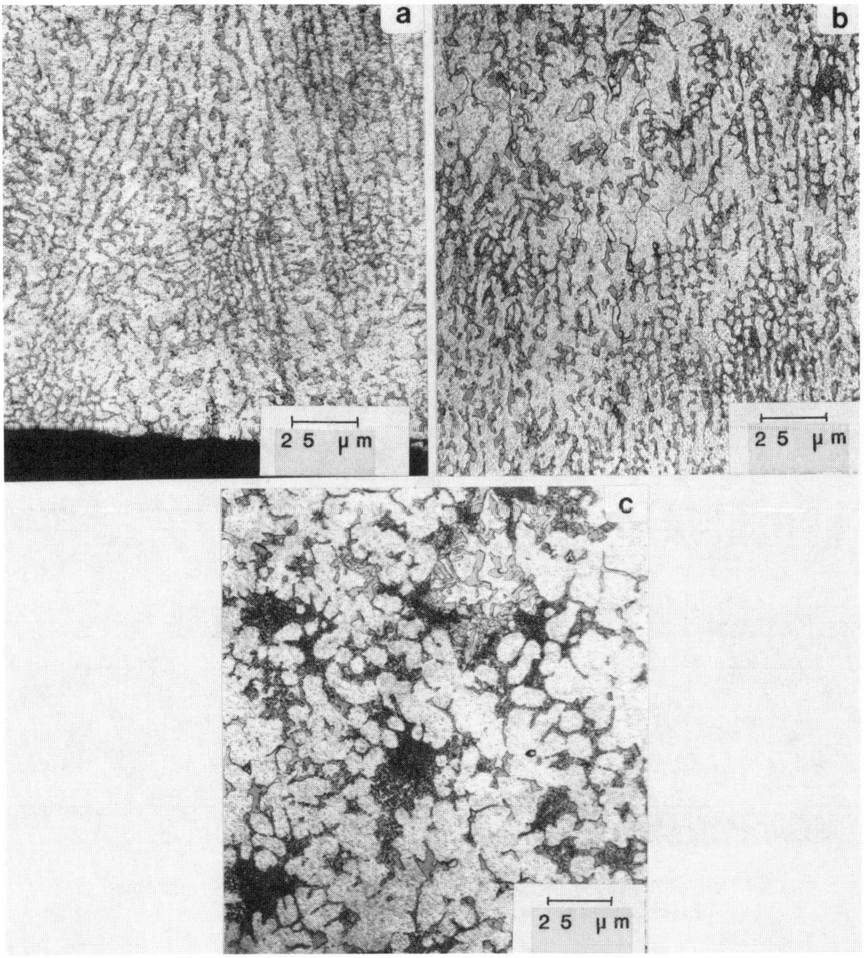

Figure 6a-c - Solidification structure of as cast A380 aluminum alloy, a) close to the wheel surface, b) at quarter thickness and c) at the centre of a strip.

solution (white areas) with a eutectic Al-Si-Al_2Cu in the interdendritic spaces. The silicon is dark and the intermetallic compound Al_2Cu is grey. The primary dendrites are parallel to the heat flow (Fig. 6a-b). In the central zone the structure is equiaxed (Fig. 6c). It covers about 0.6 mm in thickness. The secondary dendrite arm spacing (SDAS) becomes coarser as the distance from the wheel surface increases from zero to the half thickness of the strip. The SDAS varies from 3 to 6.5 μm in the region of the strip that was completely solidifed at the moment when it exited the twin rolls. In the central area, it varies from 7 to 12 μm.

It is well known that decreasing the cooling rate (increasing the solidification time) leads to coarser dendrite arm spacing. In this case the segregation at the centre of the strip becomes more significant. Also, some porosity has been observed through the thickness particularly in the central zone.

In the case of 1 mm thick strip, the characteristic structure is shown in Fig. 7a-b. As before, two zones are observed. Close to the wheel surface the dendritic structure is fine and parallel to the heat flow; the SDAS varies from 2 to 5.5 μm. At the centre, the structure is equiaxed and coarse; the SDAS varies from 5.5 to 9 μm. This zone covers about 0.2 mm. This indicates that the strip was not completely solid when it left the rolls. It is shown (Fig.13) that 0.2 mm in thickness at the centre of the strip was in the mushy zone. As opposed to the strip of 2 mm thickness, the 1 mm strip showed only little porosity.

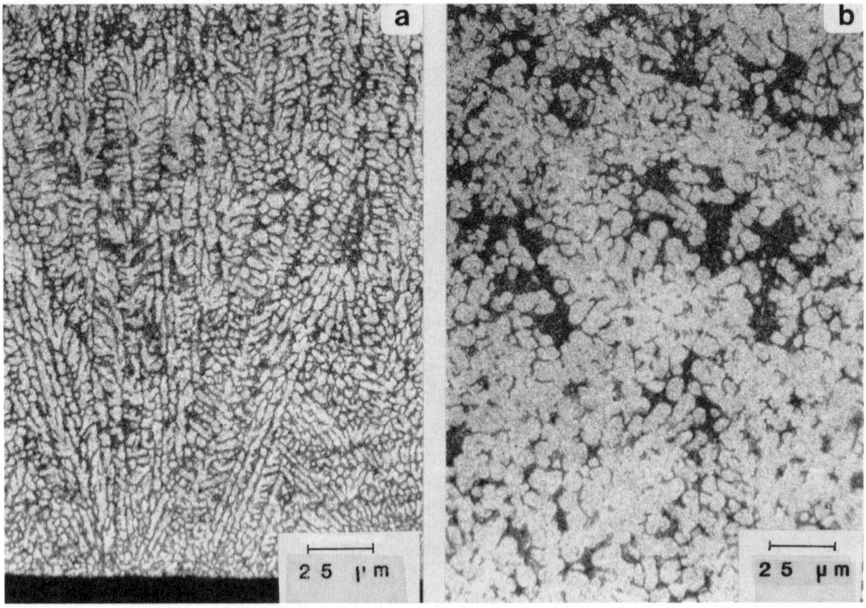

Figure 7a-b - Solidification structure of as-cast A380 aluminum alloy, a) close to the roll surface and b) at the centre of the strip 1 mm in thickness.

For 0.5 mm thick strip, the structures close to the wheel surface and the central zone are shown in Fig. 8a-b respectively. They are similar to those of 1 mm thick strip, however the strip was completely solid at the exit from the caster. The SDAS varies in the range 2 to 3.5 μm from the wheel surface to the centre of the strip.

At the exit, the mushy zone at the centre of the cast strip decreases from 1/3 of the thickness to 1/5 to zero as the thickness of the strip decreases from 2 to 1 mm to 0.5 mm. The structure is also finer, which indicates an improvement of the microsegregation with the smallest thickness.

Copper Alloys. Two zones have been observed in the 2 mm strip of brass: close to the surface an undercooled zone of about 40 μm in thickness followed by a columnar dendritic zone through almost half of the strip thickness (Fig. 9a). At the centre of the strip there is a boundary resulting from the intersection of the two solidification fronts (Fig. 9b). This indicates that the strip was almost completely solid at the exit from the twin rolls. The structure shows α dendrites (white areas). The interdendritic regions (dark areas) are rich in Zn. The secondary dendrite arm spacing varies in the range 3 to 6 μm from the wheel surface to the centre of the strip. Cu-Ni-Sn was also strip cast; the solidification structure illustrated by Fig. 10 is similar to that of the brass (7).

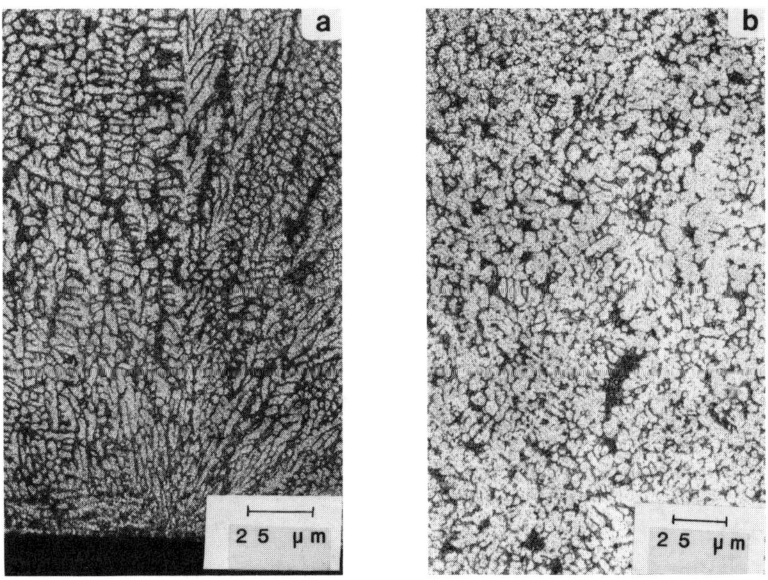

Figure 8a-b - Solidification structure of as-cast A380 aluminum alloy, a) close to the surface and b) at the centre of the strip 0.5 mm in thickness.

Moving Plate. Different materials such as 304 stainless steel, 1010 and 1020 plain C steel and Cu-Ni-Sn alloys were cast on the moving copper plate with a speed of 0.5 m/s. Fig. 11 gives a typical structure of 304 stainless steel which shows the classical three zones: undercooled, columnar and equiaxed respectively. The SDAS varies from 4 to 12 μm. The other materials present a similar solidification structure (8-10).

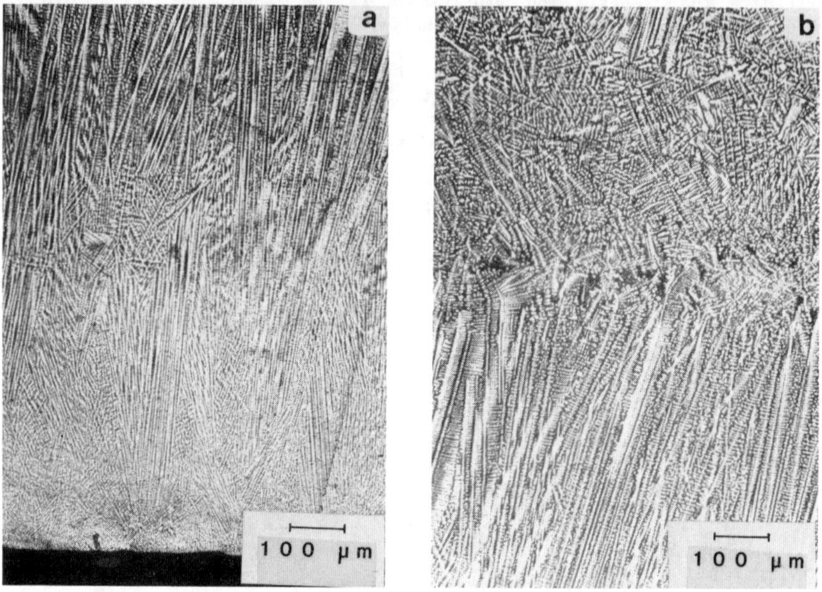

Figure 9a-b - Solidification structure of brass a) close to the wheel surface and b) at the centre of the strip 2 mm in thickness

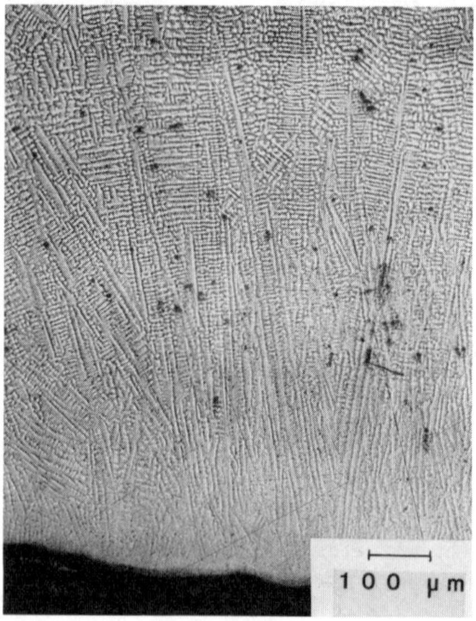

Figure 10 - Solidification structure of as-cast C72500 alloy, 1 mm in thickness.

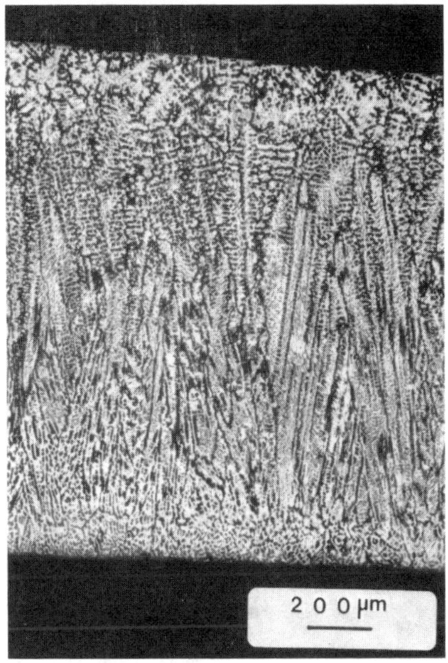

Figure 11 - As-cast structure of 304 stainless steel across the thickness, cast on the moving copper plate.

General Discussion

As stated earlier, one of the goals of the Canadian research program on strip casting is to identify strip casting technologies capable of sustained production rates of 100 tons/hour. While we have not yet produced steel strips , we can get a good idea of the capabilities of our machine by using the data from the Al - Si and Cu - Zn alloy trials.

Solidification time

A dendritic structure is characteristic of all aluminum alloy and copper alloy casting. The mechanical properties of cast alloys are influenced by the dendrite arm spacing (11-15).

A380 Alloy. The relationship between the solidification time t_f and the secondary dendrite arm spacing for 380 alloy strip cast by our twin-roll system has been established as follows (16).

$$\lambda = 10.10\ t_f^{0.43} \qquad (6)$$

where λ is the dendrite arms spacing (µm) and t_f is the solidification time (s). The solidification times t_f estimated from the SDAS measurements by using the above equation are given in Table VIII.

For the strip 2 mm in thickness, the solidification time varied from 0.06 to 0.36 s in the columnar zone, which was solid at the exit from the twin rolls. The latter value corresponds to the residence time between the rolls and thus to a speed of 5 RPM which is in agreement with the experimental conditions. The solidification time corresponding to the central zone varies from 0.43 to 1.5 s which means that the strip was completely solid one second after the exit. In the case of 1 mm thick strip the solidification times in the columnar zone (completely solid at the exit from the wheel) varies from 0.02 to 0.19 s and in the mushy zone from 0.24 to 0.76 s. The value of 0.19 s is close to the experimental residence time (0.16 s) of the strip 1 mm in thickness. For the 0.5 mm thick strip the solidification time varies from 0.02 to 0.09 s. The latter value corresponds to the residence time of the strip between the rolls. It corresponds to a speed of around 14 RPM. This value is close to the experimental one (14.5 RPM). In order to have the strip 2 and 1 mm in thickness completely solid at the exit of the twin roll, at a given contact angle, the speed must be described.

Brass. A relationship between secondary dendrite arm spacing and solidification time for brass was estimated from the literature data (17):

$$\lambda = 10.40\ t_f^{0.35} \tag{7}$$

In brass the SDAS were 3 and 6 μm close to the wheel surface and at the centre respectively. The corresponding solidification times were 0.03 and 0.20 s. The latter value is higher than the experimental residence time of the strip between the rolls (≃ 0.13 s).

304 Stainless Steel Cast on a Moving Plate. In the case of 304 stainless steel, the relationship between the SDAS and the solidification time is expressed as follows (9):

$$\lambda = 15.42\ t_f^{0.347} \tag{8}$$

The solidification times corresponding to 4 and 12 μm were 0.02 and 0.48 s respectively. Lower values of SDAS and solidification time can be expected in the case of the twin-roll process because of the higher heat transfer coefficient. It was found also, that there was no macrosegregation; however a low degree of microsegregation of Cr and Ni was observed during this process.

Table VIII. Solidification Time and SDAS for the A380 Alloy Strip Cast in Different Thicknesses.

Strip thickness	Zone	SDAS (μm)	t_f (s)
2 mm	solidified zone	3 - 6.5	0.06 - 0.36
	mushy zone	7 - 12	0.43 - 1.5
1 mm	solidified zone	2 - 5	0.02 - 0.19
	mushy zone	5.5 - 9	0.24 - 0.76
0.5 mm	solidified zone	2 - 3.6	0.02 - 0.09

Heat transfer coefficient

Table IV gives typical casting conditions for these alloys. Using our computer program we tried to find a heat transfer coefficient corresponding to our results. The main problem here is the fact that we do not know exactly how much of the exit strip is completely solidified. Micrographs taken from A380 strips clearly show that in some cases half of the strip coming out was still in a mushy state while in other cases solidification was complete.

The best example of this latter case is given by the 0.5 mm A380 strips (Fig. 12). In this case the machine produced about 5 m of solid strip at a speed of 14.5 RPM before jamming. Thus we know that this corresponds roughly to the critical speed. Using this data we were able to conclude that the heat transfer coefficient needed to complete the strip solidification before or at the roll pinch was 10 000 $W/m^2 \cdot K$. This is also consistent with simulations of tests for 1 and 2 mm thick aluminum strip. In these cases micrographs show that solidification was only complete through about 4/5 (1 mm, Fig. 13) and 3/5 (2 mm) of the strip thickness. The simulation tells us that this corresponds to a heat transfer coefficient of around 8500 $W/m^2 \cdot K$ in both cases. Thus as long as the strip is not completely solid in between the rolls the heat transfer is reduced.

Another interesting point to make is that when producing brass strip, we were able to cast at speeds of 13 RPM. In this case micrographs (Fig. 9) show that almost all of the plate was solidified between the rolls. This corresponds to a heat transfer coefficient of 10000 $W/m^2 \cdot K$. Even though the roll separation force was very low during this test, there still was good heat exchange. This leads us to believe that even higher values of H could still be obtained. After all, H values of 100 000 $W/m^2 \cdot K$ can be observed during chill block melt spinning of ribbon $\approx 1\mu$ thick (18).

Casting rate prediction. Using the 10000 $W/m^2 \cdot K$ value, we can predict the casting rate for a low carbon steel 2 mm thick strip, cast on our rolls, with a contact angle of 10°. In this case, the program gives a critical speed value of around 6.5 RPM. Given a typical industrial width of 2 meters this would yield a casting rate of 23 tons/hr.

Keeping all parameters the same and doubling the roll radius, would double the casting rate. Thus, two twin-roll machines with rolls of 1.2 meters in diameter would be able to meet the rates required by the Canadian steel makers. Since a vertical twin-roll design does not require large floor space the use of two machines in parallel would not be problematic.

Impact of surface quality

One remaining problem though is how assure that the heat transfer is uniform over the whole contact surface. If this condition is not met and the machine is operated near the critical speed, surface defects similar to those reported by aluminum (2) producers are bound to appear. Work needs to be done on that aspect. From our own experience, graphite coatings are not a very good solution because their lubricating characteristics promote slippage between the roll and the strip. Heat transfer is reduced and in the worst case the rolls can completely loose contact with the strip for short periods of time. Work is under way to look at different materials that could act as coatings. We also now have machinery to enable us to change the roll surface roughness in an effort to find the best surface condition.

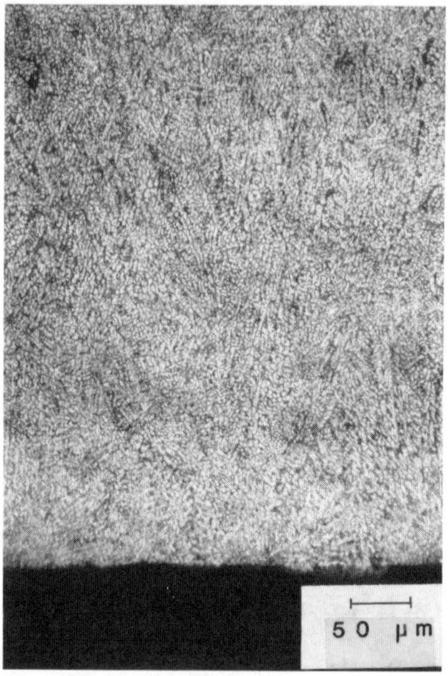

Figure 12 - As-cast structure of 0.5 mm A380 alloy strip

Figure 13 - As-cast structure of 1 mm A380 alloy strip

Conclusion

The current results of our investigations clearly show that:

- Vertical twin-roll casters are capable of casting aluminum and copper based alloys in strips of from 0.5 to 3 mm thickness.

- The heat exchange computer model we developed gives accurate predictions.

- The heat transfer coefficient at the strip - roll interface can be quite high (10 000 $W/m^2 \cdot K$) if a minimum amount of roll separation force is present.

- Manual control of the process is feasible only if the strip is not completely solid in or before the roll pinch.

- The best feedback parameter seems to be the roll separation force.

- Extrapolation from the current results indicates that a 2 m wide steel casting machine with 60 cm rolls would be able to cast steel strip at a rate of 20 tons/hr with a contact angle of 10°.

- Some effort has to be made to identify optimal roll surface conditions to promote heat exchange, avoid slipping and thus improve surface quality of the strip.

Once these steps are completed, we will be able to look at the microstructure and mechanical properties of strip-cast steels. The goals here will be to find the shortest path in terms of thermomechanical working to obtain deep drawing quality steel.

Acknowledgement

The authors would like to thank Dr. D. Briggs of CANMET for his contributions to the project and to the text and and to Dr. R. Thomson of CANMET, and Dr.G. Deep and Dr. L. Mavropoulos of Noranda Research Center for their assistance and discussions. Many thanks also go to G. St-Amand, J. P. Nadeau, P. Thibault and M. Thibodeau for their technical help.

Reference

1. Sir Henry Bessemer - US patent no. 49053 (1865).

2. M. J. Bagshaw, J. D. Hunt, R.M. Jordan, "Modeling and Control of Casting and Welding Processes", Santa Barbara, Ca, USA, Jan. 12-17, 1986, Met. Soc. of AIME, 1986, pp. 135-147

3. J.L. McCall, "Properties and Selection - Non Ferrous Alloys and pure Metals", Metals Handbook, 9th Ed. ASM, Metals Park, Ohio 44073, 2 (1979), p. 332.

4. J.L. McCall, "Properties and Selection- Non Ferrous Alloys and Pure Metals", Metals Handbook, 9th Ed. ASM, Metals Park, Ohio 44073, 2 (1979), p. 170.

5. W.C. Erickson, "Computer Simulation of Solidification", AFS Int. Cast Metals Journal, (March 1980).

6. G. Deep, Noranda Research Center, Pointe Claire, Quebec, Canada, private communication.

7. E. Essadiqi, J. Blain, L. Mavropoulos and J.Masounave, "Structure and Properties of a Strip Cast Cu-Ni-Sn and Al-Si-Cu Alloys", Proc. of TMS Conference, Chicago, Illinois, Sept. 25-29, 1988, to be published.

8. J. Masounave, J. Blain and E. Essadiqi, "La coulée directe de métaux", Séminaire sur les nouvelles applications de l'acier face à concurrence des matériaux de substitution, Commission Économique pour l'Europe, Luxembourg, (1988), 2 au 6 mai.

9. E. Essadiqi and J. Masounave, "Strip Casting of 304 Stainless Steel", NRCC/IMRI, Boucherville, Quebec, Canada, 1987, Report IGM87RT-205-621-R.

10. E. Essadiqi, J. Blain, D. Briggs and J. Masounave, "Strip Casting Simulation of 1010 and 1020 Plain C Steels", NRCC/IMRI, Boucherville, Quebec, Canada, 1988, Report IGM88RT-205-728-R.

11. M. Bamberger, B. Z. Weiss and M.M. Stupel, Met. Sci. and Tech., 3, (1987), pp. 49-56.

12. R. Elliott, "Eutectic Solidification Processing", Butterworths Ed., (1983), pp. 182-188.

13. K.J. Oswald and M.S. Misra, AFS International, Cast-Metal J., 6, (1981), p. 23.

14. G.R. Armstrong and H. Jones, "Solidification and Casting of Metals", Metal Society, London, (1979), pp. 454-458.

15. S.F. Frederick and W.A. Baily, Trans. Metall. Soc., AIME, 242, (1968), p. 2063.

16. E. Essadiqi, J. Blain and J. Masounave, "Solidification Structure of a Strip Cast Al-Si-Cu Alloys, Proc. of the CIM Conf., Vol. 11, 1988. In press.

17. T.F. Bower and M.R. Randlett, "Metallography and Microstructure", Metals Handbook, 9th Ed., ASM, 9, 1985, pp. 638-645.

18. R. Mehrabian, Intern. Metals Rev. 27, (1982), pp. 185-208.

NEW CASTING AND ROLLING TECHNOLOGY

FOR ECONOMICAL STRIP PRODUCTION

Herbert Fastert* and Hans Malinowski**

*SMS Concast Inc., Montvale, New Jersey
**SMS Schloemann-Siemag AG
Duesseldorf, WEST GERMANY

General

The CSP process is a new technology for the economic production of hot strip. CSP is the acronym for Compact-Strip Production. It represents a technical concept that revolutionizes hot-strip production (Fig. 1).

New Concept

The conventional production route for hot strip leads from the continuous caster via the slab handling area, scarfing stations and slab furnaces to the down-stream rolling mill with a roughing-stand group and a six or seven-stand finishing train.

The continuously cast sizes are up to 2,200 mm wide and 200 to 250 mm thick. These slabs are rolled down in the hot-mill roughing and finishing trains in order to achieve finishing strip thicknesses of normally 2 to 15 mm. The process requires a high degree of energy.

On the other hand, the new CSP technology allows casting of strips that are only 40 to 60 mm thick and their direct rolling in a compact CSP rolling mill. The combination of strip caster and reduced rolling work in the CSP mill shortens the production chain from liquid steel to the rolled product considerably.

Contrary to conventional plants which require high capital investment and achieve an economical production only at a high throughput rate, hot strip can be produced by a CSP plant at competitive costs starting at 500,000 t/year (Fig. 2).

Casting of Near Net Shape Products
Edited by Y. Sahai, J.E. Battles,
R.S. Carbonara and C.E. Mobley
The Metallurgical Society, 1988

Continuous Casting of Strips

In the design of the CSP caster, we have employed, to the greatest extent possible, components that have been already established and well proven in modern slab-casting plants. Consequently, the CSP-casting technology corresponds in principle with that of the conventional continuous-casting process.

The steel is supplied in the normal way, i.e., from the casting ladle via a shrouding tube, into the tundish and then through a submerged nozzle into the mold.

At the heart of this new strip-casting technology, was the development of the new mold. It is protected by worldwide patents. The top region is funnel-shaped to accommodate the submerged nozzle. It has been designed for low-stress solidification of the strand shell (Fig. 3).

The design of the taper of the funnel section permits progressive reduction in the peripheral length of the mold face from the top to the bottom of the tapered section. It balances the contraction occurring in the shell of the cast section as solidification takes place. The final cross-sectional shape of the strip is thus formed before it leaves the mold.

The parallel side walls enable rapid changing of the strip width from the control pulpit. The shape of the outlet of the submerged nozzle permits metallurgically correct distribution of hot metal.

On account of the high casting speeds, the throughput capacity is equal or greater than that of conventional slab-casting plants. The start of the casting operation is fully automatic. Fluctuations in the mold level are automatically detected and corrected. The casting powders have been especially developed for high casting speeds which, combined with high frequency oscillation and optimal mold geometry, results in a strip surface making inspection before rolling unnecessary.

Despite the high casting speeds, the rapidly solidifying thin strand is only supported for a length of 5 to 6 m below the mold instead of 20 to 40 m which is required on slab casters. Similarly, the CSP caster requires only a small bending radius of 2 to 3 m in comparison to 10 to 12 m for a conventional slab caster (Fig. 4).

Depending on the maximum casting speed and strip thickness, the strand support consists of only two or three guide segments. They can either be changed individually or simultaneously with the mold as a complete exchange unit. Finely tuned spray-water cooling in the strand-guide region provides optimal and rapid strand cooling.

A bending unit below the last strand guide segment separates the dummy-bar head from the fully solidified hot strand. This unit simultaneously forms the radius for entry into the withdrawing and straightening machine that unbends the strip into the horizontal position.

The design of the dummy-bar permits a firm connection between the starting head and hot strand within only a few seconds after the start of casting. Upon introduction of the dummy-bar, the gap between the starting head and mold walls is automatically and securely sealed by a ceramic fiber mat. Additional materials, such as cords or cooling scrap, are no longer necessary. Accordingly, there is less wear in the mold and the equipment connected with it. The dummy-bar concept results in a turn-around time of less than 20 minutes (Fig. 5).

Soaking Furnace

In order to equalize the temperature, the cast strip, which has already been cut to length by a pendulum shear, passes through a soaking furnace arranged between the CSP caster and CSP rolling mill.

The drawing shows an example of a single-strand plant with a two-zone roller-hearth furnace. The overall length of the furnace is approximately 2.5 times the length of the cut strip. The first part of the furnace serves as a heating zone and the second as a holding zone (Fig. 6).

The cut length of the cast strip enters the furnace at casting speed and passes first through the heating zone. After the tail end of the strip has left the heating zone, its complete length is within the holding zone. The entire strip has a uniform temperature. It travels parallel to an accelerating section which increases its speed to that required for entry into the first roll stand.

Rolling in the CSP Mill

The strips emerging from the furnace are rolled down to finished thicknesses in the subsequently arranged hot-strip finishing train. For this purpose, the compact CSP-rolling mill has been developed (Fig. 7).

The new design of the compact mill utilizes maximum transmittable drive torques considering an optimal roll diameter. When compared with a conventional hot-strip mill which has, for example, six stands, it has been possible to reduce the number of stands to five or four, depending on the maximum strip width and minimum final thickness. This has been the result of a constant application of the described principle. The correspondingly reduced heat radiation allows a lower entry speed resulting in less drive power.

The upstream furnace produces a strip with a constant temperature over its complete length. It is therefore not necessary to speed up the temperature as is otherwise the case in conventional mills. This procedure reduces the required drive power. Since the strip passes through the plant continuously, there are no skid marks. They usually occur on the surface in local areas of the slab-reheating furnaces due to overcooling.

The reduced rolling speed and absence of temperature speed up lead to shortening of the laminar cooling section and the runout-roller table.

Besides elimination of the roughing-stand group and fewer finishing stands, all of which are equipped with hydraulic screwdown, the reduced drive power contributes substantially to a reduction of both the capital and operating costs. The hydraulic screwdown ensures closest thickness tolerances.

Moreover, incorporation of the established and proven CVC technology (Continuously Variable Crown) and work-roll bending affects both the strip profile and flatness. This provides favorable conditions for achieving close tolerances during further processing (Fig. 8).

Pilot Plant

In order to test and verify the concept applicability under quasi-production conditions, a pilot plant was erected for continuous casting of strips in SMS' own foundry. In cooperation with experts from the steel industry, this pilot plant has proved to be a functional, efficient casting process (Fig. 9).

The width of the cast strips ranges from 1,200 to 1,600 mm and the thickness from 40 to 50 mm. After a short trial period, the planned casting speed of 6 m/min was achieved. The charged heats were between 10 and 14 t with cast strand lengths of 22 m. The type of cast steel ranged from low-carbon sheet quality to microalloyed pipe grades (Fig. 10).

The produced strips have a good microstructure and surface quality. The microstructure is, in fact, better than that in continuously cast slabs because the high solidification rate produces a fine crystalline primary structure with low central segregation (Fig. 11).

In rolling tests, 50 x 1,200 mm cast strips were rolled to a final thickness of 2.5 to 3.1 mm in only four stands of a hot-strip mill finishing train. No problems arose with either rolling or coiling. After pickling, the strips were rolled down to a 0.5 mm thickness in a reversing cold mill. They were annealed and subjected to 0.8% temper rolling.

In the test series, the surface quality of the rolled strips and various mechanically tested samples were equivalent to those from conventionally produced strips.

First CSP Production Plant

At the end of 1986, Nucor Corporation/USA placed an order with SMS for the supply of a CSP plant with strip caster and downstream hot-strip mill for the greenfield installation of a new mini-steel plant. The deciding factors in placing this order are superior economics of CSP technology as well as good casting and rolling results. Commissioning of the production plant is scheduled for 1989. It will enable Nucor to produce hot-rolled strip more economically. Both capital and operating costs are considerably lower than those with comparable conventional plant technologies.

The casting-production capacity is 820,000 t/year with a maximum casting speed of 6 m/min. The strips are up to 50 mm thick and 1,350 mm wide. The maximum coil weight is 24 t which is the equivalent of a 47 m cast strip length.

After hot strips have gone through a soaking furnace, they pass into a CSP-rolling mill where they are rolled down to final thicknesses of 2.5 to 12.7 mm through only four stands. The stands are equipped with hydraulic screwdown, work-roll bending and CVC technology.

The design of the rolling mill allows the future installation of a fifth stand for rolling of thinner finished strips. The scope of supply includes a laminar cooling section, a downcoiler for strips up to 12.7 mm thick and a coil conveyor system.

Outlook

Experience has shown that CSP technology is a functionally reliable process for economical production of hot strip. The new development considerably shortens the production chain from liquid steel to hot-rolled strip. It therefore offers steelmakers new opportunities in the reduction of costs. For the first time, CSP technology assures construction and economical operation of mini-steel plants with annual capacities of up to 1 million t of flat products.

For higher productions, e.g., 2 million t/year of hot strip, the CSP rolling mill is supplied by an upstream two-strand caster (Fig. 12). Even considering this fact, the technology is economically advantageous. As is the case with single-strand casters, the thin strand thickness and shorter solidification length lead to substantial cost savings.

The roughing-train stands in the hot-strip mill have been eliminated. Also, there is a marked reduction in processing costs in view of the shortened production chain. Even higher productions of 3 to 4 million t/year can be realized with CSP technology. Whether it be mini or high production, with the importance of its technical and economic advantages, the new technology of integrated casting and rolling of strip changes the traditional route of hot-strip production.

Features and Advantages of CSP Technology

- Small machine units with short transport routes.
- Casting of strips to transfer bar sizes.
- Same throughput capacity as conventional continuous-slab caster.
- Employment of classic components from conventional continuous casting technology.
- Very good surface and excellent internal micro-structure of the cast strips.
- Compact arrangement of caster, soaking furnace as well as rolling-mill finishing train and consequently a constant rolling temperature; elimination of temperature speedup and avoidance of skid marks.
- Improved yield, particularly within the CSP rolling mill by elimination of crop ends.
- Fewer finishing stands for rolling of the cast strip.
- Substantial energy saving through continuous operation within the same heat.
- Quality of hot strip equivalent to that of conventionally produced strip; sometimes even better.
- For the first time, economical operation of mini steelmakers for capacities of up to 1 million t/year of flat products possible.
- Capital and operating costs considerably lower than those with conventional plants.
- Revamp of existing slab-casting machines to CSP casters possible.
- Modernization of existing hot-strip mills by employment of CSP technology and elimination of slab furnaces and roughing train.

FIGURE 1

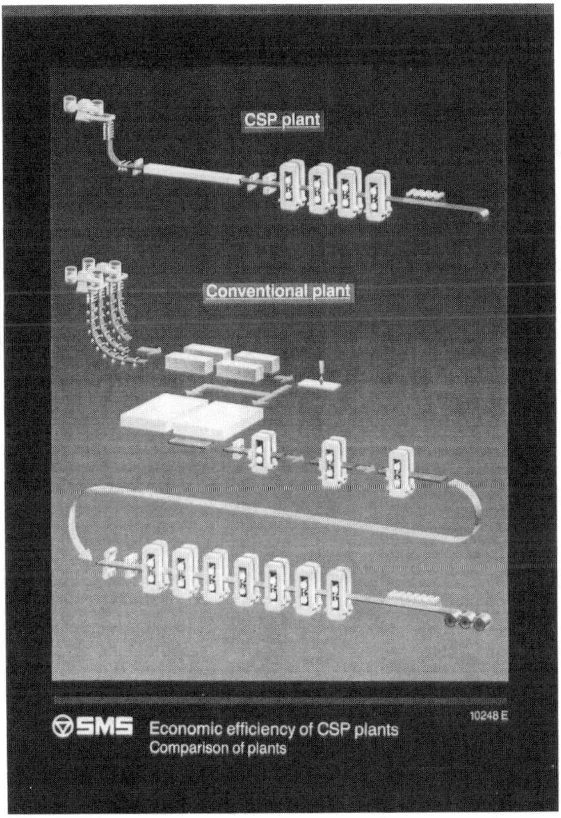
CSP plant
Conventional plant
10248 E
SMS Economic efficiency of CSP plants
Comparison of plants

FIGURE 2

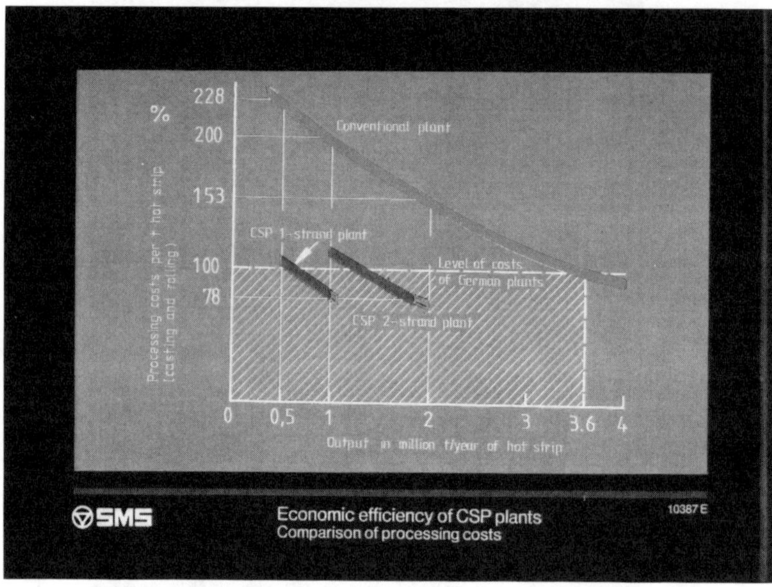
%
Processing costs per t hot strip
(casting and rolling)
228
200
153
100
78
Conventional plant
CSP 1-strand plant
Level of costs
of German plants
CSP 2-strand plant
0
0,5
1
2
3
3.6
4
Output in million t/year of hot strip
SMS
Economic efficiency of CSP plants
Comparison of processing costs
10387 E

FIGURE 3

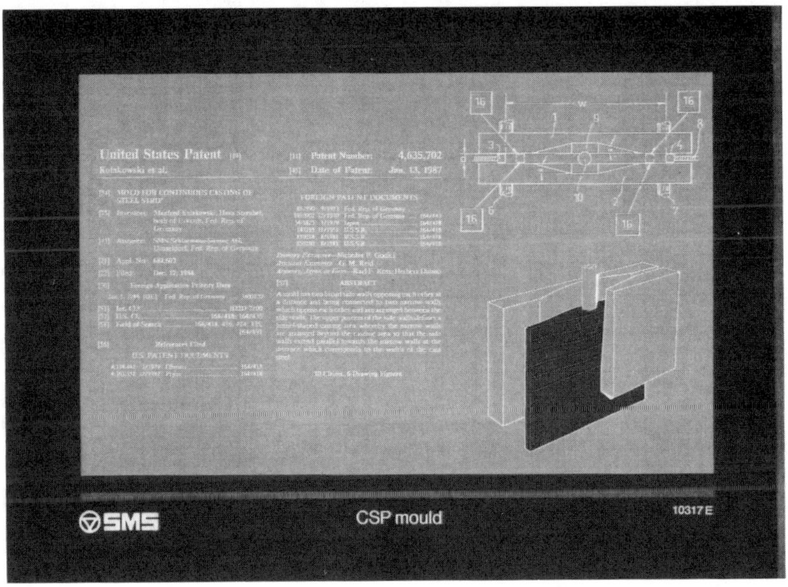
United States Patent
Patent Number: 4,635,702
Date of Patent: Jan. 13, 1987
SMS
CSP mould
10317 E

FIGURE 4

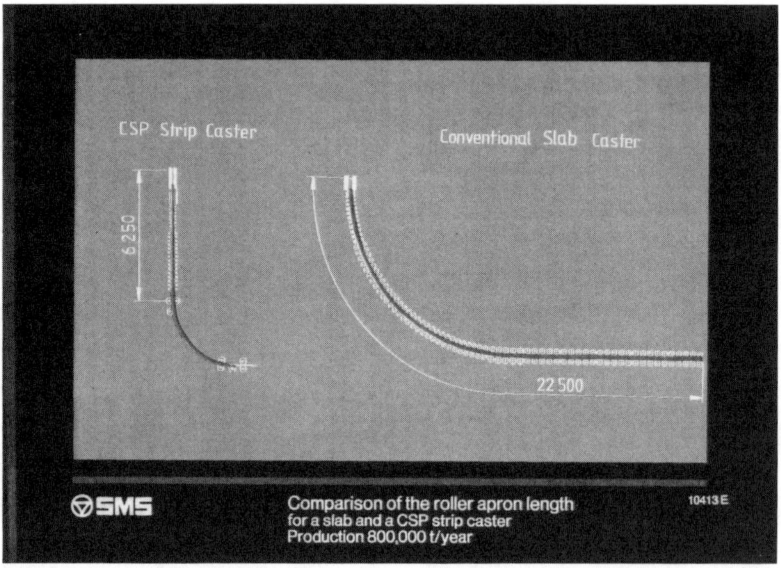

Comparison of the roller apron length
for a slab and a CSP strip caster
Production 800,000 t/year

FIGURE 5

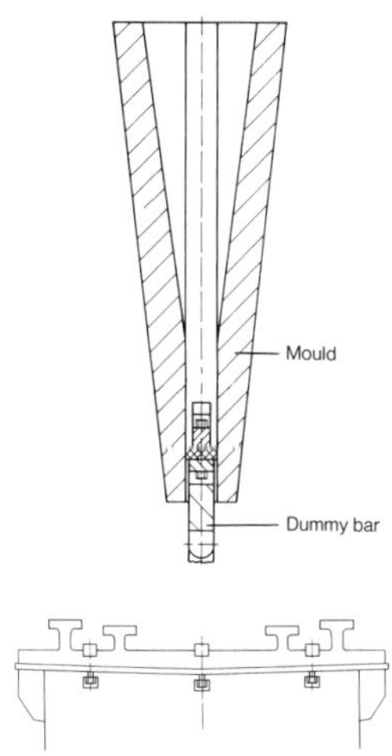

SMS **Dummy bar in starting position and shape of dummy-bar head** 10154 E

FIGURE 6

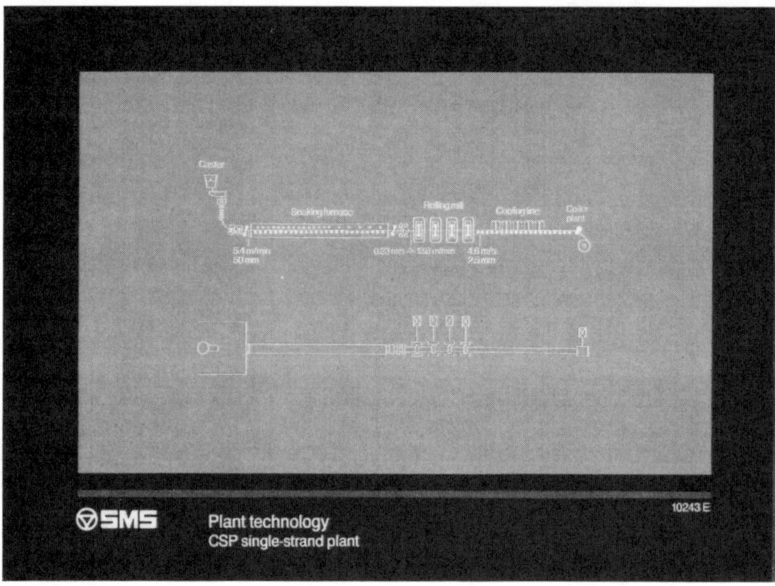
Caster
Rolling mill
Plant technology
CSP single-strand plant
10243 E

FIGURE 7

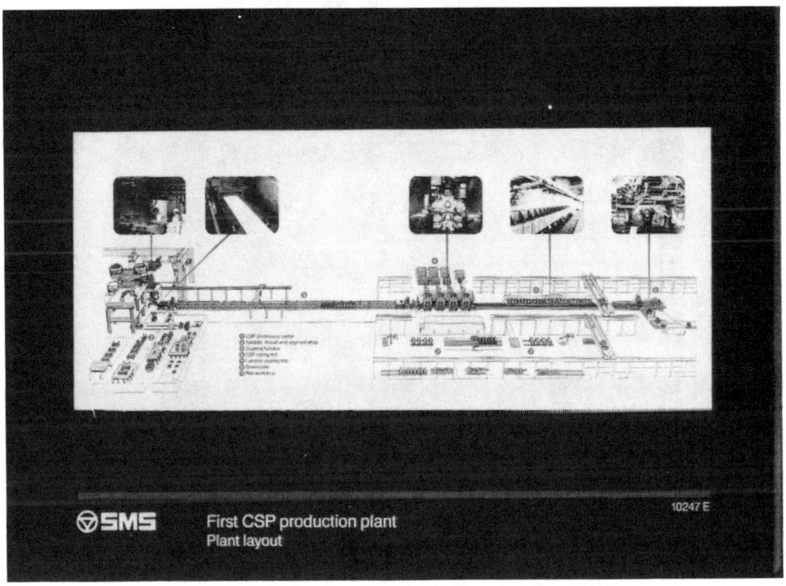

FIGURE 8

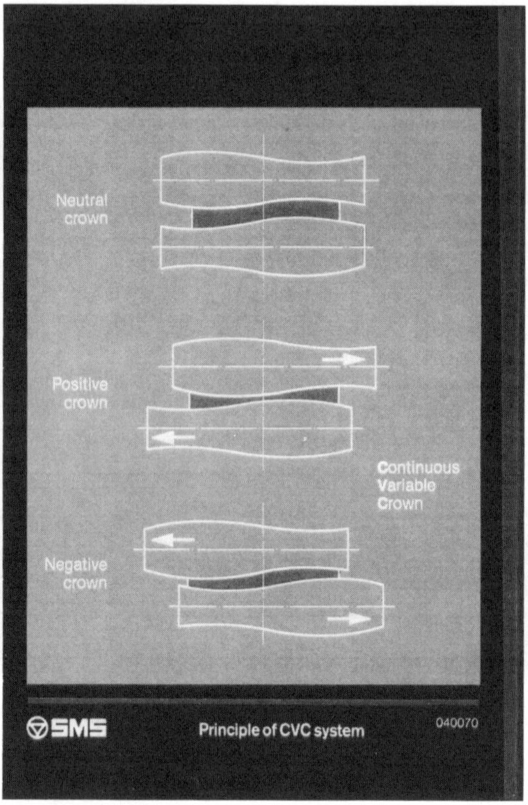
Neutral
crown
Positive
crown
Continuous
Variable
Crown
Negative
crown
SMS
Principle of CVC system
040070

FIGURE 9a

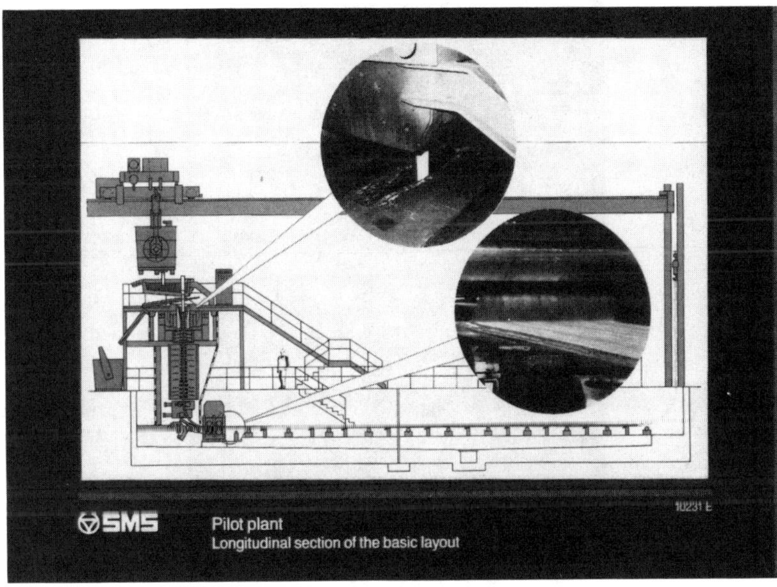
10231 E
SMS
Pilot plant
Longitudinal section of the basic layout

FIGURE 9b

SMS **Pilot plant for the casting of 50 mm thick strips**

2901-5 E

FIGURE 10

Material	C	Si	Mn	P	S	Al	N	Ti	Nb	Cu	Cr	Ni	V
Low-carbon Al killed steel	0.03–0.07	<0.02	0.15–0.25	<0.015	<0.015	0.030–0.060	0.003–0.006	—	—	—	—	—	—
Structural steel	0.10 0.14	≤0.03	0.60–0.80	<0.015	<0.015	0.005–0.025	<0.007	—	—	—	—	—	—
Structural steel with high Cu content	0.14–0.18	0.25–0.35	0.60–0.80	<0.035	<0.020	0.005–0.015	—	—	—	0.35–0.50	—	—	—
Alloyed steel	0.25–0.31	0.50–0.70	1.0–1.2	<0.045		0.025–0.035	—	—	—	—	0.60–0.80	—	—
Microalloyed steel, pipe grade (X 70)	0.07–0.17	0.20–0.40	1.30–1.45	<0.020	<0.020	0.015–0.030	<0.009	—	0.03–0.05	—	—	—	0.05–0.07
Austenitic stainless steel	<0.05	0.35–0.50	1.40–1.50	<0.035	<0.015	—	—	0.015–0.025	—	—	17.2–18.0	8.5–9.1	—

SMS — Analyses of additional cast steel grades — 10174 E

FIGURE 11

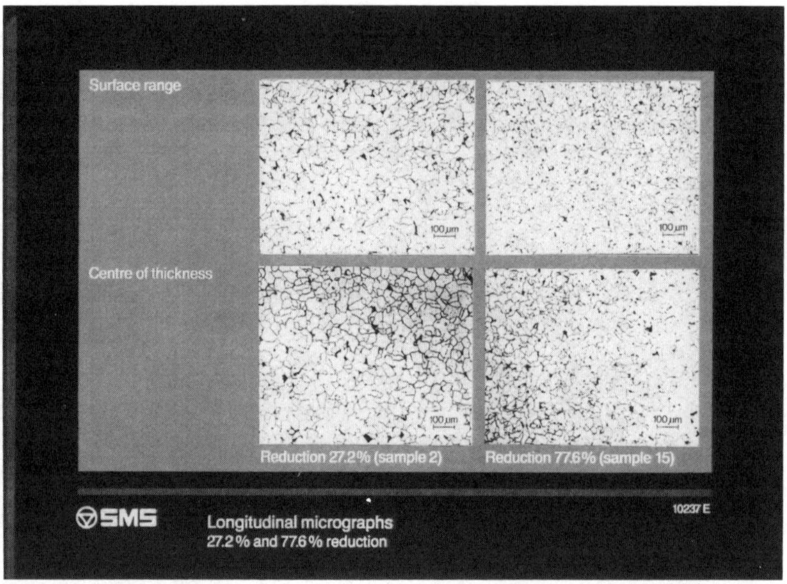

Surface range
Centre of thickness
100 µm
100 µm
100 µm
100 µm
Reduction 27.2% (sample 2)
Reduction 77.6% (sample 15)
SMS
Longitudinal micrographs
27.2% and 77.6% reduction
10237 E

FIGURE 12

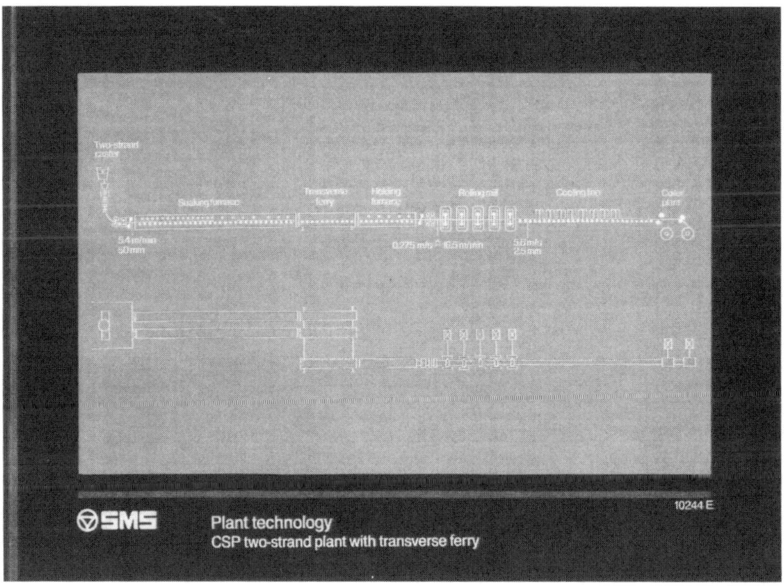
Two-strand caster
Soaking furnace
Transverse ferry
Holding furnace
Rolling mill
Cooling line
Coiler plant
5.4 m/min
50 mm
0.275 m/s ≙ 16.5 m/min
5.6 m/s
2.5 mm
10244 E
SMS
Plant technology
CSP two-strand plant with transverse ferry

CASTING TECHNIQUES V: ELECTROMAGNETIC CONFINEMENT

Session Chairman: **G. Hummert**
Westinghouse Electric Corp.
Pittsburgh, Pennsylvania

THE DEVELOPMENT OF ELECTROMAGNETIC CASTING

FOR COPPER BASE ALLOYS

B. G. Lewis and J. C. Yarwood

Olin Metals Research Laboratories
91 Shelton Avenue, New Haven, CT 06511

Abstract

Significant benefits in terms of hot mill processing and subsequent yield have been demonstrated for a range of copper alloys produced by electromagnetic casting (EMC). The macro and microstructural advantages are derived from the mold-less casting technology. This paper describes the fundamentals of the EMC process and its development for copper-base alloys. The improvements in hot working yield are described with reference to phosphor-bronze processing.

Casting of Near Net Shape Products
Edited by Y. Sahai, J.E. Battles,
R.S. Carbonara and C.E. Mobley
The Metallurgical Society, 1988

Introduction

Electromagnetic phenomena have found many applications in metallurgical processes, induction melting and electromagnetic stirring being the well-known examples. The use of electromagnetic forces to support liquid metals via levitation melting is also well established, however, the practical extension to containing and shaping molten metal during casting represents a more recent development. Electromagnetic casting processes were first described and demonstrated in rudimentary form in the early 1970's by the Soviets [1]. The aluminum industry was the first to adopt this new technology and development work by, for example Meier et al [2] of Alusuisse, has led to industry-wide EMC commercialization for vertical semi-continuous casting. The development of EMC technology for high density metals, specifically copper, was undertaken by Olin Corporation in 1978 under a Soviet license. During the last ten years the development program has moved through small scale billet casting to an automated pre-production slab casting pilot plant. Commercialization planning is underway.

Process Principles

EMC parallels standard direct chill casting (DCC) in all aspects except the use of the chill mold. In EMC, a magnetic containment cavity replaces the water cooled copper mold and solidification is afforded by direct water application to the solidifying ingot. The processes are compared schematically in Figure 1.

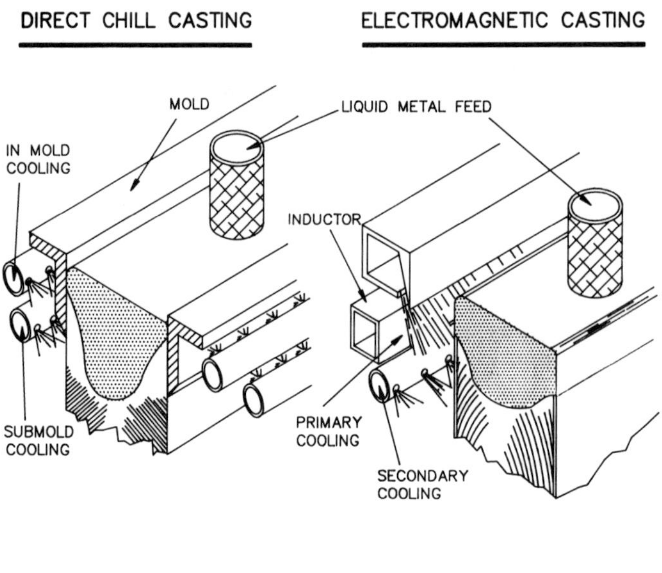

Figure 1. Comparison of normal DCC and EMC molds in operation.

The liquid metal is contained and shaped by the Lorentz force that results from the interaction between the current induced in the surface layers of the metal and the magnetic flux at the surface, Figure 2. For stable containment, the magnetic containment pressure (numerically equal to the integral of the Lorentz force through the zone of induced current i.e. skin depth) should balance the metallostatic head at all vertical positions. This criterion is achieved by shaping the magnetic profile with a metallic screen and is described by the equations in Figure 2[3].

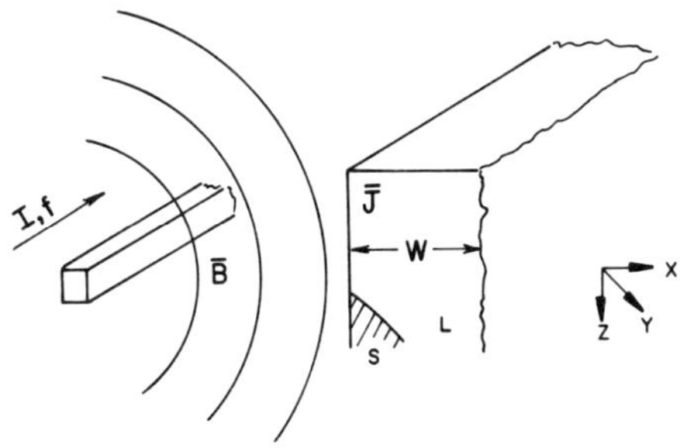

$$\bar{F} = \bar{J} \wedge \bar{B} \longrightarrow F_X = B_Z J_Y \text{ (Lorentz)}$$

$$P_X^{mag} = \int_0^{W/2} F_X \, dx \longrightarrow \frac{\mu_o}{4} H^2 \psi\left(\frac{W}{\delta}\right) \propto I^2$$

For uniform containment:

$$P_X^{mag}(z) = z\rho g$$

Figure 2. Schematic of magnetic condition for stable uniform containment in EMC.

Process Development

Relative to aluminum, density, surface energy and temperature distinguish copper-base alloys in the establishment of EMC casting practices. The greater density represents the need for increased containment power, the lower surface energies create the need for tight containment control and the higher temperatures are more demanding of sensors and materials. In combination, these factors map to the need for stringent control over metal flow and distribution.

Control by constant containment conditions adopted by the aluminum industry proved inadequate for copper alloy casting. Central, therefore, to the success of Olin's EMC program for high density metals has been the development of proprietary containment controls (4). The measurement and control practice takes advantage of the intrinsic process physics by monitoring the system inductance. A fast response power supply is then controlled to maintain ingot size and thereby, its stability. Casting is carried out at a head of 1"-2" of molten metal. At this level containment power consumption is a small fraction of the energy required for melting. Percentage variations in this metal head can destabilize the system, thus the management of flow and containment are subject to full process control from casting initiation through termination.

Process Advantages

The absence of a water cooled copper mold results in dramatic improvements in cast bar quality. Mold related damage, tears, and exaggerated meniscus markings are avoided. Further, due to the direct water application, recalescence and associated inverse segregation phenomena are precluded, Figure 3. Figure 4(a),(b) compares the surfaces of EMC and DCC phosphor-bronze (C510). Note the smooth EMC surface and the total absence of Sn rich surface exudate. Figure 5 depicts the near surface segregation profiles in EMC and DCC material as determined by microprobe.

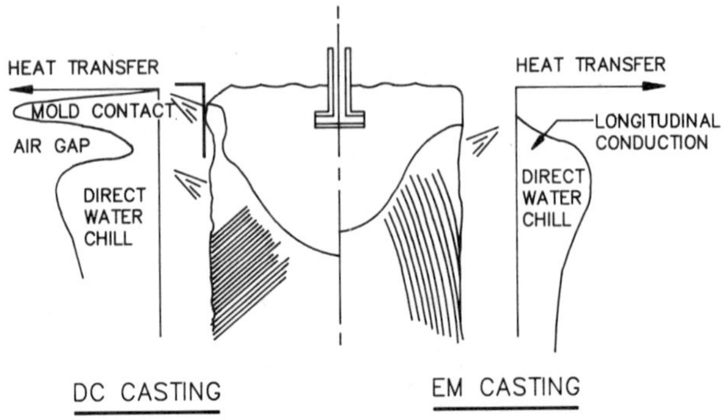

Figure 3. Heat transfer conditions in DCC and EMC compared.

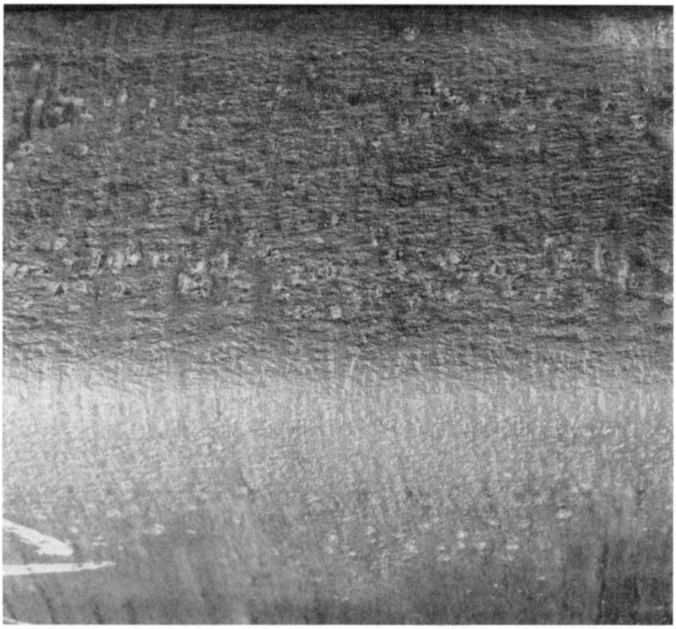

Figure 4 (a). Detail of EMC bar corner surface.

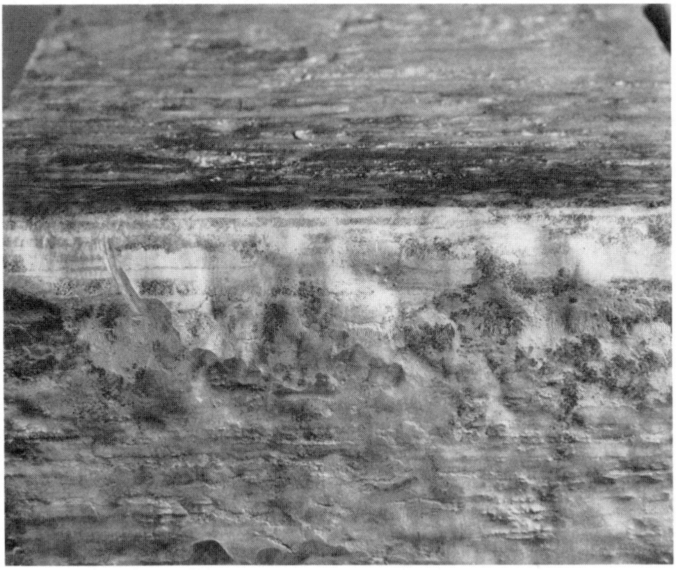

Figure 4(b). Detail of DCC bar corner surface.

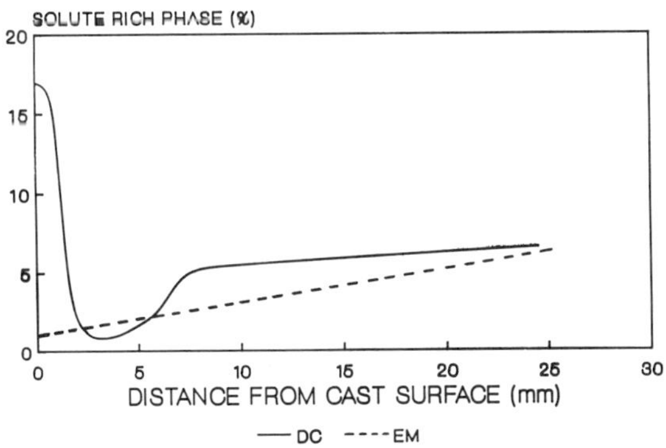

Figure 5. Typical surface solute segregation profiles.

The high near surface cooling gradients and the shape of liquid-solid interface result in modifications of both the macro and microstructure relative to DCC. The macrostructure, following the heat transfer history, is characterized by grains that initially run parallel to the cast surface before assuming growth normal to the liquid-solid interface in the sump, Figure 6 (a). The equivalent DCC structure is shown in Figure 6(b). This structure is obviously advantageous in that the cast grains are resistant to opening on hot rolling. The surface microstructure, as represented by the secondary dendrite arm spacing, is finer than for DCC, characteristic of shorter local solidification times. Typical data are shown in Figure 7. This fine structure plays an important role in potentially shortening homogenizing times for hot processing. For a given alloy, the surface macrostructure itself is retained to varying depths depending on the homogenization treatment. Thus, optimization of the hot rolling practice is important in maximizing the benefit of EMC structures.

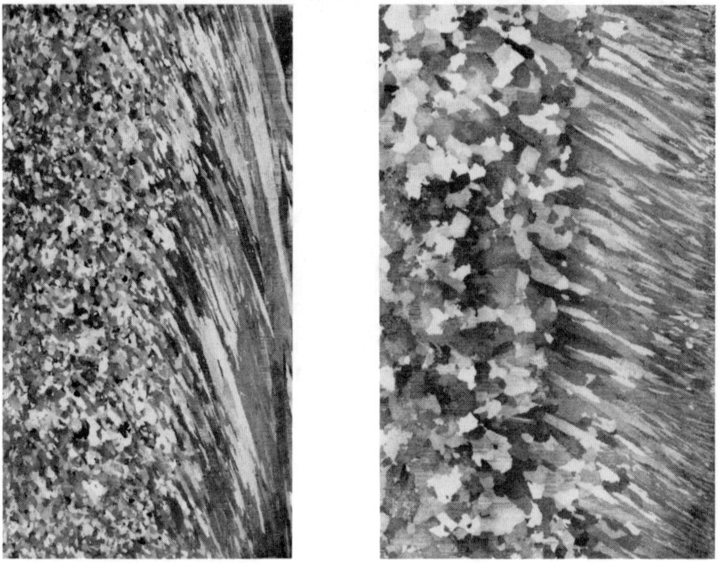

(a) (b)

Figure 6 (a,b). Typical EMC (a) and DCC (b) macrostructure of C510 compared.

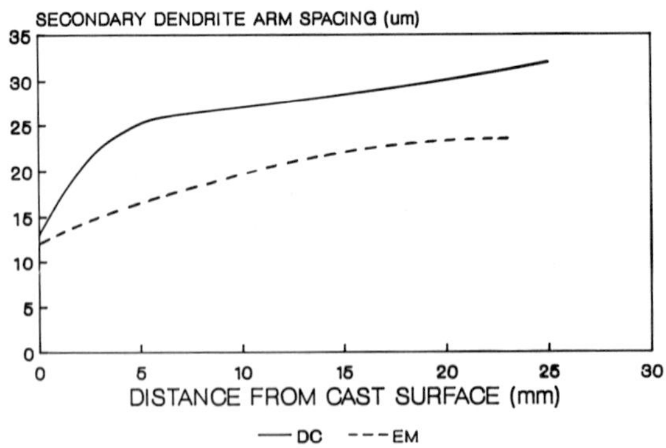

Figure 7. Near surface SDAS for EMC and DCC structures.

Economics

Copper alloy direct chill castings are normally reheated for hot working. In the strip production sequence of operations the hot rolled plate is scalped to remove surface defects and rolled in oxide. Typically, the surface and edges of the plate are milled to provide defect-free metal for subsequent cold rolling/annealing operations. Edge cracks not removed in this milling step tend to propagate upon further rolling, and necessitate edge trimming. These milling and edge trimming steps greatly reduce casting plant yield and generate scalp chips and edge trim which are troublesome and costly to recycle.

Both laboratory and commercial scale tests have demonstrated the crack resistant nature of EMC castings. The inverse segregate free surface structure, and absence of the related chill mold cast surface defects impart a rolling performance on long freezing range alloys that has previously been more typical of low alloyed coppers. Accordingly, there is need for only minimal surface restoration, mainly to remove oxides, before proceeding with cold rolling.

Laboratory hot rolling, Figure 8 (a,b), demonstrates the nature of the improvements in surface and edge integrity that can be achieved for EMC ingots. Plant rolling of pilot plant castings, Figure 9(a,b), confirms the significantly improved performance for all alloys examined.

This casting process will first find applications for those alloys that exhibit inverse segregation in standard DC casting, where yield improvements in excess of 20% have been demonstrated. However, consistency of product and the absence of all mold related defects qualify EMC for a wide range of copper-based alloys. It is further anticipated that the fundamental process disciplines that have been defined for copper alloys will be applicable to other high density, high temperature metals.

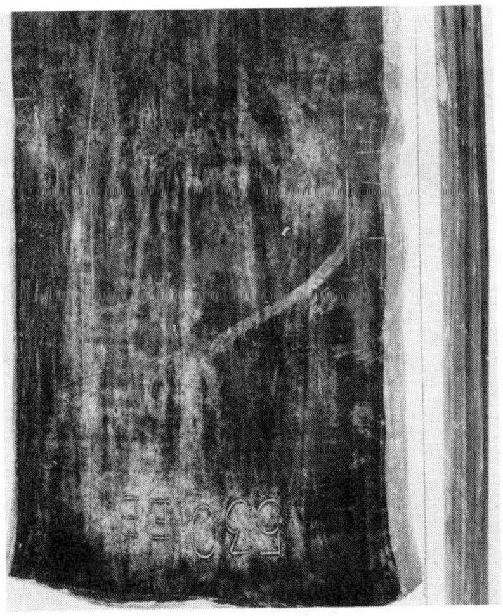

(a)

(b)

Figure 8 (a,b). Surface and edge quality if EMC (a) and DCC (b) hot rolled plates.

(a)

(b)

Figure 9(a,b). Pre-production pilot plant bars (a) and hot band (b).

Acknowledgements

The authors wish to acknowledge the technical contributions of many Olin MRL staff in furthering this program and are indebted to the Brass Group of Olin Corporation for support of this work and for permission to publish the results.

References

(1) Z. N. Getselev
"Casting in an Electromagnetic Field", J. Metals, Oct 1971, pp. 38-59.

(2) H. A. Meier et al
"Alusuisse Experience with Electromagnetic Molds", Light Metals, 1977, pp. 223-233.

(3) J. Davies and P. Simpson,
"Induction Heating Handbook", McGraw-Hill, 1979

(4) U.S. Patent 4,161,206
"Electromagnetic Casting Apparatus and Process".

DESIGN AND MAGNETOHYDRODYNAMIC STABILITY

FOR HORIZONTAL ELECTROMAGNETIC CONTINUOUS CASTING

J. R. Hull, R. J. Lari, W. F. Praeg, D. M. Rote, L. R. Turner

Argonne National Laboratory
Argonne, IL 60439 USA.

Abstract

High-frequency electromagnetic (EM) fields are proposed for the containment of thin sheets of metal during continuous casting in a horizontal orientation. Major EM components of the caster include a picture frame magnet, conducting side guards for positioning and side containment, and an eddy current return path that requires special electrodes, near the inlet outlet of the caster, that force uniform current distribution. An inductance stabilization method provides global stability, and magnetic flux compression provides local stability. Stability analysis and EM heating constraints indicate that frequencies in the range of 25-100 kHz are desirable to levitate 6-mm-thick steel. For systems without active feedback, a conducting screen is required below the metal, with a gap between the screen and the molten metal no more than twice the metal thickness. Experiments in which 10-kHz EM fields were used to statically levitate sheets of molten tin indicate that dominant magnotohydrodynamic instabilities correspond to theory.

Casting of Near Net Shape Products
Edited by Y. Sahai, J.E. Battles,
R.S. Carbonara and C.E. Mobley
The Metallurgical Society, 1988

Introduction

The ability to cast metal sheet products in near-net shape is expected to greatly reduce production costs. Although rapid cooling techniques have made it possible to cast thin (< 0.05 mm thick) sheets by means of single wheel casters, attempts to produce thicker sheets by means of belts and wheels have been only marginally successful thus far. Thin slabs have been cast with the twin-belt caster.

Electromagnetic (EM) methods have been used in a number of metallurgical applications (1) and have been used commercially in metal casting mainly for mold stirring and containment of vertical ingots in the production of aluminum. In cooperation with personnel from a number of U.S. steel companies, researchers at Argonne National Laboratory (ANL) have investigated processes for EM continuous casting (EMCC) of steel in large-aspect-ratio shapes (2). The goal of this project was to cast steel sheets about 6 mm thick. Most of the efforts focused on design of a bench-top caster capable of casting 15-cm-wide sheets. One concept involves the use of EM fields to constrain and levitate a vertical sheet of molten metal (2,3). A second concept involves EM levitation of a horizontal sheet of molten metal by the interaction of a horizontal high-frequency (HF) magnetic field and longitudinal eddy currents which are induced in the sheet by the HF magnetic field (2,4).

The horizontal casting concept requires a uniform horizontal magnetic field below the liquid metal sheet and a much reduced field or no field above. Key components of the conceptual design for the horizontal caster include

- A magnet that levitates the molten metal by the interaction of the horizontal HF field and the eddy currents generated within a skin depth of the bottom of the levitated sheet,
- A feed system in which flow is controlled by mechanical and electromagnetic means,
- A cooling system,
- Heat shields that protect the magnet system.

This paper presents general features of the magnet design for the horizontal EMCC concept, and discusses some of the major technical issues associated with the EM levitation that affect the feasibility of the concept. A more detailed discussion is given in Refs. 2 and 4.

Relevant Principles of Electromagnetic Levitation

The principles of magnetic pressure, skin depth, and electromagnetic heating (power loss) are central to EM casting.

Assume a large, non-magnetic, conducting plate as shown in Fig. 1. The bottom surface is at $y = 0$ and the top surface at $y = \infty$. An alternating magnetic field is applied in the x direction. The field inside the plate is given by

$$B_x = B_s e^{-y/\delta} \cos(\omega t - y/\delta) , \qquad (1)$$

where $B_s \cos \omega t$ is the instantaneous external field (tesla) at the surface and δ, the skin depth (meters), is equal to $(2/\omega\sigma\mu_0)^{1/2}$, where $\omega = 2\pi f$ [f is the frequency (Hz)], σ = the conductivity of the material (siemens/meter), and $\mu_0 = 4\pi 10^{-7}$ = permeability of space (henry/meter). Equation (1) indicates exponential decay and a phase shift of the field with increasing field penetration.

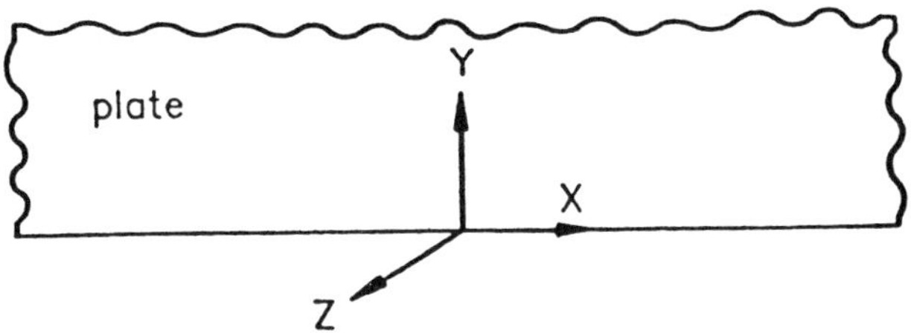

Figure 1. Coordinate system for large conducting plate.

The instantaneous current density (A/m^2) is

$$J_z = \frac{B_s}{\mu_o} e^{-y/\delta} \, 2/\delta \cos(\omega t - \phi) , \tag{2}$$

where $\phi = \frac{y}{\delta} + \frac{\pi}{4}$.

Hence, the current density is also attenuated by $e^{-y/\delta}$ and its phase is shifted by ϕ, increasing with depth.

The magnetic levitating force (N), given by the equation

$$F_y = \frac{B^2}{2\mu_o} A , \tag{3}$$

where A is the surface area (m^2) and B is the rms value of the field, must equal the gravitational force

$$G_y = mg = \rho Ahg , \tag{4}$$

where h is the conductor thickness, ρ is the density, and g is the acceleration of gravity. Hence, for a parallel field only on the bottom surface (zero field on the top surface), levitation requires that

$$\frac{B^2}{2\mu_o} = \rho hg . \tag{5}$$

For a field B_B at the bottom and B_T at the top of the conductor, levitation is achieved when

$$\frac{B_B{}^2 - B_T{}^2}{2\mu_o} = \rho hg. \tag{6}$$

The quantity $B^2/2\mu_o$ is called the magnetic pressure. If the skin depth δ is small compared to the thickness of the conducting body (a liquid metal sheet in our case), the magnetic force acts essentially as a surface force.

The average eddy current power loss per unit surface area is

$$P = \frac{B_s{}^2}{2\sigma\delta\mu_o{}^2} = \frac{I^2}{2\sigma\delta} , \tag{7}$$

where B_s and I are the rms field and current, respectively. Because $1/\sigma\delta$ is the resistance per unit length of a layer of thickness δ and unit width, the power loss is equivalent to that of an effective uniform rms current $(I/2^{1/2})$ in a layer of depth δ.

Because it is often the magnetic pressure $F/A = B_s^2/2\mu_o$, rather than the field B, that is of interest, Eq. (7) can be written

$$P = (F/A)/\mu_o\sigma\delta \qquad \text{or}$$

$$P = (F/A)\ (\omega/2\sigma\mu_o)^{1/2}\ , \tag{8}$$

which shows that for a given magnetic pressure, the power per unit area varies as the square root of the frequency and inversely as the square root of the electrical conductivity.

Caster Conceptual Design

Magnet Design

To levitate a sheet of molten metal requires a uniform horizontal field below the metal and a much smaller field above it. A conceptual magnet design to produce such a field is shown in cross section in Fig. 2. With only the ac coils, ferrite yokes, and poles present, this type of magnet is known as a "picture frame" magnet and has been used extensively in particle accelerators. As shown in Fig. 3, this magnet produces a very uniform horizontal magnetic field in the region between the bottom coil and the liquid steel, directed from pole to pole. The field falls off linearly across the coil and reverses direction in the yoke. The current is perpendicular to the plane of the paper and of opposite direction in the top and bottom ac coils, which are connected at the ends of the magnet in a suitable way.

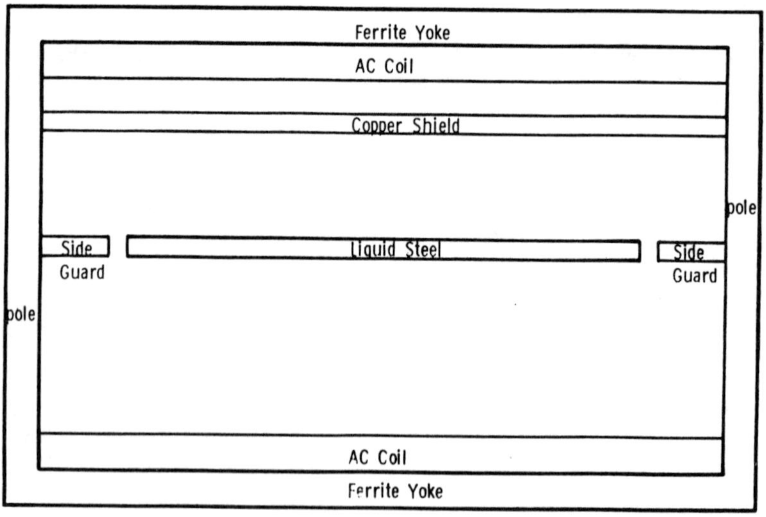

Figure 2. Cross section of conceptual magnet design for a horizontal EM caster.

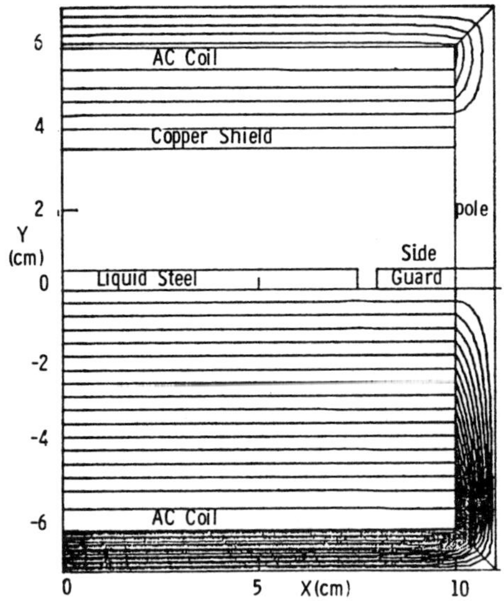

Figure 3. Magnetic flux plot for horizontal EMCC conceptual design.

The liquid steel and side guards provide parallel paths for the eddy current induced in them. They are connected to the copper shield at the ends of the magnet and form a single-turn eddy current loop in the magnet. As the ac coil tries to drive magnetic flux through the loop, the flux is almost excluded by the currents induced in the loop. Hence the magnetic field between the copper shield and the liquid steel is small when compared with the field below the liquid steel and above the copper shield. This produces the uniform vertical force on the liquid steel except near the sides. If the liquid steel would be extended from pole to pole, the vertical force would be extremely uniform. Other ways of achieving the conditions required for horizontal casting (uniform field below the metal sheet, reduced field above) have been studied (2,4).

Frequency of Operation

According to Eq. (8), EM heating increases with frequency, and the ability to remove heat places an upper bound on frequency. Calculations (2) indicate that with augmented gas-cooling methods and frequencies below 100 kHz, steel sheets may be cooled and solidified within a magnet of reasonable length (< 6 m). Analysis indicates that stability of the EM levitation becomes easier at higher frequencies and that a sheet thickness is required that is larger than the EM skin depth by a factor of 2 to 3. This places a lower limit on the frequency of about 25 kHz for a steel sheet.

Side Guards

The side guards, which should consist of a material with electrical conductivity much like that of liquid steel, essentially extend the sides of the liquid steel to the poles except for the small air gap between the sideguards and the liquid steel. Computer modeling of this geometry has shown that the vertical force on the liquid steel can be made uniform near the steel edge by adjusting the relative vertical position of the liquid steel and the side guard. Further refinements in the uniformity of the field could be made by driven side guard coils, not shown in Fig. 2.

Current Return Path and Current Contacts

For the concept of the horizontal caster, if there is to be net levitation, the currents induced in the bottom skin of the liquid metal sheet must have an external return path. The copper shield provides that return path. The liquid metal, copper shield, and the upstream and downstream current contact can be thought of as composing the one-turn secondary of a transformer, the magnet coils being the primary. The skin depth in the liquid steel is small compared to the thickness of the steel sheet. This arrangement results in a relatively high resistance of the current path in the steel. The resistance of the copper shield is made very small by using thin (less than a skin depth thick) insulated copper sheets connected in parallel. If the resistance of the current contacts is low and if the entire current loop lies inside the good-field region of the magnet, then there will be only a very small field above the steel. The value of the field is such that the voltage induced in the loop is sufficient to drive the loop current through the contact resistances and part of the copper circuit. Some field above the sheet may be useful to produce a good surface quality and to suppress wave motion.

Levitation requires uniform flux density and uniform current density over the width of the liquid steel sheet. The uniformity of the field below the sheet and the small ratio of skin depth to sheet thickness tend to force a uniform current density. However, large variations in contact uniformity might overcome this tendency. If so, uniform current density can be imposed over the width of the liquid sheet by placing many uniformly spaced contacts are placed on the solidified surface of the sheet. The distance from these contact points to where the metal is still liquid must be large enough for the current to diffuse to uniform current density. Equal current sharing between the contact points may need to be forced. This can be done by connecting, in series with each contact, a reactor with an impedance much larger than the resistance change due to contact variations. A better solution may be the use of mutually coupled reactors that force current sharing by inducing voltages in the reactor winding that is connected to a contact that does not share current equally.

The downstream multiple contacts could consist of rollers or brushes that may be placed outside the magnet. By this point in the casting process, the metal must have a skin strong enough to contain the pressure of these contacts The solidified steel sheet can pass through rollers at a temperature of 1200° C, or possibly higher (5). Oxidation of the surface would increase the contact resistance between the sheet and the rollers. The resistivity of Wustite (FeO), the major part of the oxide coating, is 25 Ωcm at 1300° C and 100 Ω cm at 1100° C (6).

The effect of the electrical contact resistance on levitation was studied with a simple circuit model. In the model, the upstream and downstream resistances are combined into a single resistance R_Q. The currents on the outside surface (bottom of steel, top of copper shield) and inside surface (top of steel, bottom of copper shield) of the loop are treated as two currents that combine to flow through the contact resistance. The time-varying flux driving each current, the resistances, and self- and mutual-inductances are determined by the geometry. The amplitude and phase (actually the real and imaginary parts) of each current are variables.

For a steel length of 5 m, all within the magnet, a width of 1 m, and a frequency of 100 kHz, the model predicts that levitation is unaffected by a contact resistance of 0.1 Ω or less, but that there is no levitation if the contact resistance is 10 Ω or more. The critical resistance is about 1 Ω. The values of Wustite resistivity above are consistent with a resistance well below 1 Ω. Predictions of the model show that the critical resistance varies directly with the length between contacts, inversely with the width of the sheet, and directly with the frequency. The contact resistance also results in Joule heating, and the need to keep the heating at the contacts manageable results in a required contact resistance of less than 1 mΩ. This value is difficult to achieve and requires that the steel at the contact be free of oxide.

The upstream contact is seen as challenging but clearly feasible; perhaps a high-melting-point metal encrusted with steel, inserted into the liquid metal of the tundish. Alternatively, it could be in the liquid metal further downstream, but far enough upstream that flow perturbations caused by the contact would die out before reaching and destabilizing the levitated liquid. The feasibility of the downstream contact, rollers or brushes, can only be determined by an experimental program.

Production-Scale Considerations

The concept for horizontal EM casting can be extended to a production-scale caster, but some additional considerations arise because of the greater width of the liquid steel sheet and because of the greater length of the magnet that accompanies a faster casting velocity.

A sheet width of up to 2 m may be desirable for a production caster, requiring a comparable coil separation. End effects could cause a transition region of 0.3 m or so between the magnet end and the beginning of the good-field region at each end of the magnet. End guards on each end of the magnet could reduce these end regions, but the end guards themselves could interfere with the feed system and the cooling system of the steel that is leaving the magnet.

For some production scenarios, the combination of conservative cooling rates and high casting velocities (0.3 to 1 m/s) could require magnets approaching 10 m in length. Such magnets are technically feasible (magnets 30 m long have been built), but, for magnet lengths greater than 10 m, the capital and operating costs (electrical power) begin to influence the total cost of casting.

The basic requirement for stability is the presence of conducting screens near the liquid steel sheet. Cooling and other apparatus can be placed within the magnet above and below these screens without affecting system stability. The additional power required with increased height of the magnet is a small fraction of the total.

Analysis of Magnetohydrodynamic Stability

Background

The stability of massive conducting liquids that are levitated by ac fields has been of interest for some time (7). Toroidal geometries and meander coils have been suggested (8,9). These geometries are essentially two-dimensional analogs of the basic circular coil method used in levitational melting (10). A magnetic hole is located between conductors, and these geometires rely on surface tension to keep the fluid from falling through the hole.

Sagardia (11) investigated the effects of an ac field in the asymptotic limit of infinite frequency (zero skin depth). He showed theoretically and experimentally that large objects could be stably levitated with two perpendicular ac magnetic fields of different frequency. Others (12,13) have also investigated such systems. The magnitude of the magnetic fields required for stability is considerably larger than that required to levitate a 6-mm-thick sheet of steel. Such fields could be used by applying a partially opposing field on top; however, the heat input would be much larger than could be removed in a practical system. Garnier and Moreau (14) discussed the effects of a finite EM skin depth on such applications. Stability is still maintained, but the required magnitude of the magnetic field increases as frequency decreases. This would not alleviate the overheating problem.

Stabilization Methods for Horizontal Electomagnetic Casting

Designing a system to overcome the magnetohydrodynamic (MHD) instabilities of the liquid and solidifying steel sheet is one of the most difficult problems associated with the development of an EM caster. No one has ever stably levitated a horizontal liquid sheet of any substantial extent, and the technical feasibility of obtaining stability is unknown at this time. Other levitational instabilities that accompany the interaction with the cooling system also affect feasibility. The production of waves by the passage of a cooling gas parallel to the steel sheet has been investigated in some detail both theoretically and experimentally at ANL. These investigations have provided constraints for the maximum gas velocity (15).

The ANL concept for a horizontal caster uses inductance stabilization for bulk stability because a uniform field, produced by a uniform current sheet, provides no stabilization. In the ANL concept, a constant voltage in a high-Q circuit drives the magnet current so that movements of the steel sheet are automatically compensated for by changes in the current. Published analyses that show a lack of stabilization for a uniform field, e.g., Ref. 8, assume a constant current for the levitating magnet.

The present concept for horizontal casting achieves local stabilization through flux compression between the solidifying steel sheet and a nearby conducting screen which may be porous for coolant flow. If the screen is too far from the sheet, it will not provide good stability, particularly against the Rayleigh-Taylor (RT) instability. If it is too close, it produces large pressure drops and inefficiencies in the gas cooling system. With the limited design and analysis to date, the feasibility for a successful design that meets all constraints is still an open question. The necessity for operating near the high-temperature environment of the molten steel may make the successful design of such a screen more difficult.

The time constants of both stabilization methods are signficiantly smaller than those associated with instabilities. The characteristic time associated with flux compression is in the microsecond range, whereas the fastest local instabilities occur in the magnetoinertial time scale, which is in the millisecond range. The inductance time constant is typically less than a second, which is less than the residence time in the caster, even at the most optimistic cooling rates.

Rayleigh-Taylor Instability

The RT instability is a general instability of an interface between two fluids of different densities, which occurs when the light fluid is "pushing" the heavy fluid. In the present context, the RT instability occurs when a slab of heavy fluid (liquid metal) lies on top of a slab of lighter fluid (air). This configuration is unstable. The opposite configuration, with air overlying the metal, is stable. In the horizontal EMCC context, the EM field must not only provide support against the downward force of gravity, it (or some other mechanism) must also provide restoring forces to prevent this basic RT instability.

Most of the work done on the magnetic RT problem has been with dc fields. Direct current magnetic fields are essentially independent of the form of the free surface. They react mainly on the velocity field in the bulk of flow or via some externally determined current flow. With a dc field, the induced current is proportional to the velocity of the liquid metal. The use of ac magnetic fields produces quite different phenomena. Alternating current fields control and react to perturbations of the equilibrium shape of the free surface. The surface and fields are closely coupled. The currents induced by an ac magnetic field are dependent on the frequency of alternation, and in many cases, the effect of fluid velocity becomes a secondary consideration.

The levitation melting technology and other metallurgical processing concepts have motivated a number of studies of the fluid motions induced internally and on free surfaces. For RT instability problems, the internal flows are not important per se. However, they play an important secondary role in providing energy to perturb the system beyond what is considered in linear stability analysis. In many cases the internal flows act as an additional destabilizing force to gravity.

For flux compression stabilization to work, the EM force needs to act in the limit of a surface force. If the EM skin depth is greater than or approximately equal to the thickness of the liquid sheet, then the behavior approaches that of the dc case, which is unstable via the kink instability (16-18). In addition, flux compression cannot stabilize against small wavelength disturbances (14). If the skin depth is larger than the wavelength stabilized by fluid surface tension, the magnetic field will diffuse into the metal rather than follow the contours of the surface.

The stabilization used in the present design concept employs passive methods only, requiring no sensing of the position of the liquid sheet or active feedback. Active feedback methods could be employed to increase stability. Successful suppression of the RT instability in a two-dimensional system has been experimentally demonstrated by means of solenoidal magnets connected by feedback to position sensors (19,20). Using electrohydrodynamics, Melcher (21) and Melcher and Warren (22) successfully demonstrated the use of feedback to stabilize against RT disturbances in a three-dimensional system.

In the discussion that follows, we first describe the influence of the effective tension that the magnetic field produces in the system, independent of any flux compression, and show that the magnetic tension stabilizes against short-wavelength disturbances, but not long-wavelength disturbances. We then discuss the influence of surface disturbances on this tension. Finally, we examine the flux compression effect of a bottom conducting shield and show that if such a shield is close enough to the levitated sheet, it can stabilize against long-wavelength disturbances.

Magnetic Tension

As in Fig. 1, we consider two fluid half-spaces, separated by a horizontal interface at $y = 0$. We ignore viscosity and surface tension to concentrate on the action of a horizontal magnetic field parallel to the interface. We assume that magnetic permeability is constant and the same for both half-spaces. We assume that, initially, a constant magnetic field permeates the space in the y direction. We also assume that the electrical conductivity is large enough that the magnetic diffusivity K_m can be ignored.

For disturbances parallel to the magnetic field, the effect of the field is the same as if there were an effective magnetic surface tension T_{mag} acting on the interface. The T_{mag} is defined by

$$T_{mag} = (B \cos\theta)^2/(\mu k) , \qquad (9)$$

where θ denotes the angle between the magnetic field B and the wave vector k of the disturbance.

Disturbances perpendicular to the direction of the magnetic field are not influenced by the presence of the field. The instability for perturbations in this direction is the same as if no field were present. Physically, this makes sense in that when the surface becomes distorted, the lines of the magnetic field are pushed in the same direction as the surface, but the field lines themselves remain flat. Without a bending of the field lines, there is no magnetic tension.

If we incorporate fluid surface tension T, let $\theta = 0$, and ignore the density of air ρ_2, relative to the density of the liquid metal ρ_1, then the condition for stability becomes

$$g\rho_1 < Tk^2 + kB^2/\mu . \qquad (10)$$

The system becomes unstable for short wave numbers (long wavelengths). A finite system can always be stabilized by making B sufficiently large.

It is of interest to calculate the effective magnetic surface tension for a practical example by using Eq. (9). Let B = 0.01 T = 100 G, λ = 1 m, and θ = 0. Then T_{mag} = 12.7 N/m , which is much larger than the 1.8 N/m fluid surface tension for steel. This suggests that the magnetic field can help to stabilize against RT disturbances.

Assume that the fluid to be levitated is 6.4 mm (0.25 inch) thick and has a density of 7000 kg/m^3. The rms magnitude of the magnetic field needed to levitate the slab is given by

$$\rho gh = B^2/(2\mu_o) , \qquad (11)$$

which yields B_{rms} = 0.0331 T = 331 G .

We now determine the critical wavelength that such a field can stabilize against. If we ignore fluid surface tension in Eq. (10), and combine Eq. (10) with Eq. (11), we have

$$\lambda < 4\pi h , \qquad (12)$$

and λ_c = 8 cm for h = 0.635 cm.

Note that in the horizontal caster, the presence of the side guards will keep the edges of the liquid metal sheet straight. Hence any disturbance in the interior of the sheet parallel to the edges (perpendicular to the field) must be accompanied by a disturbance that is perpendicular to the edges (parallel to the field) and of wavelength twice the sheet width or less. Thus, the disturbance will be stabilized by the field, provided that the sheet width is not much larger than about 4 cm.

Effect of Wavy Interface

Here we examine the effect of a sinusoidal disturbance of the interface on the behavior of the magnetic fields. We continue to examine half-spaces but caution that the results are indicative only, because the real system does not consist of half-spaces.

The magnetic levitation pressure at the bottom of the half-space is given by

$$P_{mag} = (\mu/2)[(H_o + h_x)^2 + h_y^2 + h_z^2] , \qquad (13)$$

where h_x, h_y, and h_z are the components of the field perturbation due to the wavy interface. To lowest order in the perturbed fields this becomes

$$P_{mag} = (\mu/2)(H_o^2 + 2H_oh_x) . \qquad (14)$$

Analysis shows that, under a depression, h_x is positive and thus the total magnetic field increases. Similarly, below an elevation, h_x is negative and the total field decreases. The increase (or decrease) is proportional to ε, the amplitude of the waves in the interface. Hence the field is acting like a spring to return the interface toward equilibrium. This type of behavior is often termed flux compression, but it must be distinguished from the flux compression produced by a conducting shield, as described below. We note that, under most circumstances, the field falls off as $\exp(ky)$ or $\exp(2\pi y/\lambda)$. Thus the flux is compressed within a depth λ of the interface. This suggests that the effective spring constant of the flux compression is larger for smaller λ, because a displacement ε then occupies a larger fraction of the volume available for compression. This implies a critical wavelength λ_c, such that disturbances with wavelengths greater than the critical value are unstable. This was also the result found by Garnier and Moreau (15) in a more detailed analysis that included the fluid equations.

Conducting Shield

To approach the experimental situation more closely, we must take into account the sheet of copper that is placed below the levitated metal. Sometimes this sheet takes the form of a shield, and sometimes it consists of the driving coil of the magnet. We model the copper sheet as an infinitely conducting surface.

Consider the same problem as before, but now, in addition, we have a surface of a material with infinite conductivity located at $y = -a$. We now have an air gap in the region $-a < y < 0$, and a good conductor half-space for $y > 0$.

If the conducting shield is far away, then, except for the very largest wavelengths, $ka >> 1$, the shield will have no influence. However, when we calculate the magnetic pressure for $ka << 1$, i.e., for $\lambda >> 2\pi a$ (say $\lambda > a$), we obtain

$$P_{mag} = P_o(1 + 2\varepsilon/a) , \qquad (15)$$

and we note that a has taken the place of $\lambda/(2\pi)$. This suggests that if we make

$$a < \lambda_c/(2\pi) , \qquad (16)$$

the system will be stabilized against all long-wavelength disturbances. Physically, the conducting plate forces the flux to be compressed into a volume with at most a height a. Subject to more detailed analysis, it is probably correct to assume that Eq. (16) could be directly substituted into a detailed fluid stability calculation. Thus, if, for a 6.4-mm-thick (1/4 inch) sheet of tin or steel, λ_c = 8 cm, then a_c = 1.3 cm. We note from Eq. (12) and the following that $\lambda_c = 4\pi h$, and that this leads to the equation

$$a_c = 2h \, , \tag{17}$$

a result that can be derived from much more basic considerations.

The above results apply in the high-frequency limit. If the EM skin depth is too large, then the field diffuses too far into the liquid metal and cannot stabilize against short-wavelength disturbances, even in the presence of the conducting screen.

Levitation Experiments

We realized from the beginning of the project that stable EM levitation of molten sheets was a major issue of technical feasibility, and that analysis alone would not resolve the issue. A test magnet was designed and fabricated for levitation experiments with liquid metals, such as tin, that have low melting temperatures. The test magnet was designed to have a very uniform horizontal field and to simulate the conceptual design of the horizontal caster magnet. The primary purpose of the test magnet was to investigate MHD stability over a liquid metal sheet with sufficient area to avoid domination of edge effects in the middle of the sheet.

The experimental design was constrained by available power supplies. A 30-kW, 10-kHz unit was the most powerful available supply close to the desired frequency range of 25-100 kHz. The available power limited the thickness h of the liquid tin that could be levitated to about 4.7 mm (3/16 inch), even after careful tuning of the power supply and modification of the experimental apparatus to reduce losses. At 10 kHz, the skin depth δ in liquid tin is about 3.5 mm. According to the stability analyses discussed previously, with a ratio of $h/\delta \simeq 1.4$, it was not expected that successful levitation of the liquid sheet would occur. However, the behavior of the dominant instability modes could be studied.

Experimental Apparatus

Figure 4 shows a view of the test magnet; its ferrite core is 30 cm long. The aperture for experiments is 20 cm wide by 13 cm high. The good-field region (also the maximum-field region), with a field error $|\Delta B/B_0| <$ 0.5%, is 15 cm long. The magnet coils are made from pairs of copper sheets 7.6 cm wide and 0.025 cm thick. In Fig. 5, the magnet core is shown on the left, and on the right is an experimental setup for liquid tin with the magnet coil in the background.

A closed current loop was designed as a removable insert to contain a sheet of liquid tin within the good-field region (i.e. uniform-field region) of the test magnet. The loop opened outward at one end of the magnet to allow an unobstructed view of the interior of the loop. Except for the test section, the loop consisted of a copper conductor - 6.4-mm thick plate on the bottom for rigidity, and layers of copper sheet on top to reduce resistance. The test section consisted of a rectangular solid-metal frame which contained the liquid tin. The frame served as side guards and current electrodes. For each experiment, a mass of solid tin (114 mm x 64 mm x 4.7 mm) was rolled to the appropriate thickness, weighed, and then pressed and soldered into the test section. A ceramic baseplate was placed below the opening of the test section frame to support the tin until it was fully melted by eddy current heating within the magnet and until the magnetic field was adjusted to the proper levitation strength. Many fine holes were drilled in the base plate to allow equalization of atmospheric pressure

Fig. 4. Test magnet assembly.

Fig. 5. Magnet core and experimental insert.

above and below the liquid tin.

Standard VHS-format (30 frames/s) video equipment recorded the experiments. A sloped mirror was placed inside the insert over the test section to provide a bird's eye view of the tin pool to the video camera. A remote vernier control to the power supply provided fine control of the field strength.

Solid-Plate Experiments

To test the global stability of the magnet design, a solid copper plate was levitated in the test magnet. In these experiments the test section consisted of 6.4-mm-thick copper side guards. The ends of the copper plate were bent down, and the plate was electrically connected into the closed-loop insert by means of troughs filled with a low-melting-temperature solder at both ends. The troughs were well outside the good-field region. A 3-4-mm air gap existed between the side guards and the edges of the copper plate.

The copper plate levitated easily and repeatably. As expected, its equilibrium height increased as the field strength increased. It exhibited stability in the vertical direction, even when strongly perturbed by a plastic rod, with oscillations damping out in about one second. The vertical stability continued until the field strength was increased so that the equilibrium position was well above the top of the side guards. The levitated plate also exhibited stability in the horizontal direction, although this was aided by the small viscous forces arising from the troughs of liquid solder.

Effect of Window and Side Guard Material

Experiments were conducted to determine the sensitivity of the system to MHD instabilities arising from the differences in electrical conductivity of the sideguards and electrodes and that of the liquid tin. In many of these experiments, there was no air gap between the tin and the side guards. In such a geometry, a difference in electrical conductivity between the tin and the side guards results in a vertical component of the magnetic field at the guard/tin interface. The strength of the vertical component increases with the difference in skin depth between the materials. Similarly, a difference in conductivity between the tin and the electrodes results in a vertical component of current at the electrode/tin interface. In either case, a horizontal force and nonuniformity in the vertical force are present.

Stainless steel, copper, and monel were used as side guard and electrode materials. Monel has an electrical conductivity approximately equal to that of liquid tin. For the various experimental geometries and materials, the horizontal and vertical forces throughout the sheet were calculated numerically by means of the EM computer code PE2D (23). In all cases, whenever the horizontal- or vertical-force nonuniformity was greater than a few percent of the levitational force, an instability invariably occurred at the site of the nonuniformity, at field strengths significantly below that required to levitate the sheet. At lower field strengths, such instabilities were usually preceded by significant amounts of EM stirring, observable by small random surface undulations centered around the nonuniformity.

The numerical calculations indicated that a 4.7-mm thickness of tin with a 6.4-mm thick monel window would experience a relatively uniform levitational force, and this arrangement was used in subsequent experiments. The calculations were confirmed experimentally with placid tin surfaces and instabilities that occurred in the center of the sheet.

Similar computations were also performed for geometries with air gaps or ceramic spacers between the side guards and the tin. It was also conjectured and confirmed by PE2D that both vertical and horizontal forces could be changed by modifying the shape of the copper eddy current shield that is on top of the lower field coil below the insert. It was suggested that the proper matching of the shape of the lower copper eddy current shield could compensate for nonuniformities created by the liquid metal/sideguard interface and thereby increase the stability of the system. Only preliminary experiments were performed for these types of configurations.

Behavior of Instabilities

Experiments were performed with the 6.4-mm monel test section frame, 4.7-mm tin, and an air gap below the insert less than 8.6 mm. In several instances, instabilities occurred before full levitational field strength was reached because preparation of the tin was poor. Instances in which the bottom part of the tin had even one spot that did not make good electrical contact with the window frame invariably led to instability. Upper surface imperfections caused by nonwetting at the window frame or nonconducting "dirt" embedded in the tin seemed to cause field nonuniformities that were often associated with electromagnetic stirring observable on the upper surface. When such surface stirring was observed, instability always occurred at magnetic field strengths below those required for levitation.

With proper preparation techniques, the dominant instability appeared to be the RT instability. At field strengths sufficient to levitate the sheet of molten tin, a ripple would appear in the middle of the sheet well away from the edges, as shown in Fig. 6. The wave vector of the disturbance was perpendicular to the magnetic field lines and the wavelength was on the order of 1 cm. From a placid upper surface, the disturbance develops to the finite amplitude shown in Fig. 6 [sometimes within one video frame (0.03s)] on the order of the magnetoinertial time τ_{MI} = 0.02 s, defined as

$$\tau_{MI} = h(\rho\mu/B^2)^{1/2} , \tag{18}$$

where h is the sheet thickness, ρ is the liquid density, μ is the magnetic permeability of the tin, and B is the magnetic field strength. Such behavior agrees with predictions of the analyses discussed previously, although verification would be aided by recording the experiments with higher speed video equipment.

Further evidence for the presence of the RT instability was seen in an experiment in which a thin copper sheet was placed at the bottom of the liquid tin. This sheet was free to levitate. The sheet and the fluid levitated together. No instability was observed until the solid sheet became tilted. This indicates that the basis for the observed fluid instabilities is associated with the lower free surface of the tin.

Once a disturbance reaches the finite amplitude stage, its amplitude increases within a few video frames until there is a hole in the liquid sheet. Once initiated, this tear propagates rapidly along the direction of the field lines until the sheet divides in two parts and the current loop is severed. This mode of disruption of the liquid metal sheet is referred to as the edge-tear instability. The propagation time of the edge-tear wave across the entire sheet is about eight video frames (0.27 s), approximately equal to τ_{MI} as defined by the width of the sheet. The mean propagation velocity is about 43 cm/s. The characteristic speed of the shallow fluid wave is $(gh)^{1/2}$ = 21.6 cm/s.

Pool thicknesses h of less than about 4 mm were unstable because of surface tension forces. For h < 4 mm, the pool, when perturbed, tended to pull away from the wetted interfaces with the window frame, or form a hole in its central region.

Conclusions

Several generalized observations can be made from the totality of the levitation experiments. It is expected that these conclusions will be altered for higher frequency fields (i.e., $h/\delta \gg 1$). Field nonuniformities of the order of several percent lead to instability. Avoiding this effect required a careful match between the electrical resistivity of the liquid tin and the surrounding window. It also required care in inserting the tin. In the case of a uniform field (i.e., no surface stirring observable, field strength sufficient to levitate achieved, etc.) the instability occurred as an RT instability. The initial instability occurred as a surface wave in the center of the slab, with the wave vector of the disturbance perpendicular to the magnetic field lines. The wavelength of this initial disturbance was of the order of 1 cm, and the time for the distrubance to develop was approximately equal to the magneto-inertial time.

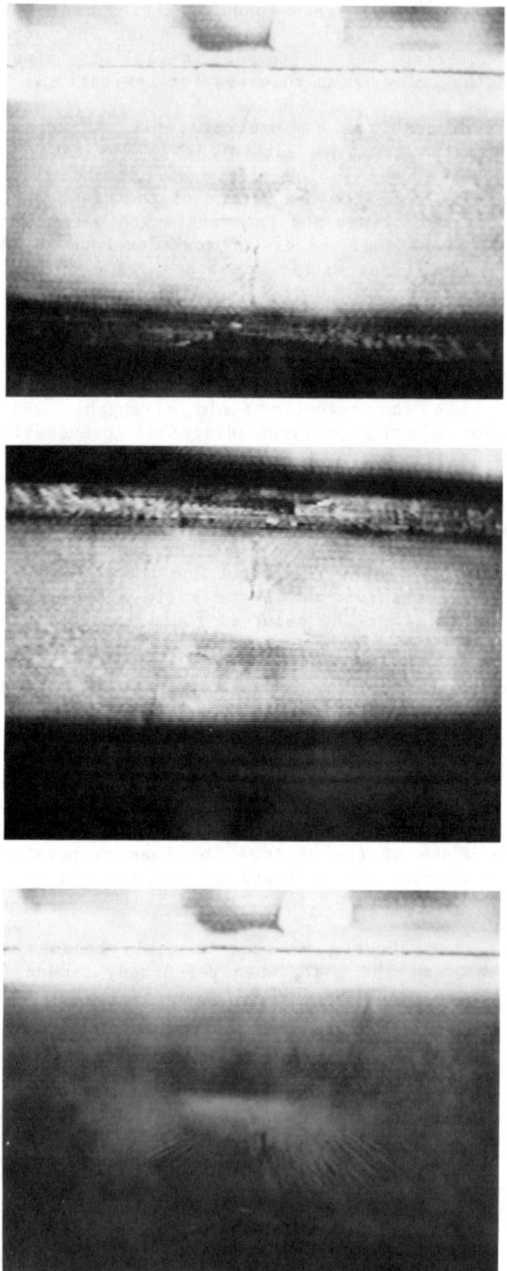

(a)

Video Frame 0

Smooth Surface. In frames 1-5 (not shown) very small surface perturbations in the center of the slab are barely visible.

(b)

Video Frame 6

Small disturbance is visible in center.

(c)

Video Frame 7

Ridge is visible. "Stretch lines" are due to oxide at tin surface.

Figure 6. Video frames (30 frames/s) of liquid tin upper surface during Rayleigh-Taylor instability. Alternating magnetic field runs left and right.

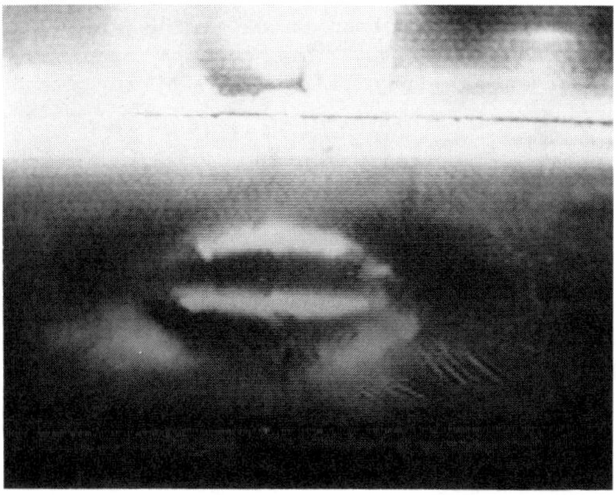

(d)

Video Frame 8

Disturbance wave vector runs up and down, perpendicular to the magnetic field.

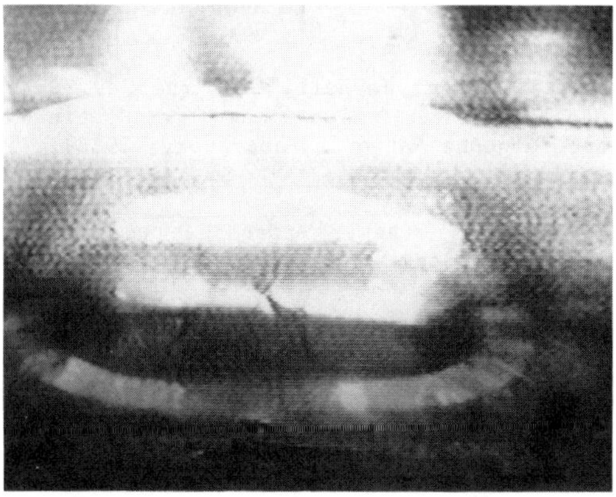

(e)

Video Frame 9

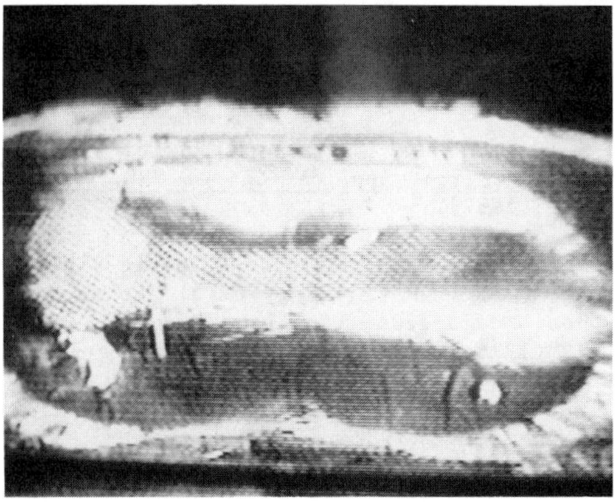

(f)

Video Frame 10

Porous ceramic cloth that supports the liquid tin is visible.

Figure 6. Video frames (30 frames/s) of liquid tin upper surface during Rayleigh-Taylor instability. Alternating magnetic field runs left and right.

Based on the analyses and experiments performed to date, the horizontal EM caster concept appears technically feasible. The concept requires a great deal more analysis and experimental study before feasibility can be conclusively demonstrated.

Acknowledgments

This work was sponsored by the U.S. Department of Energy, Office of Industrial Programs under Contract W-31-109-ENG-38, as part of the Steel Industries/Federal Laboratories Initiative. Many members of the EMCC project have assisted with the levitation experiments and interpretation of the results. The authors are indebted to K. Barnthouse, J. Battles, R. Benson, P. Clark, B. Goldman, N. Gopalsami, R. Kliss, D. McGhee, T. Merse, B. Misra, L. Rupert, V. Shah, R. Weeks, and T. Wiencek.

References

1. H. Branover, H. K. Moffatt, M. Mond, E. S. Pierson, P. S. Sulem, and A. Yakhot, "Magnetohydrodynamic Flow and Turbulence: a Report on the Fifth Beer-Sheva Seminar," J. Fluid Mech., 188 (1988) 87-106.

2. J. Battles, D. Rote, B. Misra, W. Praeg, J. Hull, L. Turner, V. Shah, R. Lari, N. Gopalsami, and T. Wiencek, Electromagnetic Continuous Casting of Steel (Final Report, Argonne National Laboratory, 1988). To be published.

3. L. R. Turner, R. J. Lari, and W. F. Praeg, "Approaches to Vertical Electromagnetic Moldless Casting of Steel" (Paper presented at this conference).

4. J. R. Hull, R. J. Lari, W. F. Praeg, and L. R. Turner, "Horizontal Electromagnetic Casting of Thin Metal Sheets," U.S. Patent No. 4,678,024 (July 7, 1987).

5. W. Wilson, private communication, Weirton Steel Company, 1987.

6. J. Cox, private communication, Bethlehem Steel Company, 1987.

7. T. F. Volkov, "Stability of a Heavy Conducting Fluid Contained by a Rapidly Varying Magnetic Field," Sov. Phys. Tech. Phys., 7 (1962), 22-27.

8. A. D. Sneyd, "Levitation of a Liquid Metal Layer by a Row of Infinite Parallel Line Currents," in Metallurgical Applications of MHD, ed. H. Moffat and M. Proctor (Cambridge, 1982).

9. A. P. Vutsens, "Freely Levitating Torus in the Electromagnetic Field of a Current-Carrying Loop and a Current-Carrying Axial Conductor," Magnit. Gidrodin., 1976, no. 4:113-120. (English translation in Magnetohydrodynamics (USSR), 12 (1976), 478-485.)

10. A. D. Sneyd and M. K. Moffat, "The Fluid Dynamics of the Process of Levitation Melting," J. Fluid Mech., 117 (1982), 45-70.

11. S. R. Sagardia, "Electromagnetic Levitation Melting of Large Conductive Loads," (Ph.D. thesis, University of Toronto, 1974).

12. V. V. Smirnov, "Levitation Melting of Large Electrically-Conducting Specimens," Russ. Metall., 1978, no. 1:80-86.

13. M. R. Harris and S. Y. Stephan, "Support of Liquid Metal Surface by Alternating Magnetic Field," IEEE Trans. Magn., MAG-11 (1975), 1508-1510.

14. M. Garnier and R. Moreau, "Stability of Molten Metal Free Surfaces in the Presence of an Alternating Magnetic Field," in Metallurgical Applications of MHD, ed. H. Moffat and M. Proctor (Cambridge, 1982) 211-216.

15. D. R. Rote, J. R. Hull, and T. Weincek, "Stabilization of a Molten Metal/Gas Coolant Interface by Electromagnetic Fields, (Paper presented at this conference).

16. J. Etay and M. Garnier, "Horizontal Mouldless Electromagnetic Continuous Casting: a Normal Mode Analysis" (Paper presented at Fifth Magnetohydrodynamic Flow and Turbulence Seminar, Beer-Sheva, Israel, March 2-6, 1987).

17. S. Asai, "Process Development and Stability Analysis of Horizontal Electromagnetic Casting," Proc. Iron Steel Inst. Japan, (1986).

18. T. Kozuka, J. Obayashi, S. Asai, and I. Muchi, "Horizontal Electromagnetic Casting of Thin Plate and its Stability Analysis," Trans. Iron Steel Inst. Japan, 27 (1987), B-216.

19. D. S. Zrnic and C. D. Hendricks, "Stabilization of the Rayleigh-Taylor Instability with Magnetic Feedback," Phys. Fluids, 13 (1970), 618-629.

20. D. S. Zrnic, C. D. Hendricks, and J. M. Crowley, "Control of an Instability on the Fluid Pendulum with Magnetic Feedback," Phys. Fluids, 13 (1970), 1875.

21. J. R. Melcher, "Continuum Feedback Control of Instabilities on an Infinite Fluid Interface," Phys. Fluids, 9 (1966), 1973-1982.

22. J. R. Melcher and E. P. Warren, "Continuum Feedback Control of a Rayleigh-Taylor Type Instability," Phys. Fluids, 9 (1966), 2085-2094.

23. Vector Fields Ltd., 24 Bankside, Kidlington, Oxford OX5 1JE, England.

A METHOD FOR VERTICAL ELECTROMAGNETIC MOLDLESS CASTING OF STEEL

L.R. Turner, R.J. Lari, W.F. Praeg

Argonne National Laboratory
9700 South Cass Avenue
Argonne, IL 60439

Abstract

Several approaches have been studied for the vertical casting of thin (3 mm - 8 mm) sheets of steel. Each approach employs electromagnetic (EM) forces, and avoids the need for contact between the solidifying steel and a solid mold. The most promising approach uses a high-frequency (HF:>100 kHz) oval solenoid magnet to provide containment of the liquid steel and a low-frequency (LF:~60 Hz) traveling field, similar to the double-sided linear induction pump, to provide levitation. The low field level of the solenoid and the low frequency of the levitation magnet result in acceptably low EM heating of the steel. The LF field penetrates the steel and provides a body force exactly counteracting the force of gravity everywhere except near the edges of the solidifying sheet. Additional HF traveling field magnets augment the levitation force near the edges but generate more EM heating. Other means of extending the levitating force to the edge and other approaches using stationary or traveling fields have also been studied.

Casting of Near Net Shape Products
Edited by Y. Sahai, J.E. Battles,
R.S. Carbonara and C.E. Mobley
The Metallurgical Society, 1988

Introduction

The ability to cast in near net shape is expected to greatly reduce the production cost of metal sheet products. Rapid cooling techniques have permitted the casting of thin (<0.05 mm thick) sheets using single wheel casters. Attempts to produce thicker sheets by means of belts and wheels have been unsuccessful.

Electromagnetic (EM) methods have been used in a number of metallurgical applications (1) and have been used commercially in metal casting mainly for mold stirring and containment of vertical ingots in the production of aluminum. In cooperation with researchers from a number of U.S. steel companies, researchers at Argonne National Laboratory (ANL) have investigated processes for EM continuous casting (EMCC) of steel in large-aspect-ratio shapes (2). The goal of this project was to cast steel sheets of about 6 mm thickness. Most of the efforts focused on design of a bench-top caster capable of casting sheets of width 15 to 30 cm. One concept involves EM levitation of a horizontal sheet of molten metal by the interaction of a horizontal high frequency (HF) magnetic field and longitudinal eddy currents, which are induced in the sheet by the HF magnetic field (2,3,4). A second concept, the one described here, involves the use of EM fields to constrain and levitate a vertical sheet of molten metal (2). Only a small effort was expended on vertical casting. Electromagnetic and liquid metal instabilities have scarcely been considered.

For vertical casting, the problems of confinement and levitation can be decoupled, and hence instabilities should be controllable in ways not available for horizontal casting. Moreover, the vertical caster has the advantage that the eddy currents all form closed loops in the steel sheet. There is no need for an external current return path and the troublesome electrical contacts described in (3).

The vertical casting concept of choice, illustrated in Fig. 1, requires three distinct kinds of magnets, namely:

(1) A high frequency, low amplitude field to provide containment of the liquid metal sheet.

(2) A low frequency (LF) traveling field that, over most of the width of the liquid metal sheet, offsets the effect of gravity.

(3) A high frequency traveling field to offset gravity near the edges of the sheet.

In another concept for electromagnetic containment and levitation, containment is provided by a stationary, high frequency, magnetic field, and levitation is provided by having the field increase downward to counteract an increasing ferrostatic head. Despite this method's advantages of simplicity and self-centering, the large electromagnetic heating it entails renders it impractical.

In a third concept, both confinement and levitation are provided by a high-frequency magnetic filed traveling upwards. With this method, the electromagnetic heating is less, but still troublesome.

High Frequency Containment Field

The high frequency field is used to define the surface of the molten metal and confine the molten metal within that surface. The frequency is chosen to produce a skin depth that is a small fraction of the thickness of

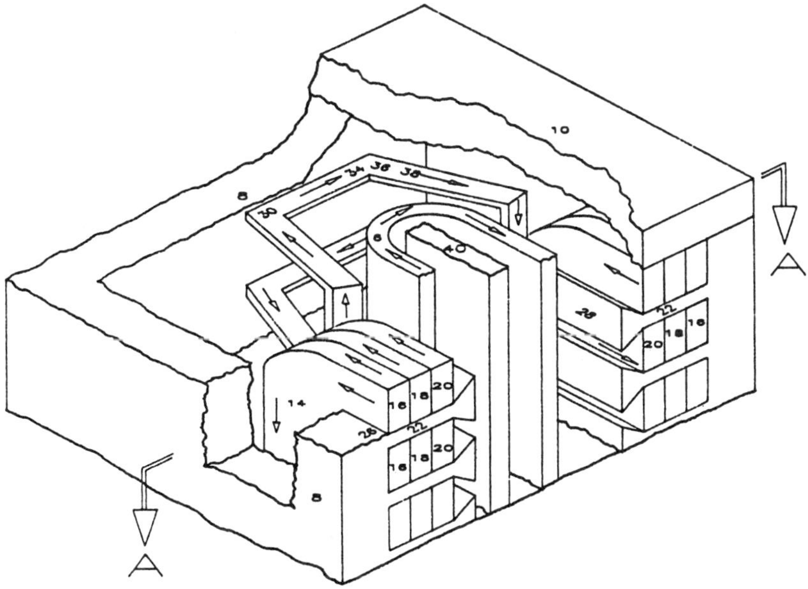

Figure 1 - Conceptual design of the vertical caster
The HF containment solenoid (6) surrounds the molten metal sheet (40).
The LF traveling field magnet for levitation consists of three layers of coils (16,18,20), steel teeth (22), tapered caps (28), and yoke (26).
The HF traveling field magnets (30) provide edge support. For clarity, the ferrite core of the edge-support magnet is not shown.

the sheet. The field is spacially uniform over the surface of the metal and parallel to it. The elongated solenoid coil completely surrounds the molten sheet and produces a vertical high frequency field at the surface of the metal, as shown in Fig. 1. The horizontal eddy currents in the surface of the metal perimeter interact with the vertical field to produce a horizontal force directed normally inward. This force contains the molten metal. A typical frequency would be in the range from 100 kHz to 400 kHz. The high frequency coil is placed inside the low frequency coil to reduce power losses.

Low Frequency Traveling Magnetic Field

The low frequency traveling magnetic field must produce a uniform body force in the molten metal. This body force is used to cancel the ferrostatic pressure head. The effect is similar to that of a linear induction motor to propel trains and other devices, or an electromagnetic pump for liquid metals.

A design study determined the geometry of magnet coils and steel required to produce a uniform body force on the liquid metal. Computer analysis of a section extending over one wavelength of the traveling wave, located far from the ends of the magnet, yielded the geometry shown in Fig. 1. The magnet section consists of 10 steel teeth with tapered caps attached to a steel yoke to provide a low reluctance path for the magnetic flux. Three layers of coils are positioned in the slots between the teeth, 120 degrees apart in phase. Each layer consists of nested race track coils, positioned to approximate a cosine distribution of current. Hence it gives a sinusoidal field distribution along the vertical length of the liquid metal. This in turn produces a uniform body force on the liquid metal.

For the magnet to have a high efficiency, the separation between the tooth caps on opposite sides of the steel sheet should be no more than two tenths of a wavelength. For larger separations, the magnetic flux tends to circle the conductors instead of passing through the molten metal, and efficiency diminishes.

In practice, that separation must accommodate the high frequency coils as well as the sheet of molten metal. (It is much better for the high frequency coils to experience the low frequency field than vice versa.) The wavelength is then determined to be at least five times that separation.

One wavelength of the magnet, far from the ends, was analyzed. Some studies were made on the effect of terminating the magnet at the end. These showed a deterioration of the uniform body force near the end of the magnet, a well-known and much-studied effect. Further analysis is needed to make the body force more uniform in the end region of the magnet. The magnet may have to be at least three wavelengths long.

As described below, the minimum width of the sheet also depends on the wavelength, and hence on the separation between the pole caps on opposite sides of the molten steel sheet.

The design study also investigated the frequency of current to use in the coils. A frequency in the range 10 to 100 Hz was found to limit the horizontal force to a small fraction of the vertical force. This range of frequencies also produced the most uniform vertical force on the liquid metal and negligible heating. Because of its ready availability, 60 Hz was chosen.

Magnet Edge Effects and Edge Support by High Frequency Traveling Fields

The LF magnets described above were analyzed for an infinitely wide sheet; i.e., the field and eddy currents are everywhere horizontal, and the force everywhere vertical and uniform. The actual eddy currents produced in the molten metal by the traveling magnetic field have the form shown in Fig. 2. The horizontal component of the eddy current density J_x interacts with the horizontal magnetic field to produce the vertical force on the molten metal except near the edge of the sheet. Near the edge of the sheet, the eddy currents are mostly vertical and tend to produce a horizontal force instead of vertical. Because J_x is zero at the edges of the liquid steel, the traveling field will provide no levitation there. Moreover, it will produce reduced levitation over a distance $k^{-1} = \lambda/2\pi$ or so from the edge, where λ is the wavelength of the traveling field. Variation of the upward magnetic force per unit volume F_m is given by

$$F_m = B_y J_{ox} [1 - \cosh kx/\cosh (kw/2)], \quad (1)$$

where B_y is the LF field, independent of x, w is the full width of the sheet, x is measured from the middle of the sheet, and J_{ox} is the LF current density induced by B_y when $kw/2 \gg 1$. In the analysis that follows, we take w = 0.3 m and λ = 0.4 m, so that kw/2 = 2.36. For larger values (wider sheets), the correction of edge effects is easier.

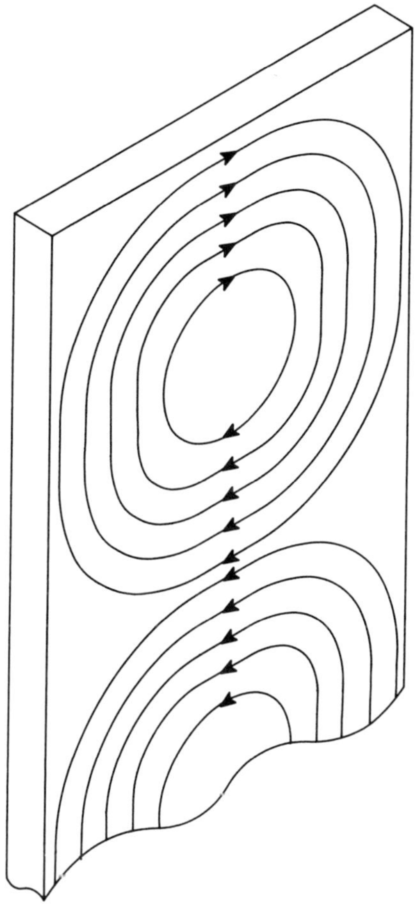

Figure 2 - Eddy currents in the molten steel sheet due to the LF levitation magnet. Near the edge, where the currents are vertical, the LF magnet does not provide levitation.

HF traveling magnetic fields at the edges of the steel sheet can provide levitation to enhance the reduced levitation there from the LF field. The combined fields produce a levitation force uniform from edge to center to within a few percent. Alternatively, a strip of high conductivity in electrical contact with the edge can lead to uniform levitation. That solution is probably impractical for production casting, but may be useful in some applications.

High Frequency Edge Coils and Net Levitation

The third magnet system is the HF traveling field coils to levitate the edges of the sheet. The system is shown in Fig. 3. The coils were modeled by a ribbon geometry, and the parameters varied to give the desired field and force pattern. The field in air was calculated by the Biot-Savart law.

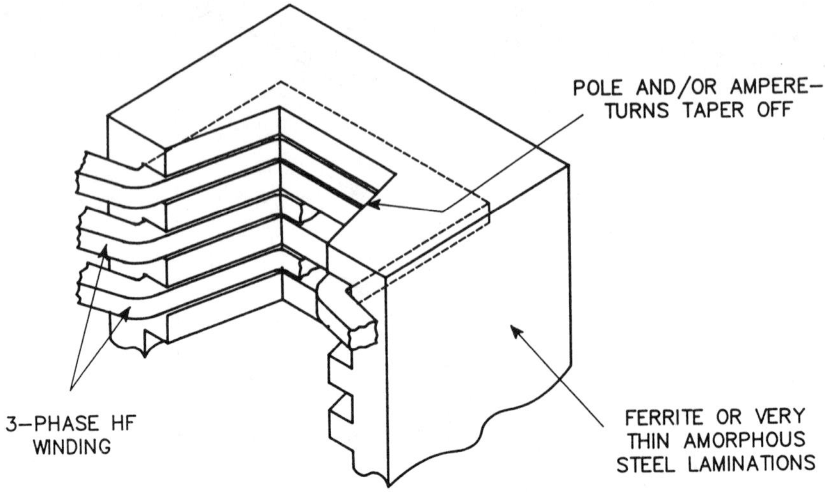

Figure 3 - Conceptual design of the edge support magnet. End guards on the magnet (not shown) could further shape the field distribution.

Consider first the field in the air. All three components, B_x, B_y, and B_z, have the z variation of a traveling field: $B_z \sim \cos(\omega t - kz)$, B_x and $B_y \sim \sin(\omega t - kz)$. In order that the force be largest near the edge (large x) and die out near the center of the steel (small x), let B_x, B_y, and B_z all have the variation $\exp(\gamma x)$. The surface of the steel, $y = 0$ is a plane of symmetry. Hence, B_x and B_z are independent of y near $y = 0$, and $B_y \sim \beta y$. The levitating force is found to be proportional to β.

Inside the steel, all components of the field (and of the current density J as well) decrease in amplitude as $\exp(-y/\delta')$, and vary in phase as y/δ. For the traveling field, these two skin depths, δ and δ', are not exactly equal. Their geometric mean is the classical skin depth δ_o, and their difference can be related to the y variation outside the steel, β.

The levitating force per unit area is found to be

$$F_x/A = -\beta\ \delta_o\ B_{zo}/8\ \mu_o, \tag{2}$$

and the pressure is

$$F_y/A = \left(B_{xo}^2 + B_{zo}^2\right)/4\ \mu_o, \tag{3}$$

in the approximation that $\delta = \delta' = \delta_o$.

Desired results for the force are that:

(1) The amplitude of the force is independent of z, and the phase varies linearly as z.

(2) The coil position should be near the edge of the liquid metal sheet.

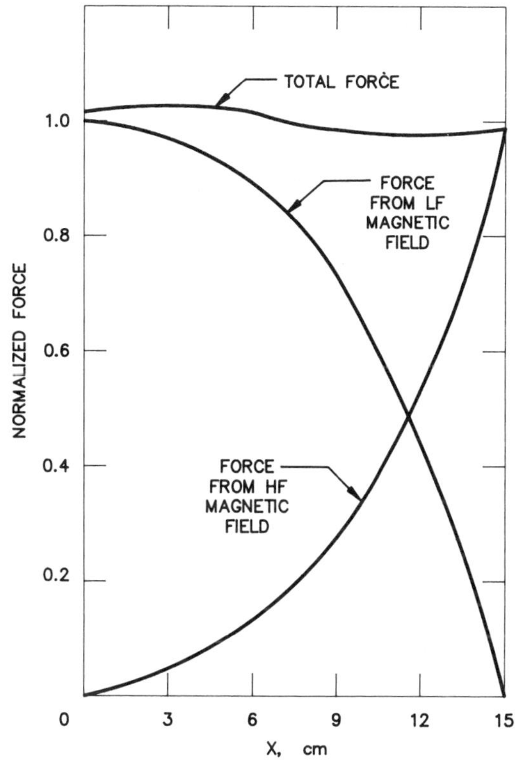

Figure 4 - Levitation force from the LF and HF magnets combine to give a net force uniform to within 2.5%.

(3) The levitating force should increase with x to offset the declining force from the LP coils as given by Eq. (1).

(4) The ratio of normal pressure to levitating force, as given by Eq. (2) and (3), should be small.

Parameter values were found that satisfied the first three of these four criteria. In particular:

(1) The maximum variation of combined LF and HF force across the sheet is only 2.5%, as shown in Fig. 4, and can be improved even more.

(2) The distance from the edge of the sheet to the coil is only 2 cm.

(3) The ratio of pressure to lifting force is less than 20 over most of the sheet and less than 70 everywhere. But that ratio is still too large.

Some other problems in addition to the force ratio remain:

(1) The best coil geometry yet found requires a peak current of 60 kA, a peak field of 7,000 G in the z direction and a peak field gradient of 30 G/mm in the y direction.

(2) The HF magnets (3 phase) and LF magnets (3 phase) must be arranged to avoid mechanical interference with each other. This is indicated in Fig. 1.

(3) The expression relating the fields and forces inside the liquid metal to the fields outside were obtained under the assumption that all three components of field varied the same way (exponentially) in x. The field components found with the computer model do not all vary the same way; further analysis is required.

Edge Support with a Conducting Strip

In some experimental situations it may be convenient to have the edges of the liquid metal in contact with a high conductivity metal strip, rather than to construct the edge support magnets described above.

With the strip, the eddy currents in the steel may flow into the strip to turn around rather than turn around in the liquid metal. In fact, the eddy currents, and hence the levitating force, can be made to increase, decrease, or remain constant near the edge, depending on the conductivity and width of the strip.

Electromagnetic Heating with the Three-Field Caster

The electromagnetic heating by the containment magnet and the LF levitation magnet is not excessive. The heating at the edges of the sheet by the HF edge support magnets must be reduced if the method is to be practical.

If the frequency of the HF containment magnet is 300 kHz and magnetic pressure needed for containment corresponds to a ferrostatic head of 1 mm, then the surface heating is 20 W/cm^2. The heating is proportional to the magnetic pressure and to the square root of frequency.

The levitation magnet design gives a satisfactory field at 60 Hz. From the viewpoint of liquid metal displacement during a cycle, 60 Hz is satisfactory as well.

For a given force density, F, the density of EM heating is given by

$$P = vF, \tag{4}$$

where v is the phase velocity of the traveling field,

$$v = \lambda\nu, \tag{5}$$

λ the pitch of the magnet (wavelength of the traveling field) and ν the frequency. For a force density equal to the gravitational force density (ρg = 69000 N/m^3), a wavelength of 0.4 m and a frequency of 60 Hz, the density of EM heating is a modest 1.65 W/cm^3.

The HF edge containment magnet described above would give a very high electromagnetic heating.

Alternative: A Fixed-Field Caster

The conceptually simplest vertical caster would be one in which the magnetic pressure at each position would just counterbalance the ferrostatic head there. To levitate a column of molten metal of height y requires a magnetic pressure, $B^2/2\mu_o$, equal to the gravitational pressure ρgy. Hence,

$$B = (2\ \mu_o \rho g)^{1/2} y^{1/2} \tag{6}$$

where ρ is the density, μ_o is the permeability of free space and g is the acceleration of gravity. The eddy current power loss per unit surface area increases linearly with y, the distance from the top of the liquid metal, when the field increases as $y^{1/2}$. The power also depends inversely on the skin depth so the frequency should be chosen to make the skin depth about 1/4 the thickness of the molten metal sheet.

In two dimensions, the condition that $B_y = ky^{1/2}$ determines the field everywhere. The conditions $B = -\nabla\phi$ and $B = \nabla x A$ then determine the scalar potential ϕ and vector potential A. In polar coordinates, the expressions are:

$$\phi = kr^{3/2} \left(\sin \frac{3\theta}{2} - \cos \frac{3\theta}{2}\right) \tag{7}$$

$$A = kr^{3/2} \left(\sin \frac{3\theta}{2} + \cos \frac{3\theta}{2}\right) \tag{8}$$

Lines of constant ϕ define the shape of the poles of the magnet. Lines of constant vector potential A define the edges of the coils.

The yoke of the magnet returns the flux from the bottom pole to the top pole. This is placed far enough from the coil to avoid flux leakage to the yoke. The cross-sectional area of the yoke must be sufficiently large to carry the flux from pole to pole and not saturate. The conductors form "race track" coils around the molten metal sheet. Horizontal eddy currents are induced around the perimeter of the molten metal sheet with a small skin depth. These interact with the vertical field to produce the needed horizontal force on the metal.

The simplicity of this kind of caster is attractive, but in casting molten steel with frequencies of 100 kHz and above, the electromagnetic heating is intolerable for ferrostatic heads of a few centimeters, much less the tens of centimeters this stationary field caster would entail.

Alternative: Casting with a High-Frequency Traveling Field

Before the edge support methods described above were thought of, vertical casting with a low frequency traveling field appeared infeasible. Under those circumstances, the method described here was considered.

For a vertical, high-frequency field, the eddy currents flow horizontally along one face of the liquid steel, around the edge, and back along the opposite face. If such a field were traveling vertically upwards and exerted an upward force on the metal, that force would be undiminished at the edge of the liquid steel sheet.

Such a scheme is indeed possible, but for reasonable values of the parameters (thickness ~6 mm, skin depth ~1 mm, wave length ~50 mm) the eddy current heating is large (375 W/cm^2). For a smaller wavelength (20 mm) and larger skin depth (2 mm), the heating is still marginal (38 W/cm^2). Also the vertical upwards force is a surface force, and the scheme depends on viscosity or effective viscosity to transfer the force to the interior.

Conclusions

The vertical caster described here, with three kinds of magnet, is a promising approach to casting thin sheets. Electromagnetic heating at the edges remains an unresolved problem.

References

1. H. Branover, H.K. Moffatt, M. Mond, E.S. Pierson, P.S. Sulem, and A. Yakhot, "Magnetohydrodynamic Flow and Turbulence: A Report on the Fifth Beer-Sheva Seminar," J. Fluid Mech., 188 (1988) 87-106.

2. J. Battles et al., "Electromagnetic Continuous Casting Project: Final Report," (Argonne National Laboratory, 1988). To be published.

3. J.R. Hull, R.J. Lari, W.F. Praeg, D.M. Rote, and L.R. Turner, "Design and MHD Stability for Horizontal Electromagnetic Continuous Casting," (Paper presented at this conference).

4. J.R. Hull, R.J. Lari, W.F. Praeg, and L.R. Turner, "Horizontal Electromagnetic Casting of Thin Metal Sheets," U.S. Patent No. 4,678,024 (July 7, 1987).

COOLANT SYSTEM CONSIDERATIONS FOR

ELECTROMAGNETIC CONTINUOUS CASTING OF STEEL

Balabhadra Misra, James E. Battles, and Vipinchandra Shah

Argonne National Laboratory
9700 South Cass Avenue
Argonne, Illinois 60439

Abstract

The success of electromagnetic continuous casting of sheet steel will depend, to a large extent, on the development of a coolant system. The problem of removing heat from the electromagnetic caster is compounded by the induced eddy current heating and limits imposed on the maximum mass flow rate of the coolant to avoid instability of EM levitation. Extensive theoretical and numerical analyses have been conducted to define the coolant requirements, to identify possible solutions to these problems, and to select reference operating conditions. The results of these analyses on the coolant system requirements in terms of casting speed, magnet length, coolant mass flow rate, and characteristics of electromagnetic levitation show that augmentation of heat transfer by a factor of 2 to 3 would be necessary for an economically viable casting system. A study of the literature shows that this requirement can be easily met by a variety of methods. The results from using nomograms to study the effect of design and operating parameters on the electromagnetic casting process are also presented.

Casting of Near Net Shape Products
Edited by Y. Sahai, J.E. Battles,
R.S. Carbonara and C.E. Mobley
The Metallurgical Society, 1988

Introduction

A key element in the development of the electromagnetic continuous casting (EMCC) process for sheet steel is a heat transfer and coolant system able to remove a large amount of heat in a short time. Designing such a system is further complicated by the fact that additional heat is imposed by the induced eddy currents. Radiative cooling is insufficient to meet the requirements and must be augmented with con-vective cooling. Only nonoxidizing gases are permissible with the EMCC process. In addition, there are restrictions on the upper boundary of the coolant mass flow rate (velocity and density) due to the possible detrimental effect of the coolant flow on the stability of EM levitation. These restrictions, which result in poor heat transfer coefficients, make the design of a coolant system more difficult. Extensive theoretical and numerical analyses have been conducted to define the coolant requirements and identify possible solutions to these problems. Based on an initial evaluation of the coolant system requirements, reference operating conditions were selected. Consistent with these operating conditions, work has been directed toward (1) parametric investigations for establishing the cooling requirements in terms of casting speed, magnet length, coolant mass flow rate, and degree of heat transfer augmentation, (2) literature survey to obtain supporting data for heat transfer augmentation, and (3) development of nomograms for studying the effect of various design and operating parameters on the EMCC process. A description of the efforts in the above three areas is presented below.

Mathematical Approach

A schematic representation of a coolant system concept for the EMCC process is shown in Fig. 1. A mathematical model to study the solidification phenomenon and the casting characteristics of the EMCC mold was developed based on (1) one-dimensional heat flow, (2) negligible convection in liquid, (3) uniform coolant velocity and casting speed, (4) negligible axial conduction, (5) EM heat flux on only one side for the horizontal caster, and (6) heat removal by convection and radiation. A generalized computer program was used to study the solidification characteristics based on the above set of assumptions, and to estimate the coolant requirements both within the electromagnetic confinement and outside the electromagnets. These results were compared with the overall coolant requirements as discussed in the next section to ensure proper physical modeling of the EM casting process.

Calculation of Heat Transfer Requirements

A set of reference conditions (see Table I) was established as the basis for calculating heat transfer requirements and as an aid in determining methodologies to obtain the required cooling. Extensive parametric studies were carried out based on these conditions. The mean outlet temperature of 2650°F (1455°C) is based on the belief that complete solidification is required before the strip can be mechanically supported. Since it may be feasible to develop a mechanical support system for a partially solidified strip, an exit temperature of 2650°F may be an overly conservative restriction.

The heat transfer coefficients estimated from empirical equations are shown in Table II for nitrogen and argon and helium gas mixtures. The heat transfer coefficients that are required to meet the operating conditions of Table I are graphically presented in Fig. 2. Comparison of

SCHEMATIC REPRESENTATION

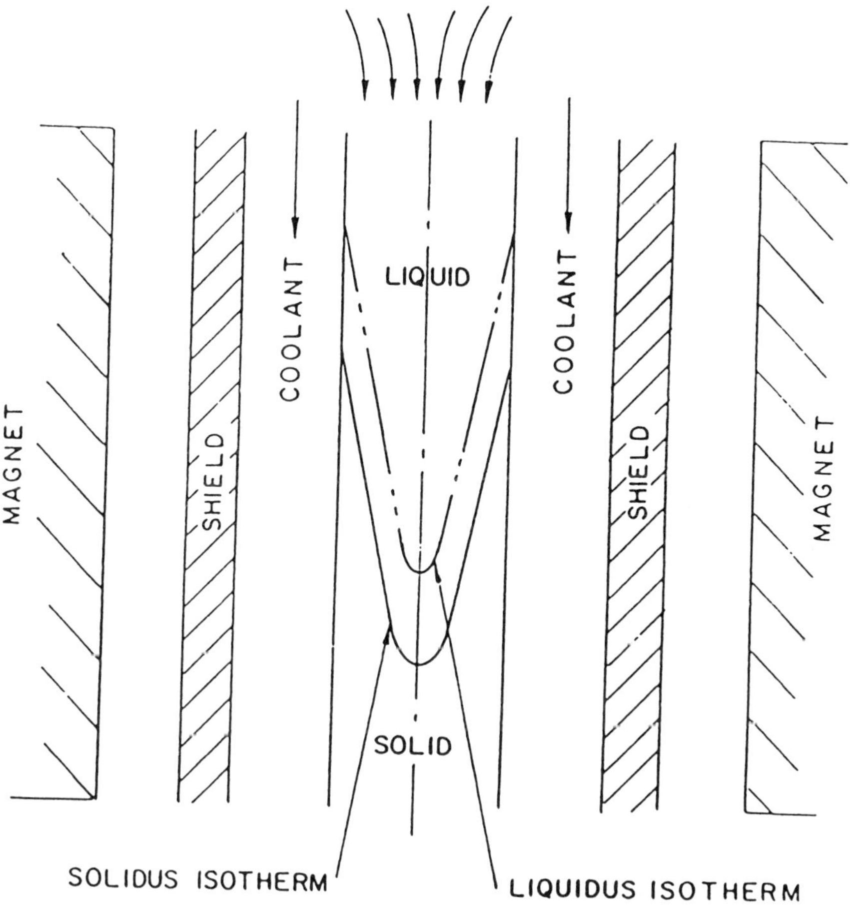

Fig. 1. Schematic of Coolant System Concept

Table I. Reference Operating Parameters

Mean Outlet Temperature	2650°F (1455°C)
Strip Size	Any width x 6-mm thick
Casting Speed	30 cm/s (67 tons/strand•hr for 48 in. wide strip)
Coolant Pressure	20, 30 psia
EM Heat Flux	50 W/cm^2
Emissivity, ϵ, of Steel	0.4, 0.6
Coolants	Nitrogen, Carbon Dioxide, or Helium and Argon
Coolant Velocity	30, 40 m/s
Magnet Length	4, 6, 8, 10 m
Coolant Channel Height	10 to 12.5 cm

Table II. Estimated Heat Transfer Coefficients

Case ID	Operating Conditions Pressure, psia	Velocity, m/s	Heat Trans. Coeff. Btu/(hr ft^2°F) N_2	10% He + 90% Ar	20% He + 80% Ar
A	20	30	15	14	18
B	20	40	19	18	22
C	30	30	21	20	25
D	30	40	26	25	31

the required heat transfer coefficients (Fig. 2) with the feasible ones (Table II) clearly shows that some heat transfer augmentation is necessary for magnets less than 10-m long.

Figure 3 illustrates the required augmentation factors for four different cases (A-D, defined in Table II) as functions of magnet length at ϵ = 0.4. As shown in Fig. 3, Case A requires an augmentation factor of 3 for a 10-m long magnet, whereas Case D requires an enhancement factor of only ~3 for a magnet length of ~6.5 m. Also, for magnet lengths that do not exceed 10 m, Fig. 3 shows that Cases B, C, and D require a heat transfer augmentation factor of only 2, which should be easily achievable.

Figure 4 presents similar calculations for the higher emissivity value (ϵ = 0.6). Even for the most-conservative coolant conditions (Case A), the required magnet length is reduced to about 8 m for an enhancement factor of 3. For the optimistic coolant system scenarios (Cases C and D), Fig. 4 shows that it is feasible to design an EMCC device that is less than 10-m long and requires little if any heat transfer augmentation. Of course, this requires verification of the velocity and pressure limits in terms of the dynamic stability of the levitated strip. Also, it should be

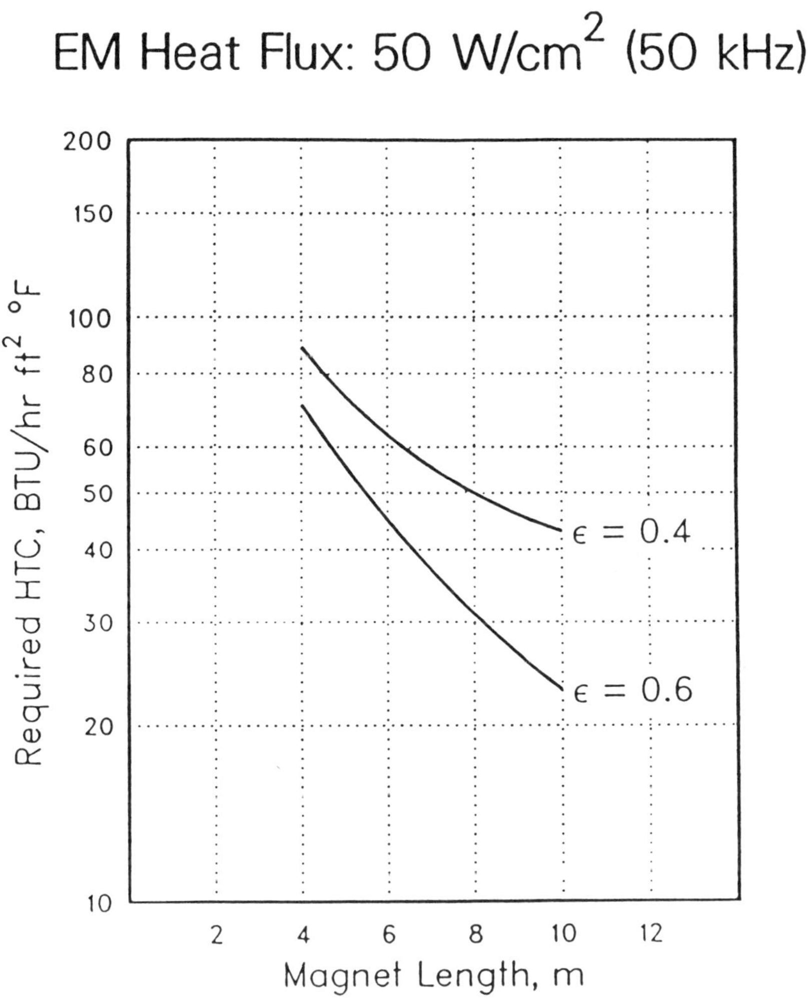

Fig. 2. Required Heat Transfer Coefficients as a Function of Magnet Length

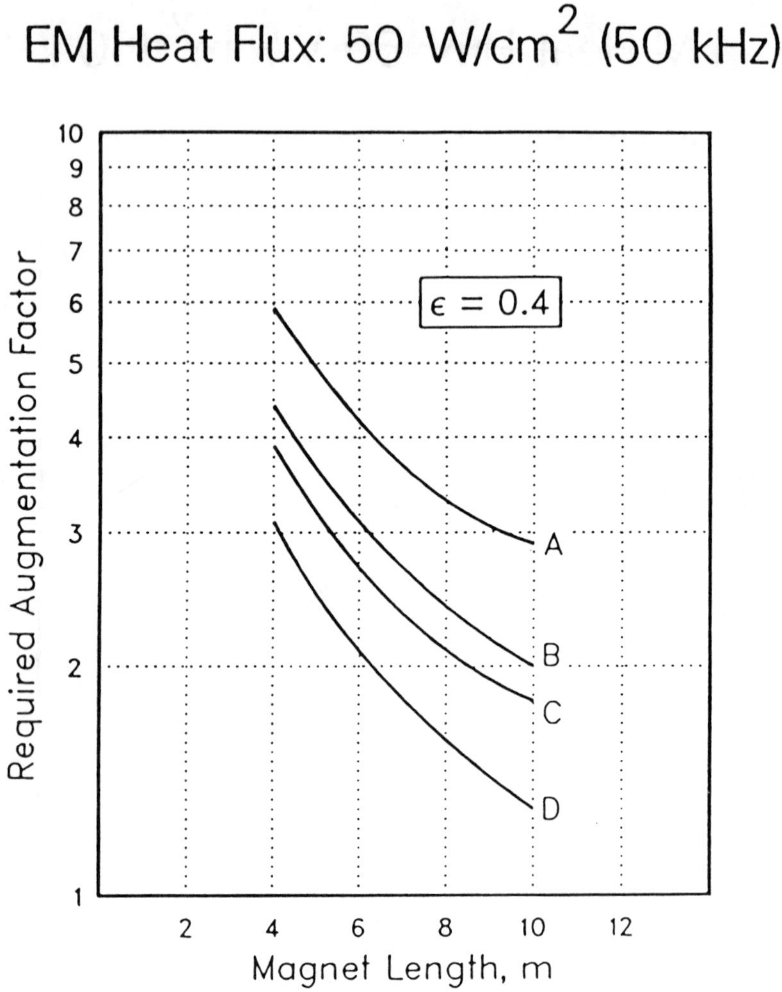

Fig. 3. Required Augmentation Factors as a Function of Magnet Length (ϵ =0.4)

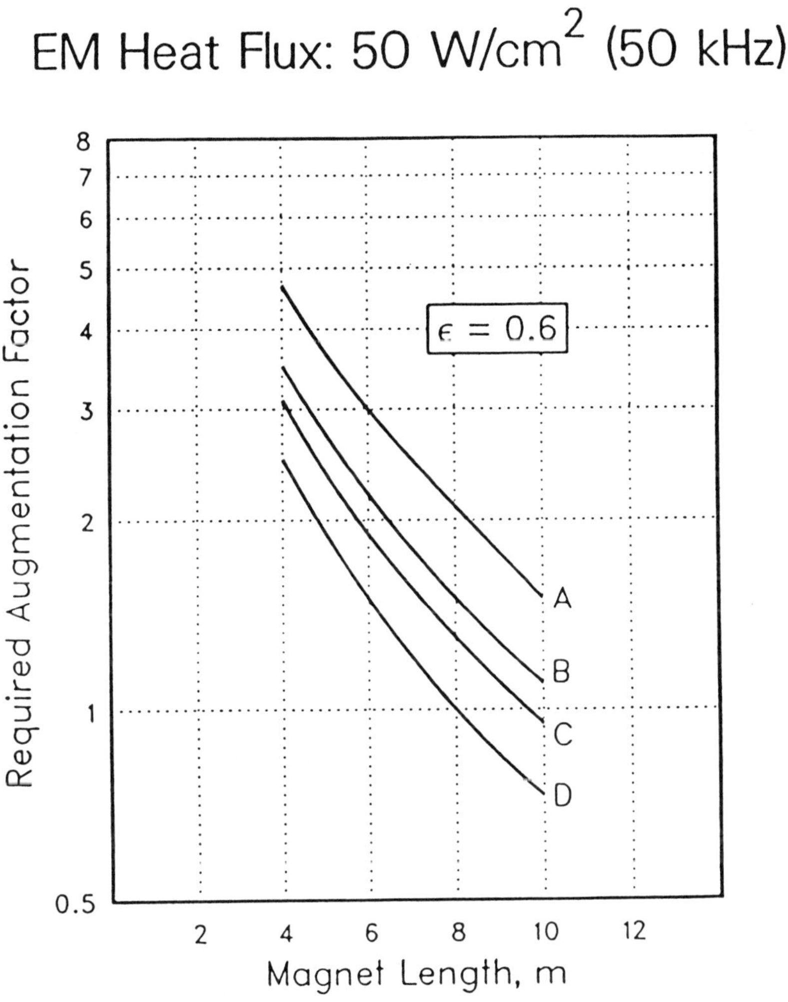

Fig. 4. Required Augmentation Factors as a Function of Magnet Length (ϵ = 0.6)

noted that the development of a mechanical support system for partially solidified strip would significantly reduce the magnet lengths discussed above.

The relative merits of gaseous coolants can be evaluated from the conventional empirical equations that form the basis for calculating the heat transfer coefficient. Since both nitrogen and carbon dioxide react with liquid steel to some extent, we considered the use of argon as a potential coolant. Because of the poor thermal transport properties of argon, we considered improving the heat transfer coefficients by using a mixture of argon and helium (which has excellent transport properties). The heat transfer coefficients for mixtures of argon and helium are presented in Table II for the same operating conditions as listed in Table I. These data show that a mixture of helium and argon, which is completely inert to liquid steel, will yield essentially the same heat transfer coefficients as for 100% N_2 or CO_2. However, a mixture of helium and argon will be more expensive to operate, even for a recirculating coolant system with very little leakage losses. Extensive parametric studies have been carried out on the heat transfer characteristics with CO_2 and a mixture of He and Ar. These results are not included here because the heat transfer characteristics of these gases are essentially similar to those of nitrogen (see Table II).

In sum, the heat transfer analyses show that heat transfer enhancement is required to design an EMCC magnet less than 10-m long with operating conditions limited to the reference values in Table I. The data presented in the next section show that achieving heat transfer enhancement by a factor of 2 to 3 is relatively easy.

Augmentation of Heat Transfer

Great technological advances have been made in heat transfer augmentation techniques that cover a wide spectrum of operating conditions and heat transfer modes. Two areas are of special interest to certain emerging industrial applications: conditions where only convective cooling by a gaseous coolant is acceptable, and conditions where no augmentation devices or turbulence promoters are permissible because they may significantly disturb the boundary layer. Augmentation of heat transfer by adding fine droplets of a liquid (e.g., water) to a gaseous coolant appears to be suitable for such applications. It is generally believed that heat transfer enhancement by mist coolants (gas + water droplets) is due to evaporation of droplets, larger heat capacity of the mixture compared with the gas alone, and micro-convection at the boundary layer; however, the exact contribution by each mechanism is difficult to assess. A survey of the literature on mist coolants was conducted, and the results found in the literature are summarized in the next section.

Mist Coolants

Augmentation of heat transfer by adding fine droplets of a liquid (e.g., water) to a gaseous coolant has been found to greatly increase the cooling potential of inherently poor gaseous coolants. Therefore, by applying a mist coolant over a partially solidified strip within the EM confinement, the total length of the magnet can be significantly reduced. Because of the high heat of vaporization of water, calculations show that evaporation of droplets can significantly enhance heat transfer (by lowering the bulk temperature), especially in situations where high surface temperatures and long residence times for the mist coolant are involved (e.g., EMCC). A study of the literature (Refs. 1-9) shows that the effectiveness of mist coolants (up to 30-fold increase in heat transfer coefficients in some cases) at low temperatures is well established. However, several experimental studies (Ref. 10-13) have indicated a sudden and significant decrease in the effectiveness of mist coolants in the temperature range of 150 to 500°C, depending on the experimental conditions. This has led many designers to conclude that mist cooling is ineffective at elevated temperatures (>600°C), even though experimental data (Refs. 14 and 15) show a monotonic increase in heat flux with rise in surface temperature. Significant progress has been made in developing analytical models for complex two-phase, two-component systems, yet no satisfactory models exist that can be utilized to design a coolant system at these high temperatures. Many experiments (Refs. 16 and 17) have been conducted to investigate the characteristics of spray nozzles and mist coolants, especially in the secondary cooling of steel, but no consensus exists on the best secondary cooling techniques. In view of the above facts, there is a need to gain a better understanding of mist coolants beyond the Leidenfrost temperature regime.

Augmentation Devices

In addition to augmentation of heat transfer by adding fine droplets of a liquid to a carrier gas, many novel techniques have been developed to modify the boundary layer in a way that increases convective transport of energy. A brief survey of the literature on heat transfer enhancement was carried out, with particular emphasis on methods that do not lead to an associated increase in frictional pressure drop. The literature (References 18-21) shows a wide selection of such techniques and devices if one requires only moderate enhancement (a factor of 2 to 3). Table III lists four methods and their experimental augmentation enhancement factors. These results show that heat transfer enhancement by a factor of two to three can be easily obtained. The augmentation method used must not interfere with the dynamic stability of the levitated strip and will require experimental verification.

Conceptual Design of Coolant System

In Fig. 1, the coolant channels are located between the thermal radiation shields that protect the electromagnet against the high-intensity thermal radiation from the liquid steel. For the purposes of thermal analysis of the steel strip within the electromagnet, the radiation shields are assumed to be cooled either by an inert gas (the same as

Table III. Methods for Heat Transfer Augmentation and their Augmentation Factors
(Many references exist, but four selected here)

Method	Augmentation Factors	Ref. No.
Swirling Flow	4	18
Displaced Device	>5	19
Scrapping Boundaries	~1.8	20
Novel Promoter	2.5	21

the EMCC system coolant) or by water. An approximate mass flow rate of the gas can be obtained based on assumptions regarding the total gas temperature rise and the coolant channel width. For the parametric studies, the coolant channel width was assumed to vary from 4 to 6 cm, and the coolant temperature rise was assumed to vary from 500 to 700°C. The coolant is assumed to be diathermic so that the total thermal energy is absorbed by the radiation shields, which are considered to be blackbodies. To prevent direct impingement of coolant jets against the liquid strip, the coolant is introduced through porous walls in the coolant manifold design. The liquid droplets are introduced into the coolant stream (mist cooling) after a solidified layer is formed on the steel strip.

Parametric Nomograms

Extensive parametric analyses have been conducted to identify the interactions of various parameters in the EMCC process. The focus of these analyses has been to (1) obtain heat transfer information, (2) determine important design and operating parameters, (3) examine how they affect cooling of liquid steel, and (4) develop design charts that will aid in the optimum design of a coolant system for the EMCC process. These analyses provide information on the total heat extraction requirements and heat transfer coefficients needed as a first step in the design of a cooling system for the EMCC process. Further, they address the feasibility of the EMCC process from a heat transfer consideration. The analyses have been performed in a very generalized fashion to provide flexibility in the selection of design and operating parameters and to aid in the design of a cooling system for EM casters ranging from bench scale to production scale.

A detailed examination of the design and operating parameters that control the EMCC process shows that extensive nomograms can be developed based on the following four major parameters:

(i) the desired outlet temperature, T_{out}
(ii) slab thickness, t
(iii) casting speed, u
(iv) magnet length, L

To cover a wide range of values of these parameters, a generalized nomogram has been developed (Fig. 5). The following case illustrates how this nomogram can be used to obtain the required heat extraction rate. The line A-B-C-D-E-F is constructed assuming an outlet temperature of 2650°F (1454°C), a slab thickness of 0.25 in. (0.635 cm), a casting speed of 10 in./sec (25.4 cm/sec), and a magnet length of 200 in. (5.08 m).

The termination of the line at F indicates the required heat extraction rate for this case, which is 1.19 x 10^5 Btu/hr ft^2 (3.75 x 10^5 W/m^2). Once the total heat extraction requirement has been determined, the convective heat transfer requirement can be calculated if one knows the electromagnetic heating and radiative heat transfer.

From an examination of the radiative and convective forms of energy and the imposed electromagnetic heat fluxes, it can be seen that the heat transfer coefficient required for cooling depends on seven parameters: $h = f(\Delta T, u, t, L, T_{out}, q_{EM}, \text{and } \epsilon)$, where h is the required heat transfer coefficient for a range of casting speed (u), magnet length (L), emissivity (ϵ), electromagnetic heat (q_{EM}), temperature difference (ΔT), and outlet temperature (T_{out}). The line ABCDEF in Fig. 6 illustrates how

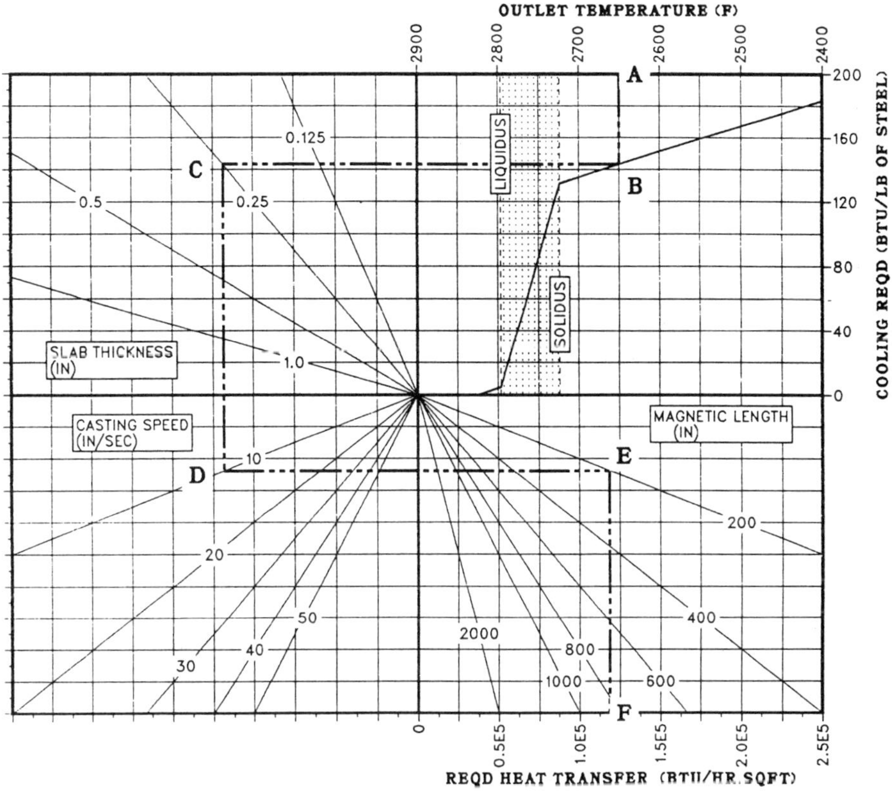

Fig. 5. Design Chart for EMCC Cooling Requirements

to use the nomograms to determine the required convective heat transfer coefficient. It shows that, for a 0.25-in.-thick slab, outlet temperature of 2650°F, casting speed of 10 in./sec, and magnet length of 200 in., the cooling requirement is 1.19 X 10^5 Btu/hr ft^2 (3.75 x 10^5 W/m^2). For this cooling requirement and with surface emissivity of 0.4, electromagnetic heat generation on one surface of 50 W/cm^2 and temperature difference between the slab and coolant of 2625°F (1440°C), the convective heat transfer coefficient required is 48.9 Btu/hr ft^2°F (277.7 $W/m^2 \cdot K$).

Conclusion

A number of gases such as nitrogen, carbon dioxide, and argon + helium mixtures were considered as potential coolants for the EMCC process. The results show that, with the imposed limits on coolant mass flow rates, augmentation of heat transfer by a factor of 2 to 3 would be necessary for an economically viable casting system. A study of the literature shows that such a modest augmentation of heat transfer can be

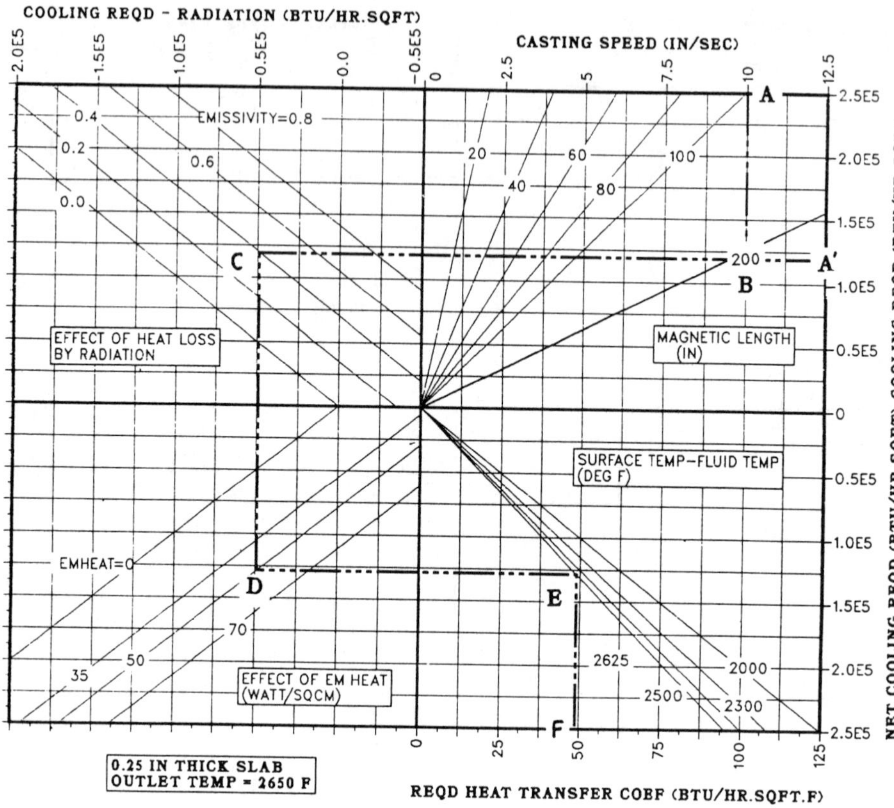

Fig. 6. Design Chart for Heat Transfer Coefficient Requirements

easily met by a variety of techniques. In addition, development of a mechanical support system within EM confinement will lead to an acceptable magnet length (<10 m).

Acknowledgment

This work was supported by the U.S. Department of Energy, Office of Industrial Programs, under Contract W-31-109-Eng-38.

References

1. T. Aihara and R. Saga, "Performance of a Compact Cooling Unit Utilizing Air-Water Mist Flow," J. Heat Transfer, 105, 18 (1983).

2. C. O. Pedersen, "An Experimental Study of the Dynamic Behavior and Heat Transfer Characteristics of Water Droplets Impinging Upon a Heated Surface," J. Heat Transfer, 13, 369 (1970).

3. J. W. Hodgson et al., "An Experimental Investigation of Heat Transfer from a Spray-cooled Isothermal Cylinder," J. Heat Transfer, 11, 457 (1968).

4. H. C. Simpson and E. K. Brolls, "Heat Transfer From a Thick Flat Plate to an Air/Water Mist in Turbulent Flow Over the Plate," Dept. of Thermodynamics and Fluid Dynamics, Univ. of Strathclyde, Glasgow (1971).

5. K. Hishika et al., "Heat Transfer from a Flat Plate in Two-Component Blast Flow," J. Heat Transfer, 102, 513(1980).

6. M. Trela, "An Approximate Calculation of Heat Transfer of an Air-Water Mist Along a Heated Flat Plate," Int. J. Heat Transfer, 24, 249 (1981).

7. W. C. Thomas and J. E. Sunderland, "Heat Transfer Between a Plane Surface and Air Containing Suspended Water Droplets," Ind & Eng. Chem., 9, 368 (1970).

8. A. N. Telford, "Exploratory Investigation of Laminar Boundary Layer Heat Transfer Characteristics of Gas Liquid-Spray System," ARL Report No. 64-136, Aerospace Research Laboratories (1964).

9. A. Acriros et al., "Research Investigation of Two-Component Heat Transfer," ARL report No. ARL-64-116, Aerospace Research Laboratories (1964).

10. H. Jacobi et al., "Heat Transfer in Cyclic Secondary Cooling During Solidification of Steel," Ironmaking and Steelmaking, 11, 132 (1984).

11. M. Mitsuttuka et al., "Recent Research, Development and Improvement of Forced Cooling Methods and Equipment at Nippon Steel Corporation," Nippon Steel Technical Report No. 12 (December 1978).

12. H. Lambert and M. Economopoulos, "Measurement of Heat Transfer Coefficients in Metallurgical Processes," J. Iron and Steel Inst., 10, 917 (1970).

13. C. J. Hoogendoorn and R. den Hond, "Leidenfrost Temperature and Heat Transfer Coefficients for Water Spray Impinging on a Hot Surface," Proc. 5th Int. Heat Trans. Conf., Washington, DC (1974).

14. H. Junk, "Warmeubergangsuntersuchungen an einer simulierten Sekundarkuhlstrecke fur das Stranggiessen von Stahl," Neue Hutte, 1, 13-18 (1972).

15. H. Muller and R. Jeschar, "Untersuchung des Warmeubergangs on einer simulierten Sekundarkuhlzone beim Stranggiessverfahrer," Arch. Eisenhuettenwes., 44 (8), 589-594 (1973).

16. J. K. Brimacomb et al., "Continuous Casting," Vol. 2, The Iron and Steel Society of AIME (1984).

17. G. F. Hewit et al., Editors, Multiphase Science and Technology, Vol. 1, Hemisphere Publishing Corp., New York (1982).

18. A. V. Sudarev, J. Eng. Phys., 15, 1049-1052 (1972).

14. H. Junk, "Warmeubergangsuntersuchungen an einer simulierten Sekundarkuhlstrecke fur das Stranggiessen von Stahl," Neue Hutte, 1, 13-18 (1972).

15. H. Muller and R. Jeschar, "Untersuchung des Warmeubergangs on einer simulierten Sekundarkuhlzone beim Stranggiessverfahrer," Arch. Eisenhuettenwes., 44 (8), 589-594 (1973).

16. J. K. Brimacomb et al., "Continuous Casting," Vol. 2, The Iron and Steel Society of AIME (1984).

17. G. F. Hewit et al., Editors, Multiphase Science and Technology, Vol. 1, Hemisphere Publishing Corp., New York (1982).

18. A. V. Sudarev, J. Eng. Phys., 15, 1049-1052 (1972).

19. S. T. Lin, L. T. Fan, and N. Z. Azer, "Augmentation of Single Phase Convective Heat Transfer with In-Line Static Mixers," Proc. Heat Transfer Fluid Mech. Inst., Stanford University Press, pp. 117-130 (1978).

20. J. K. Hagge and G. H. Junkhan, "Mechanical Augmentation of Convective Heat Transfer in Air," J. Heat Transfer, 9, 516 (Nov. 1975).

21. Saburo Maezawa and G. S. H. Lock, "Heat Transfer Inside a Tube with a Novel Promoter," Proc. 6th Int. Heat Transfer Conf., Vol. 2, p. 596 (1978).

CAD OF ELECTROHEAT DEVICES

J.C. Sabonnadière
Laboratoire d'Electrotechnique de Grenoble (CNRS UA-355) FRANCE

Ph. Wendling
MAGSOFT.CORP. - Troy, U.S.A.

O. Longeot
E.N.S.A.M. - Bordeaux, France

Summary

The purpose of this paper is to present the CAD Package FLUX (1), (2), (3) which allows, with its subsystems Flux THERM and Flux MT, a precise analysis of the electromagnetic and thermal interactions between the heated device and the inductor. The knowledge of the evolution of the temperatures allows a good prediction of the metallurgical properties of a thermally treated area (4), (5).

As an example, to compare the results given by the simulation and the values measured by test, we will use the processing of a hollow cylindrical tube heated by an annular inductor and cooled through the internal face by a pressurized water system.

Casting of Near Net Shape Products
Edited by Y. Sahai, J.E. Battles,
R.S. Carbonara and C.E. Mobley
The Metallurgical Society, 1988

Introduction

Induction heating is nowadays a well known technic, nevertheless it is often difficult for designers of induction furnaces to appreciate accurately the interactions between the inductor and the load at various temperature. However the knowledge of the thermal behaviour is necessary during the heating sequence as well during the cooling one to predict the metallurgical properties of the hardened device.

The simulation appears to be very useful for the design of the inductor but also for its operation after having been built to control the process.

Unfortunately the phenomena which occur during all the process are so complicated and strongly coupled that the classical methods of analysis become obsolete.

Numerical solution of all phenomena are generally well known but their solution is difficult because the dependance between electromagnetic and thermal phenomena is very strong.

In order to make possible for an induction furnace designer to run easily simulations of the behaviour of his designed device we have elaborated a set of packages named FLUX, FLUXTHERM and FLUXMT which are able to make a good simulation of these phenomena. These softwares have been checked to prove the accuracy of the solution, but are also designed to be very convivial using interactive C.A.D. technics.
All those softwares are based upon the finite element analysis of electromagnetic and thermal phenomena.

Inside the frame of this paper we shall describe the various capabilities of FLUX for electromagnetic problems, FLUXTHERM for thermal ones and then FLUXMT devoted to the analysis of strongly coupled electromagnetic and thermal phenomena. Finally we shall illustrate their application to the hardening of a hollow ferromagnetic cylinder and the comparison with experimental results.

II . ELECTROMAGNETIC MODELLING : FLUX-2D

Let us consider a load made of a piece of conducting material part inside a coil supplied with an alternating current density Jo.
Maxwell's equations which describe the phenomena may be written in their quasi-static form in which all the quantities depend harmonically on the time with an angular velocity :

$$\nabla \times E + j\omega B = o \qquad (j^2 = -1) \qquad (1)$$

$$\nabla \times H = J \qquad (2)$$

$$J = \sigma E \qquad (3)$$

$$H = \nu B \qquad (4)$$

$$\nabla . B = o \qquad (5)$$

(5) implies that there is a vector potential A such as $B = \nabla \times A$ which has only one component along **z** .

Then, if we introduce A into equations (1) to (4) lead to :

$$\nabla \times (\nu \nabla \times A) + j\omega\sigma A = J_o \qquad (6)$$

This last equation is solved by the finite elements method in its linear form or even in a non linear form assuming the quantities remain harmonic despite the saturation of the magnetic circuits. The results are the flux every where in the domain, the current density in the conducting materials and overall the power density losses in these materials.

At this step it is possible to use the software in order to optimize the geometry of the device to get a current density distribution in the conducting material adapted to the type of operation aimed. Any geometric or electromagnetic parameter (frequency, supplied voltage) cand be modified during the design of electromagnetic system in order to obtain better performances.

III . THERMAL MODELLING : FLUXTHERM

The fundamental phenomenon represented here is the thermal diffusion in a given media, taking into account as accurately as possible the convective and radiative exchanges.

Let us call k(T) the thermal conductivity of a material, $\rho C(T)$ its thermal capacity which both depend on the temperature. The Fourier law associated to the conservation of mass and energy lead to equation (7) which describes the time dependant evolution of temperature in the media :

$$\nabla.(-k(T)\nabla T) + \rho C(T)\frac{\partial T}{\partial t} - q = o \qquad (7)$$

and where q is the heat density brought inside the medium.

This equation must be combined with boundary conditions characteristic of heat transfer between the material and its surrounding. These conditions are either initial or boundary conditions.

$$\begin{cases} T = T_o(s) & \text{temperature known on the boundary} \\ -k\dfrac{\partial T}{\partial n} = \emptyset(s) & \text{thermal flux given on the boundary} \\ -k\dfrac{\partial T}{\partial n} = h(T-T_a) & \text{convective transfer} \\ -k\dfrac{\partial T}{\partial n} = \varepsilon\sigma(T^4-T_a^4) & \text{radiative transfer} \end{cases} \qquad (8)$$

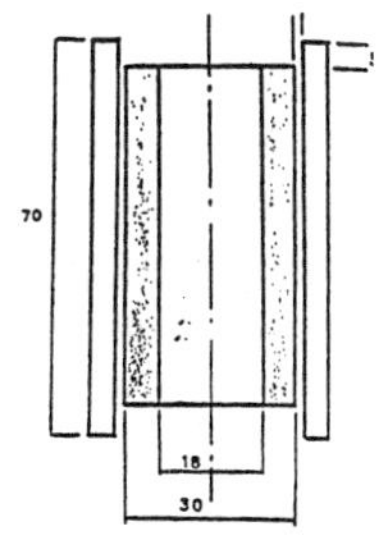

Figure 1 - Load and coil.

The software has been designed to solve equation (7) and boundary conditions of type (8) according to the kind of material and boundary described as the domain. A very important feature in FLUXTHERM is its capability to take into account the variation of parameters of equations (7) and (8) with the temperature and even very stiff variations like this of ρCfor a ferromagnetic material close to the Curie point which is modelled by an exponential type formula on the form :

$$\rho C(T) = \rho C_o + \frac{\xi}{s\sqrt{2\Pi}} \exp(-(T-T_C)/2\ s^2)$$

Another feature is its possibility to handle non linear equations and boundary conditions in order to take into account the radiative process.

The output of FLUXTHERM is the map of temperatures on the domain, the plot of temperature along a line (segment or circle) or the heat flux through any surface.

IV . COUPLING ELECTROMAGNETIC AND THERMAL EQUATIONS : FLUXMT

Engineering induction heating needs to know very precisely the interactions between electromagnetic and thermal phenomena. Indeed, the electric and magnetic parameters of the electromagnetic system are strongly dependant on the temperature, and the temperatures depends on the quantity of energy dissipated in the material which is a function of the solution of the electromagnetic problem.

In order to take these dependances into account we have built a software adapted to this coupled problem, FLUXMT in which there are software tools which allows the user to solve alternatively the electromagnetic and thermal problems with a constant updating of the various parameters.

It is thus possible to solve the electromagnetic force in the problem, extract the power density at each node as the sources of the thermal problem then solve the thermal problems during several time steps and, then, solve again the electromagnetic problem with updated parameters (σ,μ).

This software is very useful for the simulation of the heating sequence of a device and then it can be switched to FLUXTHERM for the cooling sequence in which a shower is modelled by a convection parameter and an external temperature.

V . OPERATING THE SOFTWARE

The software is operated in four consecutive steps.

V.1 Geometrical definition and discretization

The geometrical definition of the device and its inductor is made interactively on a terminal with, at each step, a possibility of graphic control of the work done. The discretization is also driven by the software which helps the designer in choosing a mesh adapted to the problem. As an example, we show on figure 2 the domain of analysis and on figure 3 the discretization of the whole domain for the electromagnetic analysis (obviously for the thermal analysis the heated device only is subdivided in finite elements).

During this step symbolic names are given to various part of the domain (BILLET, AIR, COIL...), names of region to which physical properties will be assigned eater on in the second step.

V.2 Physical properties assignment

This sub-program named PROPHY considers the data previously defined and allows the user to give a physical meaning at his problem by defining the electrical parameters (frequency, voltage) and the material of which the various part one made.

The material characteristics are stored in a data base in which the various electrical and thermal properties are assigned to a material and modelled according various possibilities. For the problem previously mentionned, the cylinder is a ferromagnetic one with the electrical, magnetic, and thermal properties depends on the temperature according to the plots of the figure 7. Any property may be isotropic or not or described as a constant or as a function of a given parameter (Temperature..., FLux density).

V.3 Solution and post-processing

Having defined completely the problem, the user runs new the solver which built the solution of the problem defined by the geometry and the physical data. The last step is the post-processor in which the results are used to define useful output data like equiflux or isothermal lines but also to compute losses in conducting sheet or the active and reactive power at the inductor ports. It is also possible to define specific isothermal lines (like on which the temperature is constant and equal at 760°C...). At this step there is a command which allows to extract from the data of the post processor the values necessary to couple electromagnetic and thermal analysis. These data are the power density at every point of the conducting material or the temperature at each node in view to update the values of the conductivity of conducting materials and the permeability of the ferromagnetic parts for the electromagnetic analysis.

All the operation of the software are driven interactively except for heavy computation which can be operated automatically.
We shall illustrate in the forecoming paragraph the application of the software to the simulation of the heating and hardening process of a hollow ferromagnetic cylinder.

VI . SIMULATION WITH THE FLUX C.A.D. SOFTWARE OF THE THERMAL CYCLE OF A HOLLOW CYLINDER

VI.1 Behaviour of the material during quinck thermal cycles

The skin heating processing by induction requires times of 2 to 10 seconds to reach a temperature of 1000 deg. C. The cooling conditions (pulverization, shower, immersion) can lead to cooling rates of the previously heated area of 50 to 400 deg. C/s. During these thermal histories, the material will be subjected to metallurgical alterations : austenitization during the heating time (reaching temperature higher than temperature of point AC3), alteration during the cooling time of the austenite in martensite or in decomposition products of austenite (gamma-alpha).

These alterations are due to the material (chemical constitution, initial structure) and to the thermal cycle (temperature rise, isothermal maintenance, cooling rate).

The supply of the modelization is to allow the provision of the metallurgical alterations (hardness of the material) regarding the parameters of the processing and the geometry of the set device inductor.

VI.2 Mathematical model of the hardness cycle

The model we used is defined by equations (9) and (10) in which the quantity Hv represents the hardness and t 300/700 the cooling criterion :

$$Hv = CM1 \tan^{-1} (S_1(-Ln(t300/700) + Ln(C1) + CE1 \quad \text{for } t300/700 \; [0, t_c] \tag{9}$$

$$Hv = CM2 \tan^{-1}(S(-Ln(t300/700) + Ln(C2) + CE2 \quad \text{for } t300/700 \; [t_c, \infty] \tag{10}$$

The coefficients CM1, S1, C1, CE1,CM2, S2, C2, CE2 must be identified. By a least square method in which we minimize the function error

$$E = \sum_{i=1}^{n} \omega_i \, (yi - Hv(t^i 300/700)) \tag{11}$$

in which Yi is the experimental value of the hardness of the value t^i300/700, we have easily computed the coefficients CE1, CE2 and CM1, CM2.

However the non linear relation between Hv and the coefficients S1, S2 or C1 and C2 make the fitting of these values more difficult. Nonetheless the method used allowed to fit the experimental plot correctly from the data of only 5 experimental points. The plots of figure 3 correspond to one value of the maximum temperature reached for different values of this maximum an interpolation must be made (figure 4).

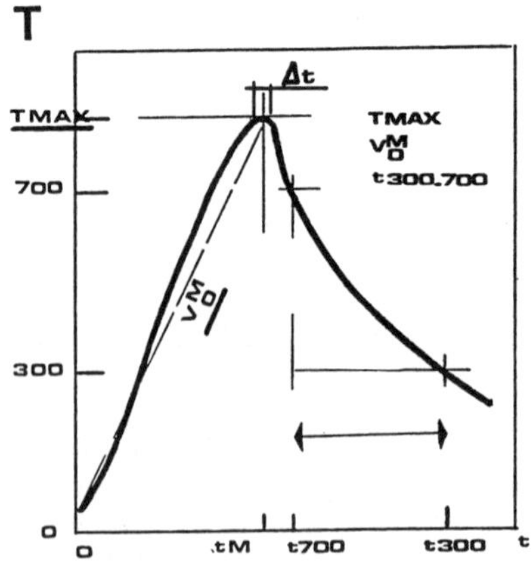

Figure 2 - Thermal cycle at a point of the device.

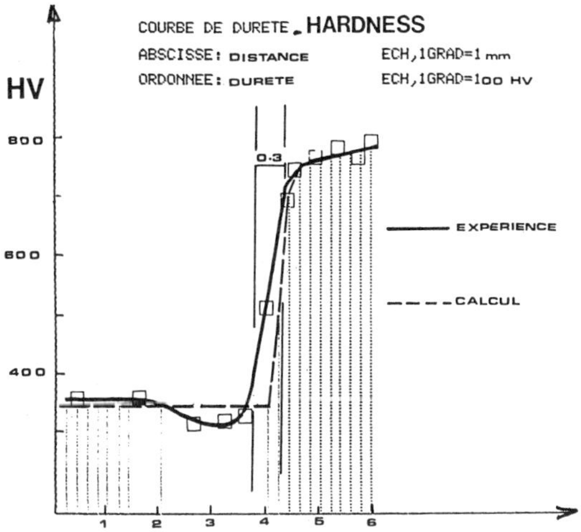

Figure 3 - Effect of the temperature to the curve "hardness criteria" (XC 42).

VI.3 Simulation of the tube heated by an annular inductor and cooled by a water circulation on the internal face

Solution :

The C.A.D. Package Flux has been used during the heating phase with a coupled magneto-thermal analysis and during the cooling phase with transient thermal analysis.

Geometry :

The subdivision used is shown in figure 5. For the chosen frequency (10 000 Hz) a three-level meshing was used.

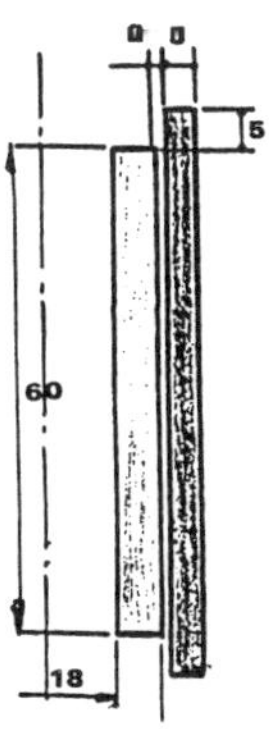

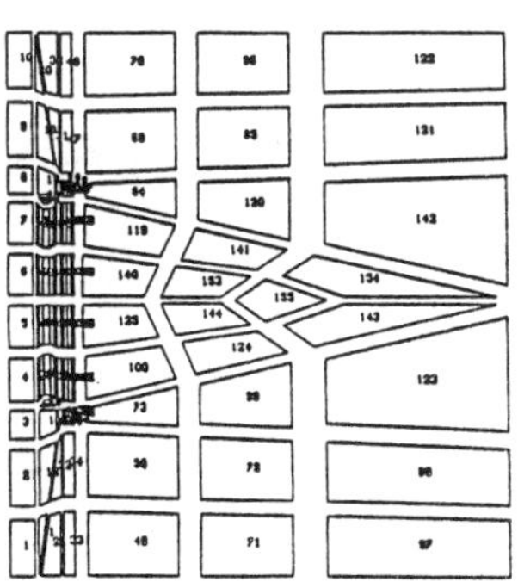

Figure 5a - Geometry (inductor-device).

Figure 5b - Three-level meshing.

$$Hv = Cm_{1,2} \tan^{-1} |S_{1,2}(-\ln t_{300.700} + \ln C_{1,2})| + C_{e_{1,2}}$$

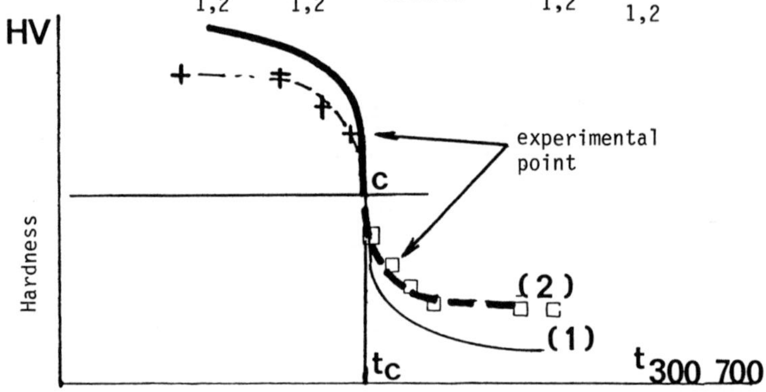

plot (1) 4 coefficients Cm_1, S_1, C_1, C_{e1}

plot (2) 4 coefficients Cm_2, S_2, C_2, C_{e2}

Figure 4 - Representation of the two models of the hardness fonction of the cooling criteria.

Boundary conditions :

Radiation on the faces (1), (2) and (3) and forced convection on the face (4). The value of the heat transfer coefficient is h = 104 W/m2.deg.C.

The control of the sources is shown on figure 6.

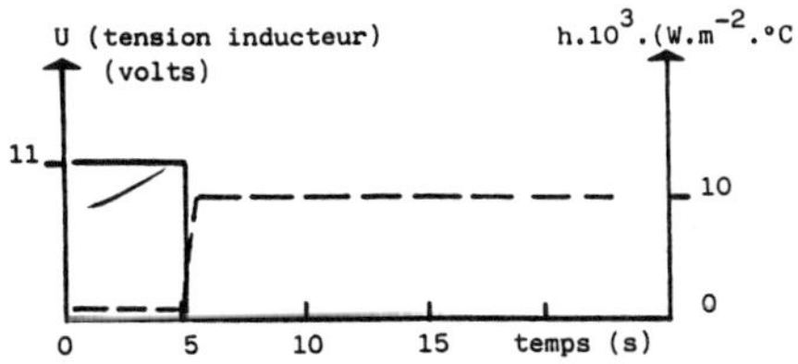

Figure 6 - ----- control of the voltage
- - - control of the water.

The material :

The material data are shown on figure 7.

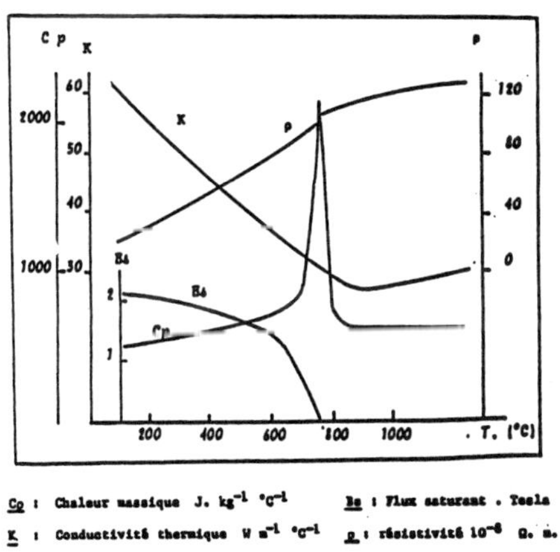

Figure 7 - Electrical, magnetic and thermal characteristic of the material.

Electrical data :

The electrical source was chosen to obtain a skin quenching, 1 to 3 mm deep, in the external face of the tube. The frequency ofthe current is 10, 10 000 Hz with a voltage of 11 volts at the terminals of the inductor.

VI.4 <u>Results of the computation</u>

Equi-flux lines (figure 8).

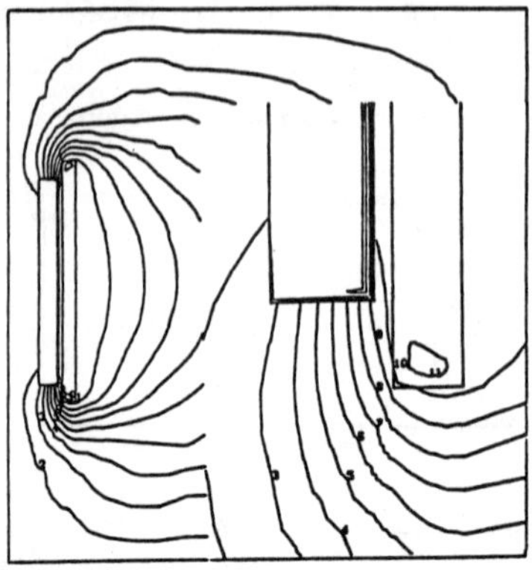

Figure 8 - (Equi-Flux).

Isothermal lines in the device with the time figure 9.

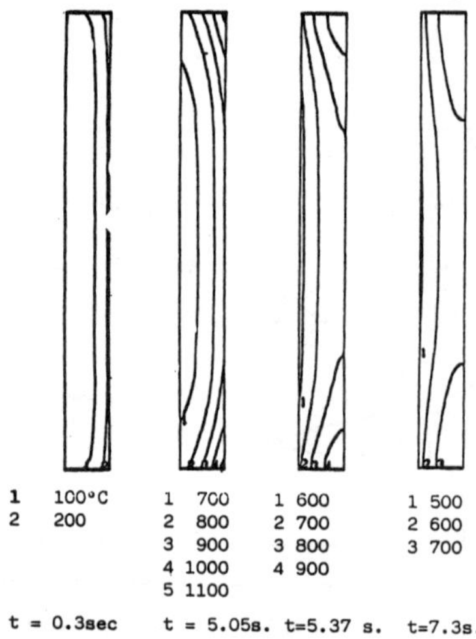

Figure 9 - Spatial repartition of the isothermal lines at different time steps (the heating phase was stopped and the cooling phase begun at 5.05 s).

It is possible to note the overheating of the corner due to the overreach of the inductor.

Distribution of eddy current in the device (figure 10, figure 11 and figure 12).

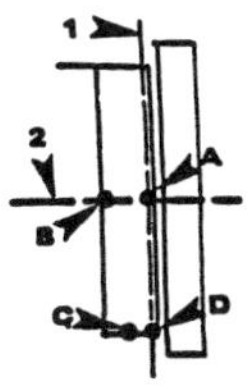

Figure 10 - Choice points
and lines

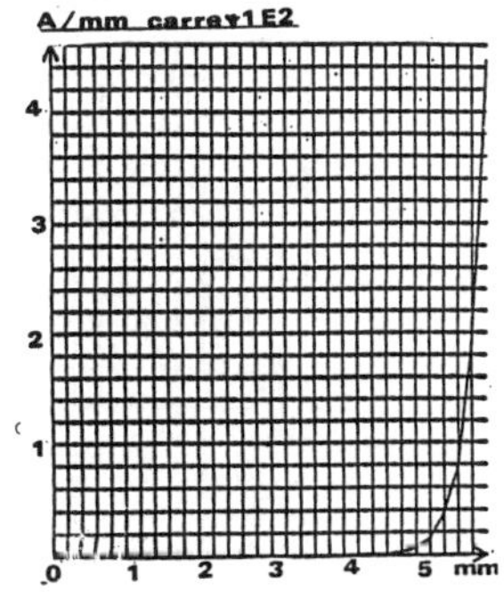

Figure 11 - Distribution of eddy current
along the line.

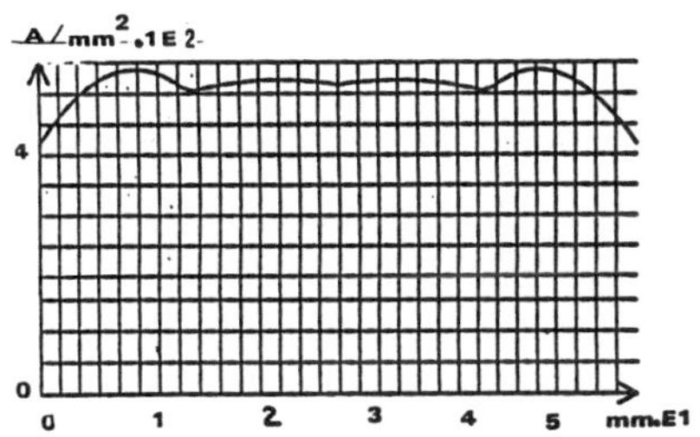

Figure 12 - Distribution of eddy current along the line.

Time variation of the temperatures (figures 13 and 14).

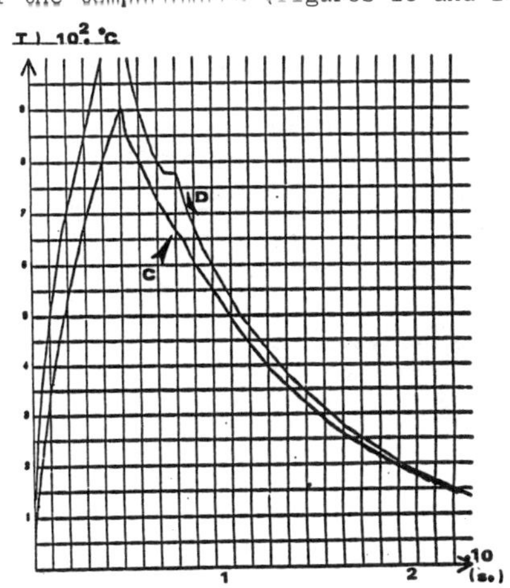

Figure 13 - Time variation of temperature at points C and D.

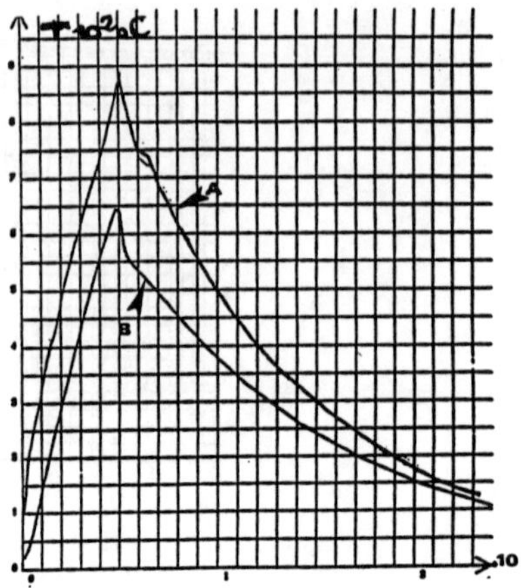

Figure 14 - Time variation of temperature at points A and B.

The time variation of the temperatures at each point of the device allows the determination of the thermal data needed to compute the hardness of the iron after processing (figure 14).

The thermal and electrical data can be modified very easily to alter the thermal cycles (figure 15).

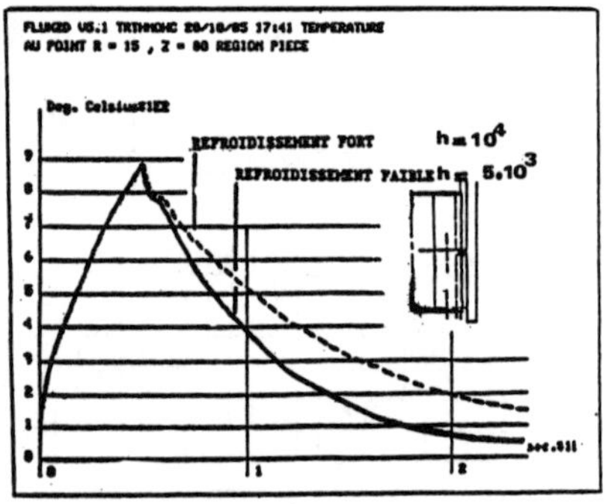

Figure 15 - Time variation of the temperatures at a point in the middle of the device for two kinds of cooling phases.

Iso-hardness lines :

The display ofthe metallurgical data allows the drawing of the iso-hardness lines (figure 16).

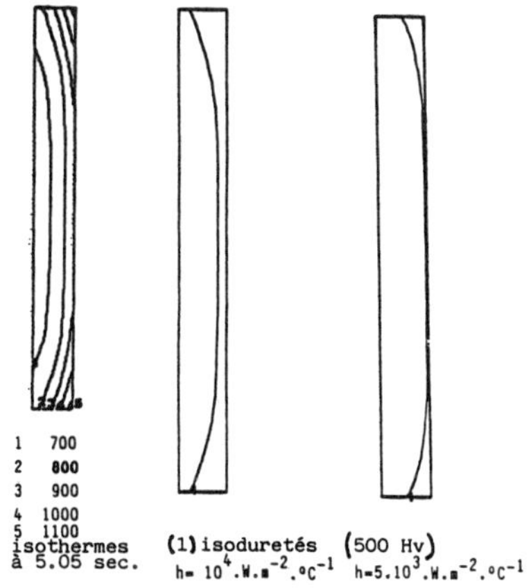

Figure 16 - Spatial representation of the hardnesses.

Alteration by Nital of the treated area iso-hardnesses 650 Hv.

VII . COMPARAISON BETWEEN EXPERIMENTAL AND COMPUTED DATA

Various experiments have been conducted in order to check the results of our software against experimental data. Some of the results are reported below on figures 17 and 18.

On figure 17 we have reported a comparaison on experimental and computed spatial distribution of hardness (for a value of the hardness equal to 650 Hv) at the extremity of the cylinder.

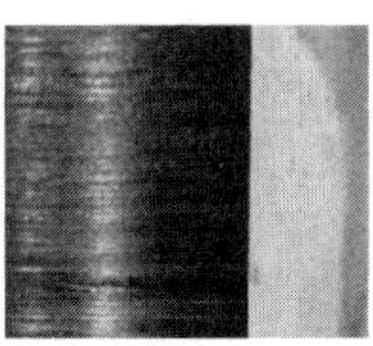

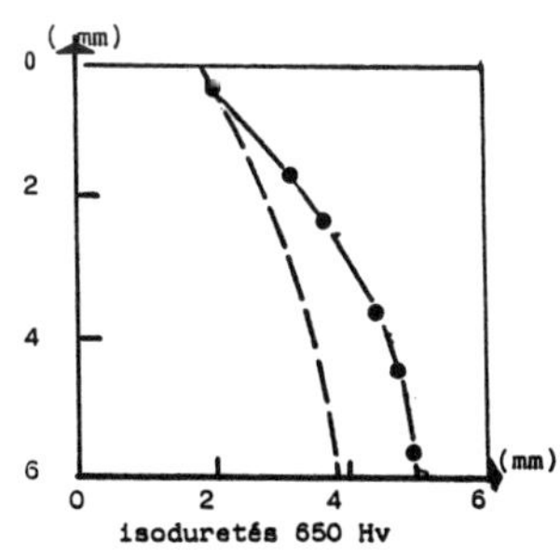

Attaque au Nital de la zone transformée (essai).

Figure 17 - Comparison (measurements •• computation — —) the annealed profil at the ends of the tube.

Time variation of the tempered area :

Figure 18 allows the comparison between the measured hardness (200 gr.) and the results of the computation along a radial straight segment.

Coupe 2

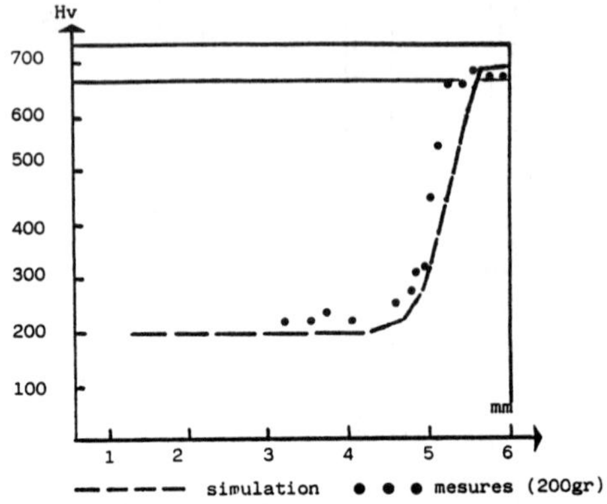

Figure 18 - Comparison between the mesured hardness and the results of computation.

CONCLUSIONS

The computation of the magneto-thermal and thermal phenomena during the heating then cooling phases by the package FLUX allows a precise analysis of the effect of each parameter on the metallurgical results. Moreover, the easy use of FLUX allows for real computer aided design ofthe thermal processing with a given device in a metallurgical aim.

REFERENCES

1. J.C. Sabonnadière, G. Meunier, B. Morel
"FLUX-2D : A general interactive finite element package for 2D electromagnetic field"
Proc. of Compumag Conference, Chicago, 1981.

2. Ph. Masse
"Analyse Méthodologie de la Modélisation Numérique des équations de la physique des milieux continus à l'aide de la méthode des éléments finis"
Thèse d'Etat I.N.P.G., Juin 1983.

3. H. Bleuvin
"Analyse par la méthode des éléments finis des phénomènes magnéto-thermiques : Application aux systèmes de chauffage par induction"
Thèse Docteur-Ingénieur I.N.P.G., Octobre 1984.

4. O. Longeot, D. Durand, J.C. Bonvalet
"Modélisation de la trempe superficielle par induction"
"Méthode de prévision de la profondeur trempée"
Rev. Traitement Thermique 190.85, pp.69-73.

5. O. Longeot
"Analyse des phénomènes magnétiques, thermiques et métallurgiques -lors de la trempe superficielle en chauffage par induction - Mesure et contrôle de la puissance active injectée à la pièce en temps réel"
Thèse Docteur-Ingénieur ENSAM Bordeaux, Juin 1985.

ELECTROMAGNETIC EDGE SHAPE CONTROL OF METAL SHEET PRODUCED BY TWIN-ROLL PROCESS

T.Yuhara, T.Kozuka, I.Muchi and S.Asai

Department of Iron & Steel Engineering
Faculty of Engineering, Nagoya University
Furo-cho, Chikusa-ku, Nagoya 464
Japan

Abstract

The Lorentz force, which is caused by the induced current, suppresses fluid flow motion under direct magnetic field.

In the production of thin sheet by a twin roll direct caster, a saw shape edge appears in the both sides. To suppress the edge quality, direct magnetic flux is applied to the composite roll made of magnetic material, such as plain steel and non-magnetic one, such as stainless steel. The strong magnetic field appears at the plain steel part between rolls and reduces the velocity along the axis of rolls. This quick reduction of the velocity at the edge part gives smooth edges of sheet.

Casting of Near Net Shape Products
Edited by Y. Sahai, J.E. Battles,
R.S. Carbonara and C.E. Mobley
The Metallurgical Society, 1988

Introduction

There is an endorsement to indicate that the study of magnetohydrodynamics (MHD) must be fruitful in electromagnetic processing of materials as well as it has been useful in the original fields of astrophysics and plasma physics. It is because liquid metal is a typical electromagnetic material and the vast knowledge of MHD which has been accumulated in fusion research and the field of astrophysics over a long period has a high potential to solve metallurgical problems.

This paper is one of the series researches concerned with the application of MHD to metallurgical problems. These days, great interest is focused on the twin-roll process as connected with production of thin sheets or amorphous metals. Since the twin-roll process has been invented by Bessemer, controlling of the edge shape at both side of sheet products has been a crucial and difficult problem. It has been a big obstacle interrupting the success of the process. In the original process, holding of molten metal at edge totally depends on the surface tension of molten metal induced by the curvature at edge corner. Thus, it gets more difficult with increasing the sheet thickness due to less curvature. In order to get a smooth edge shape, a permanent magnet is so set above the rolls to produce a strong magnetic field between them. The field can be expected to reduce the velocity of molten metal along the direction of the roll axis. This reduction of horizontal velocity suppresses surface waves on molten metal pool and brings about smooth edges. The mechanism by which surface waves produce saw edge shapes was examined by use of a high speed VTR. Based on the analysis of the recorded frames of VTR a mathematical model is developed to simulate the process producing saw shape edges.

Experimental

Apparatus

A schematic overview of the experimental apparatus is shown in Fig.1. The important difference of this apparatus with the conventional twin-roll process is that a permanent magnet is set above the rolls so as to make a magnetic circuit and to impose a direct magnetic field between them as shown in Fig.2. The intensity of the imposed magnetic field can be controlled by adjusting the distance between the pole pieces of permanent magnet and the rolls. Each of the rolls used in the process are composite ones made of a magnetic material, such as plain steel and a non-magnetic one, such as stainless steel as shown in Fig.3. A strong magnetic field is concentrated on the part of plain steel between the rolls. The two pairs of composite rolls were prepared as shown in (a) and (b) of Fig.3. (a) refers to a single ferrite roll (SFR) with one piece of ferrite, and (b) is a double ferrite roll (DFR) with two pieces of ferrite. These induce different patterns of magnetic field along the roll axis.

Procedure

Twenty grammes of tin particles are put in a nozzle made of Pyrex glass and air in the nozzle is purged out by argon gas. Then, the nozzle position is so adjusted that tin particles at the nozzle exit are located in the heater zone to be melted down and heated up to a preset temperature. The position of the nozzle is quickly so adjusted that the tip of the nozzle is in the roll gap as deeply as possible after metal temperature has reached the preset one. Then, the molten metal is injected between the rolls by using a back pressure of argon gas, which is so controlled by a sequencer unit as to open an electromagnetic valve of argon gas at the moment when a stopper located at the nozzle exit is released up.

The strength of magnetic field, H, is adjusted by changing the distance between the rolls and the pole pieces of magnet. The relation between the reciprocal values of magnetic flux at the plain steel part and

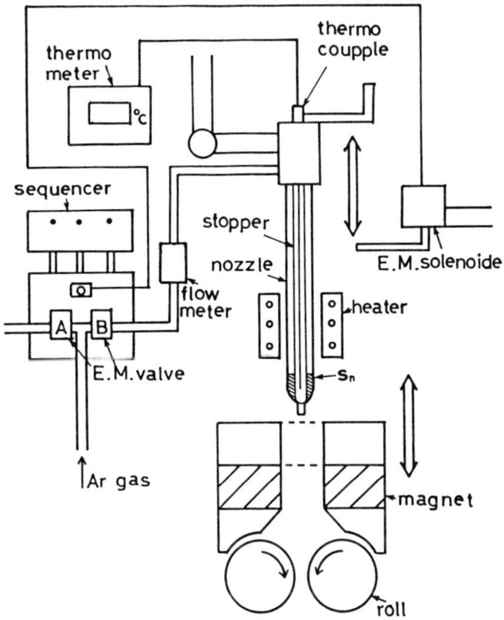

Figure 1 - Schematic view of experimental apparatus.

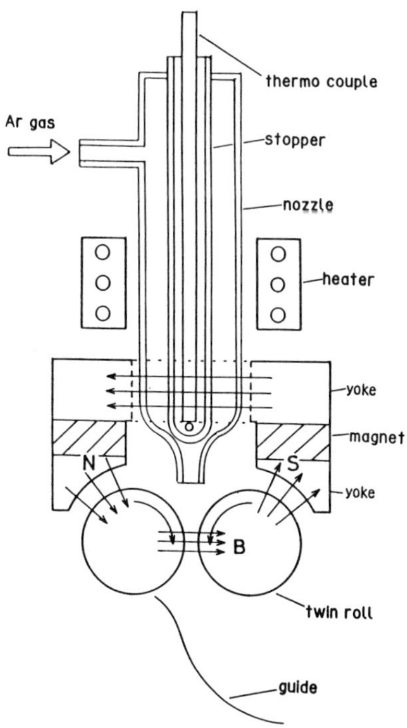

Figure 2 - Schematic view of magnetic circuit.

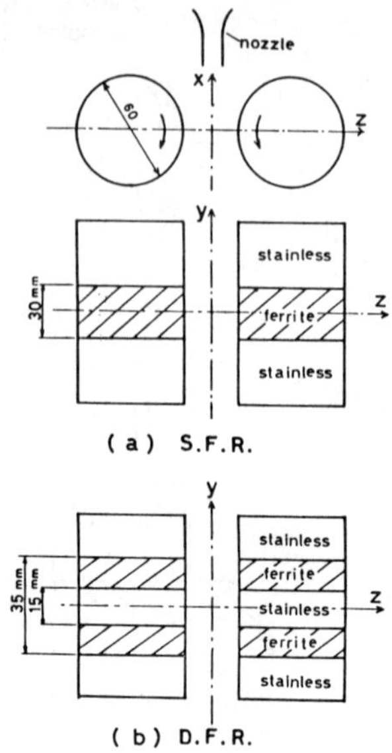

Figure 3 - Design of roll.

the roll gap is shown in Fig.4. These results were obtained under the condition of no rotation of roll. Though the magnetic flux is going to reduce with increase in rotation speed, it is not necessary to take account of the effect of roll rotation on magnetic flux under the employed rotation speed, say 120 or 130 r.p.m..

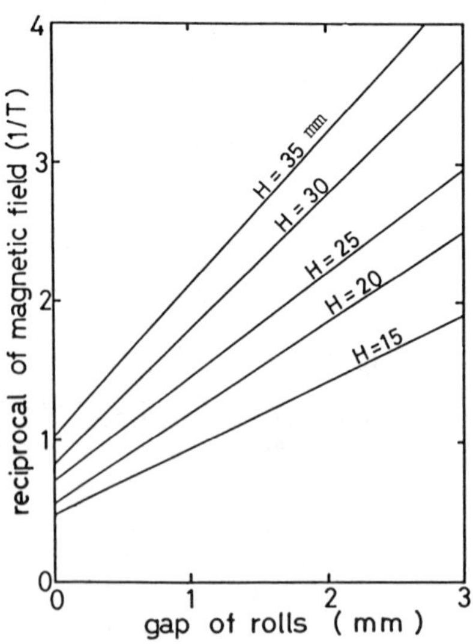

Figure 4 - The relation between the reciprocal values of magnetic flux and roll gap.

Results and Discussion

Imposition of Magnetic Field

At first, the casting condition of no magnetic field under which significant edge disturbances appear, is chosen as the reference in order to make the effect of magnetic field on the edge shape conspicuous. Figure 5 (a) shows a typical saw edge pattern which was obtained with a rotation speed of 130 r.p.m. and no magnetic field. Each of the saw edges appearing at both sides of the sheet is named a notch. The notch height is defined as the distance between the bottom and tip of the notch as shown in the side sketch of Figure 5 (b).

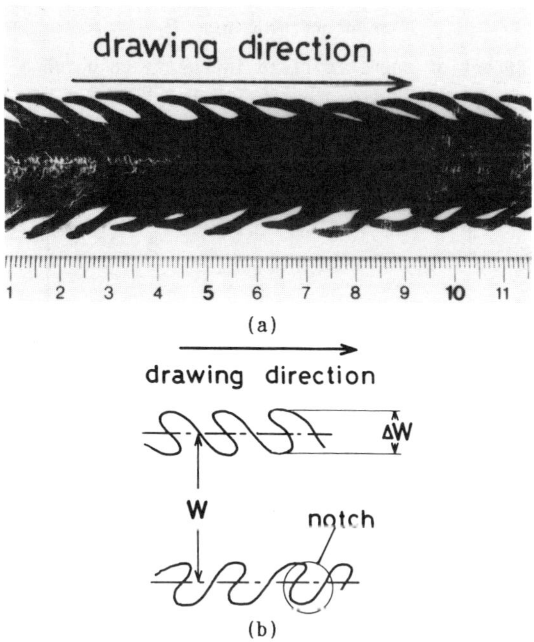

Figure 5 - Typical edge shape of thin plate produced by twin roll process.

Figure 6 shows the effect of magnetic field on the notch height. The notch height decreases with increase of magnetic flux in both cases of SFR and DFR. Accompanied with a decrease in the notch height, the number of notches per unit length also decreases as shown in Fig.7. Regarding the height and number of notches, there is no significant difference in tendency between SFR and DFR. However, the sheet width shows different trends in both cases as seen in Fig.8. That is, the sheet width in the case of SFR slightly decreases with increase in magnetic intensity, but in the case of DFR it is not affected by magnetic intensity. The reason for this is explained as will follow. The electromagnetic body force which suppresses the motion of molten metal acts over the total width of the sheet in the case of SFR. The motion of molten metal along the axis of the roll is also suppressed to reduce the sheet width. In the case of DFR the body force affects only the side edges of the sheet so that the width is not substantially influenced by magnetic field. The effects of magnetic field on the edge shape are clearly seen in Fig.9 and 10. It gets smoother with increase of magnetic flux.

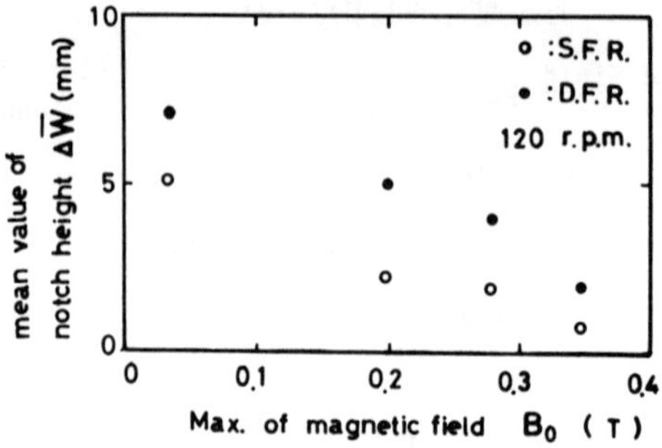

Figure 6 - Effect of magnetic field intensity on notch height.

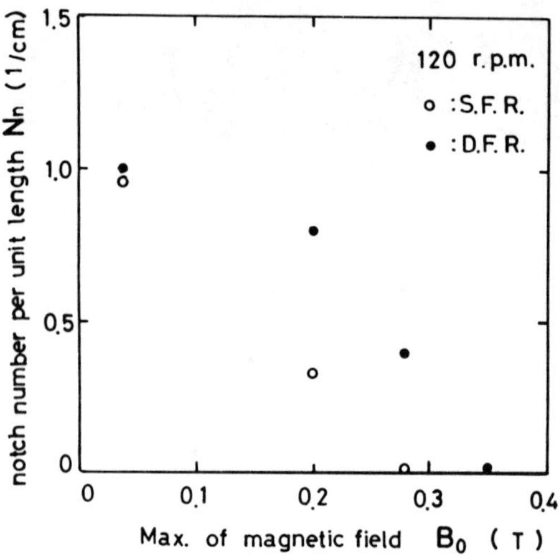

Figure 7 - Effect of magnetic field intensity on notch number.

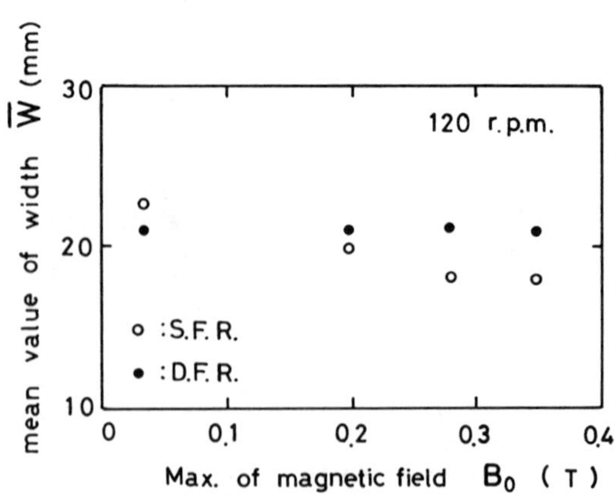

Figure 8 - Effect of magnetic field intensity on width of products.

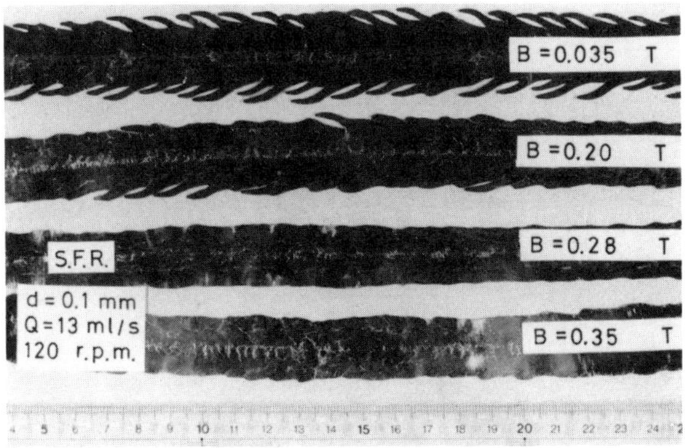

Figure 9 - Effect of magnetic field on edge shape of products in the case of SFR.

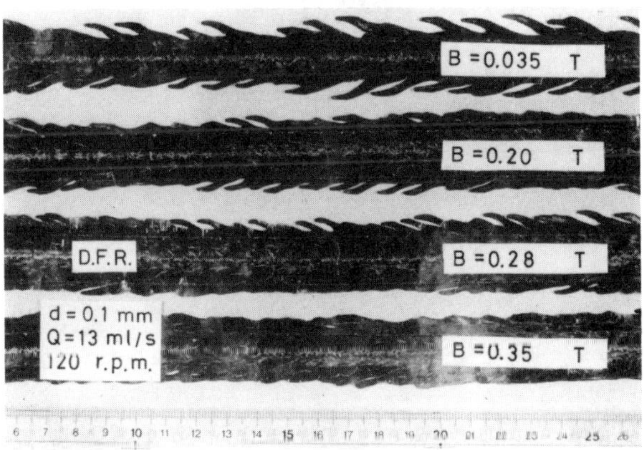

Figure 10 - Effect of magnetic field on edge shape of products in the case of DFR.

<u>Observation of Surface Wave Motion on Molten Metal Pool</u>

In order to clarify the reason why the saw edges appear on both sides of the sheet, surface wave motions on molten metal pool being held between the rolls were taken by a video tape recorder with an optical fiber scope during the injection of molten metal. Recording was conducted with a frame interval of 1/60sec and a shutter speed of 1/5000sec. Figure 11 shows a series of three frames((a),(b),(c)) sequentially taken by a camera from the upper part of the roll edge. The position of camera view is indicated in the top figure of Fig.11. Tips of waves are seen as white portions between the rolls. From the fact that the white portions shift from the center to the side and disappear in the edge part, it can be imagined that the wave introduced by injected melt moves toward the direction of the roll edge and disappears between the rolls. Based on this observation, the mechanism of saw edge formation can be drawn as Fig.12. The surface wave generated at the center part of the pool is gradually developed during traveling from the center to the lower side and the shape of the wave is fixed as a saw edge pattern by solidification at the kiss part of the rolls.

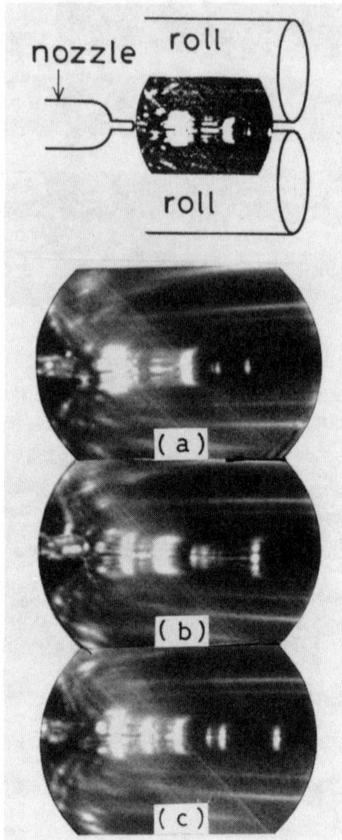

Figure 11 - Transitional behavior of surface wave of metal poll.

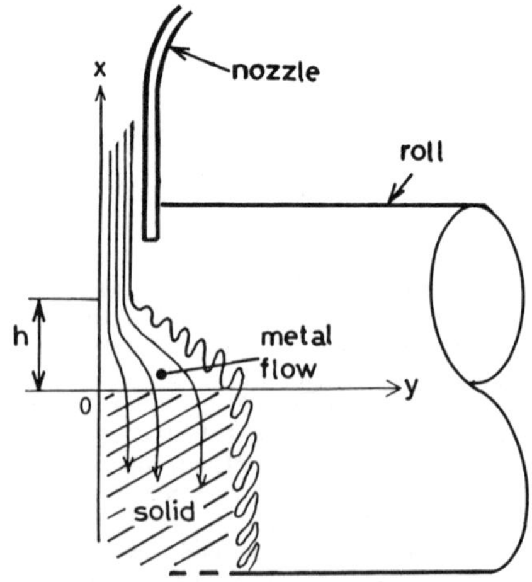

Figure 12 - Modelling of metal pool.

Simulation of Saw Edge Pattern

Single Ferrite Roll (SFR)

Based on the measurement of magnetic flux, its profile along the direction of roll axis can be expressed by a step function as follows :

$$B(y) = \begin{cases} B_0 & (0<y<l_1) \\ 0 & (l_1<y) \end{cases} \quad (1)$$

where l_1 is the distance from the roll center to the boundary between plain and stainless steel parts. Thus, $0<y<l_1$ is corresponding to the plain steel part and $l_1<y$ to the stainless part.

The surface wave was observed to be carried with the flow of molten metal. Then, the surface position of molten metal could be expressed by superposing a formulation of the travelling wave, $a \sin(ky-\omega t)$, and the metal flow, $v_x t$.

$$x = a \sin(ky-\omega t) + h - v_x t \quad (2)$$

where, a is the amplitude of the surface wave, k is the wave number, ω is the velocity of angle, h is the height of molten metal at the center of the roll and v_x is the velocity of molten metal in the x-direction.

The equation of motion of molten metal is written as Eq.(3) by considering the inertia force and the electromagnetic force induced by magnetic flux, B, and neglecting both the viscous term and the effect of surface tension.

$$\rho \frac{d^2y}{dt^2} = - \sigma B^2(y) \frac{dy}{dt} \quad (3)$$

By substituting Eq.(1) into Eq.(3) and integrating the resulting equation, Eq.(4) is obtained.

$$\frac{dy}{dt} = -\frac{\sigma}{\rho} B_0^2 y + v_{yi} \quad (4)$$

where v_{yi} is the initial velocity in the y-direction.
Again integrating Eq.(4) yields Eq.(5).

$$y = -\frac{\rho}{\sigma B_0^2} v_{yi} \exp\left(-\frac{\sigma B_0^2}{\rho} t\right) + \frac{\rho}{\sigma B_0^2} v_{yi} \quad (5)$$

where, the integral constant was determined by the condition of y=0 at t=0. By rearranging Eq.(5), the explicit expression of t is obtained as Eq.(6).

$$t = -\frac{\rho}{\sigma B_0^2} \ln \left| 1 - \frac{\sigma B_0^2}{\rho V_{yi}} y \right| \quad (6)$$

By using the condition of dy/dt=0, a critical value of y which refers to the maximum value of y, is derived.

$$y_c = \frac{\rho v_{yi}}{\sigma B_0^2} \quad (7)$$

Putting x=0 in Eq.(2) by assuming that molten metal has been solidified at the kiss point of the rolls (x=0), Eq.(8) is obtained.

$$0 = a \sin(ky - \omega t) + h - v_x t \qquad (8)$$

By considering that the cast length is $X=v_x t$ and eliminating t from Eqs.(6) and (8), Eq.(9) is derived.

$$X = v_x t = h + a \sin(ky + \frac{\omega \rho}{\sigma B_0^2} \ln(1 - \frac{\sigma B_0^2}{\rho v_{yi}} y)) \qquad (9)$$

Double ferrite roll (DFR)

Based on the measurement of magnetic flux in DFR, its profile is expressed as Eq.(10).

$$B(y) = \begin{cases} 0 & (0<y<l_0) \\ (B_0/l)(y - l_0) & (l_0<y<l_0+l) \end{cases} \qquad (10)$$

where l is the half width of the ferrite band and l_0 is the distance from the center of the roll to the inside boundary between ferrite and non-ferrite parts. In the range of $0<y<l_0$, there is no magnetic field so that the relation of $y=v_{yi}t$ is obtained from Eq.(3). On the other hand, in the range of $l_0+l>y>l_0$, the following equation is derived from Eq.(3).

$$\int_0^{l_0} \frac{dy}{v_{yi}} + \int_{l_0}^{y} \frac{dy}{\{v_{yi} - (\sigma B_0^2/3\rho l^2)(y-l_0)^3\}} = t = t_1 + t_2 \qquad (11)$$

Let's define t_1 and t_2 as follows, respectively.

$$t_1 = \int_0^{l_0} \frac{dy}{v_{yi}} = \frac{l_0}{v_{yi}} \qquad (12)$$

$$t_2 = \int_{l_0}^{y} \frac{dy}{\{v_{yi} - B_0^2/3\rho l^2 (y-l_0)^3\}}$$

$$= \frac{l_0 A}{v_{yi}} \{ \frac{1}{6} \ln \left| \frac{(Y - 1 + A)^3}{(Y-1)^3 + A^3} \right| + \frac{1}{\sqrt{3}} \tan^{-1}(\frac{2Y - 2 - A}{\sqrt{3} A})$$

$$- \frac{l_0 A}{v_{yi}} \{ \frac{1}{\sqrt{3}} \tan^{-1} \frac{-1}{\sqrt{3}} \} \qquad (13)$$

where, $Y=y/l_0$ and $A=-(3\rho l^2 v_{yi}/\sigma B_0^2 l_0^3)^{1/3}=(1/N_1)^{1/3}$, N_1 is a measure of the ratio of pondermotive force to inertia force and is well known as the interaction parameter. The critical value of y can be derived through the condition, dy/dt=0.

$$y_c = -Al_0 + l_0 = (\frac{3\rho l^2 v_{yi}}{\sigma B_0^2})^{1/3} + l_0 \qquad (14)$$

Following the same procedure as that of the single ferrite roll, the edge shape can be predicted by Eqs.(8) and (11).

Calculated results

The effects of magnetic field on the edge shape are shown in Fig.13 by use of the mathematical model. The parameters which are required in the calculation were evaluated as follows : v_x was adjusted with the velocity calculated from the rotation speed, 120 r.p.m.. v_{yi} was calculated from the width of the product obtained under no magnetic field. The wave length, λ, was obtained from the notch number. The amplitude of surface wave, a, is given as Eq.(15).

$$a = \Delta W \, v_x / 2 v_{yi} \quad (15)$$

The density, ρ, and the electric conductivity, σ, were evaluated at the melting point of tin.

With superposition of magnetic field, the edge shape gets smoother in both cases of SFR and DFR. Also, regarding the width of the sheet, the experimental results that the width in the case of SFR is more suppressed than that in DFR, was simulated qualitatively by this model.

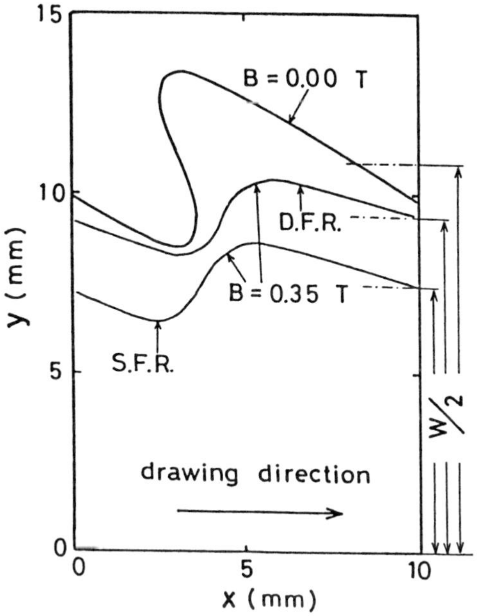

Figure 13 - Calculated results of edge shape of products.

Conclusion

It has been found that the saw edge shape appearing at both sides of a cast sheet can be suppressed by imposing magnetic field in the twin-roll process. The mechanism of the saw edge shape formation was observed by use of a high speed VTR system. Based on the observation, a mathematical model was developed to simulate the edge shape pattern. The model could qualitatively predict the edge shape and the difference of the sheet width in SFR and DFR.

CASTING TECHNIQUES VI:
STRIP CASTING

Session Chairman: **G. Houze**
Allegheny Ludlum Steel Corp.
Brackenridge, Pennsylvania

CONTROLLING THE PHYSICAL AND MECHANICAL PROPERTIES OF CAST STAINLESS STEEL BAND

David B. Love, John D. Nauman
Allegheny Ludlum Steel Division
of Allegheny Ludlum Corporation
Technical Center

INTRODUCTION

Since the early 1980's, Allegheny Ludlum has been actively developing a single-roll casting process for the production of stainless steel band. The cast material has been considered as a replacement for hot-rolled band, although it has been referred to as strip. Like hot-rolled band, the cast band still has required cold-rolling and annealing to achieve the surface and mechanical properties of conventionally produced strip.

The reason for Allegheny Ludlum interest in this new technology was to achieve the many benefits of casting materials near their final thickness. These benefits included savings in labor, energy, yield, and capital investments. Rapid solidification and the elimination of the requirement for hot-rolling has also been leading to improvements in quality and the development of new alloys.

The research and development on the band-casting process at Allegheny Ludlum has revealed that through proper control of casting variables, it has been possible to control the band's physical and mechanical properties, in terms of width, thickness, surface roughness, edge uniformity, and microstructure. The relationships between casting variables and these properties were examined in this study.

A series of samples were cast under various conditions, and analyzed for differences in width, thickness, edge quality, surface roughness, internal porosity, and microstructure. The alloy considered in this study was Type-304 austenitic stainless steel (18%Cr, 8%Ni), the largest volume stainless steel sold by Allegheny Ludlum.

Casting of Near Net Shape Products
Edited by Y. Sahai, J.E. Battles,
R.S. Carbonara and C.E. Mobley
The Metallurgical Society, 1988

Table I : General Casting Conditions

Coil #	Diameter	Substrate	Speed m/min	(ft/min)	Temp C	(F)
C1	std + 50%	C Steel	std + 3	(10)	std	
C2	std	stainless	std + 18	(60)	std	
C3	std		std + 33	(110)	std	
C4	std		std + 18	(60)	std	
C5	std		std + 6	(20)	std + 17	(30)
C6	std		std		std + 28	(50)
C7	std	C Steel	std		std + 6	(10)
C8	std	Copper	std		std + 6	(10)
C9	std	C Steel	std + 12	(40)	std + 6	(10)
C10	std		std + 12	(40)	std + 6	(10)
C11	std		std + 12	(40)	std + 6	(10)
C12	std		std + 12	(40)	std + 6	(10)
C13	std		std + 12	(40)	std + 6	(5)

std - standard conditions are proprietary

Table II : Band chemistry

Coil #	C	Mn	P	S	Si	Cr	Ni	Mo	Cu
C1	.054	1.24	.029	.004	.37	17.21	8.14	.26	.30
C2	.045	1.60	.034	.010	.44	17.69	8.79	.82	.37
C3	.026	1.65	.025	.007	.50	18.62	8.41	.26	.31
C4	.050	1.61	.030	.005	.50	18.30	8.35	.30	.30
C5	.047	1.65	.030	.005	.46	18.28	8.39	.31	.36
C6	.056	1.69	.033	.007	.42	18.37	8.24	.31	.38
C7	.053	1.70	.029	.008	.45	18.22	8.34	.32	.34
C8	.048	1.74	.031	.008	.47	18.20	8.19	.30	.32
C9	.063	1.59	.031	.004	.41	18.28	8.11	.31	.36
C10	.064	1.62	.033	.006	.44	18.97	8.25	.34	.34
C11	.056	1.58	.026	.004	.47	19.56	8.33	.37	.35
C12	.058	1.61	.030	.005	.45	18.42	9.53	.30	.32
C13	.058	1.53	.031	.007	.48	19.45	9.73	.32	.36

DESCRIPTION OF CASTING PROCESS

The steel for casting was melted in a 1350-kg (3000 lb) capacity, coreless induction furnace. The metal was then poured into a transfer tundish, which directed the metal into the casting tundish and onto the casting roll. A schematic of the water-cooled casting roll and the casting tundish is shown in Figure [1]. The width of the inside of the casting tundish in contact with a casting roll was set to obtain the desired width of the band. For the present study the width of the tundish was either 330 mm (13 in) or 560 mm (22 in). The band caster is shown in operation in Figure [2].

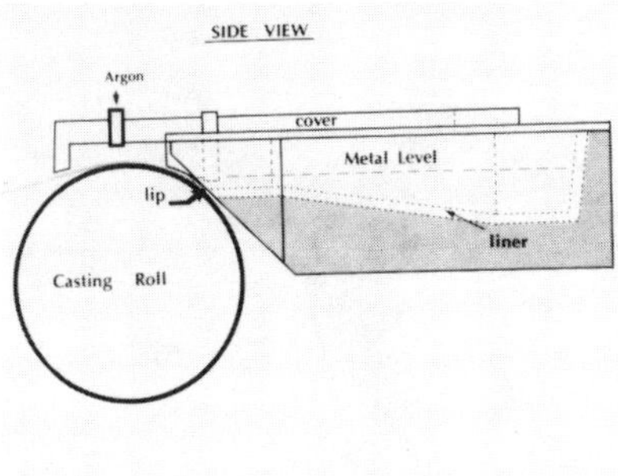

Figure 1. Side View of single roll casting tundish

Figure 2. Casting of Type 304 at Allegheny Ludlum Technical Center

A single horizontal casting roll was used to solidify the liquid steel into band, as shown in Figure [1]. The diameter of this roll was either a standard diameter or diameter enlarged by 50 percent. The thickness of the cast band could be varied from 1.4 mm (0.055 in) to 0.25 mm (0.010 in), by controlling the residence time of the liquid pool on the casting roll, as is discussed later.

The primary variables that were studied to control the shape and structure of the cast band were the level of major chemical elements, width of the casting tundish, speed of the casting roll, length of the casting pool, diameter of the casting roll, composition of the casting roll, and superheat of the melt.

ANALYSIS OF CAST BAND

The band cast under the various conditions, shown in Tables I and II, was measured and tested for shape and mechanical properties. These mechanical properties, as measured in a room-temperature tensile test, were used to evaluate the potential usefulness of the structures for conventional applications of stainless steel.

In previous studies it had been determined that the better the strength and elongation of the cast band, the better the strength and elongation of the wrought product [2,3]. Although the strength and ductility of cast band was always less than that of wrought strip, with cold-rolling and annealing the properties are improved. The results of previous studies on the cold rolling and annealing of cast band is shown in Figure [3] [2,3]. The cast band had about 30 percent elongation before rolling. After rolling and annealing, the elongation was increased to nearly 50 percent and the tensile strength was increased to about 675 MPa (90 ksi). Further rolling and annealing continued to improve the strength and ductility of the band, providing sufficient strength and ductility to meet even the most rigorous specifications.

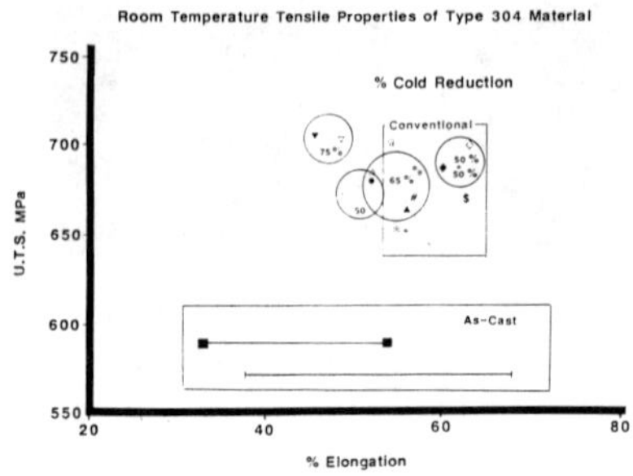

Figure 3. Plot of ultimate strength as function of elongation for cast and finished strip

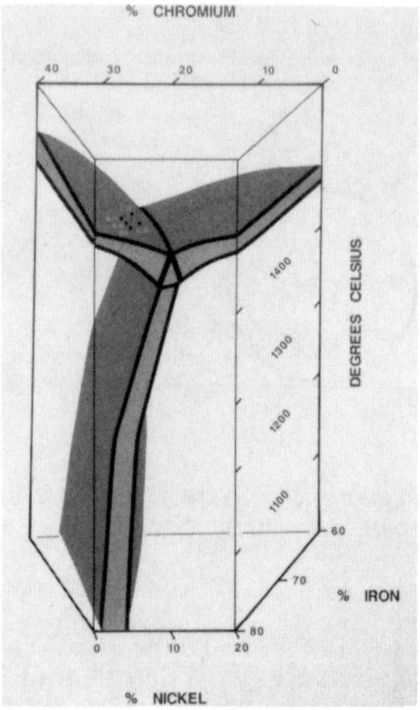

Figure 4. Three-dimensional ternary phase diagram for Fe-Cr-Ni

For this study only the tensile properties of the cast band are reported. These properties for the various coils are shown in Table V. Most of the cast band from these coils had an elongation less than 25 percent with some variability, which was typical of these cast conditions. The band from these coils has been subsequently rolled to various finished gages, but the results are not reported here.

The upper surface and lower surface (casting roll side) of the cast band was examined at low magnification to observe porosity, roughness or cracking. The surfaces near the edges were examined to observe any curling, waviness or cracking.

The thickness across the width of the band from each coil was measured with an ultra-sound detector. The results of these measurements are shown in Table III. While the thickness varied some in all cast coils, the proper control of casting could normally hold the variations in thickness, both across the width and along the length, to within 2-5%.

ANALYSIS OF THE MICROSTRUCTURE

The microstructure of the cast band was analyzed to determine the variations in the structure that led to variations in properties. The microstructure was analyzed through five different views, to observe cracks, pores, non-metallic inclusions, and metallic phases. Two sections, longitudinal and transverse to the direction of casting, were taken perpendicular to the casting roll. Three sections were taken parallel to the surface of the casting roll at different distances from the roll surface (bottom surface), Table IV. The percentage ferrite and percentage high- and low-chromium austenite was determined from these later sections.

Table III: Transverse thickness profile data

Coil #	Mean Thickness mm	(in.)	St. Dev mm	(in.)	Percentage St. Dev of Mean
C1	0.92	(0.0363)	0.05	(0.0020)	5.5%
C2	0.73	(0.0286)	0.04	(0.0014)	4.9%
C3	0.91	(0.0360)	0.05	(0.0021)	5.8%
C4	0.84	(0.0330)	0.03	(0.0012)	3.6%
C5	0.87	(0.0344)	0.09	(0.0036)	10.4%
C6	1.00	(0.0395)	0.03	(0.0013)	3.3%
C7	1.01	(0.0396)	0.06	(0.0022)	5.5%
C8	1.02	(0.0401)	0.07	(0.0029)	7.2%
C9	0.96	(0.0379)	0.03	(0.0010)	2.6%
C10	0.98	(0.0384)	0.06	(0.0023)	6.0%
C11	0.96	(0.0377)	0.10	(0.0040)	10.6%
C12	1.04	(0.0411)	0.09	(0.0037)	9.0%
C13	0.94	(0.0370)	0.04	(0.0015)	4.0%

The explanation of the microstructure had to be traced back to the solidification conditions, shown in the ternary phase diagram Figure 4. In this Fe-Cr-Ni diagram, the solidification paths of all of the alloys in this study (18-20% Cr, 8-10% Ni) are shown to lie close to the three-phase region of liquid + ferrite + austenite. Because of the proximity of the alloys to this three phase region, slight variations in chemistry or cooling conditions could radically alter the solidified structure. In general, the solidification of an 18Cr-8Ni stainless steel passes from liquid through a ferrite + liquid region, through an entirely ferritic region, and through a solid-solid transformation to form austenite. The solidified structure should have dendrites of austenite with intercellular delta-ferrite cores. The various structures of these steels when cast were more thoroughly studied by Dean [4].

Table IV : Metallographic structure data

Note : Samples perpendicular to casting direction are denoted A,B,C depending on the distance the section is from the casting wheel surface

Sample	Upper Surface
A	---------> 50um (0.020in.) from casting roll surface
B	---------> 38um (0.015in.) from casting roll surface
C	---------> 25um (0.010in.) from casting roll surface

Casting Roll Surface

Coil	% Ferrite*			% High Chrome Structure			% Low Chrome Structure		
	A	B	C	A	B	C	A	B	C
C1	4	4	8	42	65	56	58	35	44
C2	7	10	11	44	35	85	56	65	15
C3	8	6	7	83	80	100	17	20	0
C4	7	10	9	85	70	70	15	30	30
C5	14	9	7	65	85	57	35	15	43
C6	6	6	7	43	41	63	57	59	37
C7	6	15	15	63	81	52	37	19	48
C8	4	6	9	39	78	76	61	22	24
C9	9	8	14	76	63	78	24	37	22
C10	5	12	8	72	80	78	28	20	22
C11	10	13	15	63	70	74	37	30	26
C12	5	4	3	46	22	52	54	78	48
C13	6	6	6	70	72	57	30	28	43

* Determined by optical comparison methods

When this steel was cast into band, the microstructure near the surface of the casting roll was almost wholly austenitic with none of the usual dendrites or the ferritic cores. Due to the rapid, non-equilibrium, cooling of the liquid metal (2500 C/sec) [1], a metastable primary austenite appeared to form. A region of this phase was observed in the longitudinal section of the band, as shown in the sample from coil C12 in Figure 5. The phase was absent in the section from coil C11 in Figure 6, in which the composition of the alloy was further from the three-phase region.

As solidification continued further from the surface of the casting roll, three distinct phase regions were observed, as shown in Figure 7. These regions were analyzed with a chemical microprobe to determine their composition. Region A, which was relatively high in nickel, was more austenitic than Region B. Region B, which was higher in chromium and lower in nickel, was more ferritic. Region C was the primary ferrite core of a dendrite, which had not been transformed to austenite at lower temperatures.

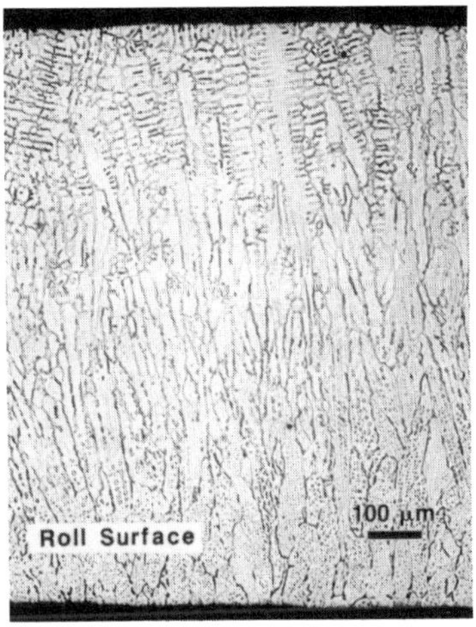

Figure 5. Severe primary austenite band near casting roll surface on high nickel coil C12 (10% oxalic etch)

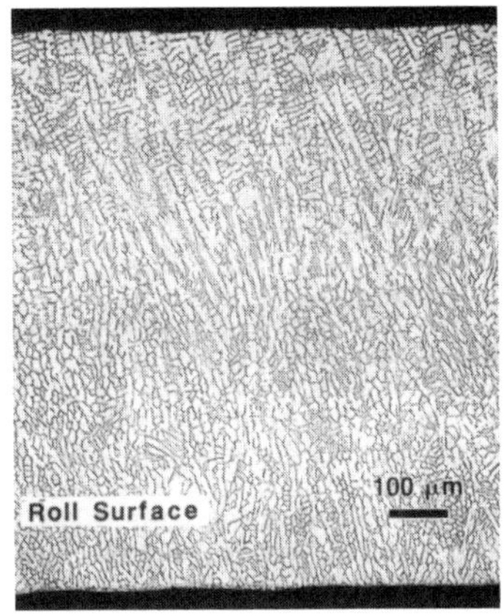

Figure 6. Absence of primary austenite band near casting roll surface on high chromium coil C11 (10% oxalic etch)

The chemical difference between Regions A and B indicated that there may have been two solidification fronts passing through the band. First, the ferrite cores (Region C) and secondary dendritic growth started, producing a high chromium structure (Region B). The remaining liquid was depleted in chromium slightly, which caused the adjacent dendrites to form from liquid closer to the Cr-Ni eutectic, producing a more austenitic structure (Region A). These three regions were present in all casting conditions investigated.

MAJOR ELEMENTS IN MELT CHEMISTRY

Four chromium levels and two nickel levels were investigated in this study, as shown in the compositions of coils C8 through C12 in Table II. Changes in the chromium and nickel content of the steel had no effect on the width of the band, the shape of the edges, or the thickness of the cast band. However, at the higher levels of chromium and nickel, the surface of the band became rougher, as indicated by the increased deviation in the thickness shown in Table III.

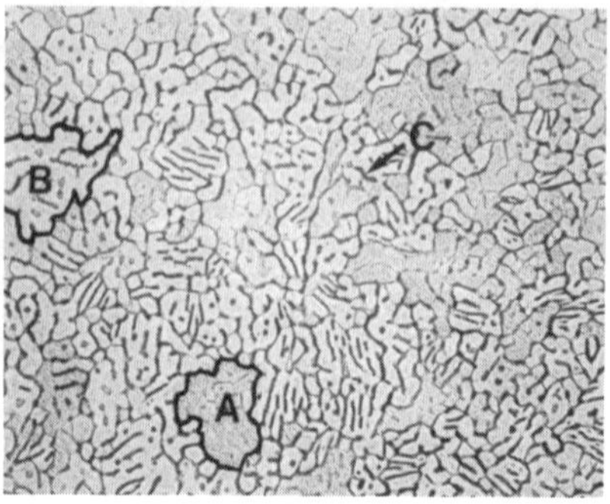

Figure 7. Three distinct chemistry regions visible in section of cast band, parallel to casting direction A - Low chrome austenite, B - High chrome austenite C - Ferrite Core

Region	Description	Fe	Cr	Ni	Si
A	Darkened grains with dendritic cored internal structure	71.62	18.86	8.88	.64
B	Light grain with uniform shading - no evidence of dendritic coring	71.31	19.54	8.36	.78
C	Dark grain boundaries	71.34	23.81	4.85	-

Some slight porosity or voids were noted in the center of band cast at 12 m/min (40 ft/min) above standard speed, but the porosity seemed to decrease as more chromium was added. No porosity was observed in the coil with 19.5 percent chromium and 8.3 percent nickel. This may have been due to the greater hot strength, as described below in the microstructure.

By varying the major elements in the melt chemistry, the path of solidification and transformation, shown in the diagram in Figure 4, was varied, particularly the path through the two-phase ferrite + austenite region. The levels of ferrite and austenite in the final band was different for each variation in composition, and the effects on microstructure, properties, and surfaces were significant.

In general, additions in chromium increased the ferrite content of the band, increased the percentage of high-chromium austenite, increased the strength, and decreased cracking and porosity of the surfaces. The band of primary austenite near the casting roll surface was nearly eliminated as the chromium content was increased, as show in Figure 6. The improvement of mechanical properties and the reduction in cracks might be expected for an austenitic alloy containing higher levels of ferrite [5].

Variations of the nickel content in the steel also had a significant effect on microstructure. As the nickel was increased, the percentage of high-chromium austenite decreased, ductility decreased, surfaces became more porous and uneven, and ferrite levels decreased. The thickness of the band of primary austenite near the casting roll surface increased, extending over half the thickness of the band, as shown in Figure 5.

Theoretically, higher nickel should have improved the ductility of an un-worked stainless product [6]. The lower strength and ductility, shown in Table V, may be attributed to small "hot-cracks". Hot cracking in low-ferrite austenitic steels has been shown to persist for rapidly solidified structures, such as weldments [5] and cast slabs [7]. The poorer band flatness for this composition may have resulted from the increased cracking and porosity of the surface.

Table V: Average mechanical properties of as-cast band

Coil #	Thickness mm (in)	.2% Yield MPa (ksi)	UTS MPa (ksi)	Elongation %
C1	1.03 (0.0405)	225 (32.6)	379 (54.9)	14.8
C2	0.99 (0.0390)	284 (41.1)	452 (65.5)	11.2
C3	0.85 (0.0335)	258 (37.4)	376 (54.5)	7.3
C4	0.81 (0.0320)	267 (38.7)	505 (73.2)	19.3
C5	0.87 (0.0344)	225 (32.6)	379 (54.9)	14.8
C6	0.91 (0.0360)	354 (51.3)	526 (76.3)	25.0
C7	0.99 (0.0390)	255 (36.9)	390 (56.6)	10.8
C8	0.99 (0.0390)	277 (40.1)	417 (60.4)	11.2
C9	0.90 (0.0355)	258 (37.4)	473 (68.5)	12.5
C12	0.98 (0.0385)	273 (39.6)	490 (71.1)	13.0
C11	0.84 (0.0330)	268 (38.8)	417 (60.4)	11.1
C12	1.02 (0.0400)	246 (35.6)	349 (50.6)	9.8
C13	0.91 (0.0360)	285 (41.3)	479 (69.5)	12.5

WIDTH OF THE CASTING TUNDISH

The inside width of the casting tundish at the point it contacted the casting roll was set at either 330 mm or 560 mm. The band produced was nearly the same width as the casting tundish. For a 330mm tundish the band was about 325mm wide at room temperature, while for a 560 mm tundish the band was about 555 mm wide. The slight decrease in width of the band was attributed to the contraction of the metal from casting to room temperature. It appeared that further increases in the width of casting tundish would result in corresponding increases in the width of the band without any limitations.

The changes in the width of the casting tundish had no

effect on the shape of the edges, the thickness of the band, or the conditions of the surfaces of the band. The shape of the typical edge of the band is shown in Figure 8. There was a slight wave along the edge, as less than 4 mm of the edge curled up periodically.

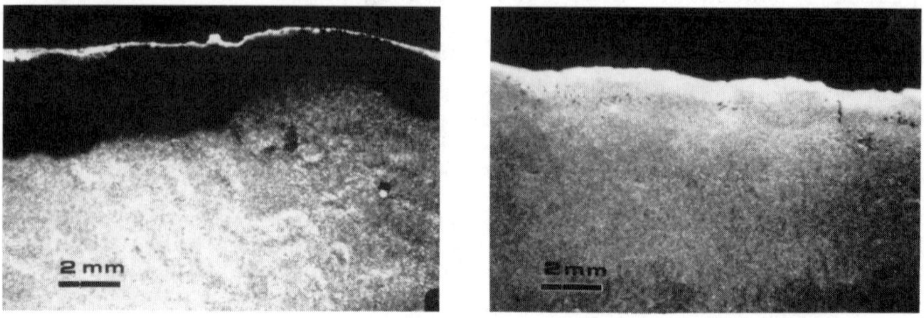

a) b)

Figure 8. Photograph of the edge of band at standard casting speed (a) and 33 m/min above the standard (b), as viewed from the surface.

SPEED OF THE CASTING ROLL

Surface speeds of the casting roll between 15 m/min (50 ft/min) to 150 m/min (500 ft/min) had been successfully used to cast band. In this particular study, a standard speed was selected and coils were cast at this speed and faster, where the maximum speed was 33 m/min above that standard, as shown in Table I.

The average width of the band was not affected by the casting speed, but the edges were slightly changed. As the speed was increased to 33 m/min above the standard, the curl of the edge decreased, as shown in Figure 8. This decrease in curling may be explained by either the decreased cooling or thinner band at the higher speed.

The thickness of the band varied, as expected, with the speed of the casting roll. The relationship between roll speed and band thickness is shown in Figure 9 for different pool lengths. This relationship was modeled mathematically by calculating the thickness of the band solidified as a function of the residence time on the casting roll and adding the boundary layer of liquid that was "dragged" along with the solidifying band. The model calculations are shown in this figure as solid lines.

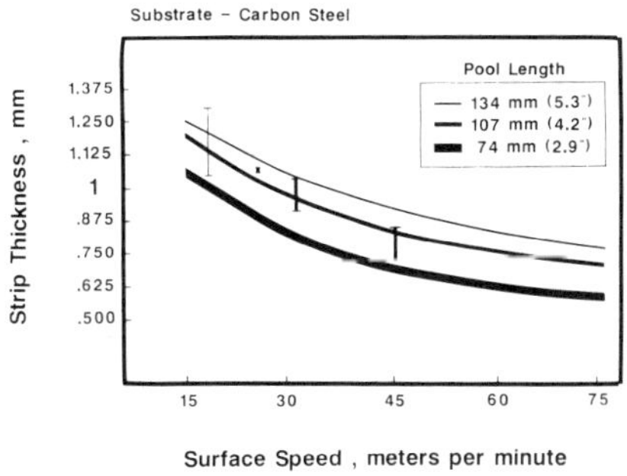

Figure 9. Type 304 band thickness vs. surface speed for various pool lengths

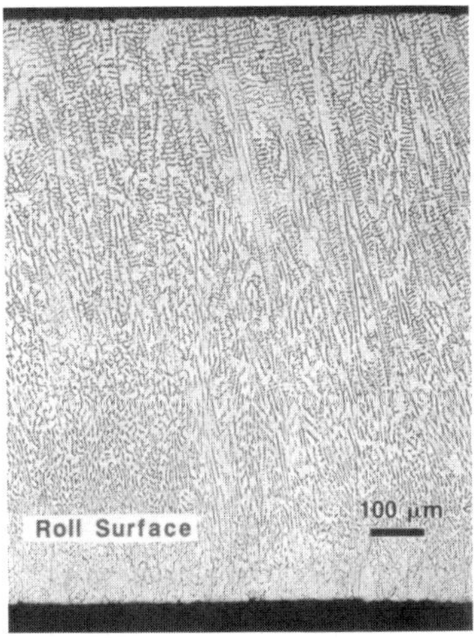

Figure 10. Primary austenite band evident at slow casting standard speed (10% oxalic etch)

The apparent porosity in the surface or the center of the band was slightly affected by changes in speed. At speeds greater than 12 m/min above the standard no porosity was observed, while at the standard speed there was slight porosity in both the center and the surface of the band. The porosity observed at the lower casting speeds may have been voids formed during solidification or during subsequent handling of the hot band.

Changes in the speed of the casting roll affected the microstructure of the band in many ways. At slow speeds, the casting roll had more time to cool before re-entering the casting tundish, allowing the surface of the roll to run cooler. The lower temperature of the casting roll may have increased the undercooling, which in turn increased the amount of metastable austenite. When casting roll was at the standard speed, the band of primary austenite in the band near the roll was quite evident, as shown in Figure 10. This austenitic band was not evident at the roll speeds greater than 14 m/min above the standard. At 33 m/min above the standard, a coarser transformed austenite-ferrite structure was observed near the casting roll surface.

Increasing speed of the casting roll also increased the amount of liquid which was "dragged" from the casting pool along the surface of the partially solidified band. This liquid was cooled by the casting roll, through the partially solidified band, as well as by radiation and convection to the surroundings. The change in dendritic growth pattern at the top of the band which resulted from these cooling conditions at different roll speeds can be seen in Figures 10.

Other effects of increasing the speed of the roll included increased surface smoothness, shown in Figure 13, better mechanical properties (e.g. coils C4 and C5 in Table V), and less evidence of low-chrome austenitic cells in the structure.

LENGTH OF THE LIQUID METAL POOL

Pool length, or the distance over which the pool of liquid metal contacted the casting roll, can be varied in a number of ways: changes in casting-roll diameter, tundish size, tundish position on the roll, and the vertical height of the pool in the casting tundish.

The average width of the band was not significantly affected by the changes in the length of the liquid pool along the casting roll. However, the edge did curl more and the surface and center porosity increased slightly as the pool became longer.

The thickness of the band was greatly affected by the length of the pool. As shown in Figure 9, as the length of the pool was increased, the thickness of the solidified band also increased. This result was predicted by previous mathematical models [1,2].

The microstructure changed slightly as the liquid pool was lengthened. As the thickness of the cast band increased, the conduction through the band to the casting roll decreased, slowing solidification and producing more ferrite and smoother surfaces.

DIAMETER OF THE CASTING ROLL

The larger diameter of the casting roll did not affect the width or the thickness of the band to any extent. The larger roll did increase the curl on the edge of the band and porosity on the surface of the band.

The effect on microstructure of increasing the casting roll diameter, in general, was similar to the effect of decreasing the casting speed. At the same speed the surface of the larger roll had more time to cool before re-entering the casting tundish, resulting a lower temperature on the roll. This lower temperature produced a larger band of primary austenite along the bottom of the band.

COMPOSITION OF THE CASTING ROLL

Casting rolls of three different compositions were evaluated for casting this alloy. These compositions were carbon steel, stainless steel, and copper. The stainless-steel roll, which has a lower conductivity than carbon steel or copper, operated at higher surface temperature. The subsequent slower cooling rates caused by the higher temperatures decreased or eliminated the band of primary austenite near the casting-roll surface.

A copper casting roll, in theory, should have resulted in high cooling rates, due to the high conductivity. Examining the band microstructure, however, showed a structure with negligible primary austenite near the casting-roll surface. The structure also contained a coarser dendritic structure which appeared to have undergone slower cooling. It was believed that the very rapid cooling caused by the cold copper roll resulted in significant, localized shrinkage of the solidifying steel, and that this shrinkage caused the solidifying band to loose intimate contact with the casting roll. With such poor contact the band continued to solidify slowly, producing a coarse structure and smooth surfaces, as shown in Figure 11. Excellent uniformity in thickness and properties were also achieved with the copper casting roll.

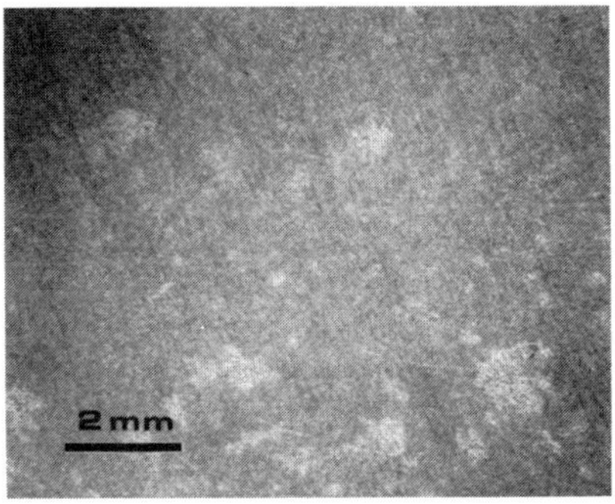

Figure 11. Surface smoothness of band cast on copper casting roll (acid cleaned).

SUPERHEAT OF THE MELT

Changes in the melt temperature affected the structure and shape of the cast band. The superheat of the melt was increased above the standard superheat by as much as 28 C (50 F), as shown in Table I for coils C5 and C6. The increase in melt temperature did not significantly affect the width or the thickness of the band, but there were some changes in the microstructure. Under similar casting conditions, the higher melt temperature produced slightly more primary austenite near the casting roll surface, coarser structure, and a moderate level of ferrite. The mechanical properties of the band were generally good, in part due to the moderate level of ferrite.

CONCLUSIONS

Using a single-roll caster, the width of the cast band was controlled completely by the width of the casting tundish, within a variation of less than 2%. The edges of the band curled up slightly by less than 4 mm. This edge variation was routinely handled by the standard trimming before subsequent processing.

The thickness of the cast band was controlled by the speed of the casting roll and the length of the liquid pool on the casting roll. There was a slight variation in the thicknesses (2-10%) that was affected by the composition of the alloy, the casting speed and the composition of the casting roll. The most uniform thickness was achieved with a higher casting speed and a copper casting roll.

The control of the microstructure in cast band was critical to the final mechanical and surface properties of the band. For Type 304 stainless steel, a minimum of primary austenite and a maximum of retained ferrite was desired. The major chemical elements that affected the retained ferrite (like chromium and nickel) could be controlled, as they have been in welding. The cooling rate of the band could also be controlled to affect the formation of austenite and ferrite in this alloy. For example, to promote the formation of ferrite, the cooling of the band was decreased by increasing the speed of the roll, using either copper or stainless steel casting roll, decreasing the diameter of the roll, or increasing the length of the liquid metal pool.

When the microstructure of the band contained sufficient ferrite cores and high-chromium austenite, without any regions of primary austenite, the band had good mechanical properties. The structure was relatively free of pores or voids in either the center or the surface. With this structure the subsequent processing would yield high strength and ductility.

The structure of the band could be further modified by conventional processing steps. The affect of various rolling and annealing cycles on the cast band has been reported previously [5,6]. Ultimately the structure of Type 304 stainless steel strip, produced from the band cast coils, was indistinquishable from strip produced by conventional processing.

While only austenitic stainless steels have been discussed here, work continues on a variety of flat-rolled products, the casting of which would have significant capital and processing cost savings over conventional processing. The unique control of structures, width and thickness of the band should allow the casting of many alloys. Some of these alloys may be produced at much lower cost, while others may have improved properties brought out by the control of structure.

REFERENCES

[1] R. K. Pitler, "Direct Strip Casting of Specialty Steels," Proceedings of the Nicholas J. Grant Symposium on Processing and Properties of Advanced High-Temperature Alloys, Cambridge, MA, June 16-18, 1985.

[2] J. F. Grubb, D. B. Love, A. Murthy, and J. D. Nauman, "Casting Specialty Steel Strip," Proceedings of the Fifth International Iron and Steel Congress, Washington, DC, April 6-9, 1986.

[3] D. B. Love, J. D. Nauman, "The Mechanical Processing of Cast Stainless Steel Strip," Proceedings of the 28th Mechanical Working and Steel Processing Conference, Greentree, PA, October 27, 1986.

[4] D. C. Dean, "The Physical Metallurgy of Cast Type 304 Stainless Steel," Iron and Steelmaker, Volume 15, March 1988, pp 12-15.

[5] T. Ogawa and E. Tsunetomi, "Hot Cracking Susceptibility of Austenitic Stainless Steels," Welding Journal, March 1982, pp. 83s to 93s.

[6] Handbook of Stainless Steels, D. Peckner and I.M. Bernstein, Mcgraw-Hill Book Company, 1987.

[7] L. Myllykoski, N. Suutala, "Effect of Solidification Mode on Hot Ductility of Austenitic Stainless Steels," Metals Technology, The Metals Society, Volume 10, December 1983, pp 453-460.

STAINLESS STEEL STRIP AND FIBER BY MELT OVERFLOW

T.A. Gaspar and L.E. Hackman

Ribbon Technology Corporation
Box 30758
Gahanna, Ohio 43230

Abstract

The Melt Overflow Process for casting metallic strip and fiber is described. The casting of Type 304 stainless steel strip is discussed. Measurements of hardness and tensile strength of as-cast strip at different thicknesses are presented. The chemical composition of the strip is also presented. The strip exhibits a fine dendritic microstructure which is compared to the microstructure of continuously cast billet. Stainless steel fibers cast by Melt Overflow and Melt Extraction are discussed. Type 304 fibers exhibit a fine, dendritic microstructure that can be observed to the fiber surface. The flexural load-deflection behavior of refractory concretes containing Type 430 fibers are compared for rapidly solidified fibers and wrought material. The rapidly solidified fibers exhibit 42% higher flexural toughness index than slit sheet fiber. The oxidation of different commercial Type 304 stainless steel fibers are measured at 1000 C for 100 hours. Rapidly solidified fibers show significantly higher oxidation resistance than slit sheet or drawn fibers.

Casting of Near Net Shape Products
Edited by Y. Sahai, J.E. Battles,
R.S. Carbonara and C.E. Mobley
The Metallurgical Society, 1988

Introduction

Ribtec's Melt Overflow Rapid Solidification Technology (MORST) is shown schematically in Figure 1. Molten metal overflows a reservoir onto the surface of a rotating chill block. The melt stream is not extruded through an orifice like melt spinning techniques, rather, the melt level is raised until liquid metal flows from a reservoir through a channel or tundish to contact the moving chill surface. More detailed descriptions of the melt overflow process have been published (1-2).

The surface geometry of the chill block determines the type of product to be cast - fiber, filament, ribbons, or strip. United States Patent 4,705,095 describes chill block surface geometries used for casting relatively thick sheet and strip with improved uniformity (3). Patents are pending for Melt Overflow process variations for casting ribbon, filaments and fiber. This paper surveys the modifications of melt overflow under development at Ribtec to cast stainless steel fibers and strip.

Stainless Steel Strip

Stainless steel strip has been cast by melt overflow for applications in reinforcing refractory concrete. Stainless steel strip as wide as 15 cm has been cast by melt overflow.

Stainless steel strip is cast form a coreless induction furnace powered by a 500 kW, 1000 Hz induction generator used for melting heats up to 100 kg of steel. A refractory plunger, measuring 15 cm diameter by 30 cm long is immersed in the liquid steel until it overflows the furnace to contact the rotating copper chill block. After the plunger is removed from the furnace, the liquid level is replenished. A 750 kg coreless induction furnace and a 1500 kg coreless induction furnace are also used for melt overflow.

Table I presents the ultimate tensile strength and hardness values measured on samples of Type 304 stainless steel strip cast by melt overflow. The variations in tensile strength can be attributed to gaseous impurities introduced during induction melting. The nitrogen content of induction melted stainless steels is typically 0.02 to 0.09 percent. The chemical

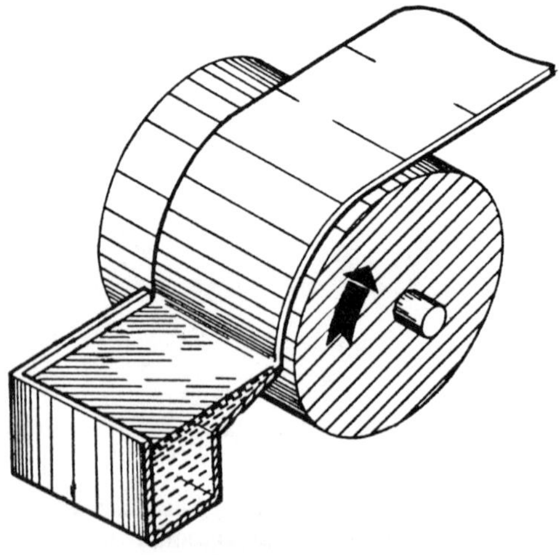

Figure 1. Schematic of Melt Overflow Process

TABLE I. TENSILE STRENGTH OF TYPE 304 STAINLESS STEEL STRIP

HEAT NO.	THICKNESS mm	THICKNESS in	HARDNESS KHN	TENSILE STRENGTH MPa	TENSILE STRENGTH psi
Fe-030	0.343	0.0135	179	565	81890
	0.362	0.0143		512	74280
Fe-031	0.349	0.0138	179	411	59650
	0.368	0.0145		422	61150
Fe-032	0.330	0.0130	168	447	64900
	0.324	0.0128		478	69350
Fe-033	0.324	0.0128	181	445	64580
	0.362	0.0143		419	60810

TABLE II. CHEMICAL COMPOSITION OF TYPE 304 STAINLESS STEEL STRIP

HEAT NO.	Cr	Ni	Si	Mn	C
Fe-030	18.16	8.56	0.63	0.76	0.08
Fe-031	18.89	8.66	0.46	0.98	0.06
Fe-032	18.67	8.66	0.62	0.92	0.07
Fe-033	18.14	8.80	0.55	0.74	0.08

compositions of the alloys are presented in Table II.

Figure 2 compares the microstructure of Type 304 stainless steel strip cast by melt overflow with Type 304 continuously cast billet. The dark regions are ferrite and the light regions are austenite. The microstructure in figure 2a of the Type 304 stainless steel strip is dendritic with the chill cast surface shown at the bottom of the micrograph. The microstructure of the continuously cast billet in figure 2b consists of significantly coarser, equiaxed austenite with intragranular ferrite that is martensitic in appearance. The fine, dendritic structure reflects the more rapid cooling rate experienced by the melt overflow strip, compared to the continuous cast billet.

Ribtec has a project underway to develop thicker steel strip by melt overflow using proprietary chill block surface geometries sponsored by the U.S. Department of Energy. Ferrous based strip has been cast thicker than 2 mm by melt overflow and we are investigating cast stainless steel and carbon steel strip of similar thicknesses.

Many other metallic alloys have been cast into strip by the melt

Figure 2. Microstructure of Type 304 Stainless steel (a) strip cast by Melt Overflow and (b) continuously cast billet

overflow process. Alloys based on magnesium, aluminum, brass and copper have been cast into strip up to 12 inches wide. Cobalt alloys have also been cast up to 4 inches wide.

Magnesium alloy strip cast to 0.005 in thick and 11 in wide was used to demonstrate the feasibility of producing graphite-magnesium composite tubes by the MCI pultrusion process. Direct cast magnesium sheet was effectively used as a cladding material for the tubes.

Aluminum alloys have been cast into strip for several applications. Ribbons or foils are comminuted into flake materials for coating applications or comminuted into particulate for aluminum P/M structural applications. Aluminum, brass and copper strip have been cast for home construction applications like flashing.

Cobalt alloy strip has been investigated for potential use as consumable welding electrodes in strip cladding operations. A program to develop direct cast cobalt alloy strip for welding applications is currently being run. Direct cast strip shows a significant cost advantage over powder rolled strip or cast electrodes.

Stainless Steel Fibers

Ribtec produces over 1000 tons per year of stainless steel fibers manufactured by both the Melt Extraction process (under licence from Battelle Development Corporation) and Ribtec's new Melt Overflow Rapid Solidification Technology (MORST). To cast fibers, circumferential grooves are machined on the surface of the chill block. When the chill block is in contact with the liquid metal, the metal solidifies on the tips of the threads to cast fibers. Fiber length is determined by longitudinal grooves that are milled along the length of the chill block surface.

Melt extraction is not well suited for casting fibers less than 0.2 mm diameter or fibers less than 18 mm long. Melt overflow casts fibers as fine as 0.05 mm diameter and as short as 3 mm. Micrographs of Type 304 stainless steel fibers 3mm long (Figure 3) and 50 mm long (Figure 4) cast by melt overflow are shown.

Figure 3a shows a scanning electron micrograph of 3 mm long Type 304 stainless steel fiber cast by melt overflow. The fibers have a crescent shaped cross section that improves fiber bonding to concrete matrix. Figure 3b shows the fine, dendritic solidification structure in the fibers. The dendritic structure can be observed on the free cast surface of fibers as shown in the electron micrograph in Figure 4a. There is some coarsening of the dendrites on the free cast surface from the section shown in the optical micrograph in Figure 4b.

Ribtec cast stainless steel fibers are added to refractory concretes. The primary purpose of using stainless steel fiber in refractories is to provide ductility to a brittle refractory cement system. This ductility is provided by the ductile steel fibers binding the refractory together subsequent to crack initiation.

The pull-out of the fibers from the brittle refractory matrix simulates a ductile stress-strain curve. Tests were performed by Lankard Materials Laboratory, Inc. in Columbus, Ohio to measure the flexural strength and toughness of refractory concretes containing stainless steel fibers (4). The concrete used in the tests is Wahl Refractories' PSM-CAST refractory concrete for applications to temperatures of 1650 C. The stainless steel fibers were added to the refractory at the rate of 4 percent by weight (dry) which is

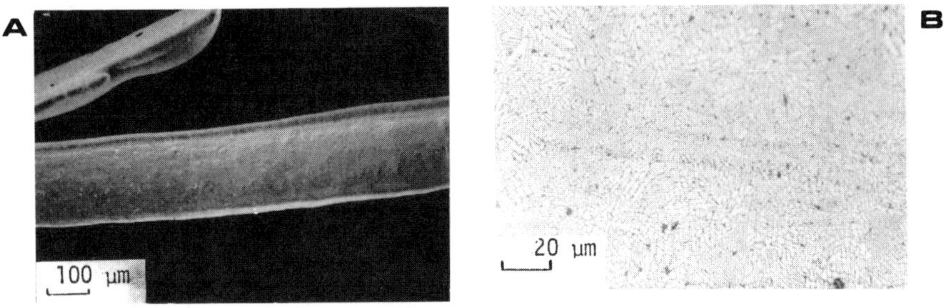

Figure 3. Micrographs of 3mm long Type 304 Stainless Steel Fibers showing (a) fiber shape and (b) dendritic microstructure

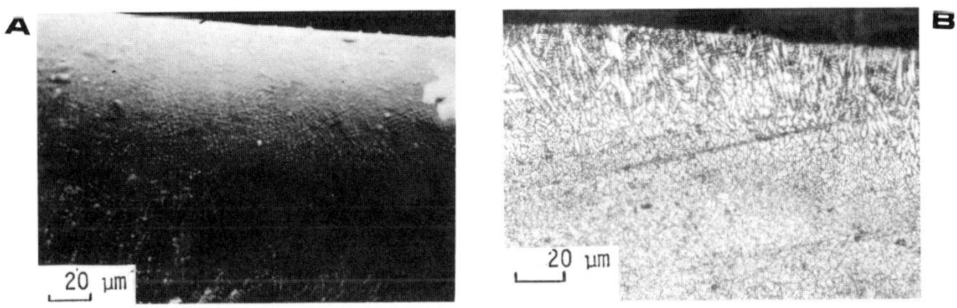

Figure 4. Micrographs of 50 mm long Type 304 Stainless Steel Fibers showing (a) dentrites at fiber surface and (b) dendritic microstructure

equivalent to 1.1 percent by volume. Three brick specimens measuring 64 mm x 114 mm x 229 mm were cast with each type of fiber, mold cured for 24 hours, then dried at 110 C for 12 hours. Flexural strength measurements were performed with the brick specimens using center point loading over a 178 mm span.

Figure 5 compares the flexural load-deflection behavior of refractory concretes containing rapidly solidified Ribtec Type 430 stainless steel fiber to those containing a commercial wrought Type 430 stainless steel fiber. Table III presents the flexural strength and flexural toughness index of the two types of fiber.

The difference in ultimate flexural strength between the two fibers is not statistically significant. However, the flexural toughness of the rapidly solidified fiber was 42 % higher than the commercial wrought fiber. Examination of the fracture surfaces following flexural strength testing showed two different modes of failure. In specimens containing the rapidly solidified fiber, fiber pullout was the principal mode of failure. In specimens containing the commercial wrought fiber, fiber fracture was the principal mode of failure.

When the refractory concrete is placed in service, hot oxidation and corrosion of the stainless steel fibers is one of the major factors in measuring it's performance. Hot oxidation tests were performed by Harrop Industries, Columbus, Ohio to compare rapidly solidified and wrought Type 304 stainless steel fibers (5). Samples of stainless steel fibers were placed into an electric kiln and heated to 1000 C in 3 hours in static air. An initial 0 hr reading was then taken. The samples were removed from the kiln and weighed twice a day until they had been exposed to 1000 C for 100 hours.

Figure 6 shows the results of the oxidation tests comparing three

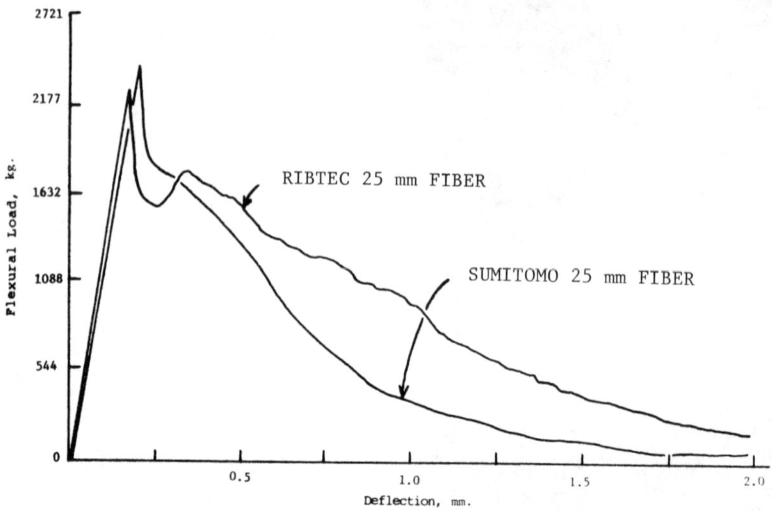

Figure 5. Flexural Load-Deflection Behavior of Refractory Concretes Tested at Room Temperature after oven drying at 110 C containing 1.1 volume percent of Type 430 Stainless Steel Fibers.

TABLE III. FLEXURAL STRENGTH AND FLEXURAL TOUGHNESS INDEX OF A 1650 C REFRACTORY CONCRETE CONTAINING 1.1 VOLUME PERCENT OF TYPE 430 STAINLESS STEEL FIBERS

	ULTIMATE FLEXURAL STRENGTH, MPa		FLEXURAL TOUGHNESS INDEX, cm^2	
	AVERAGE	RANGE	AVERAGE	RANGE
RIBTEC DIRECT CAST FIBER	13.2	12.7 - 14.1	82.9	79.9 -84.9
SUMITOMO SLIT SHEET FIBER	13.5	12.8 - 14.3	58.2	55.8 - 61.1

varieties of commercial Type 304 stainless steel fibers: Ribtec rapidly solidified fiber; slit sheet; and cold drawn wire. Rapidly solidified fibers exhibit significantly lower weight gain than either slit sheet or drawn stainless steel wire. The improved oxidation resistance for the cast fiber can be attributed to rapid solidification which imparts a fine grain size and permits greater additions of oxidation resisting elements to the alloys.

Experiments have been performed with Fe-Cr-Al alloys, Fe-Cr-Ni-Si alloys and nickel aluminides to further improve the hot oxidation reistance of the cast fibers. Strip or fiber can be cast under inert conditions using the plasma melt overflow furnaces at Ribtec.

The melt overflow and melt extraction rapid solidification processes produce stainless steel fiber that has high oxidation resistance and high toughness in refractory composites. These technical considerations combined with the lower cost of cast fibers, make them an excellent choice for reinforcing refractory concrete.

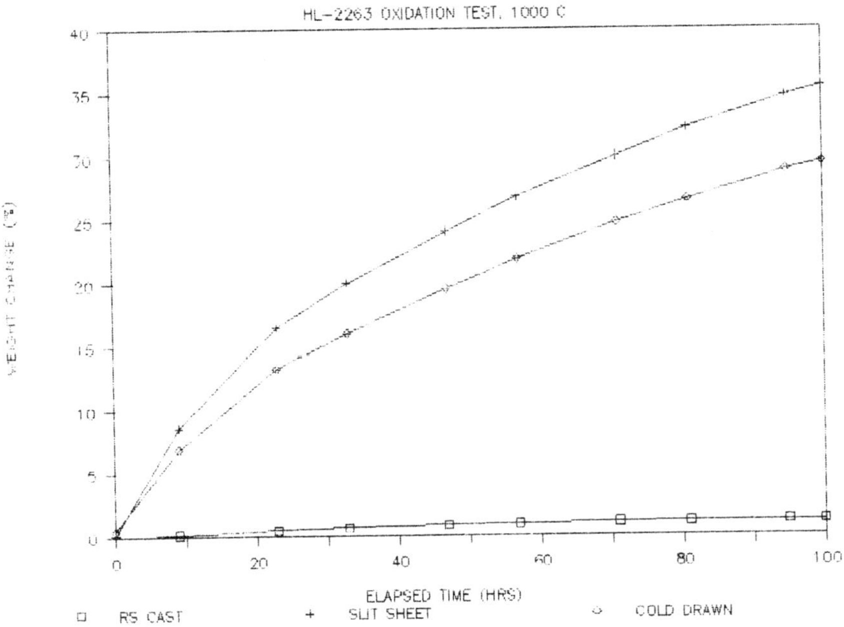

Figure 6. Oxidation of Type 304 Stainless Steel Fibers Exposed at 1000 C for 100 hours in Static Air

Conclusions

The melt overflow process has cast stainless steel strip and fiber as well as a variety of specialty metals and alloys. The stainless steel strip exhibits a dendritic microstructure and ultimate tensile strength up to 565 MPa (81,890 psi) in the as-cast condition. Strip as thick as 0.015 in was cast for applications in reinforcing refractory concretes. The use of textured chill block surface geometries for casting thicker strip is the subject of a research project sponsored by the U.S. Department of Energy.

Stainless steel fibers cast by melt overflow exhibit a fine dendritic microstructure. Additions of 1.1 volume percent of stainless steel fibers to refractory concretes significantly improve the flexural toughness of the resulting composite. Refractory concretes containing rapidly solidified fibers exhibit higher flexural toughness when compared to concretes containing wrought stainless steel fibers. The rapidly solidified cast stainless steel fibers have superior oxidation resistance when compared to either wrought stainless steel fibers or drawn stainless steel fibers.

References

1. "Rapidly Solidified Foils by Melt Overflow", Materials Research Society Proceedings, 58, ed. B.C. Giessen, D.E. Polk, and A.I. Taub (Pittsburgh, PA: Materials Research Society, 1986).

2. L.E. Hackman and T. Gaspar, Process of Making Strip by Melt Overflow, Ind. Heating, LIII, (1), January, 1986.

3. T.A. Gaspar, U.S. Patent 4,705,095, Textured Substrate and Method for the Direct, Continuous Casting of Metal Sheet Exhibiting Improved

Uniformity, November 10, 1987.

4. Newell, J., "Evaluation of Melt Extraction 430 Stainless Steel Fibers as Compared to 430 Stainless Steel Sumitomo Fibers as Reinforcement in Refractory Concrete", (LML Report No. 102, Lankard Materials Laboratories, Inc., Columbus, Ohio, 1988).

5. Whalen, E.T., "Oxidation of Type 304 Commercial Fibers" (Test # HL-2263, Harrop Industries Inc., Columbus, Ohio, 1986).

MELT FLOW CONTROL IN TWIN ROLL CASTING

S.Miyake, F.Kogiku, M.Yukumoto, M.Ozawa, T.Kan and A.Momoo

Iron and Steel Research Laboratories
Technical Research Division, Kawasaki Steel Corp.
1, kawasaki-cho, Chiba, JAPAN

Abstract

In the strip casting by twin roll, the strip thickness is principally determined by the flow rate of melt. It means that the fluctuation of flow rate, if any, inevitably causes the deviation of strip thickness. In order to control the flow rate fluctuation, mathematical analysis and practical improvement have been carried out on a twin roll caster.

The analysis has shown that two melt pools in the tundish and between the two rolls absorb the flow rate fluctuation. The degree of absorption becomes remarkable above a certain frequency of fluctuation, about 0.1 Hz for the caster in subject. On the other hand, installing a level meter and a sliding gate has been effective to minimize the fluctuation below that frequency. Consequently, the fluctuation is kept limited within ± 0.2 kg/s in the range of 2 to 10 kg/s and the strip thickness deviation in the longitudinal direction is reduced to within ± 7 %.

Due to such improvement, some of the hard workability materials have been cast to strips at a reasonable yield and productivity. For example, a new stainless steel strip with 30% Cr has been produced as the coils of 0.4 mm thickness × 350 mm width in 500 kg coil.

Casting of Near Net Shape Products
Edited by Y. Sahai, J.E. Battles,
R.S. Carbonara and C.E. Mobley
The Metallurgical Society, 1988

Introduction

The steel industry must develop a number of innovative technologies to maintain its advantage over other metal materials. Strip casting is undoubtedly counted as one of them. While several processes have been proposed and developed, the single roll(melt drag) and the twin roll are most widely adopted by investigators in the world.

At present status of development, one of the inferior items of the strip casting is strip thickness deviation. The reason is understandable especially for twin roll casting: the strip thickness is largely influenced by the feeding speed of melt, which is difficult to control or correct precisely. A single solution, therefore, may not be prepared for the thicknes deviation; instead, various approaches with endurable efforts will have to be continued.

Melt Flow Control by Gas Pressure

Fig.1 shows a configuration of the present twin roll caster of Kawasaki Steel Corp. By program-controlled-pressuring of Argon gas, the pouring furnace continues to feed melt toward the cooling rolls. The cooling rolls, in turn, solidify the melt into thin strip and send it downstream. The coiler winds the strip up to a net shape coil of 500 kg.

In terms of the flow rate, the caster has resorted at first to melt level control with no feed-back signals. It has been turned out, however, that the flow rate happens to change remarkably by melt level decreasing due to gas leakage or by clogging or erosion of the nozzle. Fig.2 is a case of the level decrease, by which the flow rate has changed more than 10 % during casting, exerting similar influence on strip thickness.

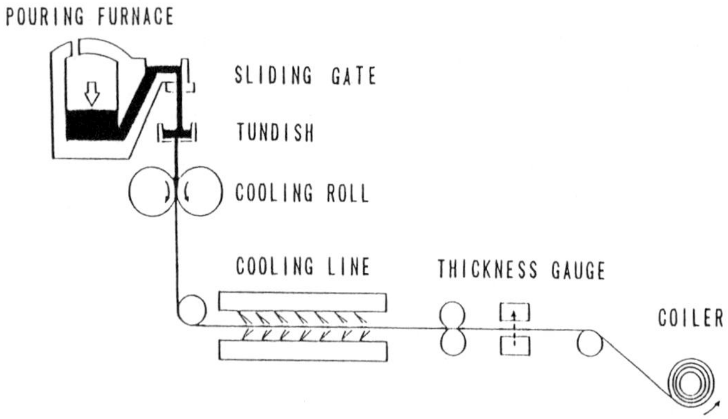

Figure 1 - Configuration of Twin Roll Caster

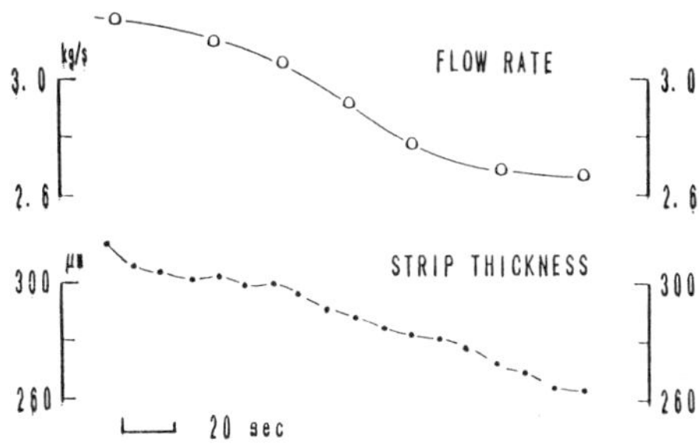

Figure 2 - Change of Flow Rate and Strip Thickness without Flow Rate Control

Monitoring and Analysis of Melt Level Fluctuation

Melt Level Monitoring

In order to monitor the melt level in the pouring furnace, a level meter has been devised. A load cell in the meter senses the change of buoyancy of a ceramic rod immersed in melt. When monitored to be out of range, the level is modified to a proper level by additional gas pressuring. Consequently, the melt level deviation has been largely reduced . Fig.3 shows a reading of the meter. It is understood that by additional pressuring, indicated by arrows, the melt level is controlled within ± 20 mm. Level monitoring has brought about another information that the level is always fluctuating with certain frequencies. The effect of the fluctuation frequency on the strip thickness deviation has been analyzed.

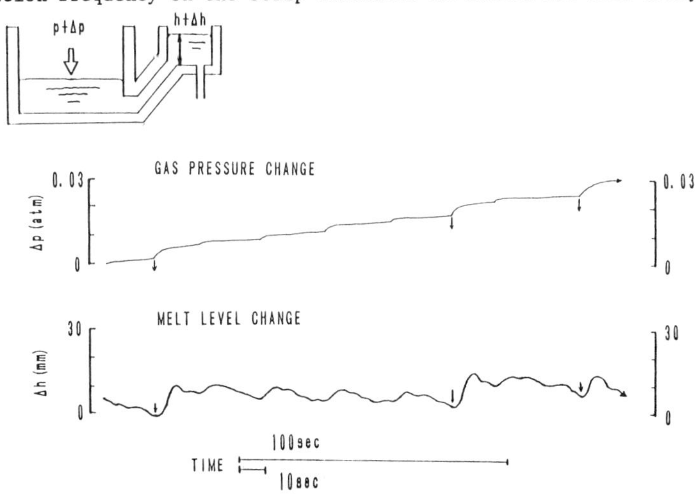

Figure 3 - Modification of Melt Level during Casting

Modeling of Melt Flow in Twin Roll Casing

To analyze the effect of melt flow fluctuation on strip thickness, the material flow of the twin roll caster has been modeled as shown in Fig.4.

At the pouring furnace:

$$q_1 = 1/R_1\sqrt{h_1} \quad \ldots\ldots\ldots\ldots\ldots\ldots\ldots\ldots\ldots\ldots\ldots (1)$$

At the tundish:

$$q_1 - q_2 = C_1\ dh_2/dt \quad \text{.......... (2)}$$

$$q_2 = 1/R_2\sqrt{h_2} \quad \text{.......... (3)}$$

At the cooling rolls:

$$q_2 - q_3 = C_3(h_3)\ dh_3/dt \quad \text{.......... (4)}$$

$$d = 2\ k\ (\ h_3/v\ )^{0.8} \quad \text{.......... (5)*}$$

$$q_3 = \rho\ d\ w\ v \quad \text{.......... (6)}$$

where q : flow rate
h : melt level
C : cross section of melt surface
d : strip thickness
w : strip width
R, k : constants
ρ : steel density, v : roll velocity

* Eq.(5) has been obtained by the authors from a heat transfer analysis of the rapid solidification process. Miyawaza et al. report a similar result (1).

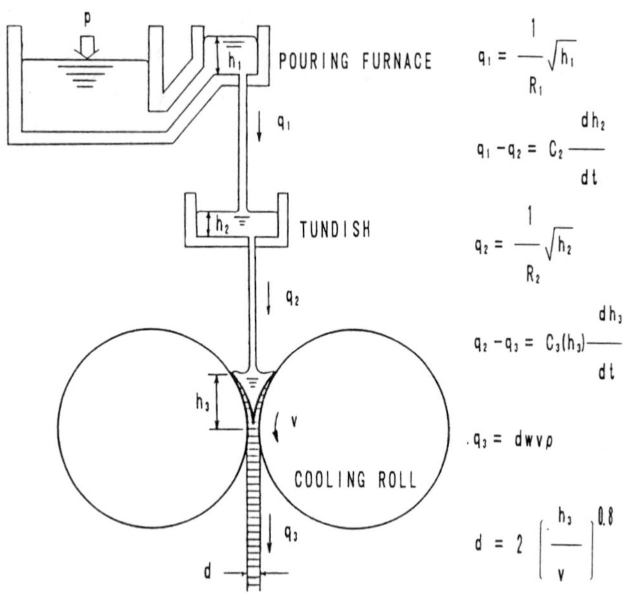

Figure 4 - Modeling of Material Flow in Twin Roll Casting

Effect of Flow Rate Change on Strip Thickness

From the analogy of the measurement, it is assumed that the flow rate fluctuation is caused by that of melt level.

$$h_1 = h_0 + A \sin 2\pi ft \quad ......(7)$$

$$h_0 = 250 \text{ mm}, \quad A = 10 \text{ mm}$$

$$f = 0.2,\ 0.1,\ 0.05,\ 0.033 \text{ or } 0.025 \text{ Hz}$$

Numerical calculation has been carried out, using the dimensions of the caster. Fig.5 shows a result when the frequency is 0.1 Hz. It is recognized that the fluctuation of flow rate is decreased by passing the tundish or the cooling rolls: from $\pm$ 0.1 kg/s, to $\pm$ 0.07 and to $\pm$ 0.04. That means the resultant deviation of strip thickness is reduced to 3.4 μm from 8.4 μm which is derived from the requirement of mass flow constancy. The relationship between the flow rate fluctuation and the resultant strip thickness deviation is shown in Fig.6. It should be noticed that above the frequency of 0.1 Hz, the thickness deviation is reduced considerably. Such dependency is reasoned by the following consideration: by approximating that q and d in the equations are proportional to h, an analytical solution is found:

$$q = ACR^2/(1 + 4\pi^2 f^2 C^2 R^2)\cdot(1/CR \sin 2\pi ft - 2\pi f \cos 2\pi ft) + Be^{-t/CR} \ldots(8)$$

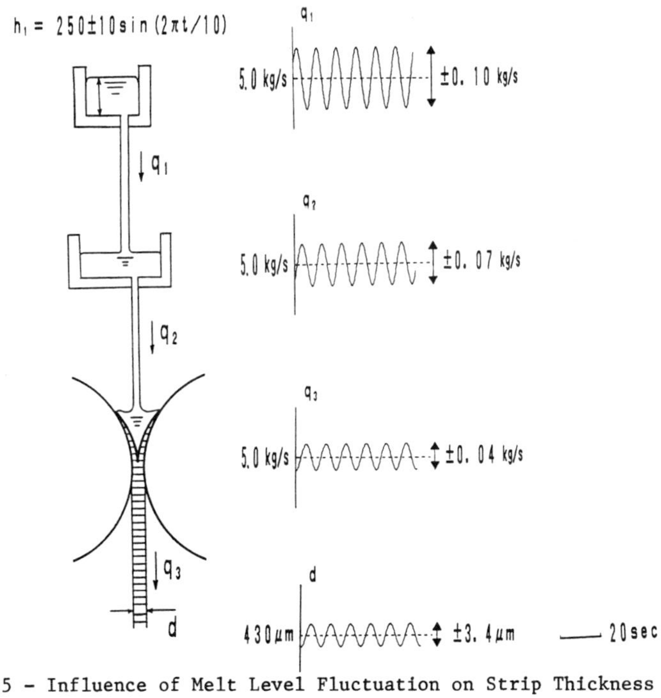

Figure 5 - Influence of Melt Level Fluctuation on Strip Thickness

Eq.(8) tells that when $fCR \gg 1$, q is proportional to $1/f^2$. In other words, two melt pools in the tundish and between the cooling rolls behave as an absorber for the fluctuation above a certain frequency which depends on the dimensions of the caster. Below that frequency, the melt pools no more behave as an absober: thickness deviation increases to the value equal to the flow rate fluctuation.

To reduce the flow rate fluctuation below that frequency and also to defend from nozzle clogging or erosion, a sliding gate nozzle has been installed on the pouring furnace. The control-period is determined to be 3.6 sec, based on the analysis above mentioned.

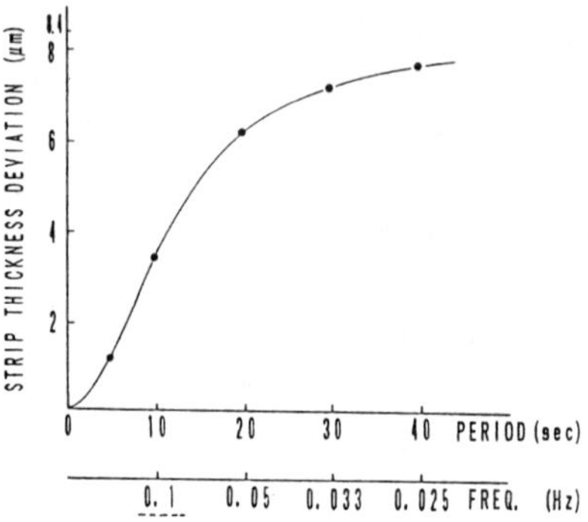

Figure 6 - Relationship between Frequency of Flow Rate Fluctuation and Resultant Strip Thickness Deviation

Flow Rate Control at Present

The current flow-control is diagrammed in Fig. 7. The flow rate is coarsely adjusted by the melt level control with the gas pressure and finely adjusted by the PI control with the sliding gate nozzle. Statistical data have shown that due to the improvement of the flow rate control, the strip thickness deviation in the longitudinal direction is reduced from ± 15 % to less than ± 7 %. Fig.8 shows an example of the control. It is recognized that in 20 sec. of casting, the flow rate becomes stable at a set value of 5.0 kg/s. The fluctuation is limited within ± 0.2 kg/s, when averaged over the control period of 3.6 sec.

The strip thickness also presents stable values at around 470 μm. A gradual change of strip thickness, from 470 μm to 480 μm and to 450 μm, cannot be explained by the flow rate change. The cause must be attributed to other factors: probably the change of melt pool profile between the rolls, which will be the next target of control.

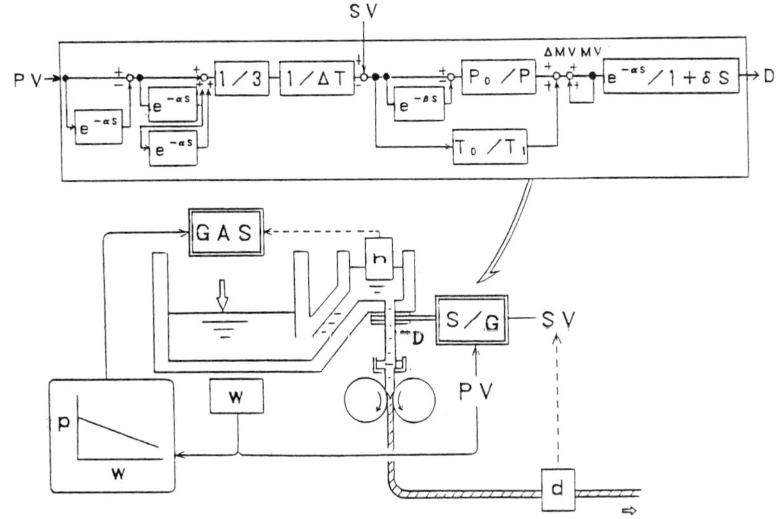

Figure 7 - Diagram of Flow Rate Control by Gas Pressure and Sliding Gate Nozzle

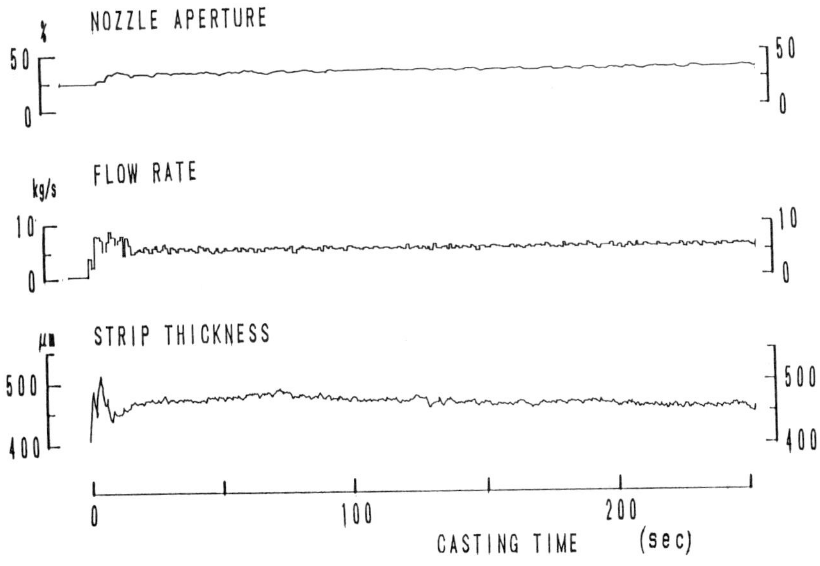

Figure 8 - Flow Rate Control by Sliding Gate and Strip Thickness Change in the Longitudinal Direction

Strip Qualities at Present

As to reducing the strip thickness deviation in the transverse direction, although not referred in the paper, setting of side dams on the roll edges and/or power up of the cooling ability of the rolls (2) are effective. Summarizing the current level, the strip thickness deviation in either direction is now controlled within ± 7%, which might be equal to the level by a single roll caster (3).

The improvement of flow rate control has increased the product-yield by the twin roll caster. As an example, a high chromium stainless steel strip (Fe-30Cr-14Ni) has been produced for the first application in over-lay welding material, the thickness of which is specified 0.4 mm ± 10 % by JIS. The strip has met the specification and also shown sufficient workability.

Summaries

(1) In order to reduce the flow rate fluctuation in twin roll casting, on-line measurement and mathematical analysis have been carried out.

(2) Due to level monitoring, the melt level in the pouring furnace has been modified within ± 20 mm of a proper level.

(3) The analysis of material flow in twin roll casting has shown that
 1) Two melt pools in the tundish and between the two rolls absorb the fluctuation of flow rate. The degree of absorption becomes remarkable above a certain frequency of fluctuation, about 0.1 Hz for the caster in subject.
 2) To control the fluctuation below that frequency, an auxiliary equipment is to be installed.

(4) By installing a sliding gate controller, the flow rate fluctuation has been reduced to within ± 0.2 kg/s in the range of 2 to 10 kg/s. Consequently, the strip thickness in the longitudinal direction is limited within ± 7 %

(5) The flow rate control has contributed to semi-commercial production of high chromium stainless steel strip.

References

1. A.Kasama and S.Mizogucti, K.Miyazawa, M.Ito and T.Sugai,"A Consideration on the Rapid Solidification of Steels - Solidification of Steels by Twin-Roll Caster," (Fifth International Iron and Steel Congress. Proceedings of the 69th Steel Making Conference) Vol.69 (1986) 850

2. M.Ozawa, F.Kogiku, M.Yukumoto, S.Miyake and T.Kan,"Heat Transfer Analysis on Twin Roll Casting," (The 4 th Conference on Modeling of Casting and Welding Processes, Palm Coast, Florida, 19 April 1988) 3-4

3. J.F.Grubb, D.B.Love, A.Murthy and J.D.Nauman,"Casting Specialty Steel Strip," (Fifth International Iron and Steel Congress. Proceedings of the 69th Steel Making Congress) vol.69 (1986) 845

THE FORMATION OF THIN STEEL STRIP

IN A LABORATORY-SCALE TWIN-ROLL CASTING

K. Miyazawa, T. Mizoguchi, M. Nakamura* and T. Ohashi**

Nippon Steel Corporation
R & D Labs.-I
1618 Ida, Nakahara-ku, Kawasaki 211, Japan

Abstract

An experiment of direct casting of stainless steel strip was done using a laboratory-scale twin-roll caster to reveal effects of operational variables on the strip thickness. It has been confirmed that the contact time between roll and metal is the most important factor to control the strip thickness under low roll supporting force. Furthermore, on the basis of the experimental data, observation of solidification structure and theoretical analysis of the heat transfer in the process, the formation mechanism of strip in the roll gap is discussed.

* Present address: Japan Technical Information Service, Sogo No.3 Building Annex, Kojimachi 1-6, Chiyoda-ku, 102 Tokyo, Japan

** Present address: Nippon Steel Corporation, Central R & D Bureau, Ohtemachi 2-6-3, Chiyoda-ku, 100 Tokyo, Japan.

Casting of Near Net Shape Products
Edited by Y. Sahai, J.E. Battles,
R.S. Carbonara and C.E. Mobley
The Metallurgical Society, 1988

Introduction

Twin-roll casting is seen as one of feasible processes to produce thin strip directly from the molten state with merits of near net shaping and rapid solidification (1, 2). In such thin strip casting processes as twin-roll (3-5, 8-10), single-roller (or melt drag) (6) and melt overflow (7) methods, it is generally important to control the strip thickness precisely and to obtain better surface and internal qualities of strip. Thus, more precise and highly controlled casting technique is now desired for the practical production of thin strip.

In this study, an experiment to produce rapidly solidified thin strips was done using a laboratory-scale twin-roll caster to reveal effects of operational variables on strip thickness. On the basis of the experimental data and theoretical analysis of the process, factors affecting the strip thickness and the solidification phenomena have been examined.

Experimental

Figure 1 shows a schematic view of the twin-roll process. Experimental conditions adopted are shown in Table I. In the experiment using a laboratory-scale twin-roll caster with copper rolls (diameter 300 mm, width 100 mm), austenitic stainless steel (SUS 304, melting weight 5-8 kg) was cast under Ar gas atmosphere to thin strips, whose thickness was around 0.5-3 mm. Springs were used for supporting the copper rolls. Main operational variables in the experiment are roll surface velocity, contact arc length between roll and cast metal, initial roll gap, roll supporting force and teeming temperature of molten metal. By changing the roll surface velocity and the contact arc length in range of 0.07-2.36 m/s and 30-85 mm, respectively, effect of these variables on strip thickness was investigated under the condition of low roll supporting force of 100-1000 kgf.

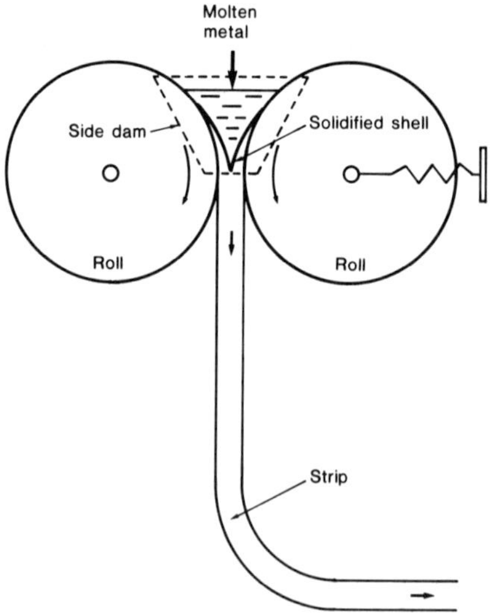

Fig.1 - Schematic view of twin-roll process.

Table I. Experimental Conditions

Cast alloy	SUS 304
Casting weight	5-8 kg
Roll material	Copper
Roll diameter	300 mm
Roll width	100 mm
Roll surface velocity	0.07 - 2.36 m/s
Contact arc length	30 - 85 mm
Initial roll gap	0.1 - 1.8 mm
Roll supporting force (by springs)	100 - 1000 kgf (usually 100 kgf)
Melt superheat	50 - 90 K

Theoretical Analysis

In the twin-roll process, molten metal is continuously poured into a space surrounded by roll surfaces and side dams. The molten metal contacting with the rolls is cooled down to a solidification temperature through removal of the superheat and undercooling of the melt and then commences to solidify. Moving downwards with roll rotation, the solidifying shells on both roll surfaces increase in thickness, and near the location of the minimum distance between rolls (namely, kiss point), the two shells are pressed to form a strip. In the case of higher roll rotation speed, the end point of solidification exists below the kiss point, while the endpoint is located above the kiss point for lower roll rotation speed. In the latter case, rolling deformation of the solidified metal takes place under higher roll supporting force.

The heat transfer in the metal between rolls is due to heat conduction and convection. According to the previous theoretical analysis (3, 4), the differential heat balance in the molten and in the solid regions is given by the following equations:

$$v_{1x}(\partial T_1/\partial x) + v_{1y}(\partial T_1/\partial y) = \alpha_1(\partial^2 T_1/\partial y^2) \quad (1)$$

$$v_{2x}(\partial T_2/\partial x) + V_{2y}(\partial T_2/\partial y) = \alpha_2(\partial^2 T_2/\partial y^2) \quad (2)$$

where v: velocity, T: temperature, α: thermal diffusivity, x,y: coordinates in casting and thickness direction, respectively, subscript 1: melt, 2: solid.

The boundary conditions are given as follows:

at meniscus

$$x = -x_4 \ : \ T_1 = T_o \quad (3)$$

at starting point of solidification

$$x = -x_3 : T_1(y=H) = T^o_{nu}, \ y_s = H \quad (4), (5)$$

at end of solidification

$$x = -x_2 : y_s = 0 \quad (6)$$

at center between rolls

$$y = 0 : \partial T_i/\partial y = 0, \quad i = 1,2 \tag{7}$$

at solidification front

$$y = y_s : T_1 = T_2 = T_{nu} \tag{8}$$

$$k_1(\partial T_1/\partial y)-k_2(\partial T_2/\partial y) = \rho \cdot \Delta H \cdot V_c \tag{9}$$

at roll surface

$$y = H : -k_i(\partial T_i/\partial y) = h(T_i-T_r), \; i = 1,2 \tag{10}$$

where T_o: teeming temperature, H: half space between rolls, T^o_{nu} : temperature when solidification starts, T_{nu}: temperature of solid/liquid interface, y_s: distance from center to solid/liquid interface, k: thermal conductivity, ρ: density, ΔH: latent heat, V_c: solidification rate, h: overall heat transfer coefficient at roll surface, T_r: roll temperature.

On the beginning of solidification, the initial undercooling of the melt depends on the rate of nucleation effective for solidification and the cooling rate of the melt. The higher the cooling rate, the larger is the initial undercooling. Once the solidification begins, the undercooling of melt at solid/liquid interface usually decreases due to the evolution of latent heat and the decrease in cooling rate. The solidification rate is related to the undercooling of melt as follows:

$$V_c = a(T_m-T_{nu})^n = a \cdot (\Delta T_{uc})^n \tag{11}$$

where a: coefficient, n: exponent, T_m: melting temperature.

The relation between the solidification rate and the position of the solid/liquid interface is expressed as follows:

$$V_y - V_x(dy_s/d_x) = V_c \tag{12}$$

where V is velocity at solid/liquid interface.

For the numerical calculations of the temperature field in the roll gap, the heat balance equations and the boundary conditions were transformed to do calculations in a computational domain by using the following new variables, according to the previous computational technique (2).

$$0 \leq y \leq y_s : \quad \eta_1 = y/y_s \tag{13}$$

$$y_s \leq y \leq H : \quad \eta_2 = (y-y_s)/(H-y_s) \tag{14}$$

The unknown temperature at the solid/liquid interface T_{nu} can be implicitly calculated to satisfy the equations (9), (11) and (12) during the computation of temperature field using the implicit finite difference method.

The velocity fields and the roll separating force are obtained by the previously proposed methods (3, 9). The velocity field in the region where the melt coexists with the solidified shell is expressed as follows:

$$v_{1x} = V_x + a_1(y^2-y_s^2) \quad (15)$$

$$v_{1y} = y \cdot d(a_1y_s^2-V_x)/dx - y^3\cdot(da_1/dx)/3 \quad (16)$$

$$v_{2x} = u_x + a_2(y^{m+1} - H^{m+1}) \quad (17)$$

$$v_{2y} = y\cdot d(a_2H^{m+1}-u_x)/dx - y^{m+2}\cdot(da_2/dx)/(m+2) \quad (18)$$

where

$$a_1 = 3(u_x-q/2H)y_s^{m-1}/2H^{m+1} \quad (19)$$

$$a_2 = (m+2)(u_x-q/2H)/(m+1)H^{m+1} \quad (20)$$

$$V_x = u_x+a_2(y_s^{m+1}-H^{m+1}) \quad (21)$$

u_x: velocity component at roll surface in x-direction, q: volumetric flow rate per unit width, m: exponent in a constitutive relationship of materials (namely, $\dot{\varepsilon} = A\,\sigma^m$, where $\dot{\varepsilon}$: strain rate, A: coefficient, σ : stress).

Results and Discussions

Effect of operational variables on strip thickness

In the experiment, strip of austenitic type 304 stainless steel with thickness of about 0.5-3 mm and width of about 100 mm was obtained. Figure 2 shows the effect of the roll surface velocity and the contact arc length between roll and metal on the strip thickness. It is seen that the thickness decreases with increasing the roll surface velocity and with decreasing the contact arc length. The increase in the roll surface velocity and the decrease in the contact arc length mean the decrease in the contact time between roll and metal, which results in the lack of time for the solid shells to grow on the roll surfaces.

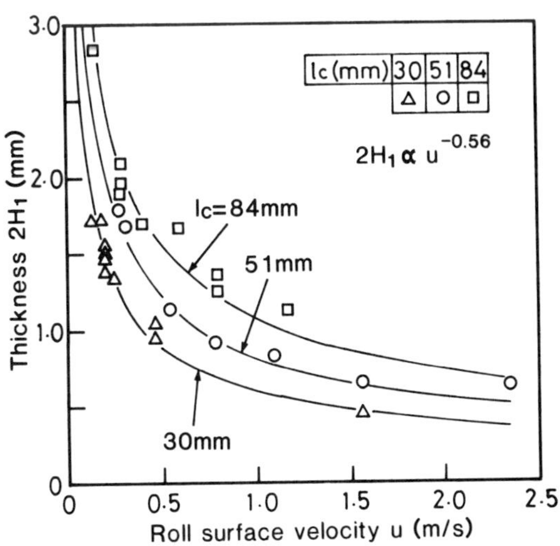

Fig.2 - Effects of roll surface velocity (u) and contact arc length (l_c) on strip thickness.

The experimental results in Figure 2 is summarized as Figure 3. Namely, Figure 3 shows the effect of the contact time (the ratio of contact arc length to roll surface velocity) on the strip thickness. The thickness increases predominatingly with the increase in the contact time and in proportion to about 0.56 power of the contact time.

In our experiment, the roll supporting force, the initial roll gap and the superheat of melt were changed in range of 100-1000 kgf, 0.1-1.8 mm and 50-90 K, respectively. As the result, it has been found that the effect of these variables on the strip thickness is minor under the adopted experimental condition and the contact time is the predominant variable to determine the strip thickness. This means that the rolling of the solidified shells scarcely took place under our experimental condition and the growth of the solid shell on the roll surfaces determined the strip thickness.

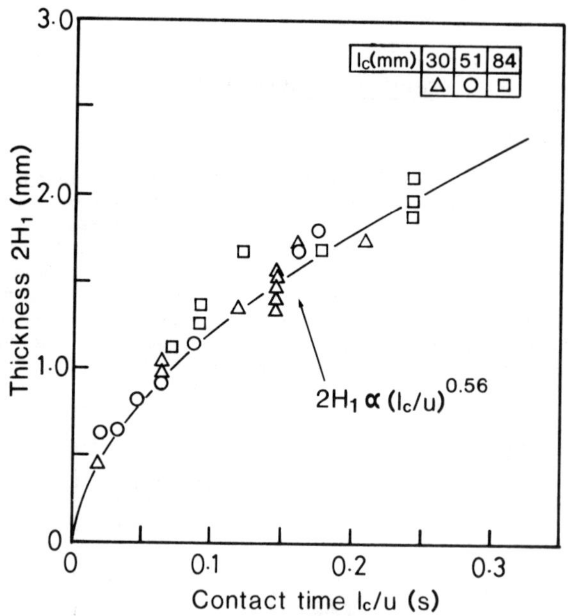

Fig.3 - Effect of the contact time between roll and metal on the strip thickness.

Solidification phenomena of strip

In the pool of molten metal between rotating rolls exist the forced convections due to the teeming stream, the roll rotation and the downward movement of solidifying shell. As mentioned above, the melt contacting with rolls begins to solidify after the removal of the superheat and the undercooling. The superheat and the latent heat of melt transfer into the rolls through the solidifying shells, which are moving downwards with the roll rotation and pressed to a strip near the kiss point of rolls.

Figure 4 shows the solidification microstructure of cast strip on the longitudinal section. In Figure 4(a), the dendritic structure is observed over the cross section. In the surface layer of the strip, bundles of dendrites with very narrow primary arm spacing are growing in random direction. The primary arm spacings increases with the distance from the strip surface and the secondary arm is clearly observed near the strip center. It is well known that, when molten metal is flowing on the solidification front, the primary arm of dendrite grows inclined to the upstream direction of the flow. Thus it is thought that the inclination of the primary arms observed in Figure 4(a) comes from the relative flow of residual molten metal on the solidifying shells.

On the other hand, Figure 4(b) shows a case where a zone of equiaxed crystals exists at the center of strip. Such a zone of equiaxed crystals was observed in our experiment when the teeming stream was fairly turbulent, or when the contact arc length was too short for the solidification front to reach the center of strip at the kiss point of roll. In the former case, if free crystals are floating in the pool of molten metal, these crystals would get between two solidifying shells near the kiss point to form a equiaxed crystal zone. In the latter case, the equiaxed structure would be formed by the preferential growth of crystals in random direction or the partial remelting of grown dendrites due to abrupt decrease in cooling rate of strip after it leaves the roll gap.

Figure 5 shows a observed distribution of secondary dendrite arm spacing in cross section of strip. The secondary arm spacing of the strips produced in our experiment is usually about 3-10 μm and shorter at the strip surface.

The relationship between the secondary dendrite arm spacing and the average cooling rate for austenitic stainless steel is shown in Figure 6. Three lines in Figure 3 have been presented by other investigators (11-13). The secondary arm spacing observed in our experiment is plotted on each line in the figure. Using the relation proposed by Esaka, et al (13), the average cooling rate of strips obtained by our experiment is estimated to be roughly in order of 10^2-10^3 K/s in our experiment.

Figure 7 shows the relationship between the average cooling rate and the strip thickness. The numerically calculated result in the figure shows that the average cooling rate increases with decreasing the strip thickness. From the comparison between the calculated and the experimentally observed relationships, the heat transfer coefficient on roll surface is estimated to be 0.1-0.5 cal/cm^2 · s · K in our experiment.

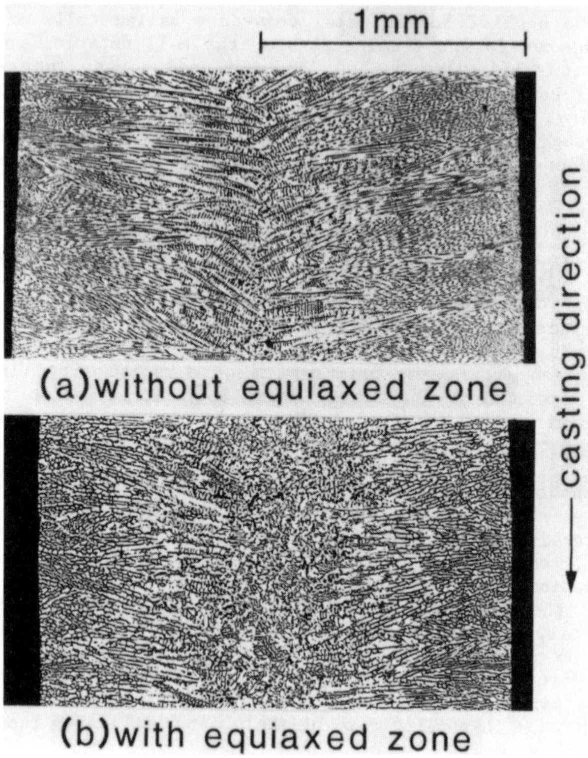

Fig.4 - Solidification structure of cast strip on the longitudinal section.

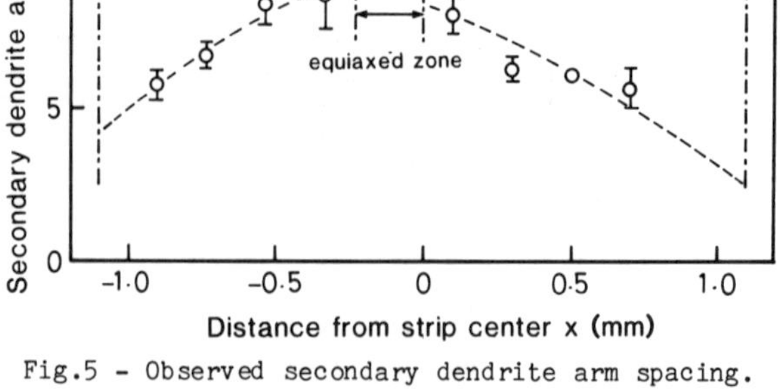

Fig.5 - Observed secondary dendrite arm spacing.

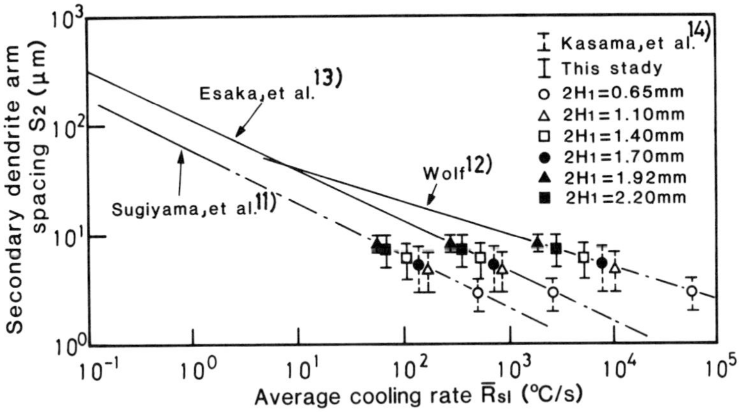

Fig.6 - Relation between the observed secondary dendrite arm spacing and average cooling rate. (SUS304)

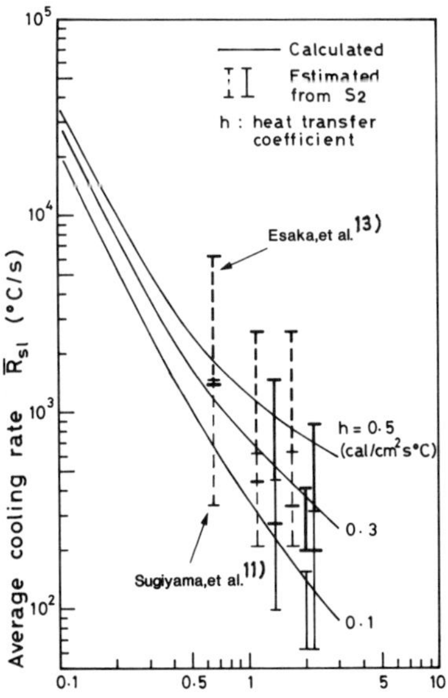

Fig.7 - Relation between the average cooling rate and strip thickness.

Main factor affecting strip thickness

Figure 8 shows the calculated thickness of solidifying shell on roll surface. After the removal of melt superheat and the undercooling, the thickness of solidified layer increases with the time elapsed from the beginning of contact with roll surface (t). The shell thickness at the end of solidification decreases with increasing the roll surface velocity. The initial undercooling (ΔT^o_{uc}) makes the starting point of solidification move downwards, but the end point of solidification is not changed by the initial undercooling, because the solidification rate is faster as the undercooling of melt at the solid/liquid interface is larger. It is noted in Figure 4 that the thickness of solidified layer is not in proportion to the square root of the elapsed time. Thus, strictly speaking, the strip thickness is not proportional to the contact time even if the rolling of the strip does not take place.

Figure 9 shows the calculated result on effect of the roll surface velocity on the distance from the meniscus of melt on roll surface to the starting point of solidification . The distance increases with increasing the roll surface velocity, the melt superheat, the initial undercooling and with decreasing the heat transfer coefficient on roll surface.

Figure 10 shows effect of the roll surface velocity on the roll gap during rolling and the average temperature of strip at roll outlet. In this calculation, the contact time between roll and metal becomes longer as the roll surface velocity is lower. Thus, with decreasing the roll surface velocity, the thickness of solidified shells at the roll outlet increases and the roll gap is enlarged. The average temperature of strip at the roll outlet decreases with decreasing the roll surface velocity.

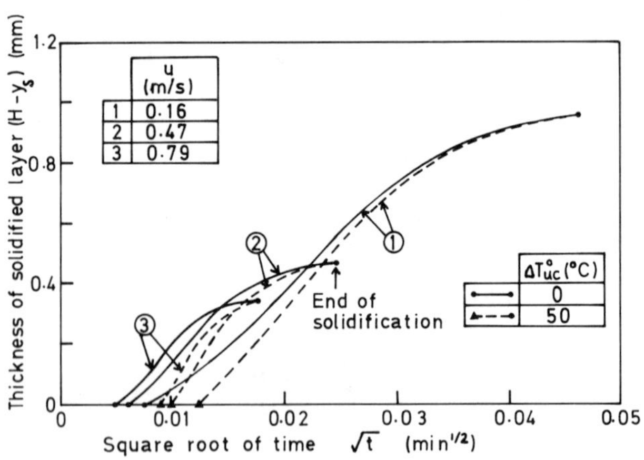

Fig.8 - Calculated thickness of solidified layer.

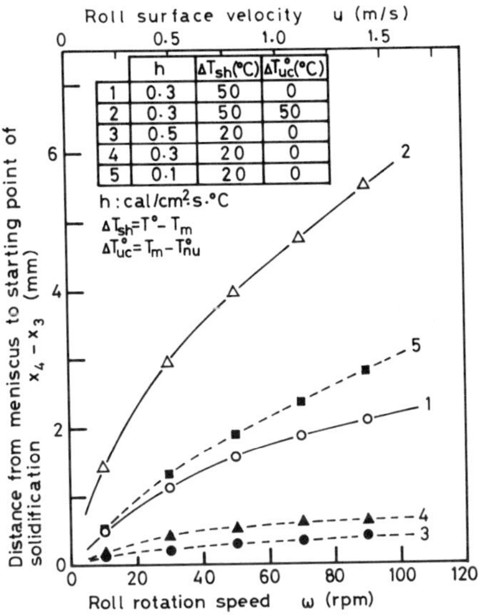

Fig.9 - Effect of roll rotation speed on the distance from meniscus to starting point of solidification.

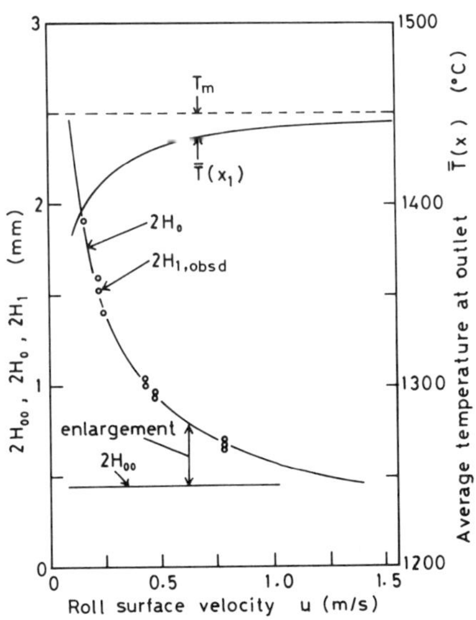

Fig.10 - Estimated roll gap during rolling ($2H_0$) and average temperature of strip at outlet. ($2H_{00}$: initial roll gap, $2H_1$: strip thickness)

Thickness range of metals so far cast by various casting processes is shown in Fig.11. In the casting processes except the continuous casting of steel slab, molten metal is solidified by the heat removal through such moving molds as roll, drum, wheel and belt. In general, for faster casting speed, the solidified metal becomes thicker and the production rate higher.

Twin-roll, single-roller (or melt drag) and melt overflow processes can produce thin strip thinner than a few mm. These thin strip casting processes are compared in Fig.12 with regard to the relation between the thickness and the roll surface velocity. For all the casting processes, the strip thickness is in inverse proportion to a power of roll surface velocity. The contact time is the main factor determining the strip thickness. It is to note that the strip thickness obtained in our experiment agrees well with the data in a melt drag casting in which the contact are length is double as long as our case.

In the twin-roll casting, the exponent n of roll surface velocity depends on mechanism to support rolls and mechanical properties of cast metal. For operational condition of relatively low roll supporting force, the relation between strip thickness and roll surface velocity is controlled by the solidification rate and the exponent n is around 0.5. In this case, the strip thickness is influenced more significantly by thermal diffusivity of metal and overall heat transfer coefficient on roll surface. On the other hand, in the case of soft metals or strong roll supporting force, the exponent n gets toward zero and the strip thickness is controlled by rolling.

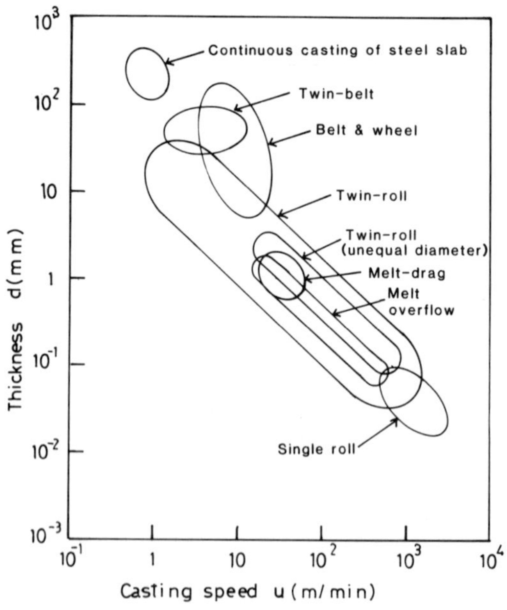

Fig.11 - Relation between thickness of solidified metal and casting speed in various casting processes.

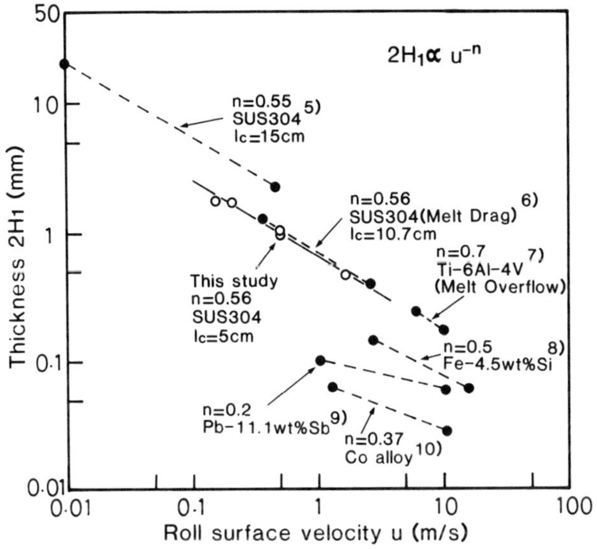

Fig.12 - Effect of roll surface velocity on thickness of rapidly solidified strip and sheet.

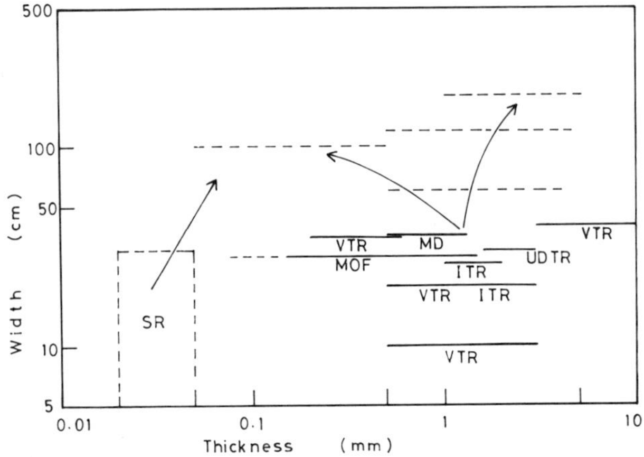

Fig.13 - Width and thickness of rapidly solidified strip and sheet in various casting processes.
(VTR: vertical twin-roll, ITR: inclined twin-roll, UDTW: unequal diameter twin-roll, MD: melt drag, MOF: melt overflow, SR: single roll).

Conclusions

Using a laboratory-scale twin-roll caster, an experiment to produce thin strip of austenitic stainless steel directly from the molten state was done. The thickness of the strips obtained was about 0.5-3 mm. It has been found that the contact time between roll and metal is the most important factor to control the strip thickness under the condition of low roll supporting force. Furthermore, on the basis of the experimental results and the theoretical analysis of heat transfer in the process, the formation mechanism of strip in the roll gap was investigated. Under low roll supporting force, the solidification rate of the shells solidifying on roll surfaces determines the strip thickness.

It was possible to cast thin steel strip with good surface quality and solidification structure and without cracks by our small-scale twin-roll caster. Such a casting technique should be developed for casting of wider strips. As shown in Figure 13, in future, research and development of thin strip casting will tend toward wider strip with higher production rate, and thinner and wider strip for more practical use of rapid solidification.

References

1. T. Kusakawa, "Outline of Today's Strip Casting of Ferrous Materials," Trans. Iron Steel Inst. Japan, 26(4) (1986), B123-125.

2. I. Ohnaka, "State of the Art of Research and Development in Rapid Solidification Technologies in Japan," Trans. Iron Steel Inst. Japan, 27 (1987), 919-928.

3. K. Miyazawa and J. Szekely, "A Mathematical Model of the Splat Cooling Process Using the Twin-Roll Technique," Met. Trans. A, 12A (1981), 1047-1057.

4. K. Miyazawa, et al., "Theoretical Analysis of Twin-Roll Rapid Solidification Process with Taking Account of Supercooling Phenomena of Melt," Trans. Iron Steel Inst. Japan, 26(7) (1986), B258.

5. S. Nishioka, et al.," Research on Twin-Roll Casting Process," Trans. Iron Steel Inst. Japan, 26(4) (1986), B147-148.

6. R.K. Pitler, "Direct Casting of Coilable Ferrous Alloy Strip," Advanced High-Temperature Alloys, ed. S.M. Allen, et al. (Metals Park, OH: American Society for Metals, 1986), 1-10.

7. L.E. Hackman and T. Gaspar, "Producing Strip by Melt Overflow Process," Industrial Heating, 53(1) (1986), 36-38.

8. M. Yukumoto, et al., "Analysis of The Solidification Process in the Roller Quenching Method," Rapidly Quenched Metals V, Vol.1, ed. S. Steeb and H. Warlimont (Amsterdam: Elsevier Sci. Pub., 1986), 91-94.

9. K. Miyazawa, T. Choh, and M. Inouye, "Rolling Characteristics in the Twin-Roll Rapid Solidification Process," Trans. Japan Inst. Metals, 24 (1983), 781-788.

10. J. Ishihara and I. Ikuta, "Non-Uniformity of the Thickness of Amorphous Ribbon Made by the Double Roller Quenching Method," J. Japan Inst. Metals, 46 (1982), 94-99 (in Japanese).

11. M. Sugiyama, T. Umeda, and J. Matsuyama, "On Microsegregation and Microstructure in 18Cr-8Ni and 25Cr-10Ni Austenitic Stainless Steels," Tetsu-to-Hagane, 60 (1974), 1094-1113 (in Japanese).

12. M. Wolf, "Strand Surface Quality of Austenitic Stainless Steels: Part 2, Microscopic Solidification Structure," Ironmaking Steelmaking, 13 (1986), 258-262.

13. H. Esaka, et al., "Relation Between Secondary Dendrite Arm Spacing and Cooling Rate for SUS304 Steel," Current Advances in Materials and Processes," (Report of the ISIJ Meeting), 1(1) (1988), 268.

14. A. Kasama, et al., "A Consideration of the Rapid Solidification of Steels - Solidification of Steel by Twin-Roll Caster -," Trans. Iron Steel Inst. Japan, 26(4) (1986), B139-140.

THERMALLY STABLE ALLOYS PRODUCED BY TWIN ROLL CASTING

I. Jin and D.J. Lloyd

Alcan International Ltd., Kingston Research and Development Centre
P.O. Box 8400, Kingston, Ontario, Canada K7L 5L9

Abstract

There has been considerable interest in recent years in thermally stable aluminum alloys, i.e. alloys which do not soften after long exposure to elevated temperatures up to 350°C. These alloys have been developed by using essentially the powder metallurgy route. The powder metallurgy route involves the production of rapidly solidified alloy powders or flakes, vacuum degassing, consolidation, and extrusion. Alloys of this general type show good thermal stability and high strength, however, the process is very complicated and expensive.

This paper discusses the possibility of producing thermally stable alloys by twin roll casting. Alloys based on the Al-Cr-Zr system can be directly cast to thin strip on a twin roll caster. With proper thermomechanical treatments, the alloy exhibits a combination of good thermal stability and strength. The casting technique, thermomechanical treatment, microstructure, and mechanical properties of the alloy are discussed.

Introduction

The purpose of this paper is to describe the current situation regarding the development of twin roll cast thermally stable aluminum alloys within Alcan, and to present the properties of the alloys developed. In general, thermally stable aluminum alloys are made by the addition of transition elements which have a low diffusion coefficient and a low solid solubility in aluminum. Because of the low solubility, the alloy development involves an inherent difficulty. The alloys must be solidified from an exceptionally high melt temperature and the cooling rate during the solidification must be sufficiently high to suppress the formation of primary intermetallic particles. To meet these conditions, the alloys have usually been produced by the powder metallurgy route.

The powder metallurgy route involves the production of rapidly solidified powders or flakes, vacuum degassing, consolidation, and extrusion. The rapid cooling rate in the powder atomizing, splat quenching or melt spinning processes makes it possible to extend the alloy solubility limits far beyond the limits dictated by the equilibrium phase diagram. A typical alloy of this type may contain 6-10 w/o Fe, 1-10 w/o Cr, 1-10 w/o Zr, 1-10 w/o Ce, 1.5-10 w/o V, and 1-2 w/o Mn (1-3). The strength of these alloys are very high (a yield strength of 420 MPa or higher), however the process is very complicated, expensive, and often results in loss of properties due to the consolidation step.

This paper presents a new family of thermally stable alloys which contain a relatively low level of solute elements. The particular advantage of the alloys is that they can be cast on a twin roll caster directly into a thin strip form, and exhibit a combination of a high strength and good thermal stability with a simple thermomechanical treatment.

Experimental Details

Alloy Composition

Initially a range of Al-Cr, Al-Mn and Al-Zr binary alloys were cast. These alloy slabs were fabricated, and their microstructures and mechanical properties were evaluated. Based on the results of the initial trials, the alloy compositions were systematically extended to multicomponent systems to obtain the best combination of castability, strength and thermal stability. The best performance was achieved with alloys which contain 0.5-1.0 w/o Cr, 0.4-0.6 w/o Zr, 1.7-2.1 w/o Mn, and 0.5-1.0 w/o Mg. Table I gives alloy compositions of two typical alloys. Alloy 1 is a low Mg/high Cr/high Zr version, and Alloy 2 is a high Mg/low Cr/low Zr version.

Table I. Alloy Compositions (w/o)

Alloy No.	Fe	Si	Mg	Mn	Cr	Zr
1	0.23	0.08	0.53	1.82	0.88	0.50
2	0.15	0.05	0.86	1.87	0.63	0.40

The alloys were prepared from superpurity Al (99.99 w/o Al), high purity Cr powder (99.9 w/o), Fe powder (99.9 w/o), Mn powder (99.9 w/o) and an Al-15 w/o Zr masteralloy.

Casting Machine and Casting Procedure

The casting machine used in the experiment was a home built twin roll caster similar to a commercial Hunter type caster. The design features included in the caster were a thin gauge/high speed casting capability, and the ability to maintain a constant strip gauge and casting speed. The caster consists of a drive motor, a speed reducer, a pair of spur gears, spindles and rolls, a hot metal tundish, and a metal feed nose tip. A recirculating water cooling system (285 L/min flow rate) and a dedicated melting furnace (a charge capacity of 350 kg and a temperature limit of 1200°C) were also installed. Table II gives the specifications for the caster.

The weighed aluminum and alloying elements were melted in the furnace. When the melt temperature was controlled to the desired level (usually 950-1000°C), the melt was thoroughly stirred, fluxed with a 90% Ar+10% Cl_2 gas mixture, and introduced into the tundish, and then to the rollbite through the metal feed tip. Once casting started this way, the casting parameters were adjusted to find the optimum speed range in which the heat flow was stable, and problems such as strip sticking to the rolls and primary intermetallic formation could be prevented.

Table II. Caster Specifications

Roll Diameter	: 356 mm (14 inches)
Roll Width	: 610 mm (24 inches)
Strip Gauge	: 2-5 mm
Strip Width	: 460 mm (18 inches)
Roll Separating Force	: 1.3 MN (150 tons) maximum
Roll Torque	: 24500 N·m (217000 in-lbs) maximum
Roll Speed	: 4 m/min (160 in/min) maximum

In general, twin roll casting is possible only in a speed range which is bound by the minimum possible speed and the maximum possible speed. Above the maximum speed, stable heat transfer from the metal to the roll breaks down and heatlines form. A detailed analysis of the heatline development mechanism is given in references (4-6). It can be stated that the maximum speed decreases as the melt temperature increases, and as the alloy freezing range increases. At too low a speed, metal freezes in the feed tip, and therefore a stable casting operation is not possible.

Another defect which commonly appears at high casting speeds is strip sticking to the rolls (6,7). When sticking occurs, the strip suddenly adheres along its width to one of the rolls. This interrupts a stable casting operation. The sticking mechanism is delicately related to the friction hill along the arc of contact. The commonly used remedies are to slow down the caster or to apply more roll coating powder (magnesium oxide, or graphite powder).

For Al-Cr-Zr alloy casting, a further consideration must be given. The alloy composition is in the peritectic range where the intermetallic liquidus temperature is very high. If the metal temperature anywhere in the tip falls below the liquidus temperature, intermetallic primary particles nucleate and grow. The primaries are harmful to the mechanical properties and reduce the solute content in the aluminum solid solution. To avoid the primary formation, the melt temperature had to be kept very high.

The optimum casting parameter values determined by a number of casting trials were as follows:

Strip thickness	= 3.0 - 3.5 mm
Tip setback	= 18 - 20 mm
Melt temperature	= 850 - 900°C
Casting speed	= 460 - 640 mm/min

Fabrication Route

The thermal stability of an aluminum alloy comes essentially from the fine dispersoids in the matrix which are able to stabilize the deformed dislocation structure, and do not coarsen extensively at elevated temperatures. Thus the as-cast alloys have to be heat treated so that the supersaturated solutes (Cr, Zr and Mn) precipitate uniformly in the aluminum matrix. To find the best possible properties in regards to strength and thermal stability, extensive TEM studies and mechanical property measurements were carried out with the as-cast binary and multicomponent alloys. From the results, the following fabrication route was determined:

As-cast Material
↓
Precipitation Anneal
At 375°C for 48 Hours
↓
Cold Roll to the Final Gauge
(Usually 75%)

Results and Discussion

As-Cast Structure

Figure 1 is an x-ray radiograph of a 3.2 mm thick Al-0.6 w/o Cr - 2.0 w/o Mn - 0.8 w/o Mg - 0.41 w/o Zr alloy strip sample which was cast with a deliberately low melt temperature (800°C). The dark spots in the radiograph are primary intermetallic particles. Figure 2 shows the microstructure of the sample shown in Figure 1. Both Al_7Cr (large angular shape) and Al_3Zr (smaller elongated shape) primaries are present in the microstructure. As discussed before, the primaries form whenever the melt temperature in the tundish or in the feed tip is lower than the equilibrium liquidus temperature of the primaries.

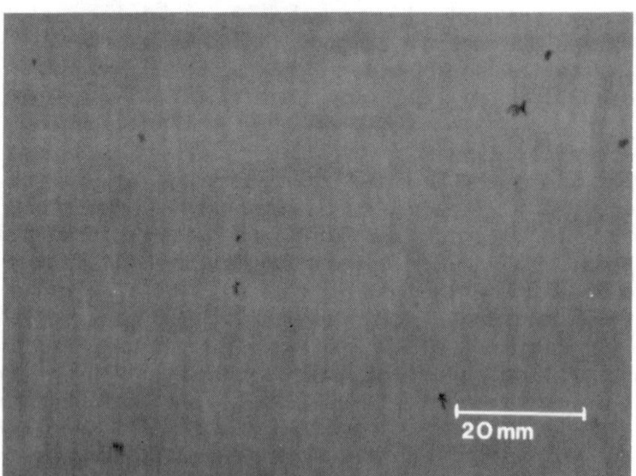

Figure 1 - X-ray radiograph of a twin roll cast Al-0.6 w/o Cr - 2.0 w/o Mn - 0.8 w/o Mg - 0.41 w/o Zr alloy sample, showing the primary intermetallic particles.

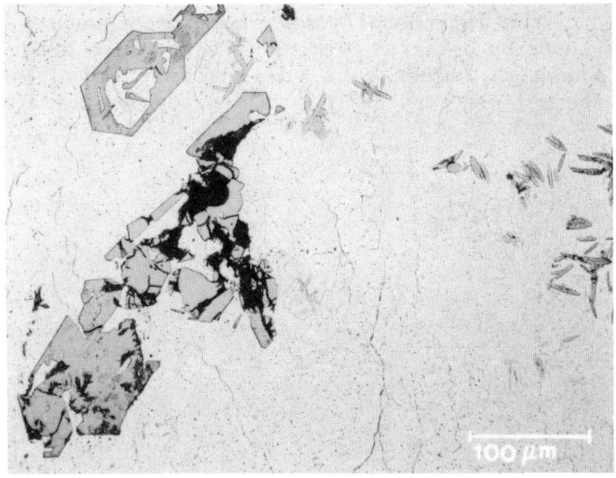

Figure 2 - Microstructure of the as-cast sample shown in Figure 1.

Figure 3 is an optical micrograph taken from a slowly cooled (2°C/s) Al-0.8 w/o Cr - 2.0 w/o Mn - 0.45 w/o Zr alloy sample, and Figure 4 is a micrograph taken from a twin roll cast sample of the same alloy. As expected, the slowly cooled sample includes large primary particles at the aluminum grain boundaries, and the twin roll cast structure is fine and uniform. The intermetallic phases in the slowly cooled sample are Al_7Cr (large angular particles), Al_3Zr (clustered small particles) and (Fe, Mn) Al_6 (dark Chinese script like network). The fine dots in the twin roll cast sample are Al-Fe-Si intermetallics. In the twin roll sample, practically all of the chromium and zirconium are in the aluminum solid solution.

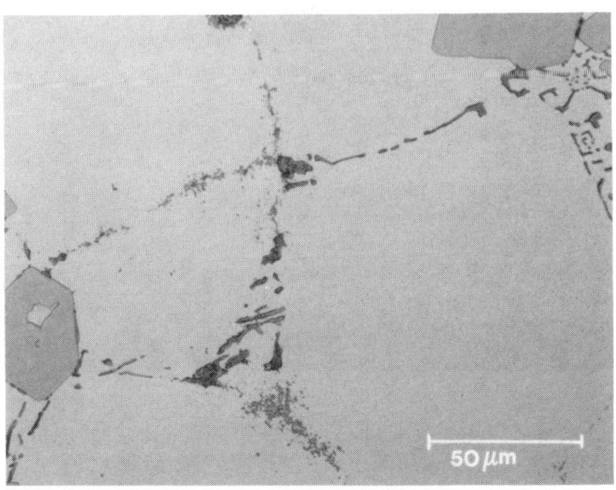

Figure 3 - Microstructure of Al - 0.8 w/o Cr - 2.0 w/o Mn - 0.45 w/o Zr alloy; slowly cooled (2°C/s).

Figure 5 is a TEM micrograph of an as-cast Al-0.77 w/o Cr - 0.45 w/o Zr alloy sample. The micrograph confirms that the structure is essentially free of primary particles of chromium and zirconium intermetallics, except for the occasional iron intermetallics. The microstructure also contains well developed sub-grains of a few microns in size. This subgrain structure is produced by the high temperature rolling associated with twin roll casting, and by thermal recovery during cooling to room temperature.

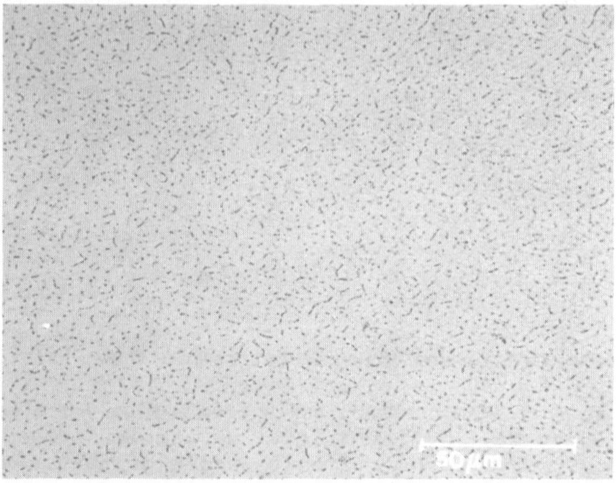

Figure 4 - Microstructure of the same alloy as in Figure 3; twin roll cast

Figure 5 - TEM micrograph showing the as-cast microstructure of Al-0.77 w/o Cr - 0.45 w/o Zr alloy.

Microstructure of Heat Treated Alloys

The as-cast alloys were heat treated at 375°C for 48 hours to induce zirconium precipitation in the matrix. The heat treated alloys, as expected, do not show any microstructural change under an optical microscope. However higher magnification TEM micrographs show some recovery of the substructure and Al_3Zr precipitates. Figure 6 shows the precipitates and the boundaries pinned by the precipitates, which slows down recovery and its resulting softening.

Figure 6 - TEM micrograph showing Al_3Zr precipitates and boundary pinning.

Mechanical Properties

To examine the thermal stability of the alloys, the heat treated and cold rolled strip samples (375°C for 48 hours followed by a 75% cold reduction) were annealed at various temperatures for 2 hours, and their mechanical properties were measured. A plot of ultimate tensile strength, yield strength and elongation vs. annealing temperature is shown in Figure 7. It is seen that the ultimate tensile strength is higher than 400 MPa, the yield strength higher than 370 MPa, and the elongation greater than 2%. Annealing at 350°C only results in a 30 MPa softening with an ultimate tensile strength of around 370 MPa. Figure 8 shows the ultimate tensile strength vs. annealing temperature curves for a number of casts. These curves show that the mechanical properties are highly reproducible.

An additional requirement for thermally stable sheet is that it have good fracture toughness, and achieve adequate fracture toughness has been a problem in sheet produced from rapidly solidified powder. To assess the fracture toughness centre cracked 400 mm wide panels were tested in the L-T orientation after cold rolling and soaking for 2 hours at 300°C. After this heat treatment the sheet had a 345 MPa yield strength and 380 MPa UTS, and the fracture toughness, K_c = 130 MPa$\sqrt{m}$. This value approaches the fracture toughness of conventional wrought Al alloys used in aerospace applications.

Conclusions

Thermally stable alloys based on Al-Cr-Zr system can be directly cast to 3-4 mm gauge strip on a twin roll caster. With a simple heat treatment and cold working, the alloys exhibit very good strength and toughness together with good thermal stability.

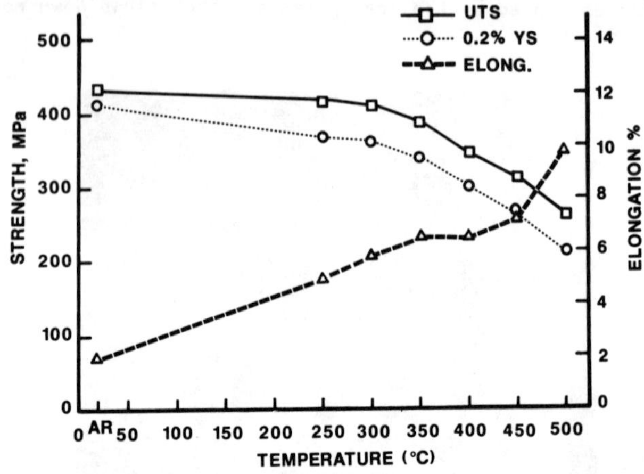

Figure 7 - Mechanical properties of twin roll cast thermally stable alloys.

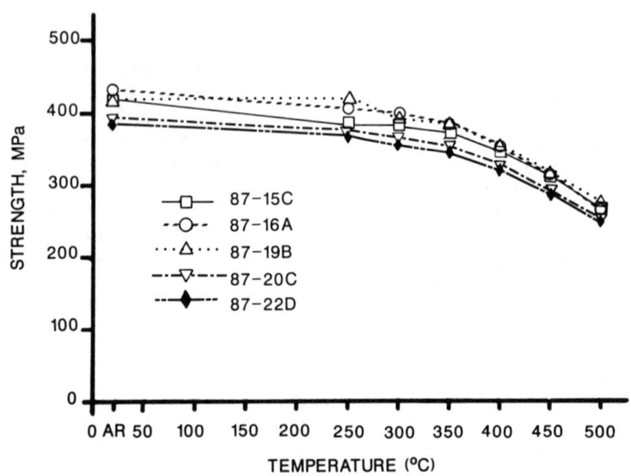

Figure 8 - Ultimate tensile strength of 5 twin roll cast alloys.

Acknowledgement

The authors are grateful to Alcan International Limited for permission to publish this paper.

References

1. D.J. Skinner, P.A. Chipko, and K. Okazaki, "Aluminum-Transition Metal Alloys Having High Strength at Elevated Temperatures," European Patent Application Publication, No. 136508, 4 October 1985.

2. W.S. Miller, and G.J. Marshall, "Thermal Stability and Tensile Properties of Al-Cr-Zr Splat Quenched Alloys," (Report No. BCR-85/39, Alcan International Limited: Banbury and Chalfont Park, 16 August 1985).

3. G.J. Hildeman, and R.E. Sanders Jr, "Aluminum Powder Alloy Product for High Temperature Application," Patent US 4379719, 20 November 1981.

4. I. Jin, "Twin Roll Casting Process: Part 8. Heat Lines and Cast Structure," (Report No. KR-82/099, Alcan International Limited, Kingston Laboratories, 17 August 1982).

5. M.J. Bagshaw, J.D. Hunt, and R.M. Jordan, "A Steady State Model for Roll Casting," Proc. Conf., Modelling and Control of Casting and Welding Processes (The Met. Soc. AIME, 1986), 135-147.

6. S.J. Bercovici, "Optimization of 3C Roll Caster by Automatic Control," Light Metals 1985 (The Met. Soc. AIME, 1985), 1285-1299.

7. R. Iricibar, and I. Jin, "Rolling Aspects of Twin Roll Casting," Light Metals 1984 (The Met. Soc. AIME, 1984), 1129-1144.

STRUCTURE CONTROL

Session Chairman: **S.M. Sorensen, Jr.**
Idaho National Engineering
Laboratory
Idaho Falls, Idaho

PREDICTING INCLUSION DISTRIBUTION IN CONTINUOUSLY CAST SLABS

T.E. Shelenberger*, D.A. Dukelow**, and H. Sato***
National Steel Corporation
*Technical Research Center
1745 Fritz Drive, Trenton, MI 48183
**Product/Process Technology
20 Stanwix Street, Pittsburgh, PA 15222
***Granite City Steel Division
20th & State Streets, Granite City, IL 62040

Abstract

To continuously cast high quality slabs, liquid steel must be delivered to the casting machine mold as free of non-metallic inclusions as possible. Those inclusions remaining in the liquid steel will tend to rise vertically during strand descent through the machine. As they rise, they will be trapped within the strand by the advancing solidification front.

A model has been developed to describe the final distribution of inclusions in conventional thickness slabs, i.e. 200-250 mm. Modeling results have been compared with actual measurements of inclusion distribution in continuously cast slabs. The model indicates that higher casting speeds will result in more inclusions being trapped in the cast slab for the same machine radius and that inclusion content will be reduced as the machine radius increases. Practical applications of the model regarding casting speed and machine radius for conventional slab casters are discussed. Application of the model to thin slabs is briefly considered.

Casting of Near Net Shape Products
Edited by Y. Sahai, J.E. Battles,
R.S. Carbonara and C.E. Mobley
The Metallurgical Society, 1988

Introduction

In the carbon steel industry, the 1980's have been characterized by customer demands for cleaner steel. Some two piece can manufacturers are specifying maximum inclusion sizes of 25 microns or less. To meet these and other demands, steelmakers have installed new equipment, such as ladle metallurgy facilities, have introduced new operating procedures for existing equipment, and have implemented statistical process control.

It is inevitable that carbon steels will contain some non-metallic inclusions of the oxide type since refining of hot metal and/or scrap is necessarily accomplished by oxidation reactions. Injection of oxygen gas is commonly used to increase reaction rates in basic oxygen and electric furnace steelmaking. Oxygen, depending primarily on the final carbon content and temperature, dissolves in low carbon liquid steel in amounts ranging from 500 to 1200 ppm. Bottom blowing with argon near the end point, e.g., with the Lance Bubbling Equilibrium (LBE) process as practiced by National Steel, results in lower oxygen contents at tap.

It is essential for further in plant processing and final application of steel to meet customer requirements that the dissolved oxygen content of steel be precisely controlled. For continuous casting, steel must be completely deoxidized, i.e. killed, so there will be no carbon monoxide evolved during solidification. Gaseous evolution can cause breakouts or create blowholes near the slab surface that can later lead to unacceptable quality.

The element commonly used for deoxidation is aluminum which is added at tap or during ladle metallurgy treatment. Soluble aluminum contents of 0.02 to 0.05% in the steel correspond to dissolved oxygen contents of less than 5 ppm in the caster tundish. For an 0.04% carbon and 1650°C endpoint, a typical resulting LBE oxygen content would be 650 ppm. Based on a 20% aluminum recovery during tap, 3.7 kg/M ton of alumina (Al_2O_3) would be generated as deoxidation and reoxidation products. Most of this alumina enters the slag phase at tap or is removed during ladle metallurgy treatment. A lesser amount is removed in the caster tundish and mold. Compared to the total alumina formed, the amount of alumina remaining in the cast slab, hence final product, is a small fraction of the original amount - about 1% or 0.04 kg/M ton.

However, depending upon size and distribution of the alumina particles or inclusions, the effect of these non-metallics can be very harmful to the forming characteristics of steel sheet and coated products. For example, cans for packing tuna fish are made by the draw/redraw (DRD) operation. In this forming operation a double reduced tin free steel (TFS) 0.18 mm thick is severely drawn. Cold working the steel in this way is an unforgiving operation that demands very clean steel to avoid forming failures.

The "starburst" failure in the side wall of a DRD can shown in Figure 1 is an illustration of what can happen when inclusions exist in a critical portion of the can. Cross sectioning the starburst defect and performing a scanning electron microscope analysis showed a series of alumina particles 2 to 9 microns in diameter in an inclusion stringer at the one-fifth line about 400 microns long (see Figure 2). (The steel used in forming the can was produced in a curved mold casting machine.) Today with cleaner steels being produced, the starburst defect is encountered much less frequently in DRD can manufacturing.

Figure 1 - Starburst defect visible on the outside upper flat section of a draw/redraw can sidewall

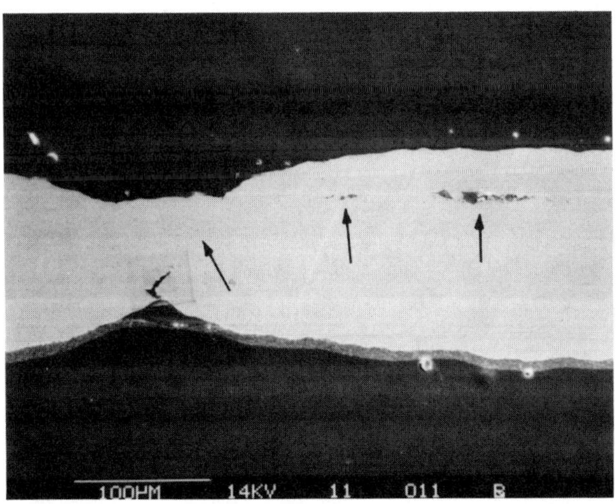

Figure 2 - Stringer of alumina particles at a starburst defect at the one-fifth line

Model Development

The flotation of solid particles in liquids is dependent on many parameters of the system, e.g. particle size and density. The flotation velocity (v)* of non-metallic inclusion particles in the static state of liquid steel would be described by Stoke's Law (Equation 1). The equation indicates that a particle of radius r would

$$v = (2gr^2 (Ds-Dp))/9n \qquad (1)$$

float to the surface of the viscous liquid steel due to the density difference (Ds-Dp) between the steel and particle. In Equation 1, g represents gravitational acceleration and n the liquid's viscosity. Continuous casting is not a static process, but a dynamic one, and thus would not be accurately described by Stoke's Law.

In an experimental study of aluminum deoxidation of liquid steel, Kawawa and Ookubo (1) developed an empirical relationship for the dynamic state of liquid steel. Their study showed that the concentration of inclusions (C) in the dynamic state would be an exponential function of the initial concentration (Co) according to Equation 2. The parameter k is an experimentally obtained characteristic of the

$$C = Co * e^{(-kt)} \qquad (2)$$

dynamic state of the liquid steel/inclusion particle system, and t is time.

Major concerns in continuous casting technology are how to predict steel cleanliness in the cast slabs produced and how to design the casting machine to obtain the cleanest steel slabs possible. Figure 3 is a schematic representation of a

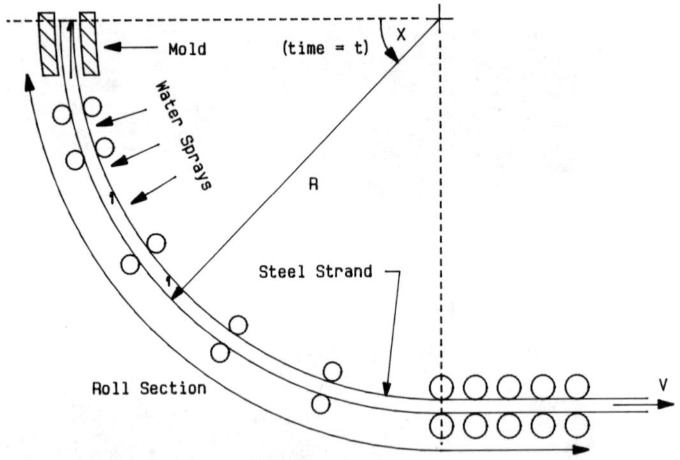

Figure 3 - Schematic of a continuous casting machine

*Additional variable description can be found in the Nomenclature section at the end of the text.

continuous casting machine. Due to operation logistics, strand solidification times, and construction costs, casting machines are built to cast the steel strand in a curved fashion and then to unbend it. In some machines, the curved portion is preceded by a short (generally less than 5m) vertical section to enhance inclusion float out.

Because of the caster design, there are geometrical restrictions to the flotation of inclusions within the strand, as indicated by the arrows in Figure 3. The inclusions have no path to float out of the liquid steel in the lower portion of the machine. Equation 3, a modification of Equation 2, has been proposed by NKK

$$C = Co * e^{(-kt)} * \sin(X) \tag{3}$$

(2) to account for these restrictions. An additional geometrical term was added to relate the angular position (X) of the cast slab from the meniscus with respect to the center point of the machine radius (R). The influence of the angle ranges from 0 to 1 depending on angle X. Since geometrically RX=Vt (where V is casting speed), Equation 3 can be rewritten as Equation 4.

$$C = Co * e^{(-kt)} * \sin(Vt/R) \tag{4}$$

In addition to the inclusion concentration decrease occurring in the liquid steel as time elapses during casting, the liquid steel solidifies forming a shell around a liquid core. A common estimate of the shell thickness (s) used in the steel industry shows it to be proportional to the square root of time (Equation 5) where K

$$s = K * t^{1/2} \tag{5}$$

is an empirical machine solidification constant. Solving Equation 5 for t and substituting the resulting expression into Equation 4 yields Equation 6. The

$$C/Co = e^{(-ks^2/K^2)} * (\sin(Vs^2/RK^2)) \tag{6}$$

equation provides a relationship for the inclusion concentration in the liquid steel just as that portion of the shell solidifies. Thus Equation 6 offers a useful estimate of inclusion concentration in the continuously cast slab.

Three limitations must be applied to Equation 6. The first is that the equation only applies to solidification from the slab's inner radius surface to the center of its thickness. The second is that due to the two dimensional solidification within a cast slab, the equation only applies to the mid-width portion of the slab. Third, it cannot be used beyond the final unbending point in the casting machine.

Discussion

Theoretical Profile Development

Design and operating information were collected from information in the literature for Bethlehem's Burns Harbor #1 Caster (3), by personal communication for NKK's Keihin #1 and #5 casters (4), and from design information at the Technical Research Center for National Steel's three casters. The collection consisted of four curved mold and two straight mold type machines. Details of these machines are shown in Table I. Using the data in Table I and appropriate casting speeds as values for the parameters in Equation 6 (and Equation 2 for the vertical section of the straight mold machines), predicted inclusion concentration profiles for slabs produced in each machine were calculated. Examples of these profiles are shown in Figure 4 for National Steel's Granite City Caster.

Curves such as shown in Figure 4 were found to be inconvenient for comparison to actual inclusion concentration data. An alternative was obtained by dividing the maximum point (Cm) of the curve into each point along the curve. Thus an inclusion concentration index (C/Cm) profile could be calculated through the slab's thickness. The results of performing these calculations on the curves in Figure 4 can be seen in Figure 5. Similar computations were made for the other five casting machines. Those results are shown in Figures 6-10. It must be noted when interpreting the profiles for the straight mold machines that the maximum inclusion concentration (Cm) used in calculating the inclusion index had two values: one for the vertical section and one for the curved section. In the vertical section, described by Equation 2, Cm was equal to Co. In the curved section, the maximum was found by differentiating Equation 6.

Table I. Casting Machine Descriptions

Caster	Mold Type	Vertical Zone (mm)	Average Outside Radius (mm)	Slab Thickness (mm)	Solidification Constant ($mm/min^{1/2}$)
Bethlehem (3)					
Burns Harbor #1	curved	none	12192	254	29.1*
National Steel					
Granite City	curved	none	12192	221	27.4
Great Lakes #1	curved	none	15240	241	27.2
Great Lakes #2	straight	2540	10058	241	27.0
NKK (4)					
Keihin #1	straight	5000	10500	210	27.0
Keihin #5	curved	none	11700	250	27.0
Thin Slab (5)					
Example	straight	4572	1219	50.8	29.3*

* Estimated from assuming complete solidification at the designated point at design operating speed.

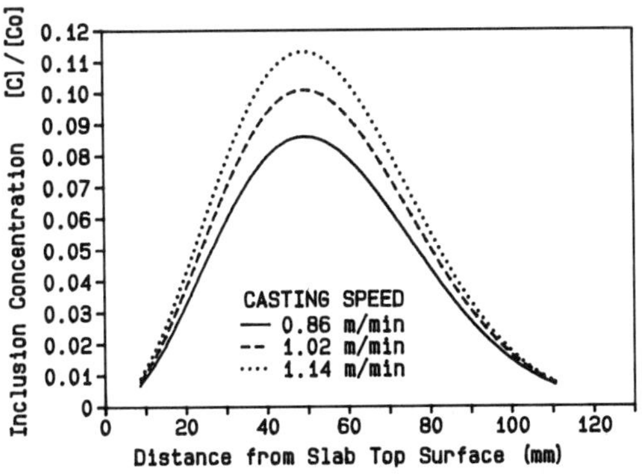

Figure 4 - Inclusion concentration profile for Granite City

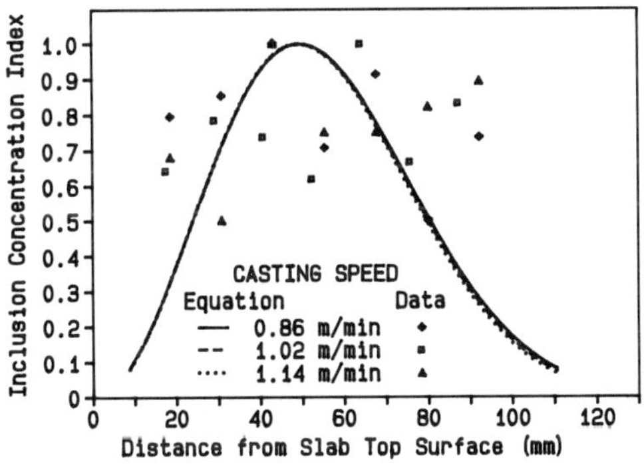

Figure 5 - Inclusion concentration index profile for Granite City

Data Collection Methodology

Attempts to collect data for comparison to the inclusion profiles such as those shown in Figure 4 were hindered in three ways. The first was in gathering data from National's casters. The initial idea was to count inclusion pits in macro-etch samples from the mid-width portion of slabs and record them on an X-Y reference grid. This method seemed reasonable based on observations of past slab macros. After collection of macro samples from the two Great Lakes casters and attempting to count inclusion etch pits, it was determined that this would not work - the slabs were too clean. The second hindrance was encountered when trying to compare data from other casting machines. Neither a study at the Burns Harbor #1 Caster (6) nor information obtained from NKK (4) reported inclusion concentrations. Only total oxygen profiles through slabs were available. Finally, initial inclusion concentrations (Co) were not available, nor easily obtainable.

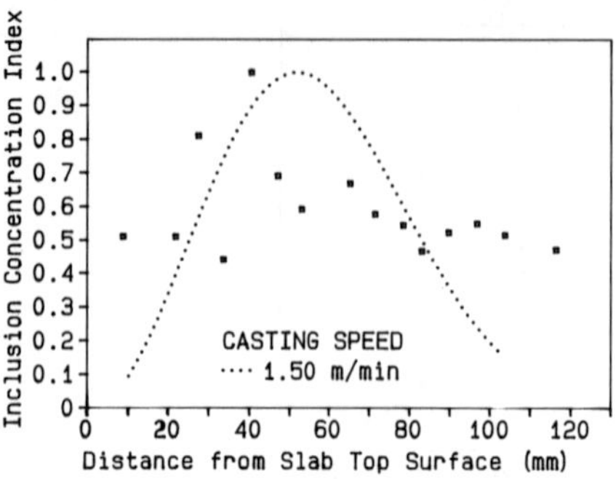

Figure 6 - Inclusion index for Burns Harbor #1 Caster

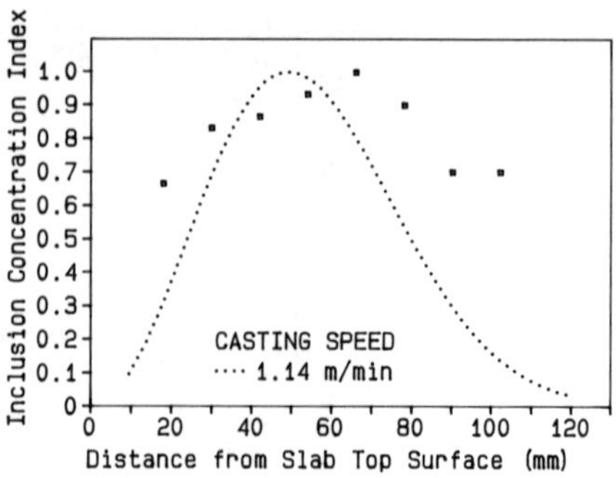

Figure 7 - Inclusion index for Great Lakes #1 Caster

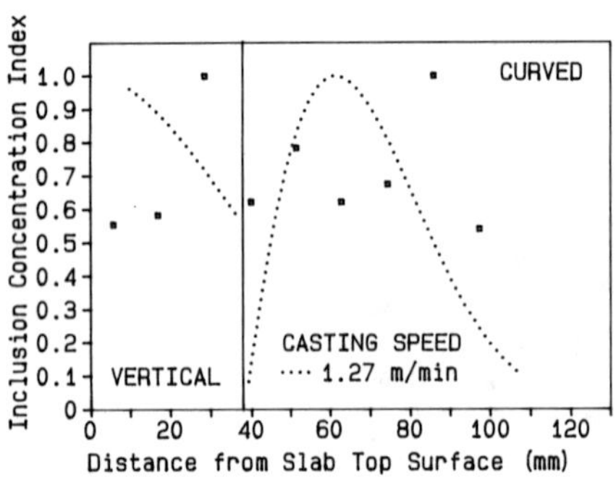

Figure 8 - Inclusion index for Great Lakes #2 Caster

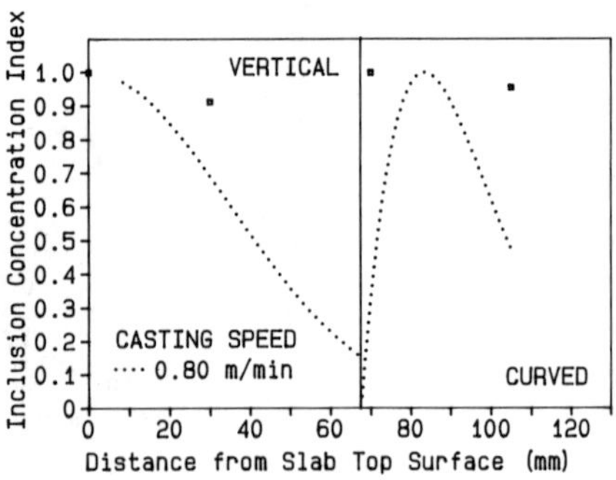

Figure 9 - Inclusion index for Keihin #1 Caster

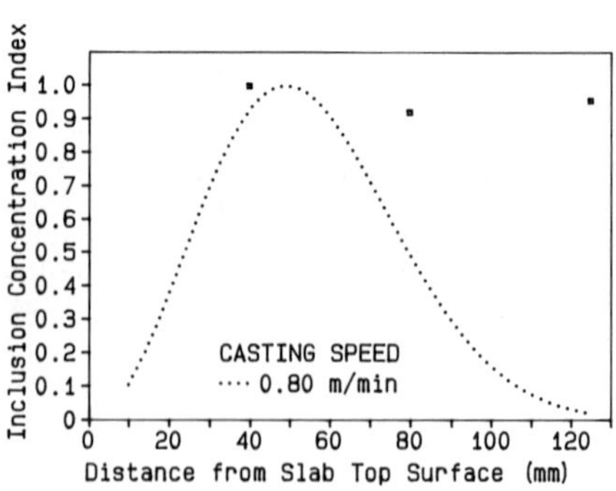

Figure 10 - Inclusion index for Keihin #5 Caster

Use of total oxygen analyses would overcome the obstacles to data comparison previously described. By assuming that the oxygen in a slab was combined with aluminum as Al_2O_3, a direct, proportional relationship between total oxygen and alumina inclusion concentration would exist. Total oxygen measurements similar to those available for the Burns Harbor and Keihin casters could be obtained for National's casters. Limiting the total oxygen data according to the restrictions of Equation 6 and dividing each measurement by the maximum measured value, an oxygen index (Equation 7) profile similar to the inclusion

$$(O)/(Om) \propto C/Cm \tag{7}$$

concentration index shown in Figure 5 was obtained. In Equation 7, O and Om are total oxygen and maximum total oxygen, respectively. Thus a means for comparing both operating and literature data to the theory was attained.

Oxygen Measurements for National Steel's Casters

Slab samples for oxygen analysis were obtained from National's three casters in the form of mid-width macro-etch cuts. These were selected from heats of low carbon aluminum killed steels. Chemistries of these samples can be seen in Appendix I. A macro was then cut into roughly 11 mm cubes (two at each position) in the direction of the slab's thickness by the Great Lakes Metallurgical Lab and analyzed for total oxygen by the Technical Research Center's Analytical Lab. The analysis consisted of cutting a 0.3g to 0.6g piece from each cube, hand filing all sides to remove air oxidation products, and determining the total oxygen in a LECO Oxygen/Nitrogen Liquid TC36 instrument. Two analyses were performed at each position through the thickness, and these two results were averaged. These analyses are reported in Appendix II. Also included are the oxygen measurements for the Burns Harbor (6) and Keihin casters (4).

Oxygen Index Calculation

After accumulating the oxygen data for the various casters, the oxygen indices for each machine were calculated. They were then overlayed on their respective inclusion concentration index plots (Figures 5-10). In cases where the top (due to slab surface oxidation) and center (due to segregation) oxygen measurements were deemed unreasonable, the values were ignored during the index calculations. Also, as noted previously, the oxygen indices for the vertical and curved sections of the straight mold machines were calculated separately.

Results

Curved Mold Machines

The model indicated that an inclusion index peak should occur approximately 50 mm from the top surface of the slab in each of the curved mold machines. For the Granite City caster (Figure 5), the data exhibited peaks at 43 mm for casting speeds of 0.86 and 1.14 m/min, and a peak at 64 mm for a 1.02 m/min casting speed. The data from the Burns Harbor #1, Great Lakes #2, and Keihin #5 (with only limited data) casters displayed peaks at 41 mm, 66 mm, and 40 mm, respectively. In four of the six cases the observed peaks were within 10 mm of the expected location, with the other two slightly farther away.

An explanation of the deviation may be found in the procedure used to obtain samples for oxygen analysis. Compared to the thickness of the slab, substantial pieces of steel were cut from the slab. They were cubes approximatly 11 mm on a side in the case of National's casters. When oxygen analyses were finally performed on the cubes, the oxygen measurement could have been taken anywhere in the 11 mm section.

The surrounding data points were not significantly less than the peaks as was predicted. This could have been due to the cleanliness of the slabs. If the entrapped inclusions were few in number, a small peak value would be expected. Since the oxygen (inclusion) index is a ratio based on the maximum value, the surrounding values would be a large fraction of the peak. Thus a more flat profile would be seen.

An interesting result of the model was that it always predicted the location of the maximum inclusion peak to be 50 mm from the top surface of the slab. Casting speed and machine radius, two parameters in Equation 6, did not significantly affect the position of maximum inclusion concentration within the slab. The modeling results, in which the curves overlapped, and the data, in which the peaks did not correlate, for Granite City (Figure 5) demonstrated this for casting speed. The peaks for the lower (0.86 m/min) and higher (1.14 m/min) speeds were at 43 mm, and the middle speed (1.02 m/min) peak was at 64 mm. As for machine radius, the 50 mm peak location was predicted by the model whether the caster was Keihin #5 (11.7 m radius) or Great Lakes #1 (15.2 m radius) caster. As with the casting speed, no correlation between peak location and radius was seen in the data.

The model predicted gathering of inclusions 50 mm from a slab's surface independent of casting speed and machine radius has provided an insight to the term "quarterline defect" often used in the steel industry when discussing cast slab quality. An operational result in curved mold slab casters is that the portion of the slab which generally has the highest inclusion content is located one-quarter of the way through the slab's thickness from its top surface. The model has shown that the inclusion concentration in curved mold machines should be heaviest 50 mm from the top surface. With most commercially cast slabs being from 200-250 mm thick, 50 mm is one-fifth to one-fourth of the way into a slab. The starburst defect previously discussed was attributed to an alumina inclusion stringer at the one-fifth line.

Straight Mold Machines

The model results for the Great Lakes #2 (Figure 8) and Keihin #1 (Figure 9) straight mold machines indicated the inclusion peak due to the curved portion of the machine would be located deeper within the slab, offset by the time it took the slab to pass through the mold and vertical section. Unfortunately the Keihin data were too limited to compare to the model. More data were available for the Great Lakes caster, showing a peak at 86 mm as compared to the predicted location of 61 mm. A 10 mm difference may be expected, as explained in the curve mold machine data, but other factors have influence in this straight mold machine. Additional study in this area is required to determine if the model accurately predicts inclusion concentration in this portion of a straight mold machine.

The inclusion concentration data available for the portion of slab solidified in the vertical section of the straight mold machines again were limited. That data which was available did not reflect the predicted curve. This was attributed to the highly turbulent flow within the caster mold and to the effect of the inlet stream shroud directing the flow of the liquid steel. Those two actual operating conditions were not accounted for in the model development, thus failure of the model in the vertical section was not surprising. A different model will be needed to describe this portion of a straight mold machine.

Practical Applications

Satisfied that the model adequately reflected operating results in curved mold machines, the effects of the continuous casting machine parameters on two features of the inclusion concentration were investigated. These features are important from both the operation (in making high quality steel with existing equipment) and the design (in specification of new machines) viewpoints.

One feature of the model is the ability of the model to estimate the location of the inclusion concentration peak (e.g. Figure 5). As previously discussed, the peak was affected by neither casting speed nor machine radius in the curved mold machines. Its location was affected by the addition of a vertical zone, such as in the straight mold machines. This phenomena was not explored further, however, since the value of a vertical section is expected to be seen in lessening the amount of non-metallics in the cast slab.

Another and more important feature of the model is the ability to estimate inclusion concentration, i.e. the relative amount of non-metallics in the slab. An application of this ability can be seen in Figure 4 for the Granite City caster. The figure shows that as casting speed increases the amount of non-metallic material entrapped in the slab increases. In this particular example, the peak amount varied from 8% (at 0.86 m/min) to 11% (at 1.14 m/min) of the initial concentration. That even at a low casting speed such a significant amount of non-metallics can remain in the cast slab emphasizes the need to provide the caster with clean steel if high quality slabs are to be produced. Another conclusion drawn from the figure is an operational drawback associated with increasing production by increasing casting speed - a larger entrapped inclusion amount. These same concepts should be considered when determining casting speeds during the design stage of new machines.

Machine radius also has an impact on the amount of inclusions that are eventually trapped in the slab as can be seen in Figure 11. The larger radius machine would be predicted to have less inclusions entrapped in the slab than the smaller radius machine - 9% and 11.5% respectively at the peak in this case. This is due to the larger radius machine having a less geometrically restrictive path for inclusion flotation. The larger the radius, the closer it approaches a machine with a vertical section which would have less restriction. A limit on machine radius would be encountered, though, due to economic considerations - the building to house the caster would be so large the cost would be prohibitive. While the radius in existing machines cannot be changed without considerable capital expense, it deserves attention during the design of a new machine.

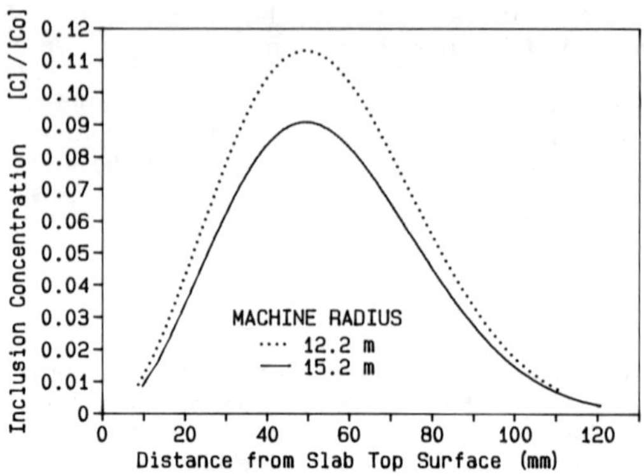

Figure 11 - Effect of casting machine radius on inclusion concentration at a casting speed of 1.14 m/min

An alternative to a machine with an extremely large radius would be one with a straight mold and a short vertical section followed by a curved section with a smaller radius. This arrangement would allow some time for the inclusions to float out of the cast strand unhindered by the geometric limitations inherent in a curved mold machine. The vertical section would thus theoretically provide for fewer inclusions in the liquid steel entering the curved section, and subsequently being trapped in the cast slab. The limited data in this study for the straight mold machines prevented quantification of this supposition, but others such as Tate, et. al. (7), have reported reduction of inclusions in cast slabs with straight mold machines compared to curved mold machines.

Lastly, the model was applied to a thin slab casting machine producing 50.8 mm thick slabs (5). A description of the casting machine can be found in Table I. According to the caster design, a slab in this machine is to be completely solidified while in a vertical position. Thus its inclusion concentration profile would not be expected to exhibit the peak which was seen in the curved mold

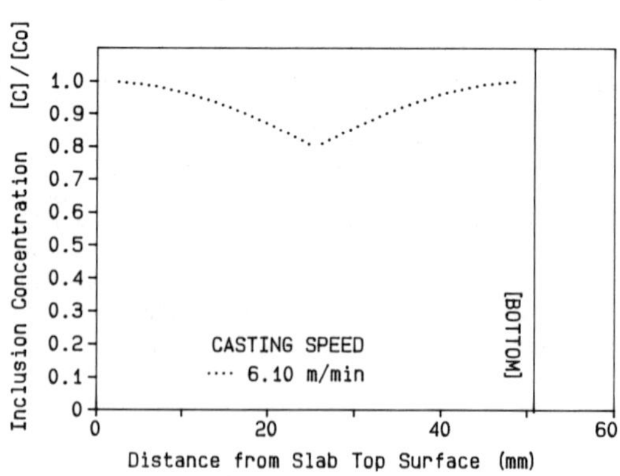

Figure 12 - Inclusion concentration profile for the thin slab caster

machines. Figure 12 shows the inclusion concentration profile predicted by the model for the thin slab caster. The most striking feature of the profile was that almost all of the inclusions initially in the liquid steel would be trapped in the cast slab. The curved mold machines (e.g. Figure 4) had predicted peak concentrations of about 11% of the initial concentration, while the minimum concentration for the thin slab machine was about 80%. Even though complete solidification is to occur in the thin slab machine without geometrical restriction to inclusion flotation, the high casting speed would prevent the inclusions from floating up through the liquid steel for removal in the mold. Thus the model indicates the need to provide the thin slab machine with even cleaner steel than conventional machines. Unfortunately, no data were available for comparison to the thin slab modeling results. Further study is required to determine if the present model accurately estimates the inclusion concentration profiles for thin slab machines.

Conclusions

Stricter quality demands placed on the steel industry by customers have motivated the industry to produce cleaner steel in order to retain its customers. One such effort has been discussed.

A model has been developed which predicts inclusion concentration profiles through the thickness of a continuously cast slab. A reasonable comparison of operational data with the model was achieved for curved mold casting machines. Additional model development remains for straight mold machines.

The model was used to evaluate the effect of casting speed and machine radius on the inclusion concentration in the final cast slab product. Higher casting speeds will result in larger amounts of non-metallics being trapped within the slab to the detriment of slab quality. This emphasizes the need to provide the caster with initially clean liquid steel and to prevent infiltration of oxygen into the liquid steel during casting. Machines with larger radii were predicted to allow for increased inclusion removal. When new machines are being designed, their radius should be closely investigated to ensure optimum inclusion removal.

From the model and data, an explanation of the "quarterline defect" observed in cast slabs was found. In typical commercial curved mold machines, the inclusion peak was predicted to be 50 mm, approximately one fourth of the way, from the slab's top surface. In straight mold machines the peak was expected to be farther from the surface, offset by the effect of the vertical section.

Some thoughts on straight mold machines were put forth, but this study was too limited to sufficiently confirm them. Additional experimental work is required in this area.

The inclusion profile in a slab produced in a thin slab casting machine was considered. The model predicted that compared to conventional casting machines, thin slab machines will require even cleaner liquid steel to ensure production of high quality slabs.

Acknowledgements

The authors appreciate the assistance of National Steel's Great Lakes and Granite City Divisions in providing sections of cast slabs representing various casting conditions. Also appreciated were the analytical services of the Trenton, MI, and Portage, IN, Technical Research Centers of National Steel.

Nomenclature

g	=	gravity (mm/sec^2)
k	=	flotation coefficient (sec^{-1})
n	=	liquid steel viscosity (g-mm/sec)
r	=	particle radius (mm)
s	=	shell thickness (mm)
t	=	time (sec)
v	=	particle flotation velocity (mm/sec)
C	=	concentration of inclusions at t=t
Cm	=	maximum inclusion concentration
Co	=	concentration of inclusions at t=0
Ds	=	liquid steel density (g/mm^3)
Dp	=	particle density (g/mm^3)
K	=	solidification constant ($mm/min^{1/2}$)
(O)	=	total oxygen (ppm)
(Om)	=	maximum total oxygen (ppm)
R	=	casting machine radius (mm)
V	=	casting speed (mm/min)
Z	=	distance from slab top surface (mm)

References

1. T. Kawawa and M. Ookubo, Tetsu to Hagane, 53 (1967), 1569-1585

2. Nippon Kokan KK, private communication with authors, December 1987

3. L. M. Harbold, R. K. Garrett and J. L. Byrnes, "Burns Harbor Slab Caster," Iron and Steelmaker, 3(6) (1976), 29-36

4. Nippon Kokan KK, private communication with authors, May 1988

5. H-F. Marten and H. Fastert, "Latest Developments and Operating Results of SMS Schloemann-Siemag AG's High-Speed Strip-Casting Machine," Iron and Steelmaker, 14(5)(1987),57-62

6. B. C. Whitmore, "Oxide Segregation in Aluminun-Killed Strand Cast Steel Slabs," Iron and Steelmaker, 5(10) (1978), 34-38

7. Masahisa Tate et. al., "Improvement of Internal Quality of Continuously Cast Slab," Nippon Kokan Technical Report, 36(1982), 1-13

Appendix I. Chemistries of Steel Samples

Element	A	B	C	D	E	F	G-H*
C (%)	0.04/0.06	0.035	0.039	0.031	0.044	0.037	NA
Mn	0.28/0.32	0.28	0.29	0.27	0.23	0.27	NA
Al	0.030/0.080	0.033	0.032	0.042	0.055	0.060	NA
P	NA	0.009	0.013	0.012	0.012	0.009	NA
S	NA	0.011	0.016	0.013	0.018	0.013	NA
Cu	NA	0.017	0.022	0.024	0.066	0.041	NA
Si	0.02 max	0.011	0.011	0.014	0.015	0.018	NA

A - Bethelehem Burns Harbor #1 Caster study (6)
B - National Steel Granite City (0.86 m/min)
C - National Steel Granite City (1.02 m/min)
D - National Steel Granite City (1.14 m/min)
E - National Steel Great Lakes #1
F - National Steel Great Lakes #2
G - NKK Keihin #1 (4)
H - NKK Keihin #5 (4)
NA - not available
* - LCAK

Appendix II. Total Oxygen Analysis Results

A		B		C		D		E		F		G		H	
Z (mm)	(O) (ppm)	Z (mm)	(O) (ppm)	Z (mm)	(O) (ppm)	Z (mm)	(O) (ppm)	Z (mm)	(O) (ppm)	Z (mm)	(O) (ppm)	Z (mm)	(O) (ppm)	Z (mm)	(O) (ppm)
9	28	6	22	6	38	6	27	6	16	6	20	0	18	0	26
22	28	18	27	17	27	18	19	18	20	17	21	30	17	40	23
28	45	31	29	29	33	31	14	30	25	29	36	70	18	80	21
34	24	43	34	41	31	43	28	42	26	40	23	105	17	125	22
41	55	55	24	52	26	55	21	54	28	52	29	140	20	170	25
47	38	68	31	64	42	68	21	66	30	63	23	180	17	210	28
53	33	80	17	76	28	80	23	78	27	75	25	210	18	250	27
66	37	92	25	87	35	92	25	90	21	86	37				
72	32	104	22	99	21	104	19	102	21	98	20				
79	30	117	20	111	21	117	16	115	24	109	28				
83	26	129	22	122	69	129	16	127	31	121	27				
90	29	141	26	134	48	141	19	139	26	132	24				
97	30	154	26	145	34	154	17	151	22	144	22				
104	28	166	32	157	26	166	28	163	22	155	27				
117	26	178	28	169	26	178	17	175	26	166	16				
		190	24	180	23	190	24	187	17	178	22				
		203	23	192	29	203	23	199	21	189	25				
		215	34	204	27	215	24	211	22	201	36				
				215	36			223	18	212	28				
								235	27	224	16				
										235	28				

A - Bethlehem Burns Harbor study (6)
B - National Steel Granite City (0.86 m/min)
C - National Steel Granite City (1.02 m/min)
D - National Steel Granite City (1.14 m/min)
E - National Steel Great Lakes #1
F - National Steel Great Lakes #2
G - NKK Keihin #1 (4)
H - NKK Keihin #5 (4)

Secondary Dendrite Arm Spacings in Continuous Cast, Thin Steel Sections

Alan W. Cramb

Dept. of Metallurgical Engineering
and Materials Science
Carnegie Mellon University
Pittsburgh, PA 15213

Abstract

The secondary dendrite arm spacing of thin section castings is often quoted as a measure of cooling rate and structural refinement. The purpose of this study was to determine the relationship between strip thickness and secondary dendrite arm spacing in low carbon, aluminum killed steels. Secondary dendrite arm spacings from low carbon, aluminum-killed strip and thin slab samples were analyzed from twin roll, Hazelett and static castings and compared to current literature values. The secondary dendrite arm spacing (λ_2) in these studies was found to be independent of caster type and related only to the distance from the chill surface (x) and a simple relationship of the form:

$$\lambda_2 = A\ x^{\frac{2}{3}}$$

was found to correlate well with the data. This relationship results because heat conduction controls shell growth and secondary dendrite arm spacing is a coarsening phenomena.

Casting of Near Net Shape Products
Edited by Y. Sahai, J.E. Battles,
R.S. Carbonara and C.E. Mobley
The Metallurgical Society, 1988

Introduction

New steel continuous casting processes are under rapid development worldwide.(1,2) Process developments are towards near-net-shape casting, where the cast dimensions are close to the final product dimensions. The major processes under development are direct strip casting, via a single or twin roll machine, thin slab casting and spray casting.(2) Solidification in all of these processes occurs more rapidly than in conventional processes due to the decrease in cast thickness and the resulting solidification structure is finer, with less microsegregation than conventionally processed material. Microsegregation is related to dendrite arm spacing and techniques which reduce dendrite arm spacing should produce more easily homogenized castings.(3,4) The purpose of this paper is to discuss the variation in secondary dendrite arm spacing which can be expected in direct and thin slab casting of low carbon, aluminum killed steels.

Experimental Details

Samples of twin roll cast strip, Hazelett cast and static cast thin slabs of low carbon, aluminum killed steel were gathered from North American Steel Companies engaged in new casting technique developments. Strip thickness ranged from 0.7 to 1.5 mm, Hazelett slab thickness was 25 mm and static cast slab sample thicknesses were 12 , 17 and 25 mm. Samples were mounted, etched to reveal solidification structure and examined by both optical and scanning electron microscopy.

Etching Practice

Samples were given a short time austenitizing anneal to aid in resolution of dendritic structure. This treatment included heating to 940°C for 1 minute followed by a quench and then reheating to 725°C for 1 minute followed by a second quench.(5) Initial attempts at resolving dendrites were made using Steads and Humphries reagent. These etches were found to be insensitive and inconsistent, therefore, a new one step etch was developed. This etch, which is a solution of 10 gm Cupric Ammonium Chloride, 7 cc Hydrochloric acid, 40 gm Magnesium Chloride and 200 cc distilled water, was found to require only short etching times, 10 to 20 seconds, followed by a rinse in 200 proof alcohol, to achieve adequate results.

Solidification Structure

The twin roll cast thin strip material which was examined had a chill zone at the surface, a columnar zone from near the surface to the center of the casting and occasionally an equiaxed zone near the center of the casting. The predominant structure was columnar with the majority of the samples having dendrites from close to the surface to the center of the sample. Equiaxed zone occurrence was sporadic and the center line, the point of contact of the dendrites from each chill surface, wandered throughout the sample rarely being in the geometric center of the casting. In certain samples, bending of the dendrites at the center line could also be seen. This condition appeared to be related to the amount of force applied to the rolls during casting.

Surface quality of strip castings was fair with evidence of most of the common casting surface

defects. However, a number of copper droplets, presumably from the rolls, could be sporadically found on the edges of the castings indicating abrasion or melting of the roll surface during casting.

The solidification structure of the static and Hazelett cast thin slabs was similar to that of the thin strip with the structure being predominantly columnar.

Dendrite Arm Spacing Measurement

Measurements were made of secondary dendrite arm spacings (λ_2) by counting the numbers of arms over a given length of primary arm (the line intercept method). All measurements were at least over 10 arms and recorded in microns as a function of distance from the chill surface. Results of the measurements are given in tables 1 - 3. Secondary dendrite arm spacing was found to be quite variable, even within the same sample, and all measurements were made from areas where resolution of the secondary arms was possible. The majority of the measurements were taken from the center portion of the thin strip material as resolution of structure within 0.01mm of the surface was extremely difficult.

Table 1: Secondary Dendrite Arm Spacings From Twin Roll Cast Strip

x (mm)	λ_2 (microns)
0.01	1.0
0.05	4.5
0.06	4.0
0.07	4.5
0.07	6.3
0.09	4.0
0.10	5.0
0.11	5.6
0.17	12.5
0.22	6.6
0.25	12.5
0.32	9.0
0.4	14.0

Results and Discussion

Secondary dendrite arm spacings in steels have been the subject of numerous studies (6 - 15), the majority of which were steady state unidirectional solidification experiments. In these experiments the solidification variables, growth rate (R) and temperature gradient (G) can be varied independently and the following empirical relationship is found to represent λ_2:

$$\lambda_2 = K R^m G^n \quad .... [1]$$

Table 2: Secondary Dendrite Arm Spacings for Static Cast Low Carbon Steel

x (mm)	λ_2 (microns)	Total Cast Thickness(mm)
1	31	25
	31	17
	32	12
2	37	25
	38	17
	37	12
3	45	17
	36	12
4	50	25
	45	17
	42	12
5	50	25
	46	17
	41	12
6	50	17
	44	12

Table 3: Secondary Dendrite Arm Spacing from Hazelett Cast Slab

x (mm)	λ_2 (microns)
8	80
10	100

The exponents *m* and *n* are usually found to be similar for secondary dendrite arm spacing measurements, their values being close to 0.4 for low carbon steels. (11 - 14) It is also normal to relate the dendrite arm spacings to local solidification time (θ_f), following the suggestion of Flemings (16), where θ_f is the freezing temperature difference between the liquidus and the solidus divided by the cooling rate:

$$\theta_f = \frac{T_l - T_s}{R \ . \ G} \quad .... \ [2]$$

T_l is the liquidus temperature and T_s is the solidus temperature. Secondary dendrite arm spacings are found to follow the following relation :

$$\lambda_2 = L \ \theta_f^r \ .... \ [3]$$

Kattamis and Flemings (17) considered secondary dendrite arm spacing to be a coarsening phenomena, similar to Ostwald ripening, and showed theoretically that the exponent r should be

0.33. Actual results in other investigations have ranged from 0.17 to 0.55 with most recent results for lower carbon steels being 0.44 (15).

The parabolic freezing law is observed to be a close approximation to actual results for most slab or strip casters. Kasama (21) has shown that, for very thin strips (< 0.05mm in stainless steel), this law is not exact and has to be modified to take into account contact resistance following the methods of Adams (18). In this study, as strip thickness of greater than 0.05 mm will be considered, a simple parabolic relationship for shell thickness was assumed. In figure 1 shell thickness is plotted verses the square root of time for strip cast material. (19 - 24) As can be seen, data fits the parabolic law where:

$$x = \alpha \sqrt{t} + C \quad .... [4]$$

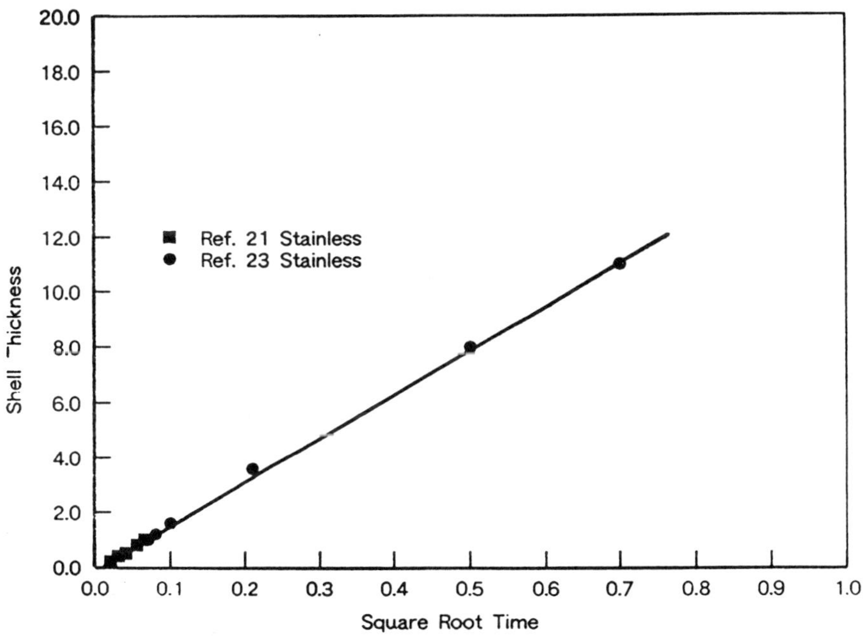

Figure 1: Shell Thickness as a Function of Square Root t

In the studies of steady state solidification, liquidus and solidus both advance in parabolic fashion, each with its own constant i. e. α_l and α_s. Thus, if G is the temperature gradient:

$$G = \frac{dT}{dx} = \frac{\Delta T}{(x_l - x_s)} = \frac{\Delta T}{(\alpha_l - \alpha_s)\sqrt{t} - D} \quad .... [5]$$

and R is the growth rate:

$$R = \frac{dx}{dt} = \frac{\alpha^*}{2\sqrt{t}} \quad .... [6]$$

where α^* is the solidification constant for a defined fraction of solid and D is a constant, then θ_f, the local solidification time, must be equal to:

$$\theta_f = \frac{[\alpha_l - \alpha_s]\ t - 2\ D\ \sqrt{t}}{\alpha^*} \quad .... [7]$$

From equation 3, λ_2 is proportional to θ_f to the one third power if coarsening of dendrites controls spacing, therefore, upon making the appropriate substitutions:

$$\lambda_2 = \frac{L\ (\alpha_l - \alpha_s)\ x^{\frac{2}{3}}}{(\alpha^*)^{\frac{5}{3}}} - \frac{2\ D\ x^{\frac{1}{3}}}{(\alpha^*)^{\frac{4}{3}}} \quad .... [8]$$

and, by rationalizing the constants, it follows that:

$$\lambda_2 = F\ x^{\frac{2}{3}} - Gx^{\frac{1}{3}} \quad .... [9]$$

If we assume D to be small as it is the difference between liquidus and solidus solidification constants, then for significant thicknesses, and, assuming superheat is small, F>>G:

$$\lambda_2 = F\ x^{\frac{2}{3}} \quad .... [10]$$

Thus it follows that a graph of $\log_{10} \lambda_2$ vs. $\log_{10}$ x should be linear with a slope of two thirds and that to a first approximation secondary dendrite arm spacing will be independent of casting machine and only dependent on cast thickness in the regime where heat transfer governs shell growth. In figure 2 the data from this study and a number of literature results from static and continuous castings are plotted for low carbon steels in the specified log-log plot. A line of slope two thirds fits the data over its complete range (four orders of magnitude in shell thickness) and suggests that the range of secondary dendrite arm spacings which will be found in strip and conventional castings can be adequately predicted in low carbon steels from a knowledge of the desired cast thickness using the following equation:

$$\lambda_2 = 22.38\ x^{\frac{2}{3}} \quad .... (11)$$

where λ_2 is measured in microns and x is measured in mm.

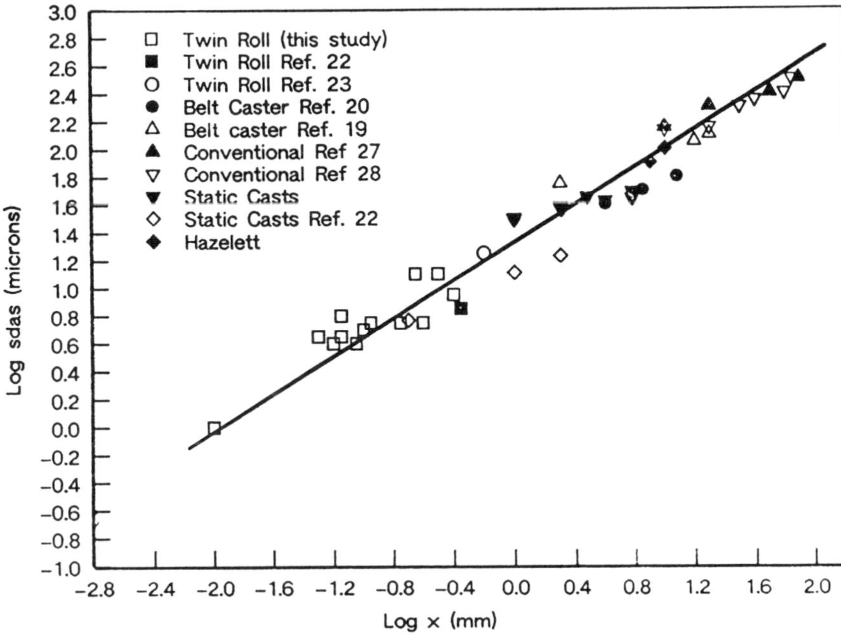

Figure 2: Secondary Dendrite Arm Spacings as a Function of Distance from the Chill Surface for Low Carbon Steels

In figure 3 the literature results from a number of studies of stainless steels is also shown to fit this model over a wide range of cast thicknesses.

In thin strip casting, the cooling rate in °C per minute is often quoted as a measure of rapid solidification. For carbon steels the data of Suzuki et al. are often quoted (8), where:

$$\lambda_2 = 710\ R^{-0.39} \quad .... (12)$$

Combining equation 12 with equation 11 allows calculation of R in °C per minute as a function of x:

$$R = 7069\ x^{-1.71} \quad .... (13)$$

Thus cooling rates for strip cast low carbon, aluminum killed steels of thickness 1.5 mm will be approximately 10^4 °C per minute near the strip center, the position of maximum secondary dendrite arm spacing and maximum microsegregation, and 10^5 °C per minute within 0.2 mm of the strip

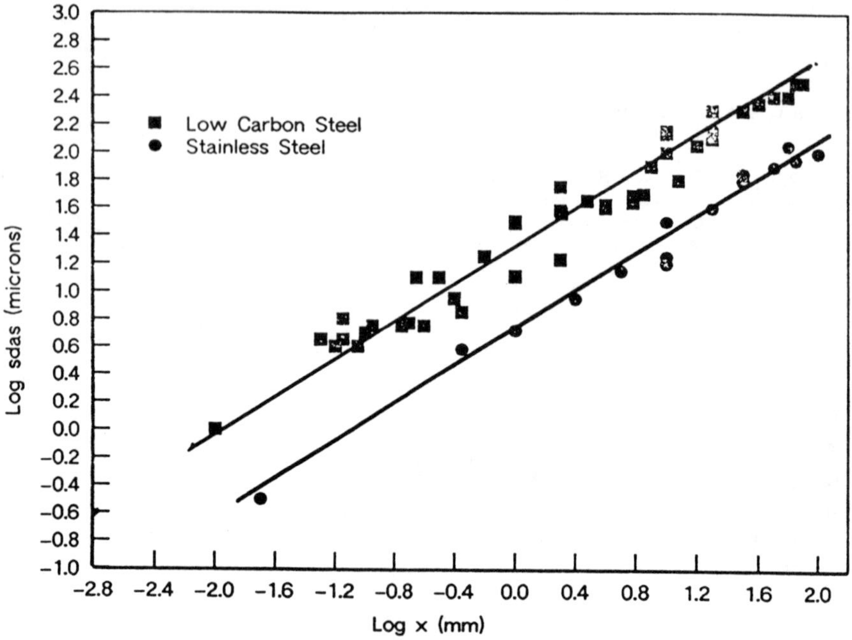

Figure 3: Secondary Dendrite Arm Spacings for Low Carbon and Stainless Steels

surface.

As previously mentioned, extrapolation of these results to lower thicknesses (< 0.1 mm) is not possible as the parabolic relation for shell thickness will not be applicable; however, for the thicknesses of sheet anticipated in the casting of low carbon, aluminum killed grades these relationships may be useful in the prediction of secondary dendrite arm spacings in strip and thin slab castings where heat conduction through the shell of the casting controls growth rate. Clearly this approach could be adapted to more complex heat transfer models for more accurate determinations of growth rate during variable rates of heat transfer. In this manner the effect of differences between operational practices in different casters could be evaluated. However, as the variation in the data from the same cast sample is quite large, this simple method allows the calculation of secondary dendrite arm spacings to within a factor of 2 for most low carbon, aluminum killed continuous castings.

Conclusions

The main conclusions of this study are that in the casting of low carbon aluminum killed steels using water cooled copper molds, the secondary dendrite arm spacing is determined simply by the distance from the chill surface in castings greater than 0.7 mm in thickness and that the solidification structure of thin strip material is similar to the surface solidification structure of thin slabs and thick slabs regardless of the casting technique.

Acknowledgement

The author would like to thank Bethlehem, Armco, USX, Inland and Weirton for their generous donation of samples from pilot casting machines and the Engineering Foundation for financial support.

The microscopy and etching techniques were developed and carried out by J. Pike and his assistance is also gratefully acknowledged.

References

1. A. W. Cramb: "New Steel Casting Processes for Thin Slabs and Strip - A Historical Perspective", Transactions of the Iron and Steel Society, Vol. 10, 1988.
2. J. Szekely: "Can Advanced Technology Save the U.S. Steel Industry", Scientific American, July 1987, Vol. 14, No.2, p 84 - 93.
3. M. C. Flemings: Solidification Processing, McGraw-Hill, 1974.
4. W. Kurtz and D. J. Fisher: Fundamentals of Solidification, Trans Tech Publications, Switzerland, 1984.
5. B. L. Bramfit: Metals Handbook Vol. 8, "Solidification Structures of Steel Ingots", p 158 - 162.
6. R. F. Polich and M. C. Flemings: "Mechanical Properties of Unidirectional Steel Castings", Trans Am Foundryman's Soc., Vol 73, 1965, p 28 - 33.
7. M. Ibarakl, T. Okamoto and H. Matsumoto: "The Structure of Unidirectionally Frozen Metastable Iron-Carbon Alloys" J. Japan Inst. of Metals, 1968, Vol. 32, p 396 - 402.
8. A. Suzuki, T. Suzuki, Y. Nagaoka and Y. Iwata: "On Secondary Dendrite Arm Spacing in Commercial Carbon Steels with Different Carbon Content:, J. Japan Inst. of Metals, Vol 32, 1968, p 1301 - 1305.
9. A. Suzuki and Y. Nagoka: "Dendrite Morphology and Arm Spacing of Steels", J. Japan Inst. of Met., 1969, Vol.33, p 658 - 63.
10. K. Gungi, K. Kusaka, E. Ishikawa and K. Sudo: "Solidification Structure of High Speed Tool Steel", Tetsu-to-Hagane, 59, 1973, p 1089, Trans ISIJ, Vol. 14 , 1974, p 257 - 266.
11. H. Jacobi and K. Schwerdtfeger: "Dendrite Morphology of Steady State Unidirectionally Solidified Steel", Met. Trans., Vol. 7A, 1976, p 811 - 820.
12. T. Edvardsson, H. Fredriksson, and I. Svensson: "A Study of the Solidification Process in Low-Carbon Manganese Steels", Metal Science, Vol. 10, 1976, p 298 - 306.
13. M. A. Taha, H. Jacobi, M. Imagumbai and K. Schwerdtfeger: "Dendrite Morphology of Several Steady State Unidirectionally Solidified Iron-Based Alloys" Met. Trans. Vol. 13A, 1982, p 2131 - 2141.
14. M. A. Taha: "Influence of Solidification Parameters on Dendrite Arm Spacings in Low Carbon Steels", Journal of Materials Science Letters, 5, 1986, p 307 - 310.
15. A Guide to the Solidification of Steels, Jerkontoret, Stockholm 1977.
16. T. F. Bower, H. D. Brody, and M. C. Flemings: Trans TMS-AIME, 1966, vol. 236, p 624 - 634.

17. T. Z. Kattamis and M. C. Flemings: Transactions of the Metallurgical Society of AIME, 233, 1965, p 992.

18. C. M. Adams, Jr., Liquid Metals and Solidification, ASM, 1958, "Thermal Considerations in Freezing", p 187 - 217.

19. K. Ushijima, T. Okazaki, M. Yoshihara, S. Shiode and M. Koide:Near Net Shape Casting, A publication of the Iron and Steel Society, 1987, "Thin Steel Slab Continuous Casting Process", p 31 - 36.

20. S. Itoyama, H. Nakato, T. Nozaki, Y. Habu, and T. Emi: Near Net Shape Casting, A publication of the Iron And Steel Society, 1987, "Development and Solidification Characteristic of Horizontal Twin-Belt Type Thin Slab Caster", p 37 - 40.

21. A. Kasama, S. Mizoguchi, K. Miyazawa, M. Ito and T. Sugai: Near Net Shape Casting, A publication of the Iron and Steel Society, 1987, "A Consideration on the Rapid Solidification of Steels - Solidification of Steel by Twin-Roll Caster", p 59 - 63.

22. C. Yoshida, K. Taniguchi, T. Nakagawa, M. Sudo and T. Nozaki: Near Net Shape Casting, A publication of the Iron and Steel Society, 1987, "Characteristics of Rapidly Solidified Cast Iron and High Carbon Steel", p 65 - 69.

23. K. Kawakami, J. Inagaki, S. Nishioka, H. Sato and M. Ohtsuki: Near Net Shape Casting, A publication of the Iron and Steel Society, 1987, "Research on Twin Roll Casting Process", p 71 - 80.

24. T. Kusakawa, H. Yamamoto, Y. Yanagi, K. Endo and K. Oka: Near Net Shape Casting, A publication of the Iron ans Steel Society, 1987 " Characteristics of Iron and Steel Thin Sheet made by Strip Casting", p 81 -85.

25. K. Bouby and J. V. Wood: "Steel Particulate Made by Rapid Solidification Processes", Powder Metallurgy, 1986, Vol. 29, No.1, p 33 - 36.

26. T. Ichiryu, I. Ohnaka, H. Tateoka and I. Yamaauchi: "Production and Properties of Rapidly Solidified Fe-Si Alloy wires by the In-Rotating-Water-Spinning Process", J. JIM, 1986, Vol. 50, p 927 - 934.

27. H. Fuji, T. Ohashi, T. Ono and K. Asano: "Fundamentals of Continuous Casting", Nippon Steel Technical Report, No. 13, June 1979, p 48 - 61.

28. K. Kawakami. T. Kitagawa, K. Murakami, Y. Miyashita, Y. Tsuchida and T. Kawawa: "Fundamental Research on Solidification Involved in Continuous casting of Steel", Nippon Kokan Technical Report, Overseas, No.36, 1982, p 26 - 41.

CONTINUOUSLY CAST FIBERS WITH

LINEARLY ALTERNATING PHASES

R.B.Pond, Sr. and J.M. Winter, Jr.

Department of Materials Science and Engineering
Johns Hopkins University
Baltimore, Md. 21218

Abstract

A method is described for continuously casting round fibers by the free flight melt spinning technique in such a way that linearly alternating phases are produced. The method involves ejecting uniform diameter oxide filmed spheres of a mechanical mixture of two alternating phases through an orifice having a diameter slightly smaller than that of the spheres in a continuous manner and solidifying the stream before breakup occurs. The "coral snake" appearance of the dual phase is portrtayed microscopically to reveal the resulting interfaces and macroscopically to reveal the phase distribution.

Although the relatively insoluble liquid phases of lead and zinc are used in these studies, a discussion of the probable use of such fibers made with soluble liquid phases such as lead and tin is presented.

Casting of Near Net Shape Products
Edited by Y. Sahai, J.E. Battles,
R.S. Carbonara and C.E. Mobley
The Metallurgical Society, 1988

Introduction

The casting of round fibers by a technique described as "free flight melt spinning" has its origin in the nineteenth century. A patent (1) issued to Edward Small in 1882 describes his method for continuously casting wire solder. In 1924 Taylor (2) described a different method of casting wire but the solidification of molten metal jets which was the technique practiced by Small was continued by Tamman and Moritz (3) in 1935 and Ulitovskii (4) in 1937. In all cases The free flight melt spinning technique takes advantage of overcoming the tendency of the fluid jet to disintegrate into droplets at some distance from the orifice by causing solidification of the stable jet before such breakup can occur. Details of this technique are spelled out in two patents (5,6) issued in 1959. In these practices an effort is made to stabilize the jet so that a continuous wire or fiber is produced.

At times it is desirable to produce a discontinuous jet. Perhaps the most beautiful descriptions of such techniques are found in the book Soap Bubbles by C. V. Boyys (7) wherein he describes the water telephone. In this device the jet is caused to become controllably discontinuous by vibrating the orifice producing the jet thereby generating a travelling array of liquid jets of various lengths. A constant frequency of vibration of the orifice produces uniform diameter spheres which, in the case of metallic materials, solidify into a product referred to a spheroidal powder or small shot.

The current experiment was devised to ascertain the possibility of generating by the free flight melt spinning technique a continuous fiber having discontinuous phases which are mutually insoluble.

Experimental Procedure

The metals selected for this experiment were 99.99% pure Zinc and 99.99% pure lead. Following the established procedures noted above spheroidal powders of approximately constant diameter were produced. The jets of metal which were disintegrated were generated by pneumatically forcing the molten metal through a 0.6 mm diameter orifice in a pyrex tube at a velocity of 223 Cm per second. The orifice was mechanically vibrated by contacting it with a sounding board on which was mounted an electric relay vibrating at 40 cycles per second. The spheroids produced were slightly larger than the orifice used in production due to the fact that the chopped jet length was considerably longer than the diameter of the orifice. After the zinc spheroids were produced the tube was chemically cleaned and then the lead spheroids produced in the same assembly.

The spheroids were allowed to oxidize by holding them in open air for several days. Equal volumes of the oxidized lead and zinc spheroids were mechanically mixed and them charged into the same tube assembly used in their production.

The tube and the mixture of particles were carefully elevated to a temperature just above the melting point of zinc. Since, at this temperature, both of the components were molten the entire mass slumped in the tube. The oxide barriers were sufficient to prevent the liquid mixing of the components or the agglomeration of the individual components which would result in a gravity segregation of the melts.

Pneumatic pressure over this molten mass caused it to be ejected from the tube. As each individual sphere passed through the orifice ita oxide film was ruptured and the molten metal allowed to interface with the adjacent phase feeding into the orifice. The composite material of this continuous jet was moving at approximately 400 cm per second and was completely solidified after travelling 120 Cm from the orifice.

Results and Discussion

The fibers produced from this operation were,continuous with alternating phases along the length of the fiber as illustrated in Figure 1 (8). This "coral snake" appearance indicates fusion between the phases as one would expect from the partial solubility of the two liquid phases. Figure 2 (9) delineates the sharpness of the interface between the zinc rich and the lead rich volumes and Figure 3 (10) shows the lead rich volume is actually a mixture of primary crystallization product lead and the eutectic.

The presence of the eutectic indicates that there had to be a transport of zinc into the lead during the interval of time between the emanation of the fiber from the orifice and solidification.This time is approximately 0.3 seconds.

This fiber is solidified during free fall and the cross section of the fiber is sufficiently small to disallow appreciable convection. Such conditions should allow experiments to be run using mutually soluble liquid components in order to establish the extent of liquid metal diffusion under microgravity conditions. In order to check this approach oxide filmed spheres of tin and lead were prepared and subjected to free flight melt spinning. However the diameter of the spheres is sufficiently small to disallow any metallographically observable concentration gradient. The entire fiber structure in these cases was identified as primary crystallization product and eutectic

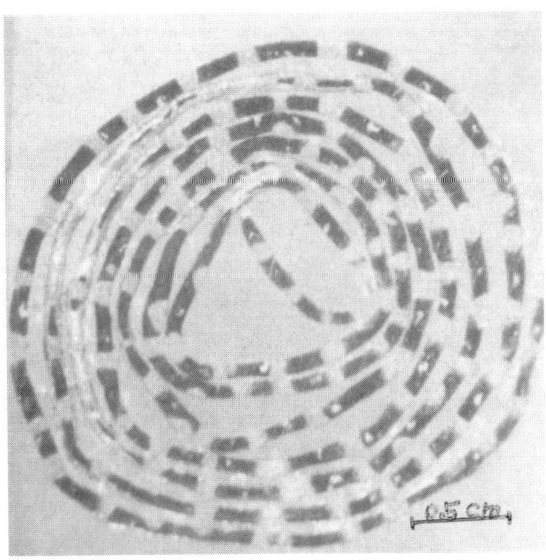

Figure 1 - Macrograph of the cross section of the lead (dark areas)-zinc (light areas) "coral snake" fiber.(8)

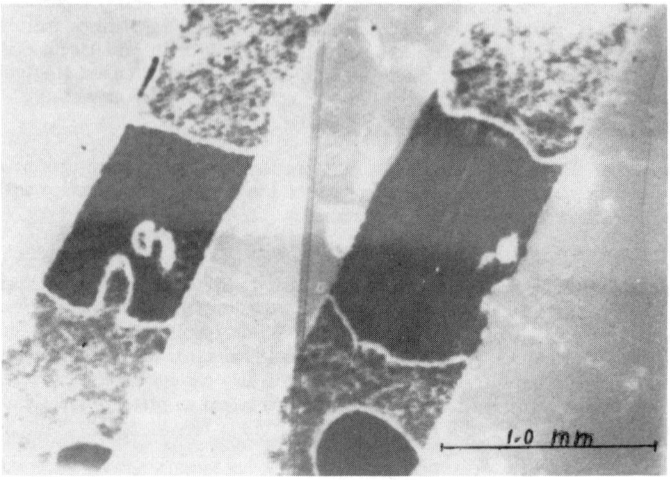

Figure 2 - Micrograph of interface os lead (mottled areas) and Zinc (dark areas).(9)

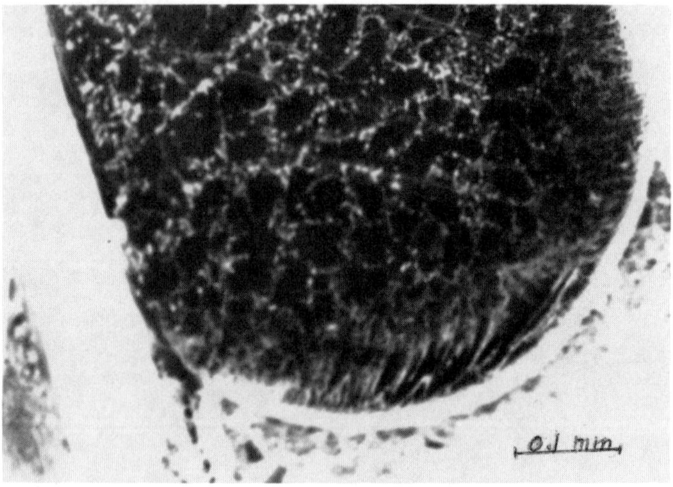

Figure 3 - Micrograph of lead rich volume showing primary lead in a eutectic newwork.(10)

References

1. Edward Small, inventor, U.S.Patent No. 262,625, "Apparatus for Making Wire Solder". filed May 21, 1882

2. G.F.Taylor, Phys Rev, 23, 655 (1924)

3. G.Tammann and G. Moritz, Z Metallk, 27, 114 (1935)

4. A.V.Ulitovskii, Stal', 7, No. 11, 111 (1927)

5. R.B.Pond, inventor, U.S. Patent No. 2,907,082, "Production of Continuous Filaments of High Vapor Pressure Metals" issued Oct. 6, 1959

6. R.B.Pond, inventor, U.S.Patent No. 2,879,566,"Method of Forming Round Metal Filaments", issued March 31, 1959

7. C.V. Boys, Soap Bubbles, Dover Inc., New York, N.Y. (1959)

MORPHOLOGY AND MECHANICAL PROPERTIES OF STAINLESS STEEL PRODUCED BY DIRECT STRIP CASTING METHOD

K. Amano, M. Tsuda, T. Tohge and K. Ebato

Nippon Yakin Kogyo Co. Ltd., Kawasaki, 210, JAPAN.

Abstract

An investigation is done on some characteristics of austenite stainless steel produced by means of a direct strip casting method. The influence of processing direct cast strips such as: 1) cold rolling—annealing, 2) annealing at high temperature—cold rolling—annealing, and 3) cold rolling—annealing—cold rolling—annealing, on the morphology and properties of products is described. Metallographical changes, recrystallization, texture and phase transformation are considered. Mechanical properties of processed strips including their tensile strength, work hardening, anisotropic properties, workability and also corrosion resistance such as pitting and stress corrosion cracking are presented.

Casting of Near Net Shape Products
Edited by Y. Sahai, J.E. Battles,
R.S. Carbonara and C.E. Mobley
The Metallurgical Society, 1988

MELT TREATMENT

Session Chairman: **R.D. Pehlke**
University of Michigan
Ann Arbor, Michigan

INTERFACIAL TENSION BETWEEN LIQUID STEEL AND MOLTEN CASTING POWDER

T. El Gammal* and P. Stracke**

*Department of Steelmaking
Institute for Ferrous Metallurgy
Aachen University of Technology
Aachen, West-Germany

**Daimler-Benz AG
Mannheim, West-Germany

Abstract

Casting powder plays an important rôle in ensuring smooth working of the continuous casting process, especially in the production of thin slabs. Interfacial phenomena between liquid steel and the molten layer of such powders influence strongly the surface quality of the slabs. In this paper the interfacial tension between several casting powders and liquid steel has been determined with the aid of the drop detachment method. The results show the effect of powder composition and steel grade on the interfacial tension for varying temperatures. The relationship between interfacial tension and the formation of oscillation marks is discussed. A meniscus of small radius is beneficial, underlining, however, the need for clean steel, particularly for casting thin strands.

Casting of Near Net Shape Products
Edited by Y. Sahai, J.E. Battles,
R.S. Carbonara and C.E. Mobley
The Metallurgical Society, 1988

Introduction

The continuous casting of thin slabs is gaining rapidly in importance. Diminishing strand thicknesses involve higher withdrawal speeds and lower interfacial contact between the metal pool and molten slag in the mold. Today strands with thicknesses as low as 50 mm and widths of 1200 mm are being produced with withdrawal speeds of up to 6 meters/min. Compared with a slab size of 1700x 200 mm, this means doubling the withdrawal speed and a reduction in contact area between the slag and the metal pool of approx. 80 %. This places extra demands on the casting powder, which, to ensure surface quality and a high degree of purity of the strand, must meet the following specifications;

a) prevent as far as possible the formation of oscillation marks,
b) absorb non-metallic inclusions and float them away from the contact area,
c) possess sufficient wettability properties,
d) lubricate the contact face mold/strand, and
e) provide an effective insulating cover.

The ability of the molten flux to penetrate between mold and strand, and consequently its lubricating power, is primarily dependent on its viscosity and the interfacial tension between molten flux and the steel. Lowering of the viscosity to maintain sufficient lubrication at high withdrawal speeds must not result in an unfavourable influence on the meniscus formation and the ability to absorb non-metallic inclusions.

Several mechanisms have been put forward to explain the formation of oscillation marks, involving extensive calculations on heat flow and solidification in the meniscus. These are referred to later in the text. However, it must not be overlooked that the interfacial phenomena play an important role in determining the shape of the meniscus, and consequently the extent of the meniscus gap, and the degree to which the molten flux penetrates this gap.

Due to experimental difficulties this important parameter has not yet been sufficiently investigated. The effect of changes in the basic composition of industrial casting powders and of alloying elements in the steel on the interfacial tension between molten powder and steel is the subject of this paper.

Experimental

Method

The method used to determine the interfacial tension between liquid steel and casting powder - the so-called drop detachment method - has been described fully elsewhere (1). In essence, it consists of extruding a drop of liquid steel out of a capillary tube into a melt of molten slag or casting powder, and recording the changes in weight of the molten slag during growth, detachment, and descent of the drop formed. This is shown schematically in fig. 1.

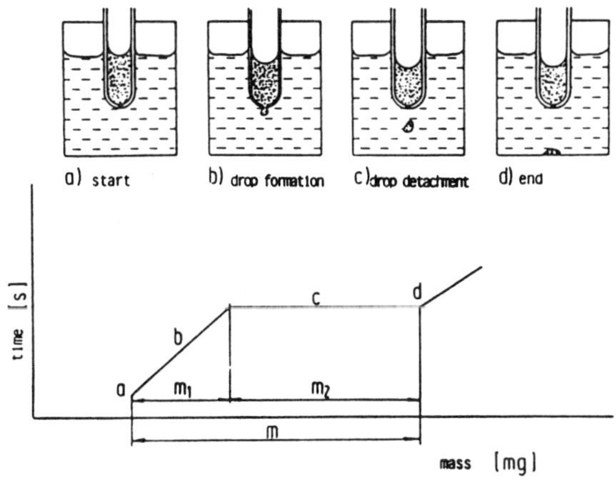

Figure 1 - Formation & detachment of drop in
in the measurement of interfacial tension (1).

The interfacial tension is then given by the relationship

$$\gamma = \frac{m_2 \cdot g}{2 \cdot \pi \cdot r \cdot \psi} \qquad (1)$$

where ($m_2 \cdot g$) is the apparent weight (due to buoyancy) of the drop after detachment.

g is acceleration due to gravity
r is the radius of the capillary
ψ is a correction factor

This method has proved very successful for determining interfacial tension at high temperatures.

Results

The compositions of the casting powders in this investigation are listed in table I. Powders 1 - 7 are commercial powders, whereas powder 8 is a synthetic mixture, based on the $CaO-Al_2O_3-SiO_2$ system. In the latter case the interfacial tension was measured against 3 different steel qualities. For the other powders steel quality CK 45 was used, the composition of which and of the other two qualities being listed in table II.

Table I. Composition of Casting Powders

No.	SiO_2 %	CaO %	MgO %	Al_2O_3 %	CaF_2 %	Na_2O %	K_2O %	Fe %	MnO %	ΣC %
1	28,90	25,28	0,05	9,08	8,66	3,73	0,14	0,82	6,83	2,10
2	43,70	31,42	0,05	7,29	9,03	3,43	0,13	0,66	0,05	0,29
3	37,60	28,80	0,05	7,62	10,45	4,93	0,17	0,45	0,02	0,30
4	39,80	35,35	0,05	6,19	7,29	5,33	0,14	0,41	0,02	0,37
5	34,50	28,97	4,05	9,55	3,39	5,90	1,18	1,26	0,05	0,29
6	38,40	26,55	1,54	16,30	4,10	3,19	1,07	0,28	0,09	0,10
7	39,10	34,19	0,05	7,13	9,07	3,86	0,22	0,82	0,02	0,46
8	45,00	45,00	-	10,00	-	-	-	-	-	-

Table II. Composition of Steels Used

	C %	Si %	Mn %	P %	S %	Cr %	Ni %	Mo %	Nb %
Ck 45	0,430	0,17	0,58	0,020	0,038	0,046	0,022	0,006	-
Continuous Cast Steel S1	0,110	0,38	1,60	0,017	0,001	0,044	0,045	0,013	0,035
X5 CrNi 18 9	0,049	0,33	1,52	0,021	0,003	17,400	8,700	0,220	<0,005

The results with powders 1-7 are shown plotted against temperature in fig. 2. Powders 3-4, which are poorer in MgO but richer in CaO + CaF_2, show the lowest interfacial tensions, whereas powders 5 & 6, which are richer in MgO + Al_2O_3, show the highest values. Powders 1,2 & 7 of intermediate composition have interfacial tensions that lie in between. All values decrease with increasing temperature, with approx. similar temperature slopes. An increase of 10°C results in a lowering of the interfacial tension by approx. 0.33 x 10^2 mN/m.

The effect of controlled additions of Al_2O_3 and MgO to powder 7 is shown in figs. 3 & 4, where it can be seen that both additions cause a rise in interfacial tension. The variation of the steel composition against a constant powder mixture (fig.5) shows that there is a drop in interfacial tension with decreasing carbon content, but with increasing Si and Mn content, from 13 x 10^2 to 11 x 10^2 mN/m at 1530°C, and that the temperature gradients for low carbon steel and Cr-Ni alloy steel are constant, being similar to those obtained when varying the powder mixture against a constant steel composition.

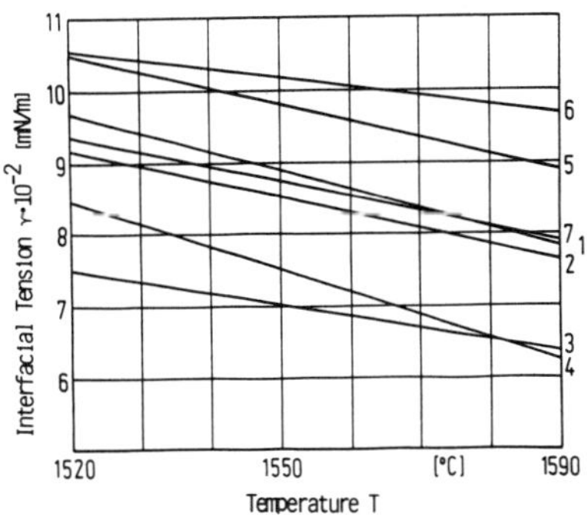

Figure 2 - Comparison of interfacial tension casting powder/steel as a function of temp.

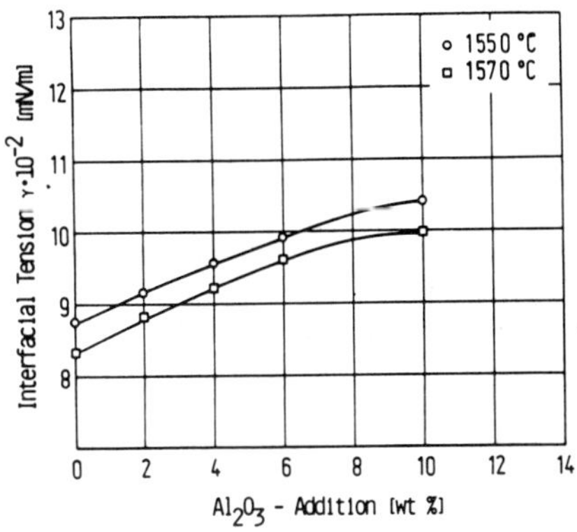

Figure 3 - Influence of Al_2O_3 addition to casting powder 7 on interfacial tension.

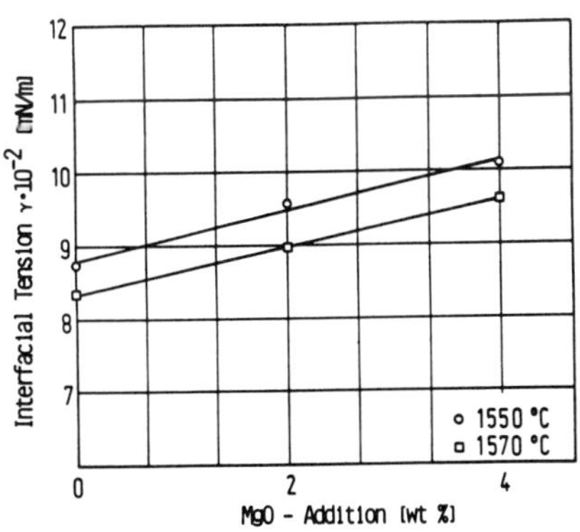

Figure 4 - Influence of MgO-addition to casting powder 7 on interfacial tension.

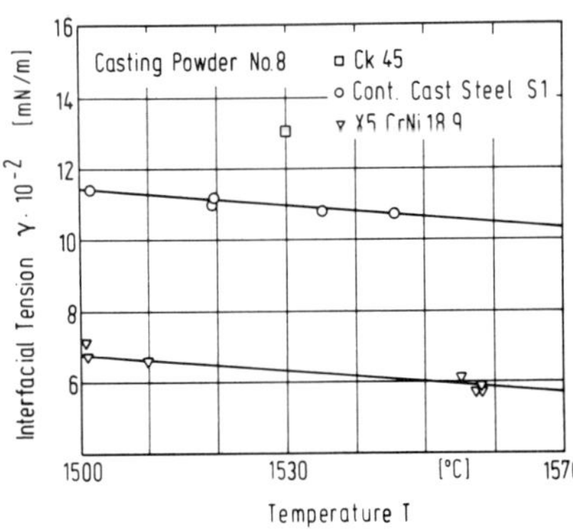

Figure 5 - Effect of varying steel composition on interfacial tension.

The interfacial tension values obtained for the various steel qualities have been applied to calculate the meniscus radius as a function of temperature.
To determine the meniscus radius the following derivation from Sato (2) has been used:

$$r = 1.699 \sqrt{\gamma_{lm/ls} / (\rho_{lm} - \rho_{ls}) \cdot g} \qquad (2)$$

where $\gamma_{lm/ls}$ = interfacial tension liquid metal/liquid slag
ρ_{lm} = density of liquid metal
ρ_{ls} = density of liquid slag
g = acceleration due to gravity

The largest radius is obtained with CK 45 steel at a value of r = 1 cm, followed by the low carbon steel with r = 0.89 cm and finally by 18/9 Cr-Ni steel with r = 0.67 cm (fig.6). The change in meniscus radius with varying powder composition is already reflected in figs. 2,3,& 4 due to the direct proportionality between interfacial tension and meniscus radius (table III). These results emphasize the rôle of the steel composition in the formation of the meniscus.

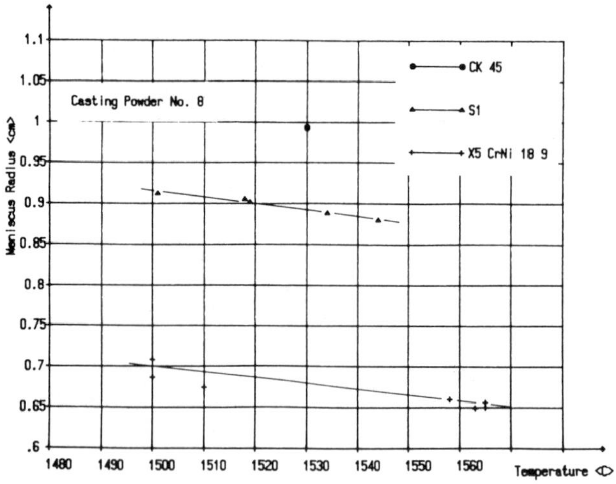

Figure 6 - Calculated meniscus radius for different steels as function of temp.

Table III. Values of the Meniscus Radius Calculated from the Corresponding Interfacial Tension Values for Different Casting Powders with Ck 45 Steel at 1550°C

Casting Powder No.	Interfacial Tension γ (mN/m)	Mensicus Radius r (mm)
1	888	7,697
2	848	7,518
3	708	6,862
4	752	7,084
5	980	8,100
6	1016	8,257
7	872	7,628

The Significance of Interfacial Phenomena in the Formation of Oscillation Marks

Previous studies on the formation of oscillation marks in continuously cast slabs have indicated three possible mechanisms:

1. According to R. Sato (2) the meniscus follows the movement of the mold oscillation, moving upwards into the slag bath during positive strip and downwards into the metal pool during negative strip (fig.7). The secondary meniscus formed below the surface of the steel is compressed during downward movement, resulting in an oscillation mark.

2. According to E. Takeuchi & J.K. Brimacombe (3) the formation of oscillation marks, both with and without sub-surface hooks, may be related to the flexibility of the solidified meniscus. A rigid meniscus skin resists being bent back fully towards the mold wall during the positive strip, allowing the liquid steel to overflow it and form a sub-surface hook (fig.8). A flexible meniscus is pulled back towards the mold during positive strip, preventing liquid steel overflow (fig.9). The oscillation marks are relatively shallow without sub-surface hooks. A similar theory has been put forward by T. Emi et al (4).

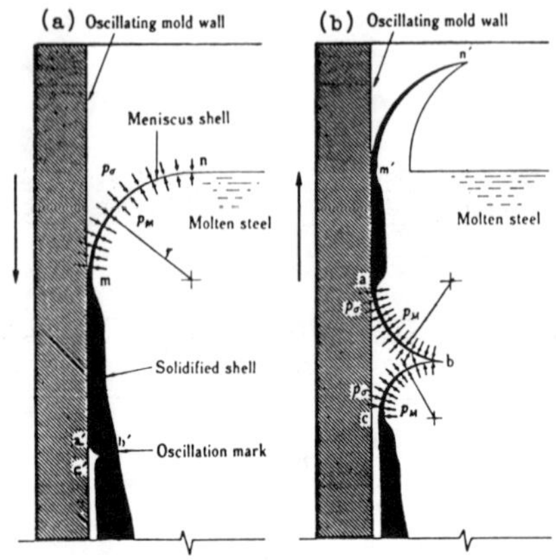

Figure 7 - Formation of meniscus shells acc. to Sato (2).

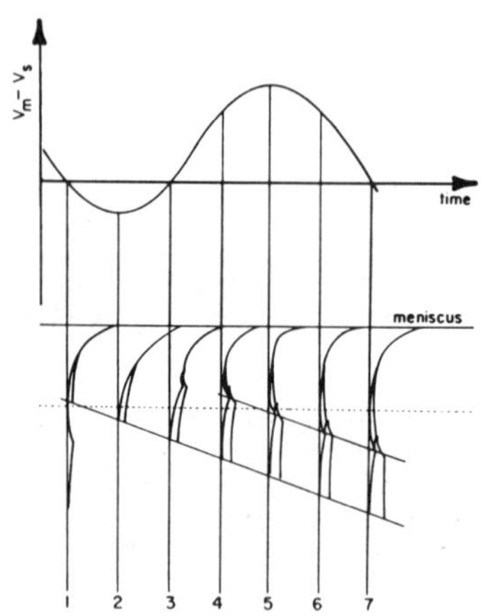

Figure 8 - Formation of oscillation marks with hooks acc.to Takeuchi & Brimacombe (3).

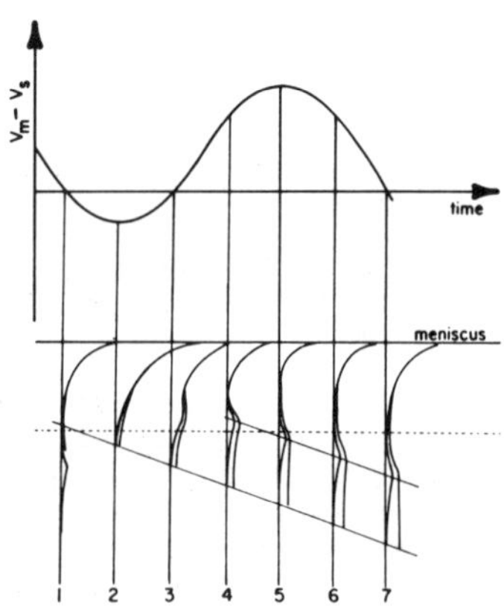

Figure 9 - Formation of oscillation marks without hooks acc. to Takeuchi & Brimacombe (3).

3. According to P. Riboud & M. Larrecq (5) the upper part of the meniscus may be remelted by liquid steel overflow on re-entering the metal pool during negative strip (fig.10). This would give more or less shallow oscillation marks.

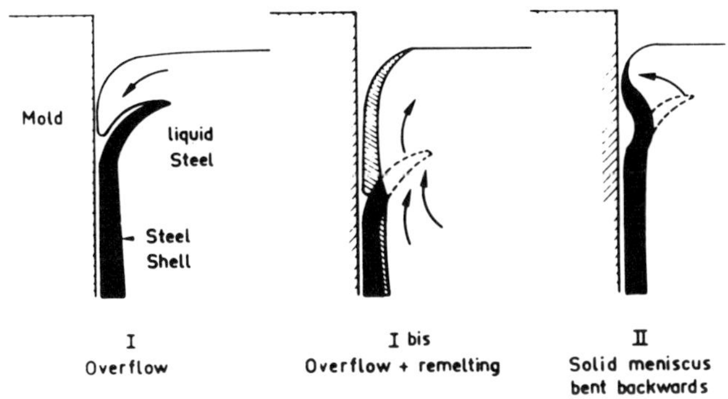

Figure 10 - Overflowing of the solid meniscus acc. to Riboud & Larrecq (5).

In all these mechanisms the interfacial phenomena between liquid steel and molten casting powder are referred to, but not investigated thoroughly, presumably due to the lack of relevant data. Clearly, however, the interfacial tension between liquid metal and molten casting powder is linked directly with the radius of the meniscus, which increases with increasing interfacial tension. Partial remelting of the meniscus (as in mechanism 3) will be shorter in the case of a meniscus of small radius, and lead to shallower oscillation marks.

As already indicated, not only the radius of the meniscus determines the shaping of oscillation marks, but also the interplay of wetting properties of casting flux and liquid steel on the solidified meniscus. P. Ackermann et al (6) have shown, both mathematically and experimentally, that, in the case of static mold casting of pure metals, the wetting angle at the 3-phase junction solid-gas, liquid-gas and liquid-solid is a major factor in the formation of surface grooves on cast slabs. Adapting this model to the casting of steel in the presence of a casting powder, the three phases become solidified metal (sm)/liquid metal (lm)/liquid slag (ls), which form a triple junction, when the solidified meniscus is drawn into the metal during negative strip. The forces acting at this point are represented schematically in fig 11.

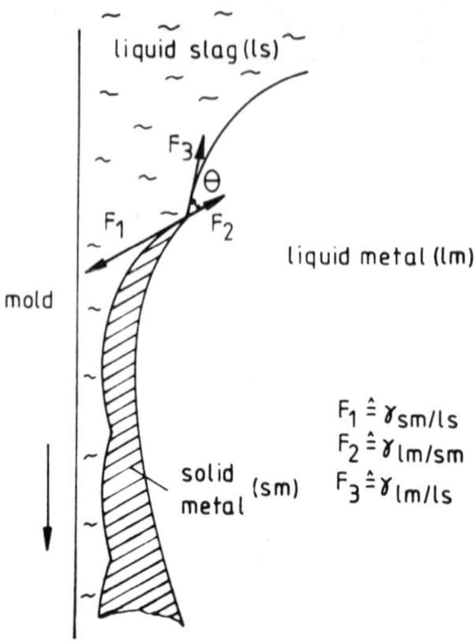

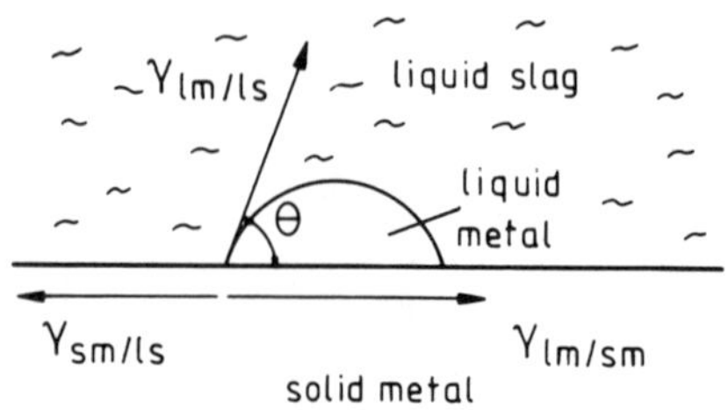

Figure 12 - Interfacial tensions acting on a sessile drop of liquid steel on solid steel in contact with liquid slag.

Figure 11 - Forces acting at the triple junction solid metal/ liquid metal/liqid slag, adapted from Ackermann et al (6).

Neglecting the ferrostatic pressure of the steel melt on the solidified shell, F_2 and F_3 may be taken as the forces due to the ferrostatic pressures and interfacial tensions of the liquid metal and slag respectively, which oppose one another. By analogy to a hypothetical sessile drop of liquid steel on a surface of a solid steel plate covered by molten slag (fig.12), the interfacial tension of solid metal/liquid slag $\gamma_{SM/LS}$ will favour the spreading of the liquid metal. This is opposed by the liquid metal/solid metal interfacial tension $\gamma_{LM/SM}$, and the force vector due to interfacial tension of liquid metal/liquid slag acts in a direction tangential to the liquid metal drop at the triple point of contact, the appropriate vector being $\gamma_{lm/ls} \cdot \cos \Theta$.

The interfacial tension between "supercooled" solid and liquid steel is not known, but it is expected to be relatively low and constant, since both are of the same composition. Furthermore, the interfacial tension between solid steel and molten casting powder has as yet to be investigated, but as the surface tension of solid steel is higher than that of liquid steel, ($\sigma_{Fe-solid}$ 2070 mN/m at 1400°C; $\sigma_{Fe-liquid}$ 1700 mN/m) (7), it may be assumed that the interfacial tension solid steel/molten slag will be correspondingly higher. The system at the triple point will be unstable as:

$$\gamma_{sm/ls} + \gamma_{lm/sm} > \gamma_{lm/sl} \qquad (3)$$

The liquid metal then creeps along the meniscus surface, displacing the slag until it reaches the mold, where it solidifies. The first meniscus (fig.13) is thus overlapped to give a secondary meniscus (fig.14). The curvature of the secondary meniscus and depth of groove are both dependent on the wetting angle Θ. The viscosity of the molten casting powder plays also a rôle here, since a viscous slag will tend to resist displacement by the liquid metal, causing the creeping liquid metal to freeze before it can reach its equilibrium position. The groove formed will then be deep.

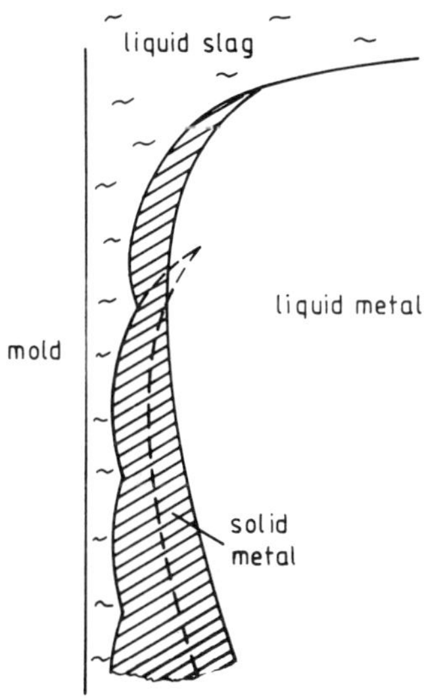

Figure 13 - Position of first meniscus on upward stroke.

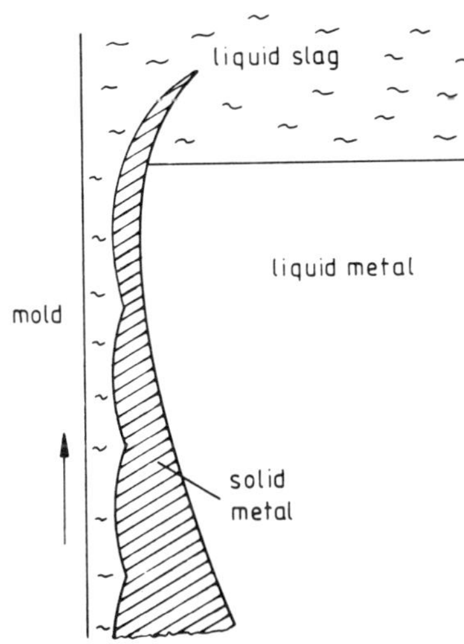

Figure 14 - Overlapping of first meniscus to form secondary meniscus.

Conclusions

In the continuous casting of thin slabs the casting speed is increased from 2 to 6 m/min, the mold oscillation frequency up to 400 cycles/min, and the stroke length reduced to 2-4 mm. To overcome the friction caused by this increased speed, additions of Na_2O, Li_2O, or CaF_2 have to be made to the casting powders to lower their viscosities. Furthermore, the properties of the casting powder have to be adjusted to suppress harmful oscillation marks, involving a lowering of the interfacial tension between the metal and the molten powder to give a meniscus of small radius, and, in addition, sufficient creep of the liquid metal over this meniscus to give a small contact angle Θ. The results show that such a match of properties would be achieved with a combination of 18/9 Cr-Ni steel and casting powder no. 8 (fig. 5).

However, the considerably smaller contact surface between casting powder and liquid steel in the casting of thin slabs, coupled with the lower rates of addition of powder, will mean increased difficulty in the removal of oxide (Al_2O_3) build-up at the interface, as compared with conventional strands. This can place undue demands on a casting powder adjusted to give low interfacial tension between metal and molten powder, since the passage of non-metallic solids through the metal/slag interface is favoured by a negative free energy change at the interface, and in turn, by a higher interfacial tension metal/slag. As indicated in fig. 3 the absorption of Al_2O_3 in the slag in itself leads to higher interfacial tensions, underlining the compromise that must be met in assessing the properties of a casting powder. It also emphasizes the need to operate with clean steel at the outset and the importance of efficient secondary refining.

References

1. T. El Gammal, R.-D. Müllenberg, "The Simultaneous Dermination of Density and Interfacial Tension of Liquid Slags and Metals". Archiv für das Eisenhüttenwesen 51 (1980) No. 6 221-226.

2. R. Sato, "Powder Fluxes for Ingot Making and Continuous Casting", Proc. 62nd Nat.Open Hearth & Basic Oxygen Steel Conf., (1979) 48-67.

3. E. Takeuchi, J.K. Brimacombe, "The Formation of Oscillation Marks in the Continuous Casting of Steel Slabs", Met.Transactions, ASM 15 B (1984) Sept. 493-509.

4. T. Emi et al, "Influence of Physical & Chemical Properties of Mold Powders on the Solidification and Occurrence of Surface Defects of Strand Cast Slabs" Proc. 61st Nat.Open Hearth & Basic Oxygen Steel Conf., (1978) 350-361.

5. P.V. Riboud, M. Larrecq, "Lubrication and Heat Transfer in a Continuous Casting Mold", Proc. 62nd Nat.Open Hearth & Basic Oxygen Steel Conf. (1979) 78-92.

6. P. Ackermann, W. Heinemann, W. Kurz, "Surface Quality and Meniscus Solidification in Pure Chill Cast Metals" Archiv für das Eisenhüttenwesen 55 (1984) No. 1 1-8.

7. L.I. Osipow, Surface Chemistry (New York, NY Reinhold Publishing Corp., 1962) 283.

VISCOSITIES OF BINARY SILICATE MELTS

Huijun Hu and R.G. Reddy

Department of Chemical & Metallurgical Eng.
University of Nevada-Reno
Reno, Nevada

ABSTRACT

The study examines the viscosity data for binary silicate melts in the composition range at 33 to 100 mole pct SiO_2 and for the temperature 1623 to 1723 K. The NO^{o}, NO^{-} and NO^{2-} ions distribution in the melts has been evaluated based on the depolymerization reaction $O^{o} + O^{2-} = 2O^{-}$. The NO^{o} values obtained by the atomic pair model agrees very well with the experimental data. An equation was derived using the hole theory and the atomic pair model to express the temperature and NO^{o} dependence of viscosity. The deduced equation

$$\eta = 3.067X10^{-9}\ T^{1/2}\ NO^{o}\ \exp(E/RT) \quad \text{(poise)}$$

is successfully applied to predict the viscosity of Li_2O-SiO_2 and Na_2O-SiO_2 binary flux melts. Excellent agreement between calculated viscosity values and experimental data for several temperatures was observed.

Casting of Near Net Shape Products
Edited by Y. Sahai, J.E. Battles,
R.S. Carbonara and C.E. Mobley
The Metallurgical Society, 1988

1. INTRODUCTION

For more than three decades, mould fluxes have been used for uphill teeming practice and continuous casting of steels. Flux usually refers to a silicate melt having a low fusion temperature caused by mainly the addition of alkali, alkaline-earth metal oxides and/or fluorides. The molten flux infiltrates into the gap between the mould and the metal shell. The infiltration of molten flux and its function as a lubricant are dependent mainly on its viscosity, fusion temperature and mode of solidification(1). If the flux is too viscous a discontinuous film may result. If the flux is too fluid, non-uniform liquid layer or droplets may form. Either of these conditions may lead to poor surface quality of casting product.

Several equations: the Andrade-Arrhenius(2), Absolute reaction rate theory and Brostow equation(3) have been developed to better express the relationship between viscosity and temperature. These equations have had limited success in predicting the viscosity of liquids, particularly silicate systems(4-7). The addition of metal oxides to silicate melt causes depolymerization and which results in a breakdown of the silicate network structure. The depolymerization reaction can be written as:

$$- Si - O - Si - \ + \ M_2O \ = - Si - O^- + 2M^+ + {}^-O - Si - \qquad (1)$$

expressing in terms of a simple electrochemical reaction:

$$O^o \ + \ O^{2-} \ = \ 2O^- \qquad (2)$$

where O^o is a bridging oxygen bonded to two silicon atoms; O^- is an oxygen bonded to only one silicon atom and O^{2-} is free oxygen. The extent to which the depolymerization reaction occurs depends primarily on the metal oxides concentrations. In general, the more the metal oxides concentration in the silicate melt, the greater the extent of depolymerization(8-11). This results in decrease in viscosity. Despite the fact that viscosity strongly depends on the composition of the melt, very little consideration has been given on incorporation of structure of melts in the previously developed viscosity temperature equations(12).

In this paper, we examine the calculation of viscosity values in a more direct manner in which we use the modified hole theory and the atomic pair model for the incorporation of rate process and structural effects, respectively, in alkali silicate melts. The predicted viscosity values thus, appear to be in an excellent agreement with the experimental data.

2. THEORETICAL CONSIDERATIONS

Viscous flow in liquids may be regarded as a rate process. The kinetic theory expression for the viscosity of a fluid can be written as(12):

$$\eta = 2\ NMVL \tag{3}$$

where M is the mass of a particle, N is the number of particles in per unit volume, V is the velocity and L is the mean free length.

In the hole theory model, randomly located and variable-sized vacancies in liquid are called holes. Thermally generated fluctuations in liquids are believed to cause holes to form new ones and old ones to be destroyed. At the same time, it is assumed that the holes are moved by the mechanism of having an ion jump into one hole and thus create a new hole in the place it has just vacated. The expression for the viscosity of the ionic liquid containing holes can be expressed as the same form as above(12):

$$\eta = 2nmv\ell \tag{4}$$

where n is the number of holes in per unit volume, m is the mass of a hole, v is the velocity and ℓ the mean free lenth.

The velocity v is given by:

$$v = \ell / t \tag{5}$$

where t is the mean lifetime between collision.

For one dimensional motion of a hole, the equipartition of energy can be expressed as:

$$1/2\ KT = 1/2mv^2 \tag{6}$$

where K is Boltzmann constant (1.38054×10^{-23} Joules/degree) and T is temperature (K).

Combining Eq.(4),(5) and (6), we obtain

$$\eta = 2\ nKTt \qquad (7)$$

where t is the mean lifetime of a hole, i.e. the average time between creation and destruction of a hole due to the thermal fluctuations, and given by:

$$t = (r/3)X(6.28m/KT)^{1/2}\ \exp(E/RT) \qquad (8)$$

where r is the average radius of the hole, m is the mass of a ionic unit and E is the activation energy of ionic unit to overcome the barrier to move to an adjacent hole.

Combining Eq.(7) and (8),we obtain equation (9):

$$\eta = (2/3)\ n\ r\ (6.28mKT)^{1/2}\ \exp(E/RT) \qquad (9)$$

Equation 9 is only a general expression and application of this to silicate melts requires the methods of evaluating n, r and E,which are described below:

CALCULATION OF E:

In the ionic liquids (i.e.liquid silicate flux), Bockris and Reddy(12) concluded that, holes are easily formed because the rate of hole formation is controlled by the thermal vibrations of the Si units relative to each other, but it is difficult to generate the jumping Si units by breaking Si-O-Si bonds holding together in a silicate network. Therefore, it was assumed that in the liquid silicates, the rate controlling process is the Si-O-Si bond rupture step. Energy of activation(E) for viscous flow depends upon the number of Si-O-Si bonds to be broken. Presence of more bonds in the liquid silicate means that more energy is required to break them. Use of atomic pair model(11) could enable us to calculate the total number of Si-O-Si bonds in the liquid silicates (i.e.the change of NO^{o} with composition of the liquid). A brief description of NO^{o} calculation method is given below:

When N moles of SiO_2 and (1-N) moles of basic oxide, MO, react to form a mole of binary silicate melt, charge and mass balance considerations, respectively,require that

$$nO^{o} = 2N - 1/2\ nO^{-} \qquad (10)$$

$$nO^{2-} = (1-N) - 1/2\ nO^{-} \qquad (11)$$

where nO^{o},nO^{-} and nO^{2-} are,respectively,the number of moles of O^{o},O^{-} and O^{2-} present per mole of binary silicate melt.

The partition function of a solution can be expressed as(11):

$$\Omega = \Omega^{o} \exp(NO^{-} \cdot \Delta G^{o}/RT) \qquad (12)$$

where Ω^{o} is partition function of pure silica and ΔG^{o} is the standard free energy change for depolymerization reaction.

At a constant composition, the equilibrium value of NO^{-} can be obtained by differentiating the partition function and setting
$\partial Ln\ \Omega / \partial NO^{-} = 0$ and solving;the following expression can be obtained:

$$(1-\exp(\Delta G^{o}/RT))(NO^{-})^{2} + (2X_{MO}-4)(NO^{-}) + 8X_{MO}(1-X_{MO}) = 0 \qquad (13)$$

Consideriding one mole of binary silicate melt,the following expression for NO^{o} can be deduced:

$$NO^{o} = (4-4X_{MO}-NO^{-})/2(2-X_{MO}) \qquad (14)$$

The calculated ionic distribution in $PbO-SiO_2$ binary melt at 1673K are shown in Fig.1. As seen from the Fig., for the range of $X_{SiO2} < 0.33$, a complete depolymerization is expected; i.e. NO^{o} tends to zero and NO^{2-} decreases with increasing SiO_2 content in the melt. For the composition range $X_{SiO2} > 0.33$, polymerization occurs,i.e.NO^{2-} tends to be zero and NO^{o} increases with increasing SiO_2 content of the melt. The calculated values of NO^{o} in $PbO-SiO_2$,Na_2O-SiO_2 and $CaO-SiO_2$ binary melts using

YOKOKAWA(11),TOOP(8) and MASSON(10) models are shown in table 1. Experimental data of NO^o at 1673K reported by Yasunari(13) are also included in table 1. For $PbO-SiO_2$ binary system,NO^o values calculated using YOKOKAWA(11) model and experimental data are plotted in Fig.2. It can be seen that an excellent agreement between the calculated values and experimental data is observed.

The experimental data(14) showed that the activation energy E can be divided into three regions. Activation energy decreases rapidly as metal oxide is added to SiO_2 until about 20 mole pct, then decreases slowly up to 60 mole pct addition;and a further increase in addition of metal oxide results in a small reduction in E. Since the E decreases with decreasing the number of NO^o in the melt, the rate of change of E with respect to NO^o is assumed to follow the fifth power to E (although other relations are tried, the fifth power gave the better correlation to the experimental data).

$$dE \,/\, dNO^o \;=\; A\,E^5 \qquad (15)$$

where A is constant. Equation (15) was integrated for the known conditions of E and NO^o. For pure liquid silicate, $NO^o = 1$ and $E^* = 519KJ$ (15), for any other composition in binary melt, $NO^o = NO^o$ and $E = E$. The expression obtained is,

$$E = E^* \,/\, [4AE^{*4}\,(1 - NO^o) + 1\,]^{1/4} \qquad (16)$$

In the above equation,A was solved assuming $E = B\,E^*$,where the value of B is 0.49 to 0.51 for Li_2O-SiO_2 and Na_2O-SiO_2 binary melt.

CALCULATION OF r and n:

Furth (16) has showed that a typical hole in a liquid is roughly the same size as an ion and can accommodate an ion. For SiO_4 tetrahedra,each silicon atom is surrounded by four oxygen atoms. The ionic radius of oxygen is 1.36A and the radius of silicon is 0.39A. Thus, the radius of SiO_4 can be calculated as

2.3Å. On the other hand, if molecular size of SiO_4 is considered, the radius of SiO_4 unit is about 2.1Å. Because the hole must hold the SiO_4 ions, the minimum radius of the hole is in the range of 2.1 to 2.3Å, and $2.1X10^{-10}$ m was used in the present calculations. For higher concentration of NO^o, existance of larger polymeric Si units (i.e. $Si_nO_{3n+1}^{2n+2}$) are expected. For the correct r value, the size of polymeric Si unit need to be used,and expected to be larger than the SiO_4 unit used here. As the data become available, correction of r value will lead to slightly higher calculated values of viscosity.

Number of holes (n) per unit volume can be related from the known values of NO^o. For one mole of binary silicate melt, n can be expressed as:

$$n = NO^o \cdot Av \qquad (17)$$

where Av is the Avogadro's number ($6.023X10^{23}$/gram mole).

CALCULATION OF $(6.28\ m\ K\ T)^{1/2}$ TERM:

The term can be rearranged as:

$$\begin{aligned}(6.28\ m\ K\ T)^{1/2} &= (6.28\ m/\ K\ T)^{1/2}\ K\ T \\ &= (6.28\ m\ Av/\ Av\ K\ T)^{1/2}\ K\ T \\ &= (6.28\ W/\ R\ T)^{1/2}\ K\ T \qquad (18)\end{aligned}$$

where R is gas constant (8.3144 Joules/degree·mole) and W is mole weight of SiO_4 unit (0.092 kg/mole).For correct polymeric Si unit, W is expected to be larger. Substituting R,W and K into Eq.(18), we can obtain

$$(6.28\ m\ K\ T)^{1/2} = 0.3637X10^{-23}\ T^{1/2} \qquad (19)$$

Substituting equations (16),(17) and (19) in Eq.(9),the following viscosity expression for binary melt is obtained:

$$\eta = 3.067X10^{-9}\ T^{1/2}\ NO^o \exp(E/RT) \quad \text{(poise)} \qquad (20)$$

3.RESULTS AND DISCUSSION

The calculated viscosity values in Na_2O-SiO_2 and Li_2O-SiO_2 binary silicate melts in the temperature range 1623 to 1723K and for pure liquid silica at 2223K are presented in this section.

A. Pure liquid silica:

The viscosity of pure liquid silica was calculated using equation (20). The calculated viscosity value is $2.27X10^5$(poise), which is in good agreement with the experimental data of $1.72X10^5$(17).

B. Alkali metal oxide-silica binary systems:

The depolymerization reaction can be written as:

$$M_2O \quad + \quad SiO_2 \quad = \quad M_2SiO_3 \qquad (21)$$

Free energy data for the reaction (21) for Li_2O-SiO_2 and Na_2O-SiO_2 silicates are shown in table 1. Using equation (13) and (14) the values of NO^o was calculated. From the known values of NO^o,the activation energy for the melts using equation (16) was obtained. The viscosity values calculated using equation (20), along with the experimental data are shown in Figs.3 to 5, and also presented in table 3 for temperatures 1623,1673 and 1723K. An excellent agreement between calculated and experimental viscosity data is observed. Fig.6 shows the comparision of calculated values and experimental viscosity data as a function of temperature at a constant NO^o. As expected, at a constant NO^o values, the viscosity of silicate melts decreases with increasing temperature. However,it should be kept in mind that when data become available, correction of the calculation of r and W will lead to slightly higher values of viscosity.

4. CONCLUSION

The method of calculation of viscosity presented in this paper appears to lead to reasonably accurate values in the composition range of $0.33 < X_{SiO2} < 1$,for binary alkali metal oxide-silica binary melts.The NO^{o},N^{O-} and NO^{2-} ions distribution in the $PbO-SiO_2$ melts has been evaluted using the polymerization theory. Calculated NO^{o} values using atomic pair model for $PbO-SiO_2$,Na_2O-SiO_2 and $CaO-SiO_2$ binary systems agrees very well with the experimental data. Calculated viscosity values for Li_2O-SiO_2 and Na_2O-SiO_2 binary melts at 1623,1673 and 1723K are in an excellent agreement with the experimental data. Further work on calculation of viscosity values in several other binary and multicomponent casting flux systems are in progress.

REFERENCES

1. E.T.Turkdogam: 'Physicochemical properties of molten slags and glasses',The Metals Society,London,1983.
2. S.Glasstone, K.J.Laidler and H.Eyring: The theory of rate processes,Mcgraw-Hill,New york,1941.
3. W.Brostow and P.Maynadier: 'High temperature science', Vol.11,1979, pp.7-21
4. R.Altan, G.Stavropoulos, K.Parameswaran and R.P.Goel: 'Physical chemistry of extrative metallurgy',TMS-AIME, New York,1985,pp.97.
5. W.L.McCauley and D.Apelian: Second international symposium on metallurgical slags and fluxes,H.A.Fine and D.R.Gaskell,eds.,TMS-AIME,Lake Tahoe,NV,1984,pp.925
6. M.W.Nichols,A.P.Lingras and D.Apellan: Second international symposium on metallurgical slags and fluxes,H.A.Fine and D.R.Gaskell,eds.,TMS-AIME,Lake Tahoe,NV,1984,pp235. pp.235
7. H.Eyring and M.S.Jhon: 'Significant liquid structures', John Wiley and Sons, New York, 1969
8. G.W.Toop and C.S.Samis,Metall.Trans, Vol.224,1962,pp.878.
9. P.L.Lin and A.D.Pelton: Metall.Trans, V.10B,1979,pp.667.
10. C.R.Masson: Can.J.of Chem.,1970,pp.1456
11. T.Yokokawa and K.Niwa: Trans.Jap.Inst. Metal, V.10,1969,pp.3
12. J. O'M.Bockris,and A.K.N.Reddy, 'Modern electrochemistry',New York,Plenum,1970, pp.574.
13. Y.Kaneko: JIM,V.41,(1977),pp.375
14. G.Hetherington: Phys. Chem. Glasses, V.5,1964,pp.130
15. R.Bruckner: Glastech.Ber.,V.37,1964,pp.413
16. R.Fürth: Proc.Cambrige Phil.Soc.,1941,pp252-281
17. R.Rossin,J.Bersan and G.Urbain: Rev.Int.Hautes Temp.Refract.,V.1,1964,pp.159.
18. D.R.Stull and H.Prophet: JANAF Thermochemical tables, 2nd ed.,NSRDS-NBS37,Supt.of Docments,Washington,DC,1971 and Supplements.

TABLE 1. The calculated and measured NO^{o} in binary system at 1623K.

X_{SiO2}	NO^{o}(Expt.)	NO^{o}(11)	NO^{o}(10)	NO^{o}(8)
		$PbO-SiO_2$ system		
0.3	0.16	0.12	0.26	0.33
0.33	0.19	0.15	0.31	0.34
0.4	0.23	0.23	0.44	0.4
0.5	0.35	0.37	0.65	
0.6	0.4	0.52	0.88	
0.7	0.63	0.65	1.14	
		Na_2O-SiO_2 system		
0.67	0.62	0.62	1.03	
		$CaO-SiO_2$ system		
0.6	0.48	0.5	0.8	
0.5	0.35	0.34	0.5	

TABLE 2. The free energy for binary systems (18).

Reaction	Free energy cal/mol
$Li_2O + SiO_2 = Li_2SiO_3$	-34400 + 0.9T
$Na_2O + SiO_2 = Na_2SiO_3$	-56800 - 0.92T

TABLE 3. Calculated and measured viscosity of binary systems.

Temperature(K)	XSiO2	NO^{o}	Viscosity (poise) Cal.	Expt.
		Na_2O-SiO_2 system		
1623	0.67	0.6	40.68	40.7
	0.65	0.58	31.36	32.4
		Li_2O-SiO_2 system		
	0.6	0.5	12.2	12.3
	0.55	0.42	5.3	3.98
	0.5	0.34	2.4	1.78
	0.45	0.25	1.	0.63
1673		Na_2O-SiO_2 system		
	0.67	0.6	28.3	28.2
	0.65	0.58	21.9	22.9
		Li_2O-SiO_2 system		
	0.6	0.5	8.71	9
	0.55	0.42	3.84	3.09
	0.5	0.34	1.8	1.4
	0.45	0.25	0.7	0.53
1723		Na_2O-SiO_2 system		
	0.67	0.6	20.4	20.4
	0.65	0.58	15.9	16.6
		Li_2O-SiO_2 system		
	0.6	0.5	6.4	7.74
	0.55	0.42	2.8	2.51
	0.5	0.34	1.35	1.15
	0.45	0.25	0.59	0.45

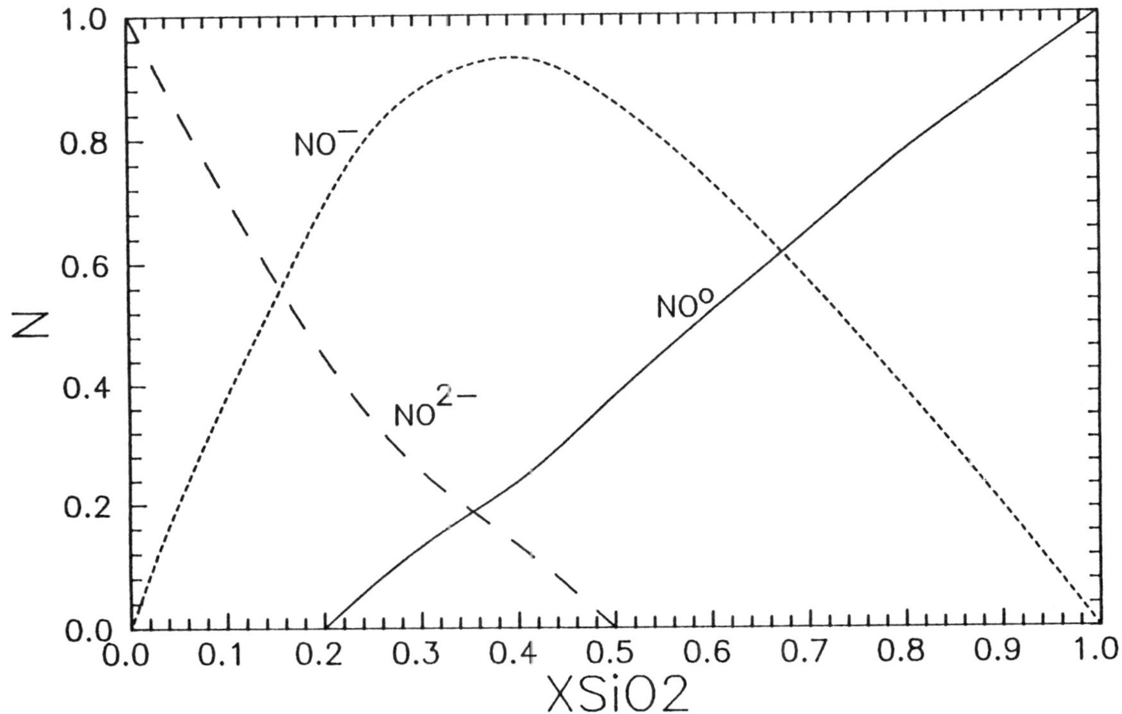

Fig. 1. The three type oxygen anions distribution in PbO–SiO2 system at 1673 K.

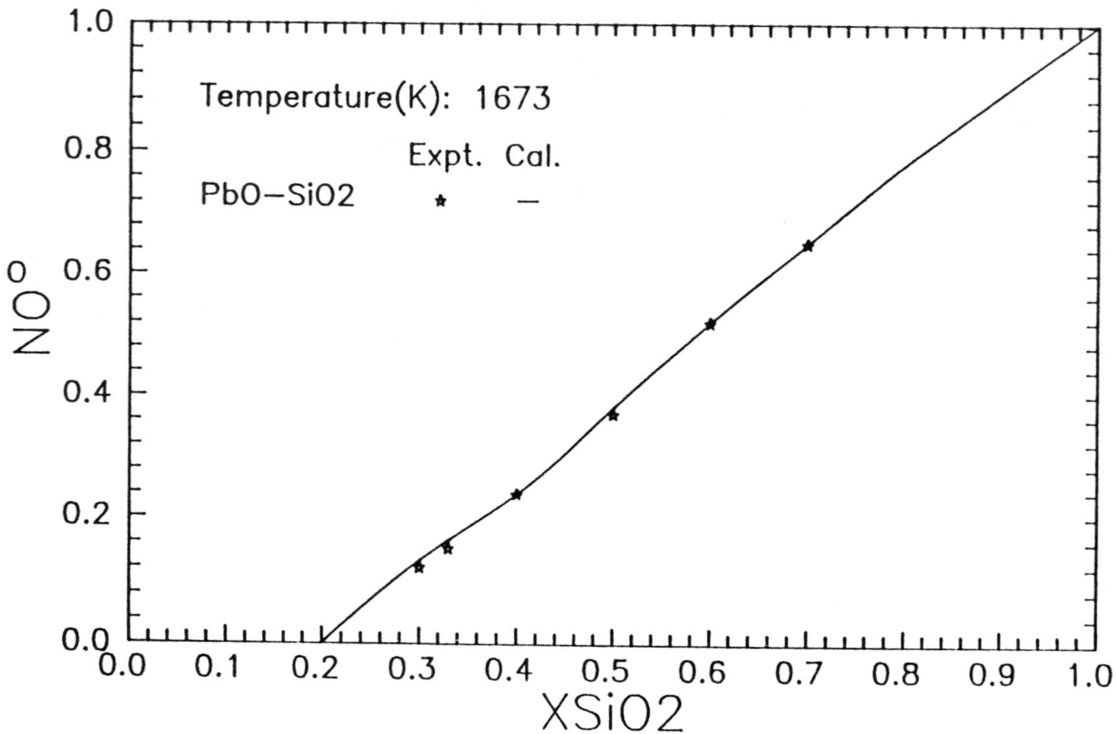

Fig. 2. Plot of calculated and measured NO^o in Binary system.

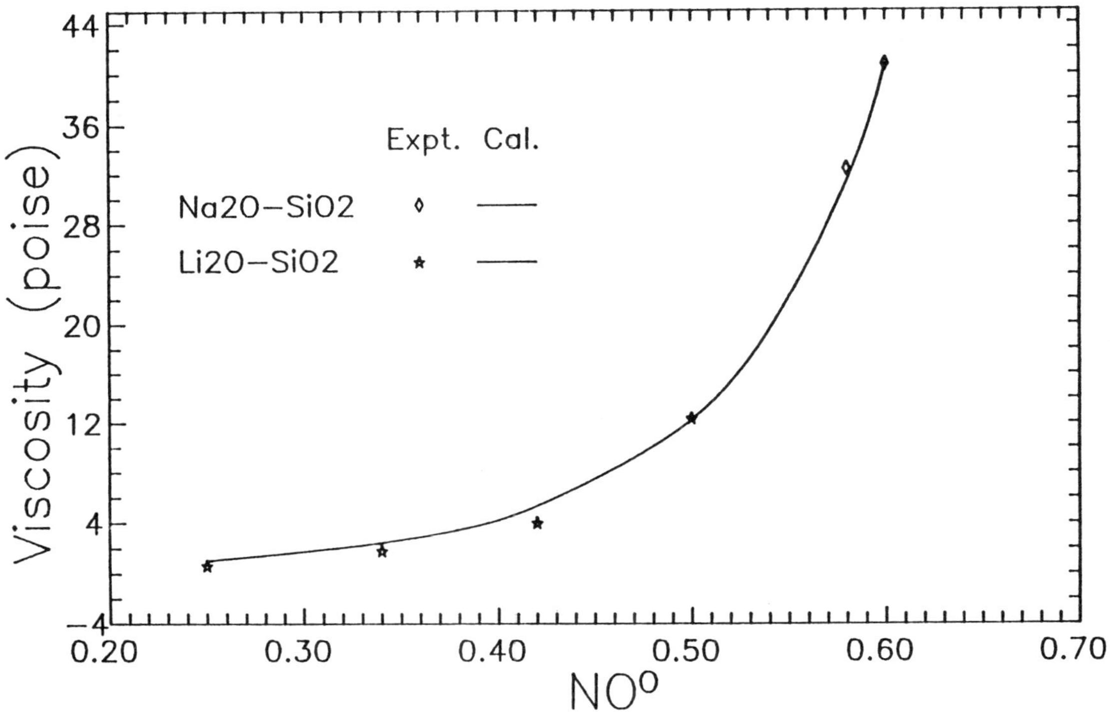

Fig. 3. Plot of calculated and measured viscosity of binary systems at 1623 K.

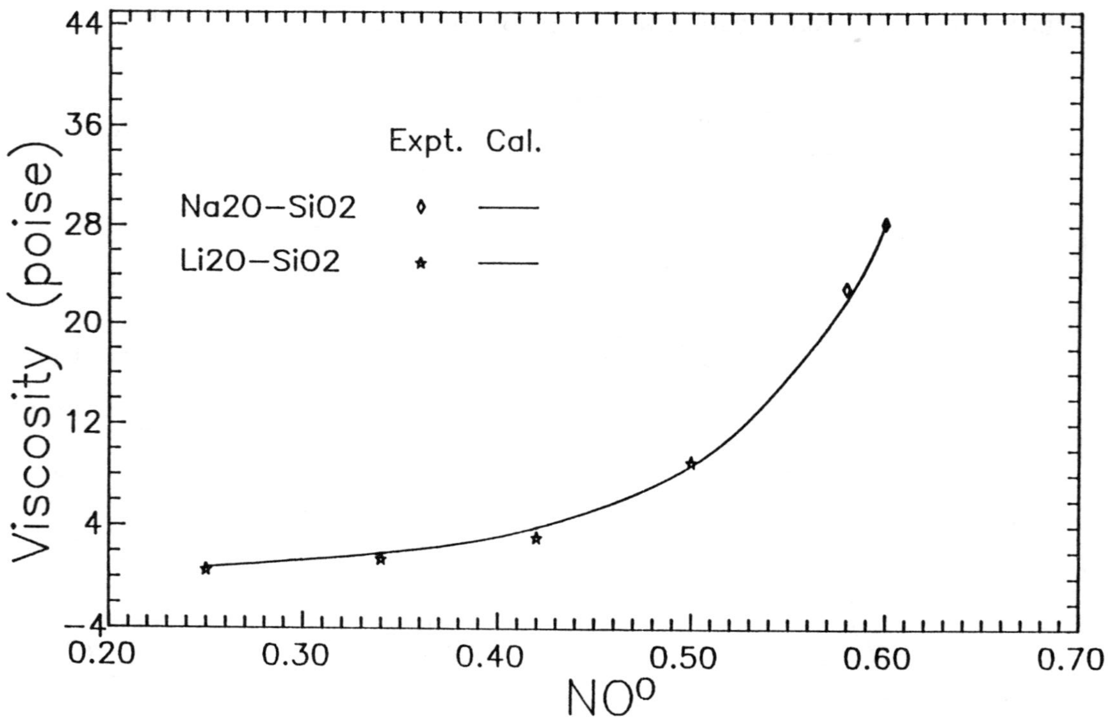

Fig. 4. Plot of calculated and measured viscosity of binary systems at 1673 K.

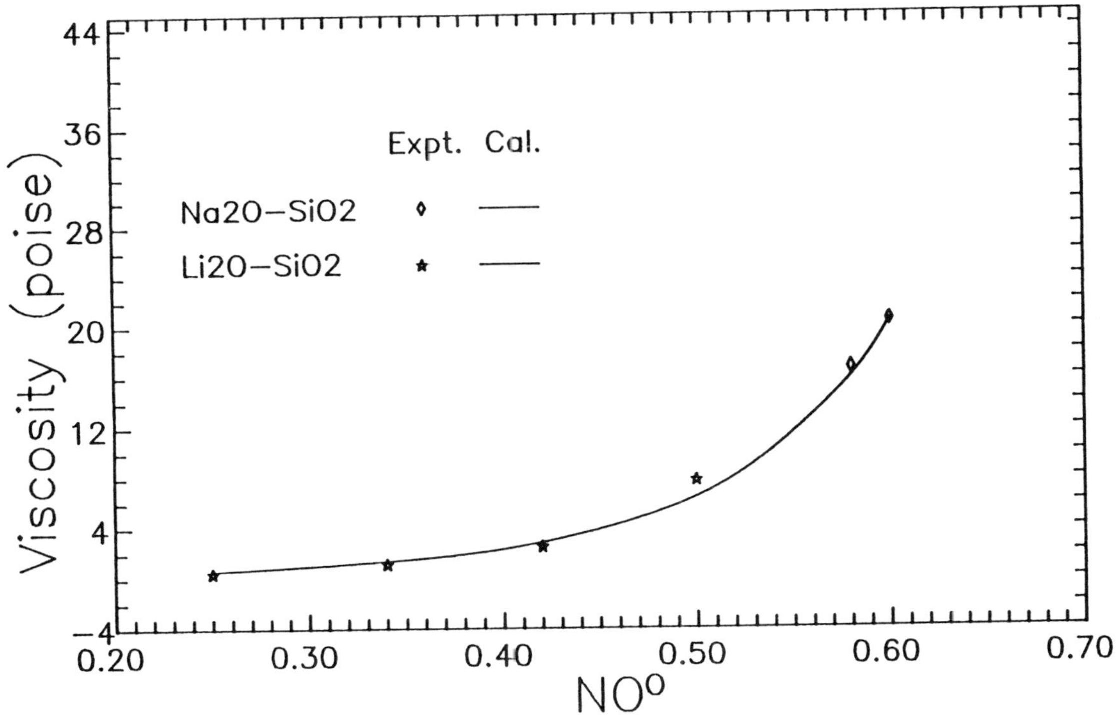

Fig. 5. Plot of calculated and measured viscosity of binary systems at 1723 K.

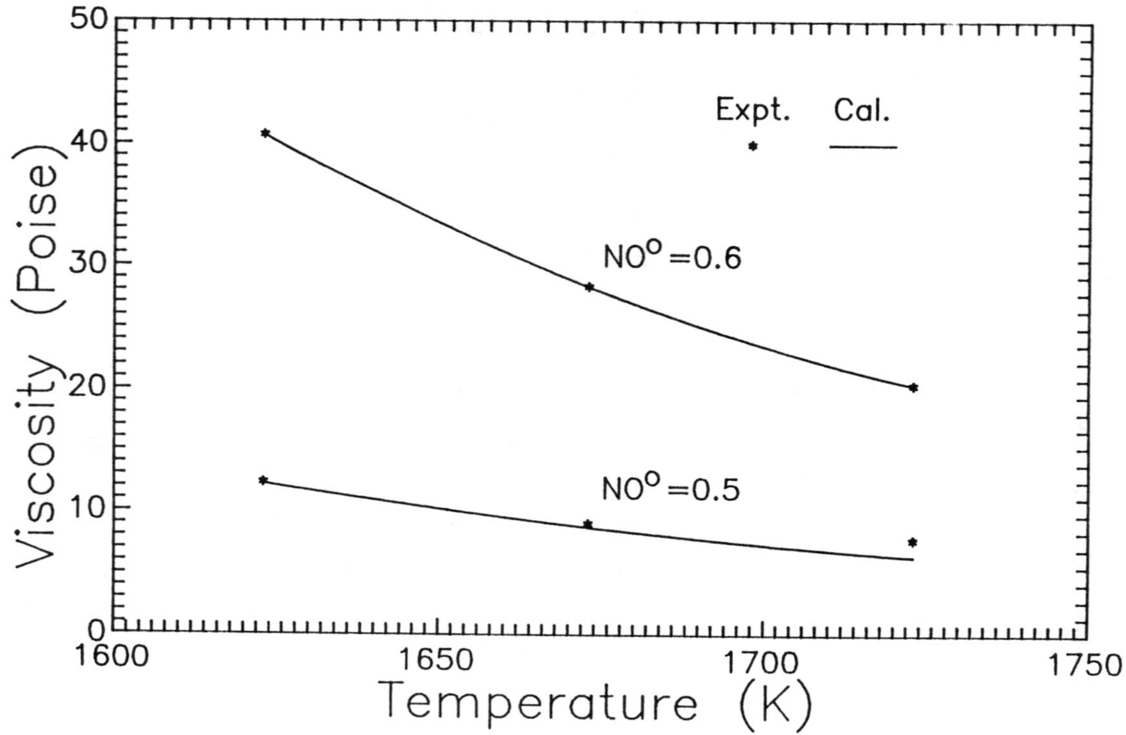

Fig. 6. Plot of calculated and measured viscosity of binary system as a function of temperature.

THE USE OF Ca-Mg ALLOY IN THE

PRODUCTION OF CLEAN STEEL

Tarek El Gammal* and Helmut Lischka**

*Inst. of Ferrous Metallurgy, RWTH Aachen,
Intzestr. 1, D-5100 Aachen, WEST GERMANY.
**SKW, D-8223, Trostbert, WEST GERMANY.

Abstract

The trend towards the casting of thin slabs, and more recently, to horizontal continuous casting has placed even more stringent demands on clean steel. Here Ladle Metallurgy plays a dominant role. This paper describes the use of metallic reagents, in particular Ca-Mg alloys, in the desulphurisation and deoxidation, and in morphology control of the inclusions. The physical and chemical properties, including thermodynamic treatment, of the reagents are presented. Test melts confirm the advantages of Ca-Mg in steel treatment, with reduction of S and O contents to low levels, conversion of inclusions to extremely fine globules and marked improvements of the mechanical properties.

Casting of Near Net Shape Products
Edited by Y. Sahai, J.E. Battles,
R.S. Carbonara and C.E. Mobley
The Metallurgical Society, 1988

ECONOMIC ANALYSIS OF THIN SECTION CASTING

J. C. AGARWAL AND MICHAEL J. LORETH

Charles River Associates Incorporated
200 Clarendon Street
Boston, Massachusetts 02116

Abstract

Thin casting of steel slab could have an enormous economic impact on the steel industry. It would allow the scrap-based minimills to compete effectively against integrated producers. This paper will examine the economics and benefits of producing thin section castings and review the strategic implications.

Casting of Near Net Shape Products
Edited by Y. Sahai, J.E. Battles,
R.S. Carbonara and C.E. Mobley
The Metallurgical Society, 1988

Background

It seems hard to believe that in 1965 only one percent of all crude steel produced in the world was continuously cast, while today the figure is more like 70 percent. The economic benefit of continuous casting has been well publicized elsewhere, so it is sufficient only to recall the major advantages:

1) Better quality,
2) Lower cost, and
3) Less home scrap and associated recycling cost.

The inherent technical and economic driving force behind the use of continuous casting has been to use the liquid steel in such a manner as to avoid as many process steps as possible. The development of near net shape casting is a logical extension of that concept, and thin section casting is the first step towards that goal. It is estimated that in the noncommunist world, 71 percent of all crude steel is now continuously cast, a percentage that is expected to go up to about 87 percent by the year 2000.

Continuous casting technology has become conventional. Slabs are typically 200 to 250 mm thick by 2,000 mm wide and are cast at rates of about 250 tonnes per hour per strand, with a withdrawal speed of roughly 1.2 m/min. Slab casters are operated principally in integrated steel mills, whereas billet machines, casting at a typical rate of 15 to 20 tonnes per hour per strand, are more widely used by minimills because the heat sizes are smaller. The advent of the minimills has contributed significantly to the widespread adoption of continuous casting of billets and rounds.

The main objectives for further development in current modes of continuous casting are:

1) Lower head room, which translates into lower capital requirement. A typical conventional continuous caster requires the tundish top to be located about 18 m above floor level;

2) Higher withdrawal speed, for in-line coupling with a rolling mill without reheating. This is unlikely to come about for slab casters because hot-strip mills operate at optimum speeds 30 to 40 times those of slab casters. Billets, however, can be rolled at much slower speeds, and it is feasible to match casting and rolling speeds for billets;

3) Lower demands for energy, operating personnel, and maintenance; and

4) Improved product quality.

The **wheel-belt billet caster** (Hitachi)(1), shown in Figure 1, has a casting speed of 3 to 6 m/min and a production rate of 30 to 60 tonnes per hour per strand. It is in-line coupled with a billet mill, where it is rolled to a 130 X 130 mm^2 square billet. The tundish top is located 4.7 m above floor level. This installation is now commercial at Daisan Steel Company, Tokyo.

The **twin-belt billet caster** (Krupp)(2), still under testing, is shown in Figure 2. It operates at high speeds, 8 to 10 m/min, and is coupled with a rolling mill. One of its interesting features, a horizontal spout, is lined with refractory and feeds the machine from the tundish. Required head room is estimated at about 5 m from floor to top of the tundish.

A **horizontal billet caster** (Kobe Steel)(3) directly casts horizontally from the side of the tundish. It is coupled with a rolling mill, and is in commercial use making stainless steel billets destined for the production of wire rod and seamless tube.

The **extra-low-head slab/bloom caster** (Sumitomo/CLECIM/Concast)(4) does not represent a radical departure from conventional design except for the small radius, 3 to 4 m, and multipoint unbending. The top of the tundish is about 7.5 m above floor level. Casting speed for a 220 X 1,600 mm^2 slab size is 2.0 m/min. The concept is still under development.

Development of Near Net Shape Casting

The objective of near net shape casting is to cast a sufficiently thin cross-section so that the product, still hot, requires only final rolling. Table I categorizes the numerous candidate solutions into three broad groups and compares them to conventional continuous slab casting. Please note that the production per unit train decreases substantially as the thickness decreases. Therefore, very thin section casting can only be justified for very high-value products.

Figure 1

HITACHI WHEEL–BELT BILLET CASTING MACHINE

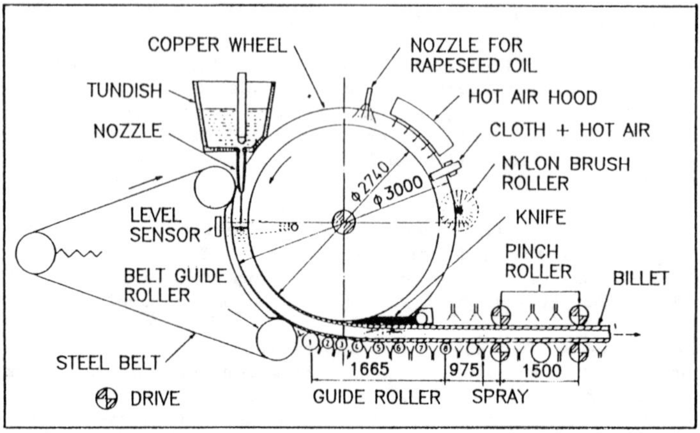

SOURCE: Iron and Steel Society, Inc. 1987. Near Net Shape Casting. P. 40.

Figure 2

KRUPP TWIN–BELT BILLET CASTING MACHINE

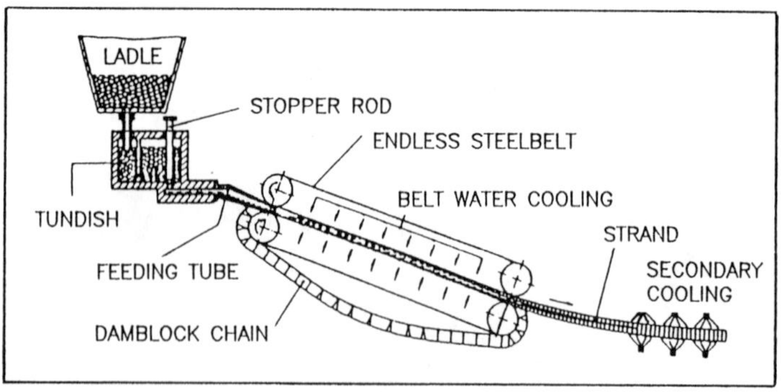

SOURCE: Iron and Steel Engineer, Vol. 58, No. 8 (August 1981), p. 18.

Table I. Continuous Casting of Thin Products

Cast Product	Typical Process	Thickness (mm)	Typical Output per Strand (tonne/hr)
Slab	Vertical C.C.	200-250	100-400
Thin slab	Twin belt	25-75	100
Strip	Twin roll	1-6	10
Thin strip	Single roll	0.02-0.5	1

SOURCE: After Marr, H.S., et al., "Horizontal Continuous Casting of Thin Slabs at the British Steel Corporation - Teesside Laboratories," in Near Net Shape Casting, Iron and Steel Society: 1987.

Due to its obvious cost and time advantages, near net shape casting is the object of major development efforts. The next ten years are likely to see large-scale commercialization in thin slab casting.

Thin Slab Casting

Twin-Belt Casting is being operated by NUCOR (5) at its Darlington Division. It produces 1,320 mm X 38 mm low-carbon steel slab at 90 tonnes per hour. The cast product is hot-rolled at a separate location for the product applications of drum stock, decking, and siding, and other such products not requiring deep drawing or forming capability; tramp elements in the scrap are probably the limiting factor.

Twin-belt casting is also being piloted actively by Sumitomo and Kawasaki. The Kawasaki caster is coupled to an in-line rolling mill.

An **Oscillating Mold - Squeeze Roll Caster** (SMS Schloemann-Siemag/Concast)(6) has completed its pilot trials and is being installed by NUCOR. End product thickness in the pilot operation ranged from 38 to 51 mm, with an adjustable width of 1,200 to 1,630 mm. Maximum casting speed was 6.1 m/min in the pilot plant. The product was rolled in-line to 3 to 19 mm, pickled, and cold-reduced. Quality was indistinguishable from that of a modern slab caster. Steel grades tested were:

- Low-C Al-killed drawing quality,
- Low-C Si/Al-killed commercial quality,
- Medium-C Si/Al-killed commercial quality, and
- HSLA.

With the exception of a revolutionary new mold that squeezes the slab, all components of the machine are standard continuous casting equipment.

K. Iverson(7) of NUCOR reported at a recent meeting that in NUCOR's pilot plant trials, the following improvements have been achieved:

1) Submerged nozzle configuration and material have been improved so that the thin section caster can operate continuously. Currently it operates for 3.5 hours, then one must change the nozzle and restart it after 45 minutes. This process improvement would increase productivity and lower costs if the improvement could be achieved on a sustained commercial operation.

2) The surface quality was excellent with no cracks in either transverse or longitudinal directions.

3) The desired mechanical properties were achieved with only 2:1 deformation in a section two inches thick.

These are impressive results from the pilot plant.

Horizontal casting (British Steel Corp.)(8) has advanced to the point where a pilot plant has consistently produced material 500 mm wide by up to 75 mm thick at speeds up to 14 m/min, equivalent to a production rate exceeding 170 tonnes per hour. The process concept is based on pouring steel into a series of trough-shaped molds located on a train that drives past underneath the tundish.

Additional development efforts are underway in West Germany (DEMAG) and Japan (Kawasaki).

Strip Casting, Thin Strip Casting

For reasonable production rates, high casting speeds are necessary, hence high cooling rates, of the order of 10^2 to 10^4 ^oC/sec. Such a high cooling rate requires very high heat transfer rates to the cooling medium, generally water. Work in this area has not moved from research to pilot plant development, nor is it likely to do so within the next five or ten years.

A variant to casting is spray forming, in which atomized molten metal is sprayed on a substrate metal strip, then compacted. It is also known as the Osprey process, or Swansea process. It has not been piloted; if piloted at all, it will probably not be with steel but with aluminum or brass, which melt at lower temperatures and are less prone to rapid oxidation.

We will confine our analysis to thin slab casting process only.

Economics

Conventional slab casting generally requires large tonnage production to be economical and practical. As the liquid steel is cast when it is very close to its freezing point, the volume and tonnage of steel handled is, by necessity, in the range of 1 million tonnes per year per strand. A normal plant capacity is about 2 million tonnes per year.

The goal of thin slab casting is to cast liquid steel directly into strip of similar thickness to hot strip produced from hot strip mills. Eliminating both the hot strip operation and a slab reheating step can achieve significant operating and capital cost savings together with higher processing yields.

The operating parameters for production of hot strip using both conventional thick slab continuous casting and thin slab continuous casting are given in Table II. In this example, capital and operating costs were determined based on a conventional operation producing 2 million tonnes per year of product, and thin strip casting of 0.4 and 0.8 million tonnes per year of product.

In the conventional process, liquid steel is continuously cast into a slab that is 225 mm thick. The slab is then reheated and sent to a hot strip mill, where the thickness of the slab is reduced to 6.3 mm. The total yield for these steps is estimated at 93 percent. Recyclable scrap is estimated at about 5 percent per tonne of product.

The thin strip casting process consists of continuously casting a 45 mm slab and reducing the slab's thickness to 6.3 mm with an in-line hot-roll machine. The expected yield is estimated at 96 percent, about a 3 percentage point increase over conventional practice.

The thin-slab technology will have major advantages over conventional practice in terms of labor, electricity, and fuel. However, at lower capacities, the labor requirement per tonne of product could exceed conventional operations due to economies of scale.

A capital and operating cost comparison is shown in Table III. The capital investment for a conventional 2 million tonne-per-year facility is $500 million, or $250 per annual tonne of product. The capital and operating costs included in Table III are based on starting with liquid steel and producing a coiled hot strip product. The cost for liquid steel production and cold rolling are not included.

For a plant producing 400,000 tonnes per year of hot strip using thin slab casting, the capital cost would be $80 million, or $200 per

Table II. OPERATING PARAMETERS FOR PRODUCTION OF HOT STRIP USING THICK SLAB CONTINUOUS CASTING AND THIN SLAB CONTINUOUS CASTING

Process	Thick Slab Casting	Thin Slab Casting	
Starting Material	Liquid Steel	Liquid Steel	
Intermediate Product	Slab	Slab	
Intermediate Product Thickness (mm)	225	45	
Final Product	Hot Rolled Strip	Hot Rolled Strip	
Final Product Thickness (mm)	6.3	6.3	
Capacity (000 Tonnes/Year)	2,000	800	400
Operating Parameters			
Steel Yield (Percent)	93	96	96
Scrap Credit (Tonnes/Tonne)	4.9	2.8	2.8
Labor (Hours/Tonne)	0.66	0.9	1.5
Electricity (kWh/Tonne)	110	40	40
Fuels (MMBtu/Tonne)	1.1	0.1	0.1

SOURCE: Charles River Associates, 1988.

Table III. Capital and Operating Costs for Production of Hot Strip Using Thick Slab Continuous Casting and Thin Slab Continuous Casting

Process			Thick Slab Casting	Thin Slab Casting	
Starting Material			Liquid Steel	Liquid Steel	
Intermediate Product			Slab	Slab	
Intermediate Product Thickness (mm)			225	45	
Final Product			Hot Rolled Strip	Hot Rolled Strip	
Final Product Thickness (mm)			6.3	6.3	
Capacity (000 Tonnes/Year)			2,000	800	400
Capital Investment ($ Million)			500	120	80
Operating Cost ($/Tonne Product)	Unit	Unit Cost			
Yield Loss	Tonne	200	14.0	8.0	8.0
Scrap Credit	Tonne	100	-4.9	-2.8	-2.8
Labor	Hour	20	13.2	18.0	30.0
Fuel	kWh	0.04	4.4	1.6	1.6
Refractories			5.0	5.0	5.0
Caster Supplies			4.0	6.0	6.0
Other Supplies			2.0	2.0	2.0
Taxes & Insurance			7.5	4.5	6.0
Allocated Technical & Admin.			2.0	2.0	2.0
Total Operating Cost			50.5	44.6	58.1
Capital Recovery ($/Tonne Product)			50.0	30.0	40.0
Total Cost ($/Tonne Product)			100.5	74.6	98.1

Note: Capital Recovery is assumed equal to 20 percent of the cost per annual tonne of capital investment.

SOURCE: Charles River Associates, 1988.

annual tonne. A plant producing 800,000 tonnes per year would have a capital cost of $120 million, or $150 per annual tonne. Under both of these capacities, the capital cost per annual tonne will be lower than conventional practice, which operates at 2.5 to 5 times the production levels of thin slab casting.

The operating cost comparison shows similar advantages. The direct operating cost of a conventional plant is estimated at about $50 per tonne of hot strip product (Table III). Yield loss and labor are major components at $14 and $13 per tonne, respectively.

For a plant producing 400,000 tonnes per year of hot strip using thin slab casting, the operating cost will exceed that of a conventional plant by about $8 per tonne. This is primarily the result of the large increase in labor cost ($30 per tonne) at this low capacity level. However, at an annual capacity of 800,000 tonnes, labor cost is reduced to $18 per tonne, resulting in a direct operating cost of about $45 per tonne. This is about $6 per tonne lower than conventional practice. Note that the twin slab casting technology is still new. It is highly likely that with more industrial experience, the labor and capital costs will decrease.

The advantages of thin strip casting are magnified when capital recovery is included. For a plant capacity of 400,000 tonnes per year, total costs are estimated at about $98 per tonne, $2 lower than conventional practice. At 800,000 tonnes per year, total costs are about $75 per tonne or an incredible $26 per tonne savings over conventional practice.

Strategic Issues

The most notable changes in steel supply to U.S. markets over the last 20 years have been the rise of the minimill and of imported steels. In addition, the increasing use of alloy steels, particularly HSLA in autos, is expected to be a significant factor in the future. Each issue is a major consideration when evaluating strategic implications of thin section casting.

Electric arc furnace production of steel has risen from about 8 million tonnes in 1960 to about 30 million tonnes in 1987. Because EAFs are virtually 100 percent scrap-charged, it appears at first glance that the rise of the EAF indicates a shift toward a higher rate of scrap recycling. Interestingly, this is not true. The rise of the EAF paralleled the replacement of the open hearth with BOFs. Since open hearths generally use a 40 to 60 percent scrap charge compared with 20 to 25 percent for BOFs, this technological change has actually led to a slight **decline** in the percentage of scrap relative to total iron units. However, the open hearth is nearly extinct. Scrap supplies cannot be expected to increase significantly in the future as the few remaining open hearths are closed. Therefore, the rise in electric furnace steelmaking capacity is ultimately limited in the

United States by the availability of scrap unless an economical source of DRI can be developed.

Minimills have concentrated on lower-value, lower-quality products. Two major factors have prevented minimills from moving into high-value products, such as flat-rolled material for automotive applications. First is scrap quality control. It is very difficult to control the level of tramp elements, such as copper, nickel, chromium, molybdenum, or cobalt, contained in steel scrap. These elements cannot be removed economically through currently available refining processes and, if present, may preclude the manufacture of certain steel products. For example, high levels of copper preclude the drawing of wire products.

The second factor that has restricted the movement of minimills into certain higher-value product lines is the economic scale of slab casting facilities. Traditional continuous slab casting requires approximately 2 million tonnes of annual production to justify the investment. Due to the generally small size of minimill operations, there has been a mismatch between the required capacity for slab casting and the steel production capacity of most minimills. For example, NUCOR, the largest and most successful minimill company, will produce approximately 2 million tonnes of steel in 1987 in three plants. Thin slab casting will change this picture drastically.

There are five major points by which the impact of thin section casting will be felt in steel industry as thin section casting technology penetrates the marketplace:

1) It will be no longer necessary to have a large integrated steel mill producing over 2 million tonnes of liquid steel to supply the flat-rolled products segment of the market. Flat-rolled products are the biggest and most profitable segment of the market.

2) Smaller steel mills, even those based on electric furnace scrap melting, will be able to produce high-quality flat-rolled products.

3) Quality control for scrap will become an even a bigger issue than now.

4) DRI-based electric arc furnace steel plants that could produce flat-rolled products could be operated economically in countries in which the domestic demand is modest and scrap supply limited.

5) International and domestic trade patterns that have normally existed between integrated steel mills and large users of flat-rolled products, such as automobile or appliance manufacturers, will be disrupted.

<u>References</u>

1. E. Niyoma et al., "Belt-Wheel Type Continuous Caster for Steel Billets," Iron and Steel Engineer, August 1981, 38-42.

2. E. Buck and D. Figge, "State of the Art in Casting Steel Billets on a Twin-Belt Caster," in Near Net Shape Casting, Iron and Steel Society: 1987.

3. T. Nozaki et al., "Horizontal Continuous Casting of Stainless at Kobe Steel," Iron and Steel Engineer, October 1985, 50-56.

4. T. Nagaoka et al., "The Basic Concept of the ELH (Extra Low Head) Caster for Slabs and Blooms," in Near Net Shape Casting, Iron and Steel Society: 1987.

5. M.D. Millet, "Development of the Hazelett Twin Belt Caster by NUCOR Steel-General Comments," in Near Net Shape Casting, Iron and Steel Society: 1987.

6. H.F. Marten and H. Faster, "Latest Developments and Operating Results of the High-Speed Strip-Casting Machine at SMS Schloemann-Siemag A.G., W. Germany," in Near Net Shape Casting, Iron and Steel Society: 1987.

7. K. Iverson, presentation at the Steel Survival Strategies Conference, New York, June 1988.

8. H.S. Marr et al., "Horizontal Continuous Casting of Thin Slab at the British Steel Corporation - Teesside Laboratories," in Near Net Shape Casting, Iron and Steel Society: 1987.

SINGLE AND TWIN-ROLL CONTINUOUS CASTING

OF STEEL STRIP AT IRSID

J. P. Birat, P. Blin, J. L. Jacquot,

P. V. Riboud, and B. Thomas

Institut de Recherches de la Siderurgie Francaise
B.P. 64, 57210 Maizieres-Lex-Metz, France

Abstract

Pilot experiments of strip casting have been conducted at IRSID, using both single and twin-roll techniques to produce strip of stainless and carbon steel ranging from 10 to 1 mm in thickness. Process variables and product properties are discussed.

Casting of Near Net Shape Products
Edited by Y. Sahai, J.E. Battles,
R.S. Carbonara and C.E. Mobley
The Metallurgical Society, 1988

AUTHOR INDEX